0332818

ROWE, JOSEPH E
NONLINEAR ELECTRON-WAVE INTERA
000332818

621.371.6   R87

KV-429-994

WITHDRAWN
FROM STOCK

of borrowing,

NONLINEAR ELECTRON-WAVE INTERACTION PHENOMENA

ELECTRICAL SCIENCE
A Series of Monographs and Texts

Edited by

Henry G. Booker
UNIVERSITY OF CALIFORNIA AT SAN DIEGO,
LA JOLLA, CALIFORNIA

Nicholas DeClaris
CORNELL UNIVERSITY,
ITHACA, NEW YORK

1 JOSEPH E. ROWE. Nonlinear Electron-Wave Interaction Phenomena. 1965

A. BLAQUIERE. Nonlinear Systems Analysis. *In preparation*

NONLINEAR ELECTRON-WAVE INTERACTION PHENOMENA

JOSEPH E. ROWE
DEPARTMENT OF ELECTRICAL ENGINEERING
THE UNIVERSITY OF MICHIGAN
ANN ARBOR, MICHIGAN

1965

ACADEMIC PRESS   New York and London

Copyright © 1965, by Academic Press Inc.
ALL RIGHTS RESERVED.
NO PART OF THIS BOOK MAY BE REPRODUCED IN ANY FORM,
BY PHOTOSTAT, MICROFILM, OR ANY OTHER MEANS, WITHOUT
WRITTEN PERMISSION FROM THE PUBLISHERS.

ACADEMIC PRESS INC.
111 Fifth Avenue, New York, New York 10003

United Kingdom Edition published by
ACADEMIC PRESS INC. (LONDON) LTD.
Berkeley Square House, London W.1

Library of Congress Catalog Card Number: 65-22768

PRINTED IN THE UNITED STATES OF AMERICA

To my parents, Anne, and the children

Foreword

The subject of nonlinear electron-wave interaction came of age in the early 1950's. The early work of A. W. Hull in 1921 suggested a need for such studies, and the velocity modulation principle was introduced in the late 1930's. After several years of successful experimentation with klystrons, traveling-wave amplifiers, and crossed electric and magnetic field devices, the microwave electron beam type of amplifier had assumed a place of importance in the electronics technology even though little was known about its efficient operation at large rf signal levels. Thus the need for a generalized nonlinear analysis became apparent and the work of A. T. Nordsieck, *circa* 1952, paved the way for later more detailed investigations.

While the bulk of the material in this book relates to the microwave electron-beam device, the basic nonlinear Lagrangian analysis is applicable to beam-plasma interactions and recently has found application in studying nonlinear ionospheric processes such as whistler-mode phenomena. It is suggested that the basic approach developed herein is directly applicable to the investigation of all charged particle-wave interaction phenomena.

It is hoped that the book will serve both as a research monograph and as a graduate textbook for a course in nonlinear interaction theory. The first four chapters provide the general framework while the following chapters consider in detail its application to various specific interaction configurations.

A great deal of the material in the book is based on the research of the author and his former doctoral students including Drs. H. Sobol, O. P. Gandhi, J. G. Meeker, G. I. Haddad, and K. L. Volkholz. Their contributions and suggestions along with those of Professors W. G. Dow and G. Hok have had a great influence on the content. A special thanks is due Professor W. G. Dow for originally interesting the author in this field and for his continued encouragement and guidance as a colleague and friend. The research support provided by the U.S. Air Force and U.S. Army is also acknowledged, since without it much of the work would never have been completed.

A particular debt of thanks is due to both Dr. G. I. Haddad and Mr. H. K. Detweiler for valuable suggestions and a careful reading of the manuscript; to Mrs. June Corkin for typing and assembling the many drafts and last but not least to my wife Anne for her many contributions.

August, 1965

JOSEPH E. ROWE

Contents

FOREWORD vii

CHAPTER I
Introduction

1. General Introduction 1
2. Scope of the Book 3
3. Classes and Description of Devices Analyzed 3
4. Necessity for a Nonlinear Analysis 11
   References 12

CHAPTER II
Eulerian versus Lagrangian Formulation

1. Introduction 16
2. Eulerian Formulation of O-TWA Equations 16
3. Lagrangian Formulation 23
4. Composite Lagrangian System 26

CHAPTER III
Radio-Frequency Equivalent Circuits

1. Introduction 28
2. Equivalence of Maxwell and Kelvin Theories 30
3. Equivalence for a Helical Wave Guiding Structure 34
4. Transmission-Line Equivalent for Surface Wave Propagation on a
   Plasma Column 51
5. Equivalent Transmission Lines for Multidimensional Propagating Structures . 54
6. Backward-Wave Equivalent Circuits 62
7. Equivalent Circuits with Spatially Varying Line Parameters 65
   References 68

CHAPTER IV

Space-Charge-Field Expressions

1. Introduction . . . 69
2. Green's Function Method for Potential Problems . . . 72
3. Potential Functions for the Cartesian Coordinate System . . . 74
4. Potential Function for a Two-Dimensional Rectangular System . . . 81
5. Space-Charge Fields for Rods of Charge . . . 89
6. Replacement of Rf Structure by an Impedance Sheet . . . 89
7. Space-Charge Potentials for Cylindrical Systems . . . 92
8. Potential Function for a Ring of Charge in an Axially Symmetric System . . . 95
9. Potential Functions for Hollow Beams . . . 102
10. One-Dimensional Disk Space-Charge Model . . . 103
11. Harmonic Method for Calculating the One-Dimensional Space-Charge Field . 106
12. Equivalence of the Green's Function and Harmonic Methods for the One-Dimensional Problem . . . 112
13. Space-Charge Fields for Specialized Configurations . . . 114
    References . . . 119

CHAPTER V

Klystron Analysis

1. Introduction . . . 120
2. One-Dimensional Klystron Analysis . . . 121
3. One-Dimensional Klystron Results . . . 130
4. Two-Dimensional Klystron Analysis . . . 144
5. Three-Dimensional Klystron Interaction . . . 150
6. Radial and Angular Effects in Klystrons . . . 155
7. Relativistic Klystron Analysis . . . 168
8. Voltage Stepping in Klystrons . . . 172
   References . . . 176

CHAPTER VI

Traveling-Wave Amplifier Analysis

1. Introduction . . . 177
2. Mathematical Analysis of the One-Dimensional TWA . . . 179
3. One-Dimensional Results . . . 193
4. N-Beam TWA Analysis . . . 217
5. Two-Dimensional TWA Analysis . . . 224
6. Three-Dimensional O-TWA Analysis . . . 231
7. Two-Dimensional Circuit, Three-Dimensional Flow . . . 234
8. Effects of Transverse Variations on TWA Gain and Efficiency . . . 235
9. Relativistic O-TWA . . . 250
10. Integral Equation Analysis . . . 253
    References . . . 261

CHAPTER VII

O-Type Backward-Wave Oscillators

1. Introduction 263
2. Backward-Wave Circuits 265
3. Mathematical Analysis 266
4. Solution Procedure 268
5. Efficiency Calculations 272
6. Relativistic Oscillator Analysis 277
7. Radial and Angular Variations in BWO's 279
   References 280

CHAPTER VIII

Crossed-Field Drift-Space Interaction

1. Introduction 282
2. Two-Dimensional Drift-Space Equations 283
3. Gap Modulation of a Crossed-Field Stream 292
4. Results for a Two-Dimensional Cf Drift Region 295
5. Three-Dimensional Drift-Space Equations 299
6. Adiabatic Motion in a Drift Region 302
   References 303

CHAPTER IX

Crossed-Field Forward-Wave Amplifiers

1. Introduction 304
2. Two-Dimensional M-FWA with a Negative Sole 306
3. Results for a Two-Dimensional M-FWA with a Negative Sole . . . 314
4. Two-Dimensional M-FWA with a Positive Sole 342
5. Adiabatic Equations for a Two-Dimensional M-FWA with a Negative Sole . 345
6. Three-Dimensional M-FWA with a Negative Sole 348
7. Effect of Cyclotron Waves 348
8. Comparison with Sedin's Calculations 350
9. Results of and Comparison of Various Nonlinear Theories for the M-FWA . . 355
   References 363

CHAPTER X

Crossed-Field Backward-Wave Oscillators

1. Introduction 365
2. Two-Dimensional M-BWO with a Negative Sole 366
3. Results for a Two-Dimensional M-BWO with a Negative Sole . . . 371

4	M-BWO with a Positive Sole	374
5	Adiabatic Equations for an M-BWO with a Negative Sole	375
6	Cyclotron Waves in M-BWO's	380
7	Theory versus Experiment	380
	References	384

CHAPTER XI

Traveling-Wave Energy Converters

1	Introduction	385
2	O-Type Traveling-Wave Energy Converter	387
3	M-Type Traveling-Wave Energy Converter	393
	References	394

CHAPTER XII

Multibeam and Beam-Plasma Interactions

1	Introduction	395
2	Nonlinear Equations for Combined One-Dimensional Beam-Plasma Circuit	398
3	Double-Beam Circuit Solutions	410
4	Interaction Equations in the Absence of a Circuit	414
5	Velocity Distributions	417
6	Two-Dimensional Effects in Beam-Plasma Interactions	419
	References	421

CHAPTER XIII

Phase Focusing of Electron Bunches

1	Introduction	424
2	Historical Background and Experimental Work	425
3	Efficiency Improvement in Traveling-Wave Amplifiers	429
4	Efficiency Improvement in O-Type Backward-Wave Oscillators	454
5	Efficiency Improvement in Crossed-Field Amplifiers	461
6	Efficiency Improvement in Crossed-Field Backward-Wave Oscillators	476
	References	485

CHAPTER XIV

Prebunched Electron Beams

1	Introduction	487
2	Mathematical Formulation of the Lagrangian Equations	488
3	Results for Klystrons	490

CONTENTS   xiii

4 Results for Traveling-Wave Amplifiers 493
5 Results for Crossed-Field Amplifiers 501
6 Rf Power Required to Bunch an Electron Beam 507
   References. 514

CHAPTER XV

Collector Depression Techniques

1 Introduction 515
2 Graphical Evaluation of Depressed Collectors 517
3 Analysis of Output Energy Distribution for Collector Depression in
     O-Type Devices 519
4 Results of Calculations for O-Type Devices 524
5 Beam Current Flow Limitation in Collector Depression 536
6 Depressed Collectors on Crossed-Field Devices 541
   References. 549

CHAPTER XVI

Modulation Characteristics

1 Introduction 550
2 Mathematical Analysis for O-Type Devices 551
3 O-Type Nonlinear Modulation Results 556
4 Mathematical Analysis for M-Type Devices 559
5 Output Spectra for Low-Frequency Modulations 562
6 Modulation by Multiple High-Frequency Signals 567
   References. 568

APPENDIX A

Rf Structure Impedance Variations

1 Helical Line for O-FWA 570
2 Helical Line for O-BWO 570
3 Tapered Interdigital Line Characteristics 575
   References. 578

APPENDIX B

O-TWA Kompfner-Dip Conditions

Text 579
References 582

APPENDIX C
M-FWA Kompfner-Dip Conditions

Text 583
References 584

AUTHOR INDEX 585
SUBJECT INDEX 589

CHAPTER

I | Introduction

1 General Introduction

The subject of this book is the interaction between drifting streams of charged particles and propagating electromagnetic waves. Of particular concern are the situations in which the wave amplitude is large and there is strong coupling between the charged fluid and the wave, for it is then that nonlinear effects are important. The interaction is considered to be nonlinear when the time-varying quantities become of a significant magnitude as compared to the corresponding steady values. Eventually all cumulative interaction systems become nonlinear if allowed to build up over a sufficient number of wavelengths.

Of particular interest here are those systems in which the drifting stream (charged fluid) is composed of electrons and/or ions and these are coupled to a slow electromagnetic wave over an extended region. The type of wave-guiding medium is not restricted and may be either some form of conventional periodically loaded waveguide system or other media such as the magnetically confined plasma in which the characteristic phase velocity of the waves is less than the velocity of light in free space.

Most of the material in the book relates to electron stream-wave configurations, although Chapter XII is specifically directed to the consideration of plasma phenomena.

The now widely familiar and much used family of linear-beam devices, represented primarily by the multicavity klystron and the traveling-wave amplifier, owe their existence to the early work of Hahn,[2,3] Ramo[9,10] and Hansen-Varian[13,16] on the principles of velocity modulation. These workers developed the fundamental concept of velocity modulation which utilizes transit-time effects and then they successfully applied their ideas to the problem of generating and amplifying high-frequency signals.

Somewhat later, other workers[8,42,51,53] investigated what is now known as the injected-beam crossed-field amplifier or so-called magnetron amplifier*.

The material in this book will be directed primarily to those devices which utilize a defined injected stream of some type. The general methods are also applicable to emitting-sole devices, although these devices are not to be discussed here. In the former case the interaction to be described is that between a directed stream and an electromagnetic wave; in the latter case the interaction process is similar although the electron stream is not well defined in the sense of a directed charged fluid. Particle and electromagnetic wave velocities both small and comparable to the velocity of light will be considered.

In particular the interaction mechanisms in klystrons, traveling-wave amplifiers (O-TWA), backward-wave oscillators (O-BWO), crossed-field amplifiers (M-FWA), crossed-field oscillators (M-BWO), multibeam devices, and electron-beam-plasma devices are discussed. Familiarity on the part of the reader with the linear (small-signal) theories is assumed, and reference to them is made only for comparison purposes and for the definition of interaction parameters. The small-amplitude theories are well covered in many articles and several books which provide a valuable introduction to this work.

An essential assumption of any linear theory, which will not be made in this nonlinear treatment, is the neglect of second-order and higher quantities, which results in the prediction of exponential growth rates with distance for fluctuating signals. As a result saturation effects are necessarily omitted and little or no information on the energy conversion process is obtainable.

If second- and higher-order terms are retained in the appropriate interaction equations, then they are highly nonlinear and generally amenable to solution only by high-speed digital computer techniques. The advent of these machines has made possible the solution of problems previously considered untractable. A great library of solutions can be obtained at relatively little cost and in little time. They are not without disadvantages, however, since the natural tendency is to substitute reams of computer data for clear thinking. Real insight into the interaction phenomena is gained through a detailed analysis and study of the digital computer solutions. Simple extrapolations of linear theory results can also provide a framework for the interpretation of large-signal calculations.

---

* It is suggested that the term magnetron be reserved for the family of crossed-field devices which utilize multicavity resonant circuits, emitting-sole electrodes or both. The term crossed-field amplifier, which may be designated M-type, or TPOM as used by the French workers, seems most appropriate for the injected-beam device.

2 Scope of the Book

It will be the objective of this book to develop physically adequate models and means of analysis to study the nonlinear interaction phenomena in the aforementioned free-electron devices. The treatment will be entirely theoretical and only a little experimental data will be introduced in support of certain calculations. The general methods to be evolved will be treated rigorously, so that they will provide a framework for the analysis of any interaction problem involving a charged fluid and an electromagnetic wave. In some cases several methods of analysis will be outlined and compared in order to point out possible alternate methods; often it is desirable to use more than one approach in order to substantiate the appropriateness of certain primordial assumptions.

In most cases the solutions of the nonlinear interaction equations will be obtained by high-speed digital computer methods, since adequate nonlinear mathematics has not yet been developed to handle these systems. However, in certain cases (Chapter XIII) extensive analytical treatments are developed which permit obtaining closed-form solutions of the nonlinear systems. This is particularly true when ideal charge bunches are postulated and this assumption is applied to the problem of phase focusing in traveling-wave and crossed-field devices.

This book is designed to serve both as an up-to-date research monograph for workers in the fields of microwave electron and plasma devices and also as a text for advanced graduate students. Much of the material has been presented by the author in various graduate courses at The University of Michigan over the past several years and has had the benefit of criticism by numerous former graduate students. Many of the author's former students have also contributed directly and indirectly as a result of their doctoral research work.

It is believed that the core material of this book could appropriately constitute a three-hour advanced graduate course in nonlinear interaction theory. Several topics such as rf to dc converters, depressed collectors and other specialized topics could easily be left to the students' independent perusal.

3 Classes and Description of Devices Analyzed

All the devices mentioned earlier have found wide application in various types of electronic equipment, each suited to a particular special purpose. Since each device has certain characteristics different from the others, no one device has dominated the science and thus we are faced with the problem of analyzing and understanding all.

At high power levels, the devices are saturated and behave in a nonlinear manner, a regime which is not tractable with a linear theory. Because we are interested in knowing the maximum efficiency of operation and how the various interaction and device parameters affect this maximum value, it is necessary to carry out a detailed nonlinear analysis. Generally this has been referred to in the literature as a "large-signal" analysis, meaning that when large-amplitude signals are present, second-order rf terms are not negligible as compared to the corresponding dc values. Additional information such as gain, phase shift, etc., are natural by-products of the analysis.

The various classes of devices to be considered are discussed briefly in the following sections.

3.1 Traveling-Wave Linear-Beam Devices

In the traveling-wave linear-beam device, sometimes referred to as an O-type device, the energy exchange process is one of converting electron beam kinetic energy into rf energy by slowing the electrons down through electromagnetic forces. Members of this class of devices include klystrons, traveling-wave amplifiers, backward-wave oscillators, double-beam amplifiers, resistive-wall amplifiers and beam-plasma amplifiers. The differences lie in the means for modulating the electron stream with rf and/or the means for extracting energy, and not in the basic interaction process. Some of the characteristics are briefly outlined below.

a. Klystrons*

The simplest form of klystron is the two-cavity version illustrated in Fig. 1, in which the first cavity is used to couple an rf wave to the beam for modulation purposes and the output cavity is used to extract rf energy from the beam. After the beam is velocity modulated in passing through the first cavity, it passes through a drift tube wherein the velocity modulation is gradually converted to density modulation, resulting in a nonlinear bunched electron beam arriving at the output. The bunches drive the output circuit and develop an rf voltage across the cavity walls.

In order to obtain maximum bunching and hence a maximum amplitude of rf current in the output, several bunching cavities may be used, as illustrated in Fig. 1. The basic principles of operation remain the same. In order to obtain wider bandwidth, the output cavities are

---

* The klystron is covered in a basic patent (U.S. 2,242,275) awarded to R. H. Varian in 1937.

frequently stagger-tuned. Generally, though, the maximum bandwidth of a high-power klystron is approximately 2–3%.

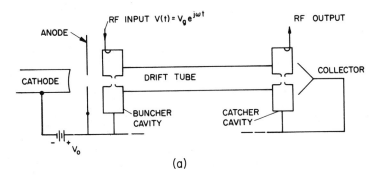

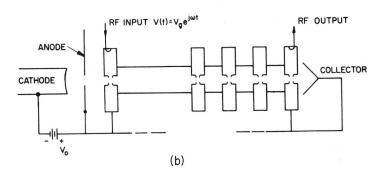

FIG. 1. Schematic diagrams of (a) a two-cavity klystron and (b) a five-cavity klystron.

b. *Traveling-Wave Amplifiers and Backward-Wave Oscillators**

These are both so-called kinetic energy conversion devices and differ from each other primarily in the mode of operation of the circuit. The forward-wave amplifier utilizes a space-harmonic circuit mode which is characterized by positive dispersion, while the backward-wave amplifier and backward-wave oscillator operate in a space-harmonic circuit mode exhibiting negative dispersion.

---

* The basic patent on the traveling-wave amplifier (*U.S.* 2,653,270) was issued to R. Kompfner on September 22, 1953. The backward-wave oscillator is covered in two patents (*U.S.* 2,880,355, March 31, 1959, and *U.S.* 2,932,760, April 12, 1960) issued to B. Epsztein of the C. S. F. Company, France.

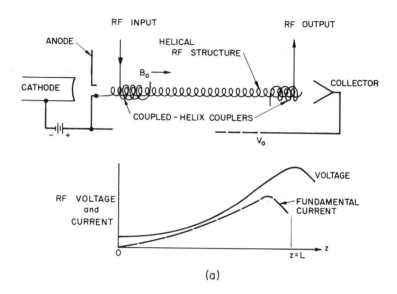

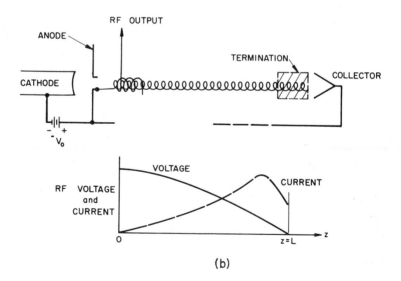

Fig. 2. Forward-wave and backward-wave devices: (a) forward-wave amplifier; (b) backward-wave oscillator.

3. CLASSES AND DESCRIPTION OF DEVICES ANALYZED

Both the forward-wave and backward-wave devices are illustrated in Fig. 2 along with their current and voltage characteristics versus distance. The rf voltage velocity modulates the electron beam all along the structure in these cases and hence the interaction is distributed, whereas a klystron has a localized velocity modulation region which is separated from the region in which the conversion to density modulation occurs. The wave modulates the beam, speeding up some electrons and slowing down others in such a way that more electrons are slowed than are accelerated per unit length along the structure and hence there is a net transfer of energy from the beam to the circuit in this region.

Various forms of periodic resonant and nonresonant circuits are used in these devices. An important example of the periodic nonresonant circuit is the helical waveguide, which is an extremely broad-band circuit for a forward-wave amplifier and also functions in a broad-band voltage-tunable backward-wave oscillator. Octave bandwidths in the amplifier and octave tuning in the oscillator are easily achievable. Other circuit forms frequently used, such as the coupled-cavity structure and the folded line, characteristically have considerably narrower bandwidths, varying between 10 and 30%. Power levels for these devices vary from the milliwatt to the kilowatt range for the backward-wave oscillator and from the milliwatt to the megawatt range for the forward-wave amplifier. Frequency coverage extends from the low megacycle region through much of the gigacycle region.

c. *Multibeam and Beam-Plasma Devices*

These two types of devices are quite similar in many respects although quite different in others. They both may be viewed as the interaction of a drifting electron beam with another charge system which itself may be either stationary or drifting.

In the case of the double-beam amplifier, a usual configuration is a solid electron beam within a hollow beam and relative motion between the two. The beams may be excited by velocity modulation through a klystron-type cavity or through a propagating circuit as in a coupled-helix coupler. Such a configuration is illustrated in Fig. 3. Little practical work has been done on this amplifier; hence little theoretical investigation, except for several small-signal theories, appears in the literature. In Chapter XII the nonlinear theory of the double-beam amplifier will be considered.

In the case of the beam-plasma amplifier the ionized plasma acts like an rf circuit. It is known that the plasma can support both forward and backward waves of propagation and thus both amplifiers and oscillators are in principle possible. A simple analysis would neglect the effects of

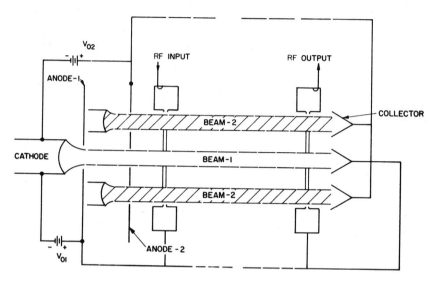

FIG. 3. Double-beam amplifier and rf coupling circuits.

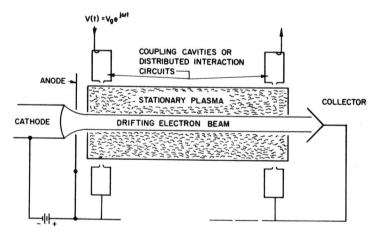

FIG. 4. Beam-plasma interaction device with coupling circuits.

the ions (viscous background medium) in the plasma and assume that the plasma electrons are stationary and thus that the interaction is similar to that in a double-beam tube. Collision effects may generally be neglected in such a nonlinear interaction process. This amplifier is shown schematically in Fig. 4, along with the modulating cavities to excite the plasma, which in turn couples the rf to the electron beam. This linear-beam interaction problem is also examined in Chapter XII.

The devices listed above constitute the principal and important types of linear beam devices and all will be treated on a nonlinear basis in succeeding chapters. The basic theory and method of approach will also apply to other possible configurations, such as drifting streams in solids and ionospheric phenomena, and hence are considered universally applicable.

3.2 Crossed-Field Interaction Devices

This class of devices differs from the previous one in that the moving charge (beam) flows in orthogonal static electric and magnetic fields. The energy conversion mechanism is one of converting potential energy to rf energy through movement of the electron beam from a low-potential cathode to a high-potential electrode which is also the rf propagating structure. The axial magnetic field in the O-type device serves to confine the electron stream so that little current is intercepted on the rf structure. However, in the crossed-field case a large fraction of the beam is collected on the structure and hence the circuit dissipation problems are sometimes paramount.

a. Forward-Wave Amplifiers* and Backward-Wave Oscillators

A schematic diagram of the interaction configuration for these devices is shown in Fig. 5 for either an amplifier or an oscillator with a so-called short-focus electron gun. The rf circuits used in practical devices have been derived from either the interdigital line or a vane type of structure. In the case of these injected-beam devices the designation M-FWA will be used to refer to a crossed-field type of forward-wave amplifier, and the designation M-BWO to correspond to a backward-wave oscillator.

The rf voltage applied to the input of the structure velocity modulates the beam, accelerating some electrons and decelerating others. Those which have been speeded up will be directed towards the sole electrode and move into a weaker rf field region. The slower electrons move

---

*Patent U.S. 2,768,328 on the linear magnetron-type amplifier was issued to J. R. Pierce on October 23, 1956.

towards the rf structure by virtue of the fact that the electric force exceeds the magnetic force and their potential energy is converted into rf energy, most of the electrons actually being collected on the structure electrode. In a later chapter a means for phase focusing and keeping the electrons off the structure will be developed. The bandwidth of such devices is generally limited by the rf structure, since those forms utilized are of a resonant periodic type. It is truly unfortunate that the helix or some other form of periodic nonresonant circuit is not available for use in a crossed-field device. The flattened helix has too low an impedance and limited power dissipation for such an application.

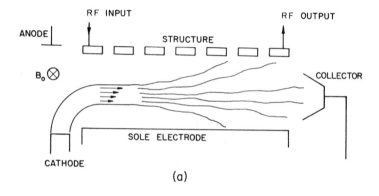

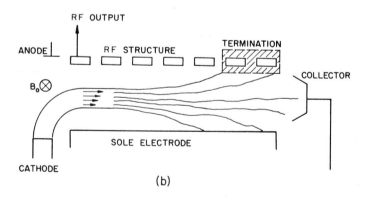

FIG. 5. Schematic diagrams of (a) a crossed-field forward-wave amplifier and (b) a backward-wave oscillator.

b. Multibeam Amplifiers

As in the case of the O-type device it is possible to devise a double- or multibeam amplifier in a crossed-field configuration. No experimental work and little analysis has been carried out on such a device, probably because of the lack of success in the O-type case. The multibeam configuration is easily handled by the nonlinear theory developed in Chapter XII.

c. Crossed-Field Plasma Amplifier

On the basis of the companion O-type device, it might be thought that such an interaction configuration might have interesting possibilities for either amplifiers or oscillators. However, plasma confinement and electron-beam focusing problems seem to preclude success with such a device. A discussion of its characteristics will, therefore, not be presented.

d. Emitting-Sole Crossed-Field Devices

A natural progression of events in injected-beam crossed-field work is to increase the beam current and space-charge density by increasing the cathode length to the point of having a continuously emitting-sole electrode. The beam is now not well defined and the device acquires magnetron characteristics. The nonlinear theory of such devices will not be treated specifically although the general methods evolved are directly applicable.

These devices may be in the form of emitting-sole amplifiers. A slightly different device is the Amplitron, which may be thought of as an injection-locked oscillator.

4 Necessity for a Nonlinear Analysis

The necessity for a nonlinear analysis arises because the small-signal theory (linear theory) is unable to give information on the saturation gain, phase shift and efficiency for high-power amplifiers and oscillators. Klystrons, traveling-wave tubes, and other linear-beam amplifiers and oscillators have found wide application in radar and other electronic systems in which it is necessary to have information on the above characteristics. All this information, along with data on the modulation characteristics and on the velocity-phase and current characteristics of the electron beam, is obtainable from the large-signal calculations.

It is most logical to proceed with the development of a nonlinear theory along the same lines used in the development of the linear theory, only retaining the second-order terms which were previously neglected.

It will be shown in Chapter II that this process leads to difficulty in view of the fact that crossing of electron trajectories results in multivalued velocity and charge density. If the fluid flow analysis is transformed to a particle analysis, this difficulty can be overcome.

The nonlinear Lagrangian equations to be developed in succeeding chapters are not readily amenable to analytical solution and hence must be solved by digital computer techniques. Several numerical solution methods will be outlined and computer solutions of most systems are given. It is possible, under certain conditions, to solve the nonlinear equations in closed form and thus obtain checks on the numerical solutions. These solutions also add to the fundamental understanding of the interaction mechanism.

The subject of phase focusing in both O-type and M-type devices is discussed in Chapter XIII and the nonlinear interaction equations are solved for the focusing of ideal hard-kernel charge bunches.

REFERENCES

A. General Theory

1. Beck, A. H. W., *Space-Charge Waves*. Pergamon Press, London, 1958.
2. Hahn, W. C., Small-signal theory of velocity modulated electron beams. *Gen. Elec. Rev.* **42**, No. 6, 258-270 (1939).
   Hahn, W. C., Wave energy and transconductance of velocity-modulated electron beams. *Gen. Elec. Rev.* **42**, No. 11, 497-502 (1939).
3. Hahn, W. C., and Metcalf, G. F., Velocity modulated tubes. *Proc. IRE* **27**, No. 2, 106-116 (1939).
4. Hamilton, D. R., Knipp, J. K., and Kuper, J. B. H., *Klystrons and Microwave Triodes*. McGraw-Hill, New York, 1948.
5. Hutter, R. G. E., *Beam and Wave Electronics in Microwave Tubes*. Van Nostrand, Princeton, N. J., 1960.
6. Llewellyn, F. B., *Electron Inertia Effects*. Cambridge Univ. Press, London and New York, 1941.
7. Llewellyn, F. B., and Bowen, A. E., Production of uhf oscillations by means of diodes. *Bell System Tech. J.* **18**, No. 2, 280-291; April, 1939.
8. Pierce, J. R., *Traveling Wave Tubes*. Van Nostrand, Princeton, N. J., 1950.
9. Ramo, S. I., The electronic-wave theory of velocity modulation tubes. *Proc. IRE* **27**, No. 12, 757-763 (1939).
10. Ramo, S. I., Space charge and field waves in an electron beam. *Phys. Rev.* **56**, 276-283 (1939).
11. Slater, J. C., *Microwave Electronics*. Van Nostrand, Princeton, N. J., 1950.

B. Klystrons

12. Condon, E. U., Electronic generation of electromagnetic oscillations. *J. Appl. Phys.* **11**, No. 7, 502-506 (1940).
13. Hansen, W. W., and Richtmyer, R. D., On resonators suitable for klystron oscillators. *J. Appl. Phys.* **10**, No. 3, 189-199 (1939).

14. Harrison, A. E., *Klystron Tubes*. McGraw-Hill, New York, 1947.
15. Heil, A., and Heil, O., Generation of short waves. *Electronics* **16**, No. 7, 164-178 (1943); [translation of article in *Z. Physik* **95**, Nos. 11 and 12, 752-762 (1935)].
16. Varian, R. H., and Varian, S. F., A high frequency oscillator and amplifier. *J. Appl. Phys.* **10**, No. 5, 321-327 (1939).
17. Warnecke, R. R., Chodorow, M., Guenard, P. R., and Ginzton, E. L., Velocity modulated tubes. *Advan. Electron.* **3**, 43-81 (1951).
18. Webster, D. L., Cathode ray bunching. *J. Appl. Phys.* **10**, No. 7, 501-508 (1939). Webster, D. L., The theory of klystron oscillations. *J. Appl. Phys.* **10**, No. 12, 864-872 (1939).
19. Webster, D. L., Velocity modulation currents. *J. Appl. Phys.* **13**, No. 12, 786-787 (1942).

C. Traveling-Wave Amplifiers, Backward-Wave Oscillators, etc.

20. Bernier, J., Essai de théorie du tube électronique à propagation d'onde. *Ann. Radioelec.* **2**, 87-101 (1947); *Onde Elec.* **27**, 231-243 (1947).
21. Birdsall, C. K., Brewer, G. R., and Haeff, A. V., The resistive-wall amplifier. *Proc. IRE* **41**, No. 7, 865-875 (1953).
22. Birdsall, C. K., and Whinnery, J. R., Waves in an electron stream with a general admittance wall. *J. Appl. Phys.* **24**, No. 3, 315-323 (1953).
23. Brillouin, L., Wave and electrons traveling together—A comparison between traveling wave tubes and linear oscillators. *Phys. Rev.* **74**, 90-92 (1948).
24. Chu, L. J., and Jackson, J. D., Field theory of traveling-wave tubes. *Proc. IRE* **36**, No. 7, 853-863 (1948).
25. Doehler, O., and Kleen, W., Sur l'influence de la charge d'espace dans le tube à propagation d'onde. *Ann. Radioelec.* **3**, pp. 184-188 (1948).
26. Haeff, A. V., The electron wave tube. A novel method of generation and amplification of microwave energy. *Proc. IRE* **37**, 4-10 (1949).
27. Heffner, H., Analysis of the backward-wave traveling-wave tube. *Proc. IRE* **42**, 930-937 (1954).
28. Johnson, H. R., Backward-wave oscillators. *Proc. IRE* **43**, No. 6, 684-698 (1955).
29. Kompfner, R., On the operation of the traveling wave tube at low level. *Brit. J. IRE* **10**, Nos. 8-9, 283-289 (1950).
30. Kompfner, R., Traveling-wave tube—Centimetre wave amplifier. *Wireless Eng.* **24**, 255 (1947).
31. Kompfner, R., Traveling-wave valve—New amplifier for centimetric wavelengths. *Wireless World* **52**, 369-372 (1946).
32. Kompfner, R., and Williams, N. T., Backward-wave tubes. *Proc. IRE* **41**, 1602-1612 (1953).
33. Muller, M., Traveling-wave amplifiers and backward-wave oscillators. *Proc. IRE* **42**, 1651-1658 (1954).
34. Nergaard, L. S., Analysis of a simple model of a two-beam growing-wave tube. *RCA Rev.* **9**, 585-601 (1948).
35. Pierce, J. R., Theory of the beam-type traveling-wave tube. *Proc. IRE* **35**, No. 2, 111-124 (1947).
36. Pierce, J. R., and Hebenstreit, W. B., A new type of high frequency amplifier. *Bell System Tech. J.* **28**, 33-51 (1949).
37. Rydbeck, O. E. H., Theory of the traveling wave tube. *Ericsson Technics* No. 46 (1950).

38. Shulman, C., and Heagy, M. S., Small-signal analysis of traveling-wave tube. *RCA Rev.* **8**, 585-611 (1947).
39. Warnecke, R., Guenard, P., and Doehler, O., Phénomènes fondamentaux dans les tubes à ondes progressives. *Onde Elec.* **34**, 323-338 (1954).

D. Crossed-Field Devices

40. Benham, W. E., Electronic theory and the magnetron oscillator. *Proc. Phys. Soc. (London)* **47**, Pt. 1, No. 258, 1-53 (1935).
41. Brillouin, L., Electronic theory of the plane magnetron. *Advan. Elec.* **3**, 85-144 (1951).
42. Buneman, O., Generation and amplification of waves in dense charged beams under crossed fields. *Nature* **165**, 474-476 (1950).
43a. Doehler, O., On the properties of tubes in a constant magnetic field—Part I, Characteristics and trajectories of the electrons in the magnetron. *Ann. Radioelec.* **3**, 29-39 (1948).
43b. Doehler, O., On the properties of tubes in a constant magnetic field—Part II, The oscillations of resonance. *Ann. Radioelec.* **3**, 169-183 (1948).
43c. Doehler, O., On the properties of tubes in a constant magnetic field—Part III, The traveling-wave tube in a magnetic field. *Ann. Radioelec.* **3**, 328-338 (1948).
43d. Doehler, O., Brossart, J., and Mourier, G., On the properties of tubes in a constant magnetic field—Part IV, Extension of the linear theory, the effects of non-linearities and the efficiency. *Ann. Radioelec.* **5**, 293-307 (1950).
44. Gould, R. W., "A Field Analysis of the M-Type Backward Wave Oscillator." Calif. Inst. Technol. Electron Tube and Microwave Lab. Tech. Rept. No. 3 (September 1955).
45. Gould, R. W., Space charge effects in beam-type magnetrons. *J. Appl. Phys.* **28**, No. 5, 599-604 (1957).
46. Guenard, P., Doehler, O., Epsztein, B., and Warnecke, R., Nouveaux tubes oscillateurs à large bande d'accord électronique pour hyperfréquences. *Comp. Rend. Acad. Sci.* **235**, 236-238 (1952).
47. Guenard, P., and Huber, H., Etude expérimentale de l'interaction par ondes de charge d'espace au sein d'un faisceau électronique de déplaçant dans des champs électrique et magnétique croisés. *Ann. Radioelec.* **7**, No. 30, 252-278 (1952).
48. Hull, A. W., The effect of a uniform magnetic field on the motion of electrons between coaxial cylinders. *Phys. Rev.* **18**, 31-57 (1921).
49. MacFarlane, G. G., and Hay, H. G., Wave propagation in a slipping stream of electrons: Small amplitude theory. *Proc. Phys. Soc. (London)* **63**, Sect. 6-B, 407-427 (1950).
50. Warnecke, R., Doehler, O., and Bobot, D., Les effects de la charge d'espace dans les tubes à propagation d'onde à champ magnétique. *Ann. Radioelec.* **5**, 279-292 (1950).
51. Warnecke, R., Doehler, O., and Kleen, W., Amplification d'ondes électromagnétiques par interaction entre des flux électroniques se déplacant dans des champs électrique et magnétique croisés. *Compt. Rend. Acad. Sci.* **229**, 709-710 (1949).
52. Warnecke, R., Huber, H., Guenard, P., and Doehler, O., Amplification par ondes de charge d'espace dans un faisceau électronique se déplacant dans des champs électrique et magnétique croisés. *Compt. Rend. Acad. Sci.* **235**, 470-472 (1952).
53. Warnecke, R., Kleen, W., Lerbs, A., Doehler, O., and Huber, H., The magnetron type traveling-wave amplifier tube. *Proc. IRE* **38**, No. 5, 486-495 (1950).

E. Beam-Plasma Interactions

54. Ash, E. A., and Gabor, D., Experimental investigations on electron interaction. *Proc. Roy. Soc. (London)* **A228**, 477-490 (1955).
55. Boyd, G. D., Field, L. M., and Gould, R. W., Excitation of plasma oscillations and growing plasma waves. *Phys. Rev.* **109**, 1393-1394 (1958).
56. Boyd, G. D., Field, L. M., and Gould, R. W., Interaction between an electron stream and an arc discharge plasma. *Proc. Symp. Electronic Waveguides, Brooklyn Polytech. Inst. 1958*, Vol. VIII, pp. 367-375. Wiley (Interscience), New York, 1958.
57. Buneman, O., How to distinguish between attenuating and amplifying waves. In *Plasma Physics* (J. E. Drummond, ed.), pp. 143-164. McGraw-Hill, New York, 1961.
58. Crawford, F. W., and Kino, G. S., Oscillations and noise in low-pressure d-c discharges. *Proc. IRE* **49**, No. 12, 1767-1788 (1961) (extensive bibliography).
59. Filiminov, G. F., Growing wave propagation in a plasma. *Radio Eng. Electron.* (USSR) (Engl. Transl.) **4**, 75-87 (1959).
60. Gould, R. W., and Trivelpiece, A. W., Electro-mechanical modes in plasma waveguides. *Proc. IEE (London)* **B105**, Suppl. 10, 516-519 (1958).
61. Gould, R. W., and Trivelpiece, A. W., Space charge waves in cylindrical plasma columns. *J. Appl. Phys.* **30**, 1784-1792 (1959).
62. Hernqvist, K. G., Plasma ion oscillations in electron beams. *J. Appl. Phys.* **26**, 544-548 (1955).
63. Kislov, V. J., and Bogdanov, E. V., Interaction between slow plasma waves and an electron stream. *Proc. Symp. Electromagnetics and Fluid Dynamics of Gaseous Plasma, Brooklyn Polytech. Inst., 1961*, Vol. XI, pp. 249-269. Wiley (Interscience), New York, 1961.
64. Langmuir, I., The interaction of electron and positive ion space charges in cathode sheaths. *Phys. Rev.* **33**, 954-990 (1929).
65. Smullin, L. D., and Chorney, P., Wave propagation in ion-plasma waveguides. *Symp. Electronic Waveguides, Brooklyn Polytech. Inst., 1958*, Vol. VIII, pp. 229-247. Wiley (Interscience), New York, 1958.
66. Tchernov, Z. S., and Bernashevsky, G. A., Amplification of microwaves by means of plasma. *Proc. Symp. Electromagnetics and Fluid Dynamics of Gaseous Plasma, Brooklyn Polytech. Inst., 1961*, Vol. XI, pp. 31-37. Wiley (Interscience), New York, 1961.
67. Tidman, D. A., and Weiss, G., Two-stream instabilities with collisions. *Proc. Symp. Electromagnetics and Fluid Dynamics of Gaseous Plasma, Brooklyn Polytech. Inst. 1961*, Vol. XI, pp. 111-121. Wiley (Interscience), New York, 1961.

CHAPTER II | Eulerian versus Lagrangian Formulation

1 Introduction

Since the invention of the traveling-wave amplifier, numerous methods have been developed to analyze its operation on a small-signal basis. Some of these are the equivalent circuit-ballistic analysis, the coupled-mode analysis, and the field analysis. All have their particular merits and advantages, although the equivalent circuit-ballistic analysis first developed by Pierce is the one most referred to and that which is possibly most useful. A basic postulate of this analysis is the description of the electron beam as a drifting charged fluid, characterized by single-valued velocity and charge density functions at each displacement plane from the input. This treatment of the electron beam is called an Eulerian formulation and is most appropriate to a small-signal analysis. An alternate description is obtained by subdividing the entering beam charge into representative "charge groups" and then carrying these charge groups through the interaction region. This particle type of analysis is called a Lagrangian analysis and is used extensively in the nonlinear treatments.

The characteristics of the Eulerian formulation are developed in this chapter and its limitations when applied to the nonlinear problem are illustrated. Also the basis of the Lagrangian method is discussed and its application to the nonlinear interaction problem is considered. The appropriateness of this method is recognized when electron overtaking occurs and as a result the beam velocity and charge density become multivalued functions of the displacement from the input plane. The transformation from the Eulerian to the Lagrangian system involves no assumptions and is outlined in detail.

2 Eulerian Formulation of O-TWA Equations

The essential elements of a traveling-wave amplifier are a drifting electron beam surrounded by an rf transmission line which supports the propagation of an electromagnetic wave having approximately the same

2. EULERIAN FORMULATION OF O-TWA EQUATIONS

phase velocity in the direction of electron flow. An axial magnetic field to keep the cylindrically shaped beam stable is incidental to the basic operation of the device.

The close proximity of the electron beam to the rf structure gives rise to mutual coupling through the rf electric field, resulting in a velocity modulation of the electrons by the rf wave. Gradually this modulation is converted into density modulation as the electrons drift through the interaction region. Electron bunches are formed and for a net transfer of energy from the beam to the wave the electron stream must slow down. Electron bunches are formed in both the accelerating and decelerating phases of the rf wave, initially there being a much smaller percentage in the accelerating phase. Eventually the favorably phased bunches in the decelerating phase begin to slip back in phase relative to the rf wave and the phase focusing condition is lost, resulting in a saturation condition

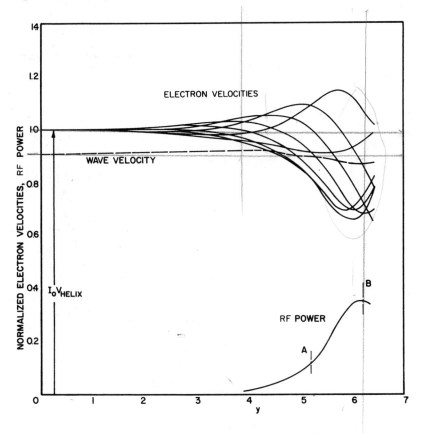

FIG. 1. Power level and electron velocity versus distance.

which limits the rf output. This loss of synchronism between the electrons and the wave limits the output and conversion efficiency.

Another important factor in limiting the conversion efficiency is the inherent debunching of the charge groups as a result of the velocity spread in the bunch. This velocity spread occurs because the rf circuit field does not act uniformly on all parts of the bunch and also due to the space-charge coulomb forces in the bunch.

The interaction process is illustrated graphically in Fig. 1, where the rf

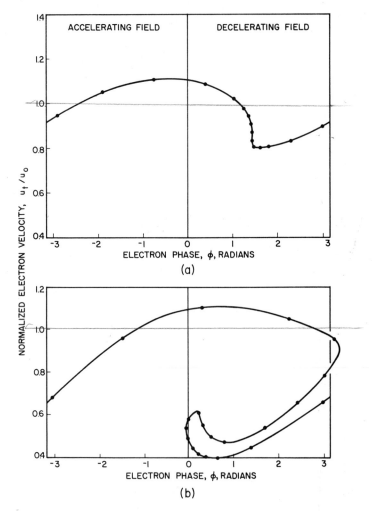

Fig. 2. Normalized electron velocity versus phase position relative to the rf wave. (a) Position A of Fig. 1; (b) position B of Fig. 1.

2. EULERIAN FORMULATION OF O-TWA EQUATIONS

power level and electron velocity are shown as functions of distance. The normalized electron velocity versus electron phase position relative to the rf wave is shown in Fig. 2 for a typical situation to illustrate the division between fast and slow electrons along with their respective phase positions in either the accelerating or decelerating regions. It is noticed that before saturation a strong electron bunch is being formed in the decelerating phase of the rf wave, while a smaller percentage of the electrons are distributed through the accelerating phase. As saturation is approached the tight bunch in the decelerating phase begins to slow down, since it has given a significant portion of its kinetic energy to the rf wave. Eventually this bunch slips back into an accelerating phase and thus takes energy from the rf circuit, limiting the output of the device. The conversion efficiency clearly could be improved if the favorably phased bunch could be held in a decelerating phase as it slows down further. This subject will be treated in detail in Chapter XIII under phase focusing, where criteria for continued focusing of the bunch are developed.

A straightforward approach to the development of the Eulerian analysis of the traveling-wave amplifier, backward-wave oscillator, or klystron is to write the Lorentz force, charge continuity, and rf circuit equations, assuming the existence of an equivalent circuit. This procedure was first utilized by Pierce in treating the traveling-wave amplifier. The force and continuity equations are

$$\frac{d\mathbf{v}}{dt} = \frac{\partial \mathbf{v}}{\partial t} + (\mathbf{v} \cdot \nabla)\mathbf{v} = -|\eta|\,[\mathbf{E}_c(r,t) + \mathbf{E}_{sc}(r,t)] \tag{1}$$

and

$$\nabla \cdot \mathbf{i}(r,t) + \frac{\partial \rho(r,t)}{\partial t} = 0, \tag{2}$$

where

$|\eta|$ = charge-to-mass ratio for an electron,
r = a generalized coordinate,
v = electron beam velocity,
E_c = rf circuit field,
E_{sc} = space-charge field,
ρ = beam charge density, and
i = beam convection current density.

A transmission-line type of equivalent circuit is assumed in which the elements of the line are adjusted so that the characteristic phase velocity

II. EULERIAN VERSUS LAGRANGIAN FORMULATION

d impedance match those of the rf interaction circuit for the desired iteraction mode.

The development of the equivalent circuit transmission-line concept is outlined in detail in Chapter III and thus will not be discussed further here. The form of this equivalent transmission line for the forward-wave amplifier is shown in Fig. 3. The following second-order differential

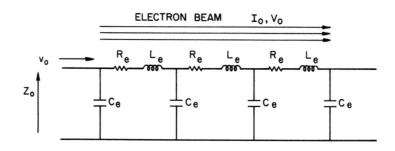

FIG. 3. Forward-wave amplifier transmission-line equivalent circuit.
$v_0 \triangleq 1/\sqrt{L_e C_e}$ meters/sec; $\quad Z_0 \triangleq \sqrt{L_e/C_e}$ ohms;
$R_e/L_e \triangleq 2\omega Cd$; $\quad C \triangleq (Z_0 I_0/4V_0)^{\frac{1}{3}}$.

equation for the rf voltage along the line is developed in Chapter III:

$$\frac{\partial^2 V(z,t)}{\partial z^2} - \frac{1}{v_0^2}\frac{\partial^2 V(z,t)}{\partial t^2} - \frac{2\omega Cd}{v_0}\frac{\partial V(z,t)}{\partial t}$$
$$= -\frac{Z_0}{v_0}\left[\frac{\partial^2 \rho(z,t)}{\partial t^2} + 2\omega Cd\frac{\partial \rho(z,t)}{\partial t}\right], \quad (3)$$

where

d = a measure of the attenuation per unit length of line,
C = a beam-circuit coupling parameter,
v_0 = the characteristic line velocity, and
Z_0 = the characteristic line impedance.

Equations (1), (2), and (3) are perfectly general, as nonlinear terms and terms depending upon transverse coordinates are retained. Numerous

2. EULERIAN FORMULATION OF O-TWA EQUATIONS

attempts have been made to solve this nonlinear system on a general basis but little success has so far been achieved. The difficulty lies not with the linear circuit equation but with the nonlinear force and continuity equations. A detailed discussion of solution methods is given in Chapter XIII. Under small-amplitude conditions the force and continuity equations may be linearized by neglecting the second-order terms in ac quantities and products of ac quantities.

If the beam diameter is small so that the circuit field does not vary across it, then a further simplification results from making a one-dimensional assumption. The resulting second-order ballistic and circuit differential equations are reduced to algebraic equations after assuming wave-type solutions of the form $e^{j\omega t - \Gamma z}$, where $\Gamma \triangleq$ the wave propagation constant.

$$2V_0(j\beta_e - \Gamma)^2 i = jI_0\beta_e \Gamma V \tag{4}$$

and

$$\Gamma V = \left[\frac{\Gamma^2 \Gamma_0(E^2/\beta^2 P)}{2(\Gamma_0^2 - \Gamma^2)} - \frac{j\Gamma^2}{\omega C_1}\right] i. \tag{5}$$

The above equations are simply combined to give the system determinantal equation.

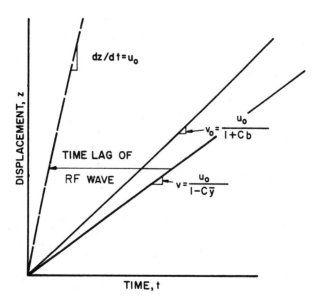

FIG. 4. Electron flight-line diagram for small-amplitude operation. $v_0 \triangleq$ undisturbed circuit phase velocity; $v \triangleq$ actual rf wave velocity.

It is seen that the real basis for the linear theory is the existence of wave solutions varying exponentially in distance and time and the assumption that quantities such as the charge density and electron velocity are single-valued functions of distance. Such assumptions are not justified in the nonlinear regime, since growth rates are not exponential and electron overtaking does occur.

Linear theory independent variables are taken as axial displacement and time, and no difficulty arises in the analysis due to the absence of multivalued flow. The rf wave velocity and phase shift per unit length are therefore constant throughout the interaction region. These characteristics are summarized in Fig. 4, where several typical electron flight lines are shown. The actual wave is shown lagging behind the stream and since the wave velocity is constant in the linear theory the time lag or corresponding phase lag $\theta(z) = C\beta_e \bar{y} z$ increases with distance.

Figure 4 may be replotted to show the phase lag directly by a simple change of variables as shown in Fig. 5. Thus it is seen that the important quantities in a linear analysis are the beam-wave relative velocity as measured by $b \triangleq (u_0 - v_0)/Cv_0$ and the departure of the actual wave velocity from v_0 as measured by $\bar{y}$, a negative number. It has been convenient to normalize the displacement in terms of C and β_e. The use of these parameters in the nonlinear theory is examined in the next section.

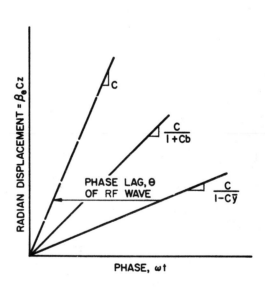

FIG. 5. Phase diagram for linear operation.

3 Lagrangian Formulation

As has been pointed out above, electron flight-line crossings do occur, resulting in multivalued current and velocity functions, and thus the use of distance and time as independent variables is limited to the small-amplitude regime. In view of the inapplicability of the fluid flow analysis to the large-amplitude regime, the beam description is altered to follow individual charge groups through the interaction region, the totality of which constitutes the beam charge. Thus, in addition to the displacement, a new independent variable indicating the relative time of entry of the charge group into the interaction region is used in the nonlinear region. Of course, the entering time may easily be transformed to the charge group entry phase relative to the applied rf wave.

The equivalent circuit and Lorentz force equations remain essentially the same in the Lagrangian frame and only the continuity of charge or conservation equation takes on a different form. The charge at any displacement plane is related to the entering stream charge as follows (unmodulated beam):

$$\rho(z, t)\, dz = \rho(z_0, 0)\, dz_0 . \tag{6}$$

Since

$$\rho(z_0, 0) = \frac{I_0}{u_0}, \tag{7}$$

then Eq. (6) is written as

$$\rho(z, t) = \frac{I_0}{u_0} \left| \frac{\partial z_0}{\partial z} \right|_t , \tag{8}$$

where the absolute value signs are introduced to indicate that all branches of the multivalued charge-density function must be considered at any displacement plane. The conservation law is satisfied by a counting operation on the entering charge groups.

The Lagrangian independent variables* are illustrated in appropriate flight-line diagrams of Figs. 6a and 6b. The new independent variables to be used in the large-amplitude theory are normalized distance and entry phase as defined below.

$$y \triangleq \frac{C\omega}{u_0} z = C\beta_e z = 2\pi C N_s \tag{9}$$

---

* In the Eulerian system, the independent variables are generally taken as z, t whereas in the Lagrangian system y, ϕ_0 are used. Dependent variables such as voltage and velocity will be written as $V(z, t)$ or $V(y, \phi_0)$ and $u(z, t)$ or $u(y, \phi_0)$ depending on the system being used. This slight sacrifice of mathematical rigor is justified on the basis of the ensuing simplicity.

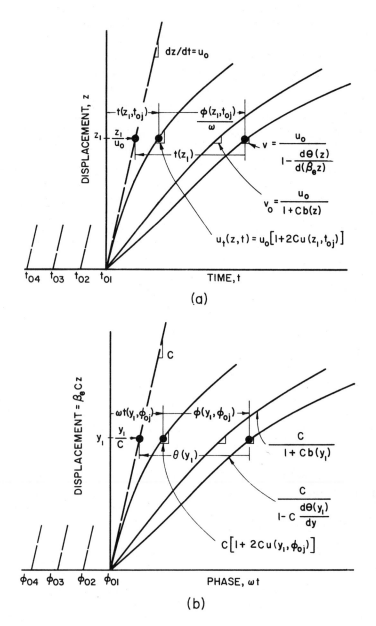

Fig. 6. Lagrangian system flight-line diagrams. (a) Distance, time system; (b) normalized distance, phase system.

3. LAGRANGIAN FORMULATION

and
$$\Phi_{0j} \triangleq \omega t_{0j}, \tag{10}$$

where

$\beta_e \triangleq$ stream phase constant, and

$N_s \triangleq z/\lambda_s$, the number of stream wavelengths.

Note that Φ_{0j} is an arbitrary initial condition denoting the phase position of the individual entering charge group relative to the rf wave.

In the more general multidimensional treatments of Chapters V–X the same normalization basis is utilized. For the sake of generality b has been written as $b(z)$ in order to include variable parameter circuits and spatially varying beam voltages.

In defining dependent variables it is convenient to refer quantities to the initial average electron velocity u_0. A result of the small-amplitude approximations is the invariance of u_0 and hence no energy conversion information can be obtained. In the nonlinear theory the average stream velocity is a function of displacement and the dependent velocity variable is introduced through the following definition of the total charge-group velocity:

$$\left. \frac{dz}{dt} \right|_{z,t_{0j}} \triangleq u_0[1 + 2Cu(z, t_{0j})]. \tag{11}$$

For an unmodulated entering beam $u(0, t_{0j}) \equiv 0$, and a nonzero value arises due to the circuit and space-charge forces acting on the charges. The quantity $2Cu_0u(z, t_{0j})$ has an rf component and a negative average component which grows with distance, accounting for the reduction in the average beam energy. Another dependent beam variable is the phase $\Phi(z, t_{0j})$, which indicates the phase position of the individual charge groups relative to the rf wave as a function of displacement. Clearly both $u(z, t_{0j})$ and $\Phi(z, t_{0j})$ are multivalued quantities, one pair for each charge group.

Following the form of the linear theory solutions, a wave-type solution for the rf circuit voltage is written in terms of the product of two slowly varying, singly periodic functions of distance and time. Floquet's theorem, or the generalization due to Bloch, suggests the following simplified form:

$$V(z, t) = \text{Re}\left\{V(z) \exp\left(j\left[\omega t - \int_0^z \beta(z)\,dz\right]\right)\right\}$$
$$= \text{Re}\{V(z) \exp -j\Phi\}, \tag{12a}$$

where

$$-\Phi \triangleq \omega t - \int_0^z \beta(z)\,dz. \tag{12b}$$

It is apparent from Eq. (12b) that the phase lag in the nonlinear case does vary with displacement. Referring to the phase diagram of Fig. 6b, we define the phase lag $\theta(y)$ of the actual rf wave relative to the hypothetical wave traveling at u_0 as

$$\theta(y) \triangleq \frac{y}{C} - \omega t - \Phi(y, \Phi_{0j}). \tag{13}$$

The construction of Eq. (13) from Fig. 6b is evident.

Thus it is seen that the dependent variable $\Phi(y, \Phi_{0j})$ has a dual nature, indicating the phase position of the individual charge group at any displacement plane and also giving the phase of the traveling rf wave at any y-plane. As a consequence of these definitions the undisturbed circuit phase velocity and the actual wave phase velocity are given by

$$v_0(z) = \frac{u_0}{1 + Cb(z)} \tag{14}$$

and

$$v(z) = \frac{u_0}{1 - \dfrac{d\theta(z)}{d(\beta_e z)}} \tag{15}$$

respectively. Thus we see that the velocity parameter $b(z)$ is still useful in the nonlinear theory, where

$$\frac{u_0}{v_0(z)} = 1 + Cb(z). \tag{16}$$

However, instead of the linear phase constant we now have a wave phase variable $\theta(z)$, which accounts for changes in actual wave velocity:

$$\frac{u_0}{v(z)} = 1 - \frac{d\theta(z)}{d(\beta_e z)}. \tag{17}$$

4 Composite Lagrangian System

The charge conservation equation (Eq. (8)) takes on the following form after the introduction of the above defined variables.

$$\rho(y, \Phi) = \frac{I_0}{u_0} \left| \frac{\partial \Phi_0}{\partial \Phi} \right| \frac{1}{[1 + 2Cu(y, \Phi_0)]}. \tag{18}$$

As noted above the circuit and force equations retain essentially the same form after introduction of the Lagrangian variables. In order to

4. COMPOSITE LAGRANGIAN SYSTEM

illustrate the exact nature of the nonlinear equations in the new system the one-dimensional traveling-wave amplifier and backward-wave oscillator equations are given below assuming that the stream-circuit coupling as measured by C is weak, i.e., $C \ll 1$.

Circuit Equations

$$\frac{dA(y)}{dy} = \pm \frac{1}{2\pi} \int_0^{2\pi} \sin \Phi(y, \Phi'_{0j}) \, d\Phi'_{0j} \tag{19}$$

$$\frac{d\theta(y)}{dy} + b = \mp \frac{1}{2\pi A(y)} \int_0^{2\pi} \cos \Phi(y, \Phi'_{0j}) \, d\Phi'_{0j} \tag{20}$$

Lorentz Equation

$$\frac{\partial u(y, \Phi_{0j})}{\partial y} = -A(y) \sin \Phi(y, \Phi_{0j}) \tag{21}$$

Phase Equation

$$\frac{\partial \Phi(y, \Phi_{0j})}{\partial y} + \frac{d\theta(y)}{dy} = 2u(y, \Phi_{0j}) \tag{22}$$

The upper of the double signs indicates the amplifier equation and the lower the oscillator equation. $A(y)$ denotes the normalized rf voltage amplitude defined by $V(y) = (Z_0 I_0/C) A(y)$. Space-charge fields have also been neglected. It should be apparent that Eqs. (21) and (22) are each m in number, where $j = 0, 1, 2, ..., m$, and m denotes the number of entering charge groups. Each of these representative charge groups is followed through the interaction region.

The full use of the new variables is outlined in succeeding chapters, which deal with specific interaction configurations. Fortunately the basic Lagrangian approach as developed above applies to both O- and M-type devices as well as various beam-plasma systems.

CHAPTER III

Radio-Frequency Equivalent Circuits

1 Introduction

The analysis of charge-electromagnetic wave interaction problems is considerably simplified by the use of an old theorem frequently used in network analysis, namely the superposition theorem. With the aid of this well-known theorem the total field or voltage associated with the electromagnetic wave may be written as the linear combination of that field or voltage due to currents flowing in the circuit and that due to induced currents flowing as a result of a modulated or bunched charged beam flowing in proximity to the circuit, independent of the exact form of the wave propagating structure. The circuit may assume any one of a number of various lumped element or distributed element arrays, all of which can support the propagation of an electromagnetic wave. In view of this separability of the problem we may address ourselves to the circuit problem and delay consideration of the charge dynamics to a later chapter.

In view of the generality and breadth of applicability of Maxwell's[8] fundamental equations the first resort in the search for an exact and appropriate treatment of the field aspect of the problem is to consider direct application of these fundamental laws in calculating the necessary fields. The basic equations which are the starting point for such an analysis are

$$\nabla \cdot \mathbf{E} = \rho/\epsilon_0, \qquad (1)$$

$$\nabla \cdot \mathbf{H} = 0, \qquad (2)$$

$$\nabla \times \mathbf{E} = -\mu_0 \frac{\partial \mathbf{H}}{\partial t} \qquad (3)$$

and

$$\nabla \times \mathbf{H} = \mathbf{J} + \frac{\partial \mathbf{D}}{\partial t} \qquad (4)$$

in free space and where the symbols E, H, D, ρ, and J refer to electric intensity, magnetic intensity, electric flux density, space-charge density,

1. INTRODUCTION

and convection current density respectively. μ_0 and ϵ_0 are the free-space permeability and permittivity respectively.

The absence of such well-known constant factors as 4π and c occurs as a result of the author's preference for the meter-kilogram-second rationalized system of units. As might be expected, adherence to this system will be maintained throughout the treatise and we hereby bid a fond adieu to all other systems.

For completeness, the simple divergence operator applied to Eq. (4) yields the well-known equation

$$\nabla \cdot \mathbf{J} + \frac{\partial \rho}{\partial t} = 0, \tag{5}$$

expressing the conservative nature of the system in support of the First Law of Thermodynamics. Elementary manipulation of the above relations yields, in the E field, the inhomogeneous wave equation,

$$\nabla^2 \mathbf{E} - \frac{1}{c^2} \frac{\partial^2 \mathbf{E}}{\partial t^2} = \frac{\nabla \rho}{\epsilon_0} + \mu_0 \frac{\partial \mathbf{J}}{\partial t}. \tag{6}$$

The equation in the E field has been chosen because of a later need to substitute this term into the Lorentz equation. Other field components may easily be found by backwards substitution through the divergence and curl equations. Equation (6) is perfectly general, and its solution will give the electric field arising from an arbitrary charge distribution and movement past the wave guiding system. It is a vector wave equation and of course may be resolved into component form for ease of handling in a particular situation.

The next step in this direct approach is to incorporate the other portion of the problem through the Lorentz equation

$$\frac{\partial \mathbf{v}}{\partial t} + (\mathbf{v} \cdot \nabla)\mathbf{v} = -\eta[\mathbf{E} + \mathbf{v} \times \mathbf{B}], \tag{7}$$

where

$\eta \triangleq$ particle charge-to-mass ratio,

$v \triangleq$ particle velocity, and

$B \triangleq$ magnetic induction.

Equations (6) and (7) govern the complete problem, subject to the imposition of the appropriate and necessary boundary conditions. This marriage is made not without difficulties and generally one finds that Eq. (6) is not altogether convenient for the purpose. Thus other equivalent but exact procedures are sought.

A noble and esoteric application of this technique was carried out by Chu and Jackson[5] in their "Field Theory of Traveling Wave Tubes". Solution was possible in the linearized sense, although the task of obtaining subsequent solutions under desired parameter variation would be a truly formidable one. In the nonlinear regime this approach is completely intractable without the introduction of multitudinous limiting assumptions.

2 Equivalence of Maxwell and Kelvin Theories

The problem which besets us is thus one of finding a more tractable equivalent method of handling the circuit problem and yet retaining the rigor of the Maxwell equation approach. Such a method has its roots in the well-established and well-known telegraphist's equations introduced by Lord Kelvin[21] in 1884. It is interesting to note that Lord Kelvin's equations preceded those of Maxwell by some 10 years. It is well known that for many problems in electromagnetic field theory the engineering method of solution as typified by the use of Kelvin's classical telegraphist's equations can lead to not only useful results but greater physical insight, whereas the Maxwell field approach may not lead to useful results.

This equivalent circuit or transmission-line approach forms the basis for the treatment of the role of the wave guiding structure in the nonlinear interaction problems characteristic of the devices studied. Schelkunoff[16] has studied the equivalence of the two methods in detail for numerous cases and finds that Maxwell's generalized field equations (including boundary conditions) can in fact be transformed into a set of equivalent coupled transmission-line equations. Thus the unassailable problem outlined above becomes readily solvable when freed of the complications introduced by the transverse boundary conditions and the difficulty in separating dependent variables.

Schelkunoff's generalized telegraphist's equations differ from the earlier, less general system due to Kelvin both for parallel wire lines separated by a homogeneous medium and for waveguides filled with an inhomogeneous dielectric medium. In the former instance, which is of chief concern to this problem, the difference lies in the single mode of propagation obtained by Kelvin and the infinite set of modes obtained by Schelkunoff. The infinite set of modes arises because of distributed coupling along the line between the fundamental and higher-order modes. Neglect of this coupling, which in many cases of interest is extremely small, reduces the generalized set to the results of Kelvin.

2. EQUIVALENCE OF MAXWELL AND KELVIN THEORIES

It is straightforward to think of an electromagnetic wave guiding system including the boundary conditions in terms of voltage and currents along a simple two-conductor transmission line. In a simple "gedanken" experiment, visualize the construction or representation of a multimode guiding system by the continuous coupling of an infinite set of two-conductor systems such as illustrated in Fig. 1.

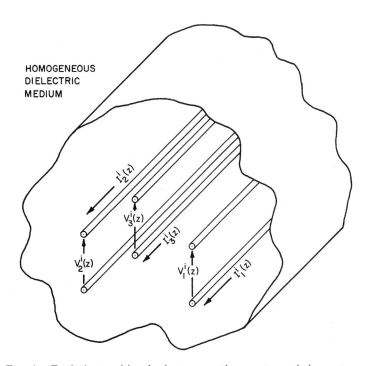

Fig. 1. Equivalent multimode electromagnetic wave transmission system.

The telegraphist's equations for such a system as illustrated in Fig. 1 are written as

$$\frac{\partial V_m}{\partial z} + \sum_{n=1}^{\infty \text{ or } N} \left[R_{mn} I_n + L_{mn} \frac{\partial I_n}{\partial t} \right] = 0 \tag{8}$$

and

$$\frac{\partial I_m}{\partial z} + \sum_{n=1}^{\infty \text{ or } N} \left[G_{mn} V_n + C_{mn} \frac{\partial V_n}{\partial t} \right] = 0, \tag{9}$$

where voltages and currents associated with the nth and mth lines are taken as instantaneous values. The equivalent line parameters R_{mn}, L_{mn},

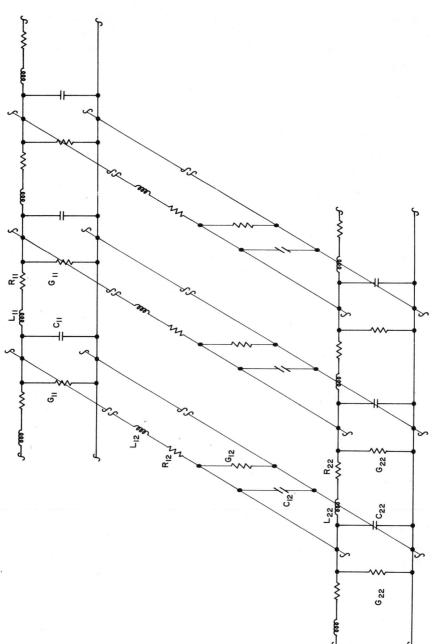

FIG. 2. Equivalent circuit representation of the infinite system of Fig. 1.

G_{mn}, and C_{mn} are the equivalent resistance, inductance, conductance, and capacitance associated with each line separately and all pairs of lines. For equal indices m and n, the parameters may be thought of as self-line parameters and for $m \neq n$ they represent mutual coupling parameters.

The equivalent circuit concept may be used to construct Fig. 2 for the system depicted by Eqs. (8) and (9), where the line parameters are used in the sense given above. The line parameters are conveniently defined on a per unit length basis to facilitate their use in either lumped or distributed element transmission systems. In the case of a single-mode system (or weak coupling to higher-order modes) such as considered by Lord Kelvin, the indices m and n are equal and the summations of Eqs. (8) and (9) disappear, resulting in

$$\frac{\partial V}{\partial z} + RI + L\frac{\partial I}{\partial t} = 0 \tag{10}$$

and

$$\frac{\partial I}{\partial z} + GV + C\frac{\partial V}{\partial t} = 0. \tag{11}$$

These equations are familiar to the undergraduate. The equivalent circuit of Fig. 2 is correspondingly simplified. In both the general (multimode) case and the single-mode line, the line parameters may indeed be functions of distance, denoting an inhomogeneous wave guiding system. Such lines are considered in Section 7 of this chapter.

Since we are interested generally in the steady-state behavior of the guiding systems we proceed to assume a harmonic nature for the voltages and currents along the line and take the voltages and currents as real parts of complex quantities defined by

$$V_m e^{j\omega t} \tag{12a}$$

and

$$I_m e^{j\omega t}, \tag{12b}$$

where ω denotes the angular frequency of the impressed electromagnetic wave. Application of Eq. (12) to Eqs. (8) and (9) respectively results in a system of ordinary differential equations with either constant or variable coefficients, depending upon the constancy, or lack thereof, of the line parameters with distance. This implicitly assumes that R, L, G, and C are time independent, which is not always valid. The results are

$$\frac{dV_m}{dz} + \sum_{n=1}^{\infty \text{ or } N} Z_{mn} I_n = 0 \tag{13}$$

and

$$\frac{dI_m}{dz} + \sum_{n=1}^{\infty \text{ or } N} Y_{mn} V_n = 0. \tag{14}$$

Z_{mn} and Y_{mn} are respectively the complex impedance and complex admittance per unit length of the line, defined as

$$Z_{mn} = R_{mn} + j\omega L_{mn} \tag{15a}$$

and

$$Y_{mn} = G_{mn} + j\omega C_{mn}. \tag{15b}$$

Generally the leakage conductance G_{mn} may be neglected.

The derivation of the transmission-line equations assumed that the line parameters were dc quantities, which limits the use of Eqs. (8)–(11). This apparent difficulty is, however, overcome in Eqs. 13 and 14, where Z_{mn} and Y_{mn} are the complex admittances. It will be shown later that Z_{mn} is related to the wave electric field and Z_{mn} and Y_{mn} together give the wave (mode) propagation constant. These relationships correctly denote the function of the equivalent circuit representation in that the line elements are selected so as to represent the electric field (Z_{mn}) acting at the beam position and the proper value of wave propagation constant $\sqrt{Z_{mn}Y_{mn}}$.

The above form of the circuit equations may be combined with the Lorentz equation and the system solved in a straightforward manner. The equivalence of the Maxwell and Kelvin theories is thus established theoretically. Further evidence of the validity of the approach exists in terms of experimental results which lend credence to the method even for very high frequencies, where such equivalent circuit concepts might not be thought to be applicable.

3 Equivalence for a Helical Wave Guiding Structure

The equivalence of the Kelvin and Maxwell theories has been carried through in general and Schelkunoff has also evaluated the equivalent line parameters for a number of particular types of waveguide systems wherein these parameters are expressed in terms of the structure dimensions and field dependencies.

One of the most important wave propagating structures of all those used in the various microwave electron devices is the helical wire transmission line, which may be viewed as a distorted two-conductor line in which the principal electric field is axial as compared to the transverse field characteristic of the two-wire line. The prominence of the helix is due to the fact that it has a reasonable impedance while being periodic but nonresonant; the latter property accounts for its inherent broad bandwidth. Bandwidths in excess of one octave are possible. In

3. EQUIVALENCE FOR A HELICAL WAVE GUIDING STRUCTURE

view of its wide range of applicability in traveling-wave amplifiers, oscillators, plasma amplifiers and parametric systems, it is appropriate to consider its equivalent circuit.

The electromagnetic field problem for the finite-wire-size helix, including attenuation, has not been solved rigorously although many noble attempts have been made [4,17]. The usual treatment of the helix waveguide has been to consider a helically conducting sheath of infinitesimal thickness[13], for which the field problem is easily solved. The impedance of an actual helix is lower than that predicted by the sheath helix model due to numerous effects such as wire size, harmonic fields, loss, and dielectric loading. These have been partially accounted for by Tien[19] and Chu[4].

The object here is only to show by example on the helical waveguide (following Schelkunoff's general method) that periodic propagating structures can be represented by equivalent transmission lines and that the line elements per unit length are derivable in terms of the structure dimensions and the electromagnetic field forms.

Special cases such as the single helical line, coupled helical lines, electron beam, etc., have been treated in this manner by several workers including the author in unpublished work. Some of these problems have been discussed by Paik[11] and Kino-Paik[7].

In view of the time dependent nature, $e^{j\omega t}$, of the electromagnetic fields being considered we are obliged to define the magnetic induction, **B**, in terms of the magnetic vector potential **A**, due to the solenoidal nature of **B**, as

$$\mathbf{B} = \nabla \times \mathbf{A} \tag{16}$$

and the electric induction as

$$\mathbf{E} = -\nabla \Phi - \frac{\partial \mathbf{A}}{\partial t} \tag{17}$$

where Φ is the scalar potential. If **B** is time independent, then **E** is derivable directly from the scalar potential. Proper choice of $\nabla \cdot \mathbf{A}$ leads to similar differential equations for all components of the general four-vector potential (A_x, A_y, A_z, Φ).

As both Φ and **A** are still arbitrary, we choose a relationship between them as given by the Lorentz condition, namely

$$\nabla \cdot \mathbf{A} + j\omega\mu_0\epsilon_0 \Phi = 0. \tag{18}$$

The result is

$$\nabla^2 \begin{Bmatrix} \mathbf{A} \\ \Phi \end{Bmatrix} - \mu_0\epsilon_0 \frac{\partial^2}{\partial t^2} \begin{Bmatrix} \mathbf{A} \\ \Phi \end{Bmatrix} = \begin{Bmatrix} -\mu_0 \mathbf{J} \\ -\rho/\epsilon_0 \end{Bmatrix}. \tag{19}$$

The interaction problems of interest to us here are those between a propagating wave and an axially (z) drifting charged beam. Thus, we are interested in both A_z and E_z fields. Assuming that the fields are spatially as well as time harmonic and taking only the z-components, we obtain the following expression for the axial electric field from Eqs. (17) and (18):

$$E_z = -\left(1 - \frac{k^2}{\beta^2}\right)\frac{\partial \Phi}{\partial z} = j\beta\left(1 - \frac{k^2}{\beta^2}\right)\Phi$$

$$= \frac{j\gamma^2}{\beta}\Phi, \qquad (20)$$

where $k^2 \triangleq \omega^2 \mu_0 \epsilon_0$. Invoking the slow-wave assumption and taking the value of the scalar potential at the sheath helix radius, we have for slow waves

$$E_{za} \approx j\beta V;$$

$$\gamma^2 = \beta^2 - k^2 \approx \beta^2. \qquad (21)$$

The field problem of the sheath helix may be solved in a variety of ways. Both TE and TM waves may be included by starting from the appropriate Hertzian vector potentials as given below.

For TE Waves

$$\mathbf{H} = \nabla \times \nabla \times \mathbf{\Pi}_1 \qquad (22a)$$

and

$$\mathbf{E} = -\mu \frac{\partial}{\partial t} \nabla \times \mathbf{\Pi}_1. \qquad (22b)$$

For TM Waves

$$\mathbf{H} = \left(\epsilon \frac{\partial}{\partial t} + \sigma\right) \nabla \times \mathbf{\Pi}_2 \qquad (23a)$$

and

$$\mathbf{E} = \nabla \times \nabla \times \mathbf{\Pi}_2. \qquad (23b)$$

The TM waves are excited by axial currents as I_z, and TE waves arise from azimuthal currents I_θ. It is necessary to assume the existence of both TE and TM modes in order to develop expressions for both the series and shunt elements.

In view of the difficulty in solving the field problem when loss is considered we shall concern ourselves here only with the reactive elements of the equivalent circuit. An approximate derivation of an expression for the series dissipative element can be made using the derived field components.

3. EQUIVALENCE FOR A HELICAL WAVE GUIDING STRUCTURE

The sheath helix being considered is shown in Fig. 3. Since the details of obtaining the field expressions from Eqs. (22) and (23) with the appropriate boundary conditions have been worked out by Sensiper[17] and many other authors, only the results of the manipulations are given here.

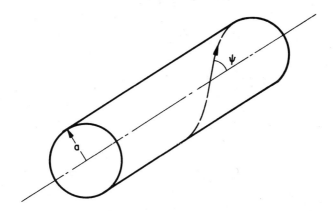

FIG. 3. Sheath helix model.

A. TE Modes Due to I_θ Currents

(1) Inside the helical sheath, $r \leqslant a$:

$$H_z^i = A_1 I_0(\gamma r), \tag{24a}$$

$$H_r^i = \left(\frac{j\beta}{\gamma}\right) A_1 I_1(\gamma r), \tag{24b}$$

$$E_\theta^i = \left(\frac{-j\omega\mu_0}{\gamma}\right) A_1 I_1(\gamma r). \tag{24c}$$

(2) Outside the helical sheath, $r \geqslant a$:

$$H_z^0 = A_2 K_0(\gamma r), \tag{25a}$$

$$H_r^0 = \left(\frac{-j\beta}{\gamma}\right) A_2 K_1(\gamma r), \tag{25b}$$

$$E_\theta^0 = \left(\frac{j\omega\mu_0}{\gamma}\right) A_2 K_1(\gamma r). \tag{25c}$$

B. TM Modes Due to I_z Currents

(1) Inside the helical sheath, $r \leqslant a$:

$$E_z^i = B_1 I_0(\gamma r), \tag{26a}$$

$$E_r^i = \left(\frac{j\beta}{\gamma}\right) B_1 I_1(\gamma r), \tag{26b}$$

$$H_\theta^i = \left(\frac{j\omega\epsilon_0}{\gamma}\right) B_1 I_1(\gamma r). \tag{26c}$$

(2) Outside the helical sheath, $r \geqslant a$:

$$E_z^0 = B_2 K_0(\gamma r), \tag{27a}$$

$$E_r^0 = \left(\frac{-j\beta}{\gamma}\right) B_2 K_1(\gamma r), \tag{27b}$$

$$H_\theta^0 = \left(\frac{-j\omega\epsilon_0}{\gamma}\right) B_2 K_1(\gamma r). \tag{27c}$$

The boundary conditions to be applied occur at $r = 0$ since the field is to be finite there, at infinity where it tends to zero, and at the radius $r = a$ where the fields must be continuous across the delta-thickness sheath. The first two conditions, i.e., at $r = 0$ and ∞, have already been applied in writing Eqs. (24)–(27). It remains to invoke those at $r = a$.

The boundary conditions at the sheath for the two sets of modes are

A. *TE Modes*

$$E_\theta^i(a) = E_\theta^0(a) \tag{28a}$$

and

$$H_z^i(a) - H_z^0(a) = \frac{I_\theta}{2\pi a}. \tag{28b}$$

B. *TM Modes*

$$E_z^i(a) = E_z^0(a) \tag{29a}$$

and

$$-H_\theta^i(a) + H_\theta^0(a) = \frac{I_z}{2\pi a}. \tag{29b}$$

Equations (28a) and (29a) express the continuity of tangential electric field across the boundary and Eqs. (28b) and (29b) denote the fields arising from the assumed currents. We also have the following conditions to be satisfied at $r = a$:

$$E_z(a) \tan \psi = -E_\theta(a) \tag{30a}$$

and

$$I_z \cot \psi = I_\theta. \tag{30b}$$

3. EQUIVALENCE FOR A HELICAL WAVE GUIDING STRUCTURE

Application of the above conditions to the earlier field equations permits expression of the axial and angular electric fields in terms of the corresponding currents. The results are

$$E_\theta^i = \left[\left(\frac{-j\omega\mu_0}{2\pi}\right) I_1(\gamma r) K_1(\gamma a)\right] I_\theta \qquad r \leqslant a, \tag{31a}$$

$$E_\theta^o = \left[\left(\frac{-j\omega\mu_0}{2\pi}\right) I_1(\gamma a) K_1(\gamma r)\right] I_\theta \qquad r \geqslant a, \tag{31b}$$

$$E_z^i = \left[\left(\frac{j\gamma^2}{2\pi\omega\epsilon_0}\right) I_0(\gamma r) K_0(\gamma a)\right] I_z \qquad r \leqslant a, \tag{32a}$$

and

$$E_z^o = \left[\left(\frac{j\gamma^2}{2\pi\omega\epsilon_0}\right) I_0(\gamma a) K_0(\gamma r)\right] I_z \qquad r \geqslant a. \tag{32b}$$

Having obtained the field components one may now proceed with the development of the transmission-line parameter expressions. Equations (32) for the axial E field may be written in terms of the circuit potential using Eqs. (20) and (21), with the result (for slow waves)

$$\frac{\partial I_z}{\partial z} + \left(\frac{\beta}{\gamma}\right)^2 \left[\frac{2\pi\epsilon_0}{I_0(\gamma a) K_0(\gamma a)}\right] \frac{\partial V}{\partial t} = 0. \tag{33}$$

The corresponding relation for the voltage gradient along the line is obtained using Eqs. (31), subject to Eqs. (28) and (30), with the following result:

$$\frac{\partial V}{\partial z} + \left[\frac{\mu_0 \cot^2 \psi}{2\pi} I_1(\gamma a) K_1(\gamma a)\right] \frac{\partial I_z}{\partial t} = 0. \tag{34}$$

Comparison of Eqs. (33) and (34) with Eqs. (8) and (9) or (10) and (11) indicates that the incremental inductance and capacitance per unit length of line may be defined as follows:

$$L_e \triangleq \frac{\mu_0 \cot^2 \psi}{2\pi} I_1(\gamma a) K_1(\gamma a) \quad \text{henrys/meter} \tag{35}$$

and

$$C_e = \left(\frac{\beta}{\gamma}\right)^2 \frac{2\pi\epsilon_0}{I_0(\gamma a) K_0(\gamma a)}$$

$$\approx \frac{2\pi\epsilon_0}{I_0(\gamma a) K_0(\gamma a)} \quad \text{farads/meter.} \tag{36}$$

The impedance for this lossless line, illustrated in Fig. 4, may be obtained from

$$Z_0 \triangleq \sqrt{\frac{Z}{Y}} = \sqrt{\frac{L_e}{C_e}} = \left(\frac{\mu_0}{\epsilon_0}\right)^{\frac{1}{2}} \frac{\cot \psi}{2\pi}$$
$$\cdot [I_0(\gamma a) K_0(\gamma a) I_1(\gamma a) K_1(\gamma a)]^{\frac{1}{2}}, \qquad (37)$$

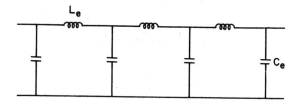

FIG. 4. Equivalent circuit for a sheath-helix transmission line.

$$L_e = \frac{\mu_0 \cot^2 \psi}{2\pi} I_1(\gamma a) K_1(\gamma a) \quad \text{henrys/meter},$$

$$C_e = \frac{2\pi \epsilon_0}{I_0(\gamma a) K_0(\gamma a)} \quad \text{farads/meter},$$

$$v_0 = \frac{c}{\cot \psi} \left[\frac{I_0(\gamma a) K_0(\gamma a)}{I_1(\gamma a) K_1(\gamma a)}\right]^{\frac{1}{2}} \quad \text{meters/sec},$$

$$Z_0 = \left(\frac{\mu_0}{\epsilon_0}\right)^{\frac{1}{2}} \frac{\cot \psi}{2\pi} [I_0(\gamma a) K_0(\gamma a) I_1(\gamma a) K_1(\gamma a)]^{\frac{1}{2}}.$$

and the characteristic axial phase velocity is

$$v_0 \triangleq \frac{1}{\sqrt{ZY}} = \frac{1}{\sqrt{L_e C_e}} = \frac{c}{\cot \psi} \left[\frac{I_0(\gamma a) K_0(\gamma a)}{I_1(\gamma a) K_1(\gamma a)}\right]^{\frac{1}{2}}. \qquad (38)$$

The dispersion equation for this transmission line is easily obtained as

$$\gamma^2 = k^2 \cot^2 \psi \frac{I_1(\gamma a) K_1(\gamma a)}{I_0(\gamma a) K_0(\gamma a)}. \qquad (39)$$

The characteristic inductance and capacitance of the line of Fig. 4 are shown versus γa in Fig. 5. As expected, the inductance decreases with increasing frequency while C_e increases, resulting in a decreasing impedance with γa and a near constant phase velocity. The impedance function $Z_0 = \sqrt{L_e/C_e}$ is shown in Fig. 6 versus γa and $\cot \psi \approx c/v$.

3. EQUIVALENCE FOR A HELICAL WAVE GUIDING STRUCTURE

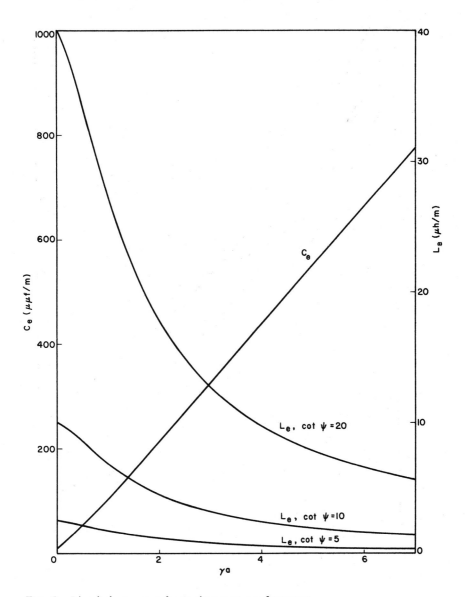

FIG. 5. Line inductance and capacitance versus frequency.
$$L_e = \mu_0 \cot^2 \psi / 2\pi [I_1(\gamma a) K_1(\gamma a)], \qquad C_e = 2\pi\epsilon_0 / I_0(\gamma a) K_0(\gamma a).$$

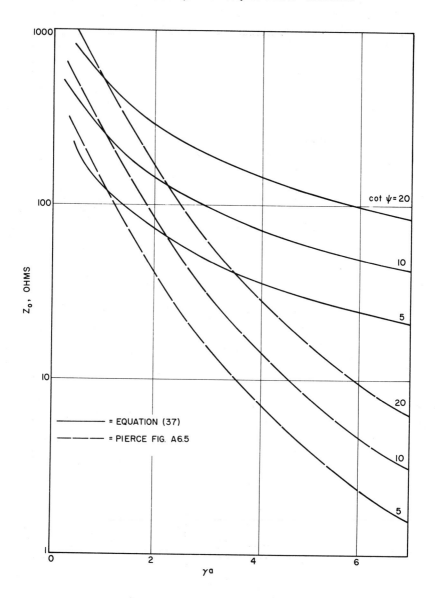

Fig. 6. Z_0 and K_s versus frequency.

3. EQUIVALENCE FOR A HELICAL WAVE GUIDING STRUCTURE

Also shown for reference is the sheath-helix impedance of Pierce calculated on a power transfer basis:

$$\frac{E^2}{\beta^2 P} = \left(\frac{\gamma}{\beta}\right)^4 \left(\frac{\beta}{\beta_0}\right)\left(\frac{60}{\gamma a}\right)$$

$$\approx \left(\frac{c}{v}\right)\left(\frac{60}{\gamma a}\right). \tag{40}$$

The impedance Z_0 is to be associated with the transverse impedance of a helix obtained as follows from Eq. (40):

$$K_t = \frac{1}{2}\left(\frac{\beta}{\gamma}\right)^2 \frac{E^2}{\beta^2 P}$$

$$\approx \frac{30}{\gamma a} \cot \psi. \tag{41}$$

A plot of this function on Fig. 6 is seen to coincide exactly with the Z_0 curves.

The phase velocity curves calculated from Eq. (38) are shown in Fig. 7. Dispersion occurs at low frequency and for $\gamma a > 2$, $v/c \approx \tan \psi$.

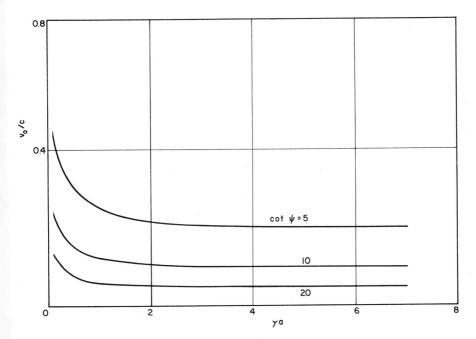

FIG. 7. Helix phase velocity versus frequency.

The above field theory treatment of the helical waveguide has assumed a lossless structure. All such transmission lines exhibit series loss, usually represented by a resistor R (ohms/meter) in series with the inductive element. The exact derivation for R requires the solution of the electromagnetic field problem for a finite-wire-size helix, which does not exist. A very accurate but still approximate development can be made, however, using the results (field expressions) for the lossless case.

Pierce[13] has calculated the total power flow along the helical sheath from the Poynting vector

$$P_t = \tfrac{1}{2} \operatorname{Re} \int (\mathbf{E} \times \mathbf{H}^*) \cdot d\mathbf{S}$$

with the result

$$P_t = B^2 \frac{\beta \beta_0}{\gamma^4} \frac{1}{F^3(\gamma a)}, \tag{42}$$

where B is determined from the input boundary conditions,

$$\gamma^2 = \beta^2 - \beta_0^2,$$

and

$$F(\gamma a) = \left\{ \frac{\pi \gamma a}{2k} \left(\frac{I_0}{K_0} \right) \left[\left(\frac{I_1}{I_0} - \frac{I_0}{I_1} \right) + \left(\frac{K_0}{K_1} - \frac{K_1}{K_0} \right) + \frac{4}{\gamma a} \right] \right\}^{-1/3}$$
$$\approx 7.154 e^{-0.6664 \gamma a}.$$

All Bessel functions are of argument γa unless otherwise indicated.

The magnetic field components H^i and H^o generate the sheath-helix currents. In general

$$\mathbf{J} = \mathbf{n} \times \mathbf{H}, \tag{43}$$

where $\mathbf{n}$ represents a unit vector normal to the sheath helix model. Hence the current densities on the inside and outside of the sheath are calculated from

$$J^i = H_z^i \cos \psi - H_\varphi^i \sin \psi \tag{44}$$

and

$$J^o = -H_z^o \cos \psi + H_\varphi^o \sin \psi. \tag{45}$$

A convenient model of an actual helical line to use is the tape helix which is illustrated in unwrapped form in Fig. 8. The tape thickness is assumed small but greater than the skin depth given by

$$\delta_s = \frac{1}{\sqrt{\pi f \mu \sigma}}.$$

3. EQUIVALENCE FOR A HELICAL WAVE GUIDING STRUCTURE

For constant total current the current densities inside and outside the tape, obtained from the sheath helix expressions, are

$$J_t^i = -jB\left(\frac{p}{w}\right)\left(\frac{\gamma}{k\beta_0}\right) I_0 \left[\frac{I_0}{I_1} + \left(\frac{\beta_0}{\gamma}\right)^2 \frac{I_1}{I_0}\right] \sin\psi\, e^{j(\omega t - \beta z)} \quad (46)$$

and

$$J_t^o = -jB\left(\frac{p}{w}\right)\left(\frac{\gamma}{k\beta_0}\right) I_0 \left[\frac{K_0}{K_1} + \left(\frac{\beta_0}{\gamma}\right)^2 \frac{K_1}{K_0}\right] \sin\psi\, e^{j(\omega t - \beta z)}, \quad (47)$$

where

$$J_t = \left(\frac{p}{w}\right) J_{sh}.$$

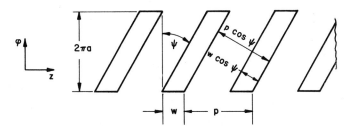

FIG. 8. Model of a developed tape helix.

The power loss along the direction of the tape is

$$\hat{P}_L = \frac{R_s}{2} \oint |J|^2\, dl \quad \text{watts/meter}, \quad (48)$$

where the integration is carried out around the tape contour and $R_s \triangleq$ the skin effect surface resistivity per square. Thus

$$\hat{P}_L = \frac{R_s}{2} [|J_t^i|^2 + |J_t^o|^2]\, w \cos\psi, \quad (49)$$

where $R_s = (\pi f\mu/\sigma)^{\frac{1}{2}} = (\omega\mu/2\sigma)^{\frac{1}{2}}$. Hence along the axial direction

$$P_L = \frac{\hat{P}_L}{\sin\psi} \quad (50)$$

and

$$P_L = R_s\left(\frac{w}{2}\right) \cot\psi [|J_t^i|^2 + |J_t^o|^2]. \quad (51)$$

Substituting for J_t^i and J_t^0 from Eqs. (46) and (47) and simplifying gives

$$P_L = \left(\frac{\omega\mu}{2\sigma}\right)^{\frac{1}{2}} B^2 \frac{p^2}{w} \left(\frac{\gamma}{\beta_0 k}\right)^2 I_0^2 \frac{\sin 2\psi}{4} F_0, \tag{52}$$

where

$$F_0 \triangleq \left[\frac{I_0}{I_1} + \left(\frac{\beta_0}{\gamma}\right)^2 \frac{I_1}{I_0}\right]^2 + \left[\frac{K_0}{K_1} + \left(\frac{\beta_0}{\gamma}\right)^2 \frac{K_1}{K_0}\right]^2.$$

The waveguide attenuation constant is written as

$$\alpha = -\frac{1}{2P}\frac{dP}{dz}, \tag{53}$$

where dP represents the power decrement per unit length and P the total power. Define $P_L \triangleq -(dP/dz)$; then

$$\alpha = \frac{P_L}{2P_t}, \tag{54}$$

where $P_L \triangleq$ power lost per meter of length, and $P_t \triangleq$ transmitted power. The voltage decay along the line is $\exp(-\beta C dz)$ and hence

$$\alpha = \beta C d$$

or

$$d = \frac{v_0 \alpha}{\omega C} = \frac{v_0}{2\omega C}\left(\frac{P_L}{P_t}\right). \tag{55}$$

The parameters α and d, expressed in terms of field quantities, are

$$\alpha = \left(\frac{p}{w}\right)\frac{\sin^2 \psi}{(2\omega\mu\sigma)^{\frac{1}{2}}}\left(\frac{\gamma^5}{\beta\beta_0^2}\right) F_1 \quad \text{nepers/meter} \tag{56}$$

and

$$d = \left(\frac{v_0}{\omega C}\right)\left(\frac{p}{w}\right)\frac{\sin^2 \psi}{(2\omega\mu\sigma)^{\frac{1}{2}}}\left(\frac{\gamma^5}{\beta\beta_0^2}\right) F_1, \tag{57}$$

where

$$F_1 = I_0 K_0 \frac{\left[\frac{I_0}{I_1} + \left(\frac{\beta_0}{\gamma}\right)^2 \frac{I_1}{I_0}\right]^2 + \left[\frac{K_0}{K_1} + \left(\frac{\beta_0}{\gamma}\right)^2 \frac{K_1}{K_0}\right]^2}{\left(\frac{I_1}{I_0} - \frac{I_0}{I_1}\right) + \left(\frac{K_0}{K_1} - \frac{K_1}{K_0}\right) + \frac{4}{\gamma a}}. \tag{58}$$

The expressions (56) and (57) agree extremely well with experimental results; in fact the correlation is much better than with the results obtained using more elaborate theories.

3. EQUIVALENCE FOR A HELICAL WAVE GUIDING STRUCTURE

Recalling that the loss factor d is related to the line elements, $R_e/L_e = 2\omega C d$, we now have a means of obtaining R_e in terms of field quantities. The result is

$$\frac{R_e}{L_e} = 2v_0 \left(\frac{p}{w}\right) \frac{\sin^2 \psi}{(2\omega\mu\sigma)^{\frac{1}{2}}} \left(\frac{\gamma^5}{\beta\beta_0^2}\right) F_1, \qquad (59)$$

which for slow waves, $\gamma \approx \beta$, reduces to

$$\frac{R_e}{L_e} \approx 2\left(\frac{p}{w}\right) \frac{\omega\beta}{(2\omega\mu\sigma)^{\frac{1}{2}}} F_1. \qquad (60)$$

The line resistance parameter is then

$$R_e \approx \frac{\mu_0}{\pi}\left(\frac{p}{w}\right) \frac{\omega\beta \cot^2 \psi}{(2\omega\mu\sigma)^{\frac{1}{2}}} \bar{F}_1, \qquad (61)$$

where $\bar{F}_1 \triangleq I_1(\gamma a) K_1(\gamma a) F_1$.

Thus the complete equivalent circuit of Fig. 9 has been obtained. As expected, the attenuation is dependent upon the fields, circuit dimensions, tape properties and phase velocity. The attenuation and resistance functions are shown in Figs. 10 and 11, in which the slow-wave approximation, $\gamma \approx \beta$, is used.

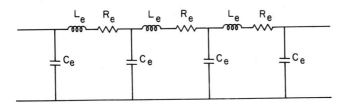

FIG. 9. Transmission-line equivalent circuit including series loss.

$$L_e = \frac{\mu_0 \cot^2 \psi}{2\pi} I_1(\gamma a) K_1(\gamma a), \qquad R_e = \frac{\mu_0}{\pi}\left(\frac{p}{w}\right) \frac{\omega\beta \cot^2 \psi}{(2\omega\mu\gamma)^{\frac{1}{2}}} \bar{F}_1, \qquad C_e = \frac{2\pi\epsilon_0}{I_0(\gamma a) K_0(\gamma a)}.$$

Similar processes may be carried out for any type of wave-propagating structure or medium after the field problem has been solved.

One other result which it is well to mention here, since electron beams may be considered as propagating structures,[2] is the similarly determined equivalent circuit with the elements indicated in Eqs. (62) and (63)

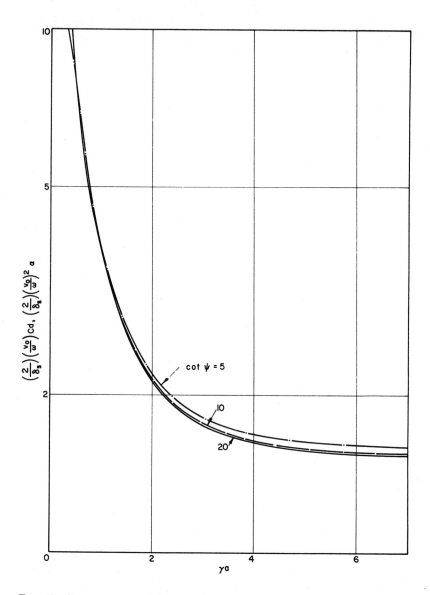

Fig. 10. Parameters α and d versus frequency.

$$\left(\frac{2}{\delta_s}\right)\left(\frac{v_0}{\omega}\right) Cd = \left(\frac{p}{w}\right) F_1, \qquad \left(\frac{2}{\delta_s}\right)\left(\frac{v_0}{\omega}\right)^2 \alpha = \left(\frac{p}{w}\right) F_1; \qquad \frac{p}{w} = 3.$$

3. EQUIVALENCE FOR A HELICAL WAVE GUIDING STRUCTURE

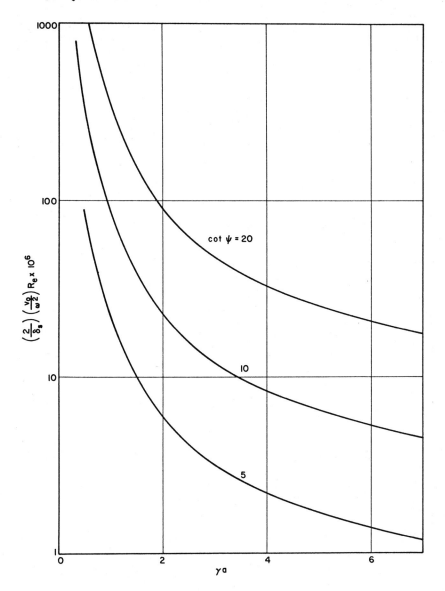

FIG. 11. Line resistance versus frequency.

$$\left(\frac{2}{\delta_s}\right)\left(\frac{v_0}{\omega^2}\right) R_e = \frac{\mu_0}{\pi}\left(\frac{p}{w}\right)\cot^2\psi\,\overline{F}_1, \qquad \frac{p}{w} = 3.$$

and illustrated in Fig. 12. The line elements are derivable directly from the Newton force law and continuity equation for the drifting beam.

$$L_e = \frac{1}{I_0}\left(\frac{2V_0}{\eta}\right)^{\frac{1}{2}}\left(\frac{\omega_q}{\omega}\right)^2 \quad \text{henrys/meter} \tag{62}$$

and

$$C_e = \frac{\sigma\epsilon_0}{\tilde{R}^2}\left(\frac{2\pi}{\lambda_q}\right)^2 = \left(\frac{\sigma\epsilon_0}{2\eta}\right)\frac{\omega_p^2}{V_0} \quad \text{farads/meter,} \tag{63}$$

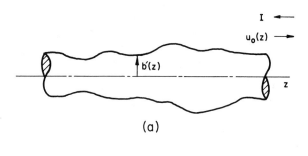

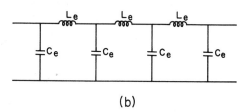

FIG. 12. (a) A one-dimensional finite-diameter drifting electron beam, and (b) its transmission-line equivalent circuit.

$$L_e = \frac{1}{I_0}\left(\frac{2V_0}{\eta}\right)^{\frac{1}{2}}\left(\frac{\omega_q}{\omega}\right)^2, \quad C_e = \left(\frac{\sigma\epsilon_0}{2\eta}\right)\frac{\omega_p^2}{V_0}.$$

where

$I_0 \triangleq$ dc beam current,

$V_0 \triangleq$ dc beam voltage,

$\eta \triangleq$ electron charge-to-mass ratio,

$\omega_p^2 \triangleq \eta\rho_0/\epsilon_0$, radian free-space plasma frequency,

$\tilde{R} \triangleq \omega_q/\omega_p$, plasma frequency reduction factor, and

$\sigma \triangleq$ beam cross-sectional area.

4. EQUIVALENT FOR SURFACE WAVE PROPAGATION

The impedance and propagation or phase constant associated with the equivalent transmission line are derived in the usual manner.

$$Z_0 \triangleq \sqrt{\frac{Z}{Y}} = \sqrt{\frac{L_e}{C_e}} = \frac{2V_0}{I_0}\left(\frac{\omega_q}{\omega}\right)$$

and

$$\gamma = \sqrt{ZY} = j\frac{\omega_q}{u_0} = j\beta_q.$$

The dc parameters of the electron beam may in fact be functions of distance and thus the equivalent line impedance and phase constant will vary with distance (Section III. 7). The inductance of the beam decreases with frequency as expected, and the capacitance appears as an equivalent parallel-plate capacitor whose plates have the cross-sectional area σ. The effective plate separation of such a capacitor is

$$s = \frac{1}{\beta_p{}^2 l},$$

where l denotes the length of a particular section of the electron beam.

4 Transmission-Line Equivalent for Surface Wave Propagation on a Plasma Column

The highly ionized plasma has received extensive theoretical study in view of its ability to propagate slow electromagnetic waves. The field problem for a surface-mode propagation on an ideal plasma column provides a basis for developing the appropriate telegraphist's equations (Paik[11]). Collision effects are ignored and a cold stationary plasma is considered. One proceeds in a manner similar to that for the electron beam treated in the previous section. It is necessary to identify voltage and current in the stationary plasma as in the drifting electron beam. The circuit current is taken as the convection current in the plasma caused by an impressed electromagnetic wave and the circuit voltage is related to the field by $E_z = j\beta V$.

The plasma column and its equivalent circuit are shown in Fig. 13. The field components for the lowest-order transverse plasma mode on the plasma column of Fig. 13 are

A. $r < b$

$$E_z{}^i = A_1 J_0(\tau r), \tag{64}$$

$$E_r{}^i = \left(\frac{j\beta}{\tau}\right) A_1 J_1(\tau r), \tag{65}$$

and
$$H_\theta^i = \left(\frac{j\omega\epsilon_0}{\tau}\right)\left[1 - \left(\frac{\omega_p}{\omega}\right)^2\right] A_1 I_1(\tau r). \tag{66}$$

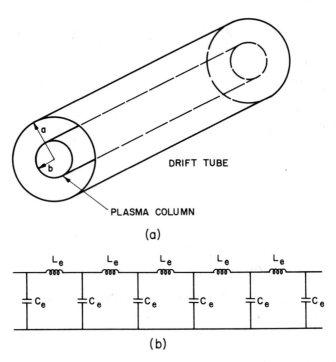

Fig. 13. (a) Ideal plasma column and drift tube and (b) its transmission-line equivalent circuit.

$$L_e = \frac{\tau}{2\pi b} \frac{1}{\omega_p^2 \epsilon_0} \frac{I_0(\tau b)}{I_1(\tau b)}, \qquad C_e = \frac{2\pi b \gamma \epsilon_0 K_1(\gamma b)}{[1 - (\omega/\omega_p)^2] K_0(\gamma b)}.$$

B. $r > b$

$$E_z^0 = A_2 I_0(\gamma r) + A_3 K_0(\gamma r), \tag{67}$$

$$E_r^0 = \left(\frac{j\beta}{\gamma}\right)[A_2 I_1(\gamma r) - A_3 K_1(\gamma r)], \tag{68}$$

and
$$H_\theta^0 = \left(\frac{j\omega\epsilon_0}{\gamma}\right)[A_2 I_1(\gamma r) - A_3 K_1(\gamma r)] \tag{69}$$

4. EQUIVALENT FOR SURFACE WAVE PROPAGATION

with the boundary conditions

$$E_z^i(b) = E_z^o(b),$$
$$E_r^i(b) + \frac{\sigma}{\epsilon_0} = E_r^o(b), \qquad (70)$$
$$\frac{H_\theta^o(a)}{E_z^o(a)} = Y_s,$$

and

where

$Y_s =$ the admittance of the drift tube,

$\tau^2 = (\beta^2 - k^2)[1 - (\omega_p/\omega)^2]$, and

$\gamma^2 = \beta^2 - k^2$.

The convection current for the rippled column with a surface-charge variation (transverse mode) is expressed as

$$i_z = \int_0^{2\pi} \int_0^b J_z r \, dr \, d\theta,$$

where

$$J_z = -j\left(\frac{\omega_p^2 \epsilon_0}{\omega}\right) E_z^i \quad \text{amps/meter.}^2$$

In terms of the electric field for this transverse mode, i_z becomes

$$i_z = \frac{2\pi b \epsilon_0 \omega_p^2}{j\omega\tau} \frac{I_1(\tau b)}{I_0(\tau b)} E_{zb}. \qquad (71)$$

Combining Eqs. (71) and $E_{zb} = j\beta V$ gives one of the familiar telegraphist's equations:

$$\frac{\partial V}{\partial z} + L_e \frac{\partial i_z}{\partial t} = 0, \qquad (72a)$$

where

$$L_e \triangleq \frac{\tau}{2\pi b \omega_p^2 \epsilon_0} \frac{I_0(\tau b)}{I_1(\tau b)} \quad \text{henrys/meter.} \qquad (72b)$$

The plasma column displacement current is given by

$$i_{zD} = \int_0^{2\pi} \int_0^b j\omega\epsilon_0 E_z^i r \, dr \, d\theta$$
$$= 2\pi b \left(\frac{\omega\epsilon_0}{\tau}\right) \left[\frac{I_1(\tau b)}{I_0(\tau b)}\right] E_{zb}.$$

Inserting the field expressions and applying the continuity equation leads to

$$\frac{\partial i_z}{\partial z} + C_e \frac{\partial V}{\partial t} = 0, \qquad (73a)$$

where

$$C_e = \frac{2\pi b \beta^2 \epsilon_0}{\gamma \left[1 - \left(\dfrac{\omega}{\omega_p}\right)^2\right]} \frac{K_1(\gamma b)}{K_0(\gamma b)}. \qquad (73b)$$

As seen in the preceding section, the inclusion of loss in the equivalent circuit is difficult whatever the propagating medium. The impedance, phase constant and dispersion equation are found for the above system in the usual manner.

5 Equivalent Transmission Lines for Multidimensional Propagating Structures

In Section 2 the subject of coupled multipaired transmission lines was discussed, and allusion made to the correspondence of the appropriate generalized telegraphist's equations to the relevant system of field equations. A great deal of interest exists in numerous areas of science relative to wave propagation in multidimensional anisotropic lattices; in particular, these multidimensional propagating structures have invaded the microwave device art in the form of biperiodic rf structures [6,10] for use in high-power O- and M-type amplifiers and oscillators. The desire to analyze such structure configurations interacting with charged beams leads to many mathematical dilemmas and hence to a desire to develop equivalent circuit methods *au* Kelvin. As Brillouin[3] points out in his treatise on wave propagation, some of the difficulties arise out of the fact that it is not always possible to separate triply periodic functions, say voltage on the structure, into the product of three singly periodic functions. Each physical case must be examined in its own right. In the case of multiply periodic structures for electron devices, the design results in a separable potential function as a result of mutually orthogonal directions of propagation.

In the development of the equivalent circuit technique it will be assumed for convenience that the charge stream is directed along one axis of the lattice. The equivalent circuit form taken for the two-dimensional anisotropic line is illustrated in Fig. 14. In developing a uniform-line equivalent circuit, valid at a single frequency, the periodicity of the structure is ignored and hence no information about the dispersion

5. EQUIVALENT FOR MULTIDIMENSIONAL STRUCTURES

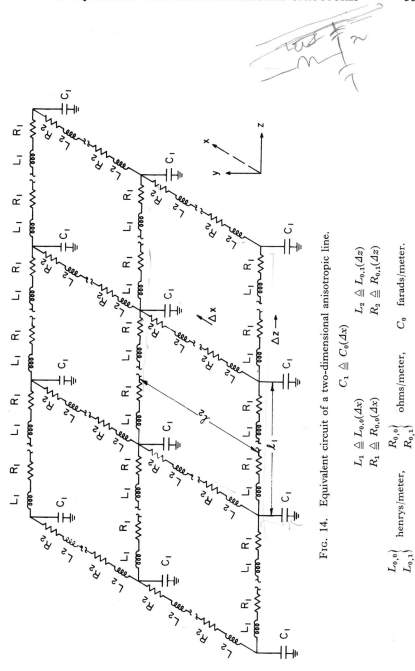

Fig. 14. Equivalent circuit of a two-dimensional anisotropic line.

$$L_1 \triangleq L_{0,0}(\Delta x) \qquad L_2 \triangleq L_{0,1}(\Delta z)$$
$$R_1 \triangleq R_{0,0}(\Delta x) \qquad R_2 \triangleq R_{0,1}(\Delta z)$$
$$C_1 \triangleq C_0(\Delta x)$$

$\left. \begin{array}{l} L_{0,0} \\ L_{0,1} \end{array} \right\}$ henrys/meter, $\left. \begin{array}{l} R_{0,0} \\ R_{0,1} \end{array} \right\}$ ohms/meter, C_0 farads/meter.

equation (ω-β diagram) can be obtained from the equivalent circuit. The uniform-line parameters are chosen to give the proper field and phase velocity for one mode and do not reflect the periodic nature of the structure. The individual line parameters are defined in the figure and it is assumed that l_1 and $l_2 \ll \lambda_g$. As illustrated for the sheath-helix guiding structure, the various line elements are related to the structure dimensions and the field dependencies.

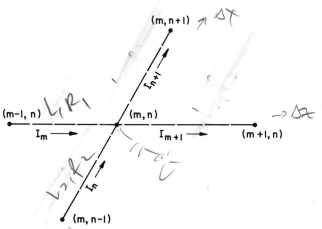

FIG. 15. Typical node of the circuit shown in Fig. 14.

Consider a typical node, illustrated in Fig. 15, of the general lossy anisotropic line of Fig. 14. The following fundamental equations are written for the voltages and currents of Fig. 15:

$$(I_m - I_{m+1}) + (I_n - I_{n+1}) = \frac{\partial q_{m,n}}{\partial t}, \tag{74}$$

$$V_{m-1,n} - V_{m,n} = L_1 \frac{\partial I_m}{\partial t} + R_1 I_m \tag{75}$$

and

$$V_{m,n-1} - V_{m,n} = L_2 \frac{\partial I_n}{\partial t} + R_2 I_n. \tag{76}$$

The total charge on the capacitance C is given by

$$Q_{m,n} = q_{m,n} + q'_{m,n} = q_{m,n} + \Delta z \rho_{m,n} \tag{77}$$

and

$$V_{m,n} = \frac{Q_{m,n}}{C_1}, \tag{78}$$

5. EQUIVALENT FOR MULTIDIMENSIONAL STRUCTURES

where

$q_{m,n}$ = the charge on the capacitance due to currents in the line,

and

$q'_{m,n}$ = the electric charge of the stream, coulombs.

In writing Eq. (77) it is assumed that an electron stream of ρ(coulomb/meter^3) passes between the biperiodic structure and the ground plane. The stream moves only in the z-direction and hence induces current in only the z-direction. It is not necessary to assume that all flux lines emanating from the stream terminate on the circuit. The quantity $\rho_{m,n}$ in Eq. (77) is written in terms of the space-charge density in the stream as follows:

$$\rho_{m,n} = \int_{x=x_1}^{x=x_2} \int_{y=y_s}^{y=y_d} \psi(y)\rho(x, y, z, t)\, dy\, dx, \tag{79}$$

where $\psi(y)$ is the coupling function for a charge element in the stream. This is the coupling function of Ramo's induced current theorem.

Equations (74)–(79) may be combined by taking the appropriate derivatives to give ($\Delta z = \Delta x$)

$$\frac{V_{m-1,n} - 2V_{m,n} + V_{m+1,n}}{(\Delta z)^2} + \left(\frac{L_1}{L_2}\right) \frac{V_{m,n-1} - 2V_{m,n} + V_{m,n+1}}{(\Delta x)^2}$$

$$-L_{0,0}C_0 \frac{\partial^2 V_{m,n}}{\partial t^2} = -L_{0,0}\frac{\partial^2 \rho_{m,n}}{\partial t^2} - \frac{R_{0,0}}{\Delta z}(I_{m+1} - I_m)$$

$$- \left(\frac{L_1}{L_2}\right)\frac{R_{0,1}}{\Delta x}(I_{n+1} - I_n). \tag{80}$$

In order to simplify Eq. (80) for the anisotropic two-dimensional line it is necessary to introduce a linear operator into Eqs. (75) and (76).

$$V_{m-1,n} - V_{m,n} = (R_1 + sL_1)I_m$$

and

$$V_{m,n-1} - V_{m,n} = (R_2 + sL_2)I_n, \tag{81}$$

where $s \triangleq \partial/\partial t$.

By analogy the following equations are written:

$$V_{m,n} - V_{m+1,n} = (R_1 + sL_1)I_{m+1}$$

and

$$V_{m,n} - V_{m,n+1} = (R_2 + sL_2)I_{n+1}. \tag{82}$$

Substituting Eqs. (81) and (82) into Eq. (80) and letting Δz and Δx approach zero gives, in the limit,

$$\left[1 - \frac{R_1}{(R_1+sL_1)}\right]\frac{\partial^2 V(x,z,t)}{\partial z^2} + \left(\frac{L_1}{L_2}\right)\left[1 - \frac{R_2}{(R_2+sL_2)}\right]\frac{\partial^2 V(x,z,t)}{\partial x^2}$$

$$-L_{0,0}C_0\frac{\partial^2 V(x,z,t)}{\partial t^2} = -L_{0,0}\frac{\partial^2 \rho(x,z,t)}{\partial t^2}. \quad (83)$$

The following definitions are made for the characteristic phase velocities and impedances of the lattice in the principal directions:

$$v_{0,0} = \frac{1}{\sqrt{L_{0,0}C_0}}; \quad v_{0,1} = \frac{1}{\sqrt{L_{0,1}C_0}}, \quad (84)$$

$$Z_{0,0} = \sqrt{\frac{L_{0,0}}{C_0}}; \quad Z_{0,1} = \sqrt{\frac{L_{0,1}}{C_0}}. \quad (85)$$

The quantities R_e/L_e in Eq. (83) may be replaced by $2\omega C d_n$, where d_n is the loss factor[13] for the principal direction in question; i.e.,

$$\frac{R_{0,0}}{L_{0,0}} = 2\omega C d_z$$

and

$$\frac{R_{0,1}}{L_{0,1}} = 2\omega C d_x, \quad (86)$$

where $C =$ the gain parameter. Substitution of Eqs. (84)–(86) into Eq. (83) and subsequent simplification and rearrangement yield

$$\left(\frac{Z_{0,0}}{Z_{0,1}}\right)(v_{0,0}v_{0,1})\left(\frac{s}{s+2\omega Cd_x}\right)\frac{\partial^2 V}{\partial x^2} + v_{0,0}^2\left(\frac{s}{s+2\omega Cd_z}\right)\frac{\partial^2 V}{\partial z^2}$$

$$-\left(1 + \frac{v_{0,0}}{v_{0,1}}\frac{Z_{0,0}}{Z_{0,1}}\right)\frac{\partial^2 V}{\partial t^2} = -v_{0,0}Z_{0,0}\frac{\partial^2 \rho}{\partial t^2}. \quad (87)$$

Integrate Eq. 87 with respect to time letting $dz = dx$, and at $t = 0$ assume that the stream enters the circuit and that the rf is simultaneously applied. Then (no integration constants)

$$\frac{Z_{0,0}}{Z_{0,1}}(v_{0,0}v_{0,1})\left(\frac{\partial^2 V}{\partial x^2}\right) + v_{0,0}^2\left(\frac{\partial^2 V}{\partial z^2}\right) - \left(1 + \frac{v_{0,0}}{v_{0,1}}\frac{Z_{0,0}}{Z_{0,1}}\right)\left(\frac{\partial^2 V}{\partial t^2}\right)$$

$$-2\omega Cd_x\left(1 + \frac{v_{0,0}}{v_{0,1}}\frac{Z_{0,0}}{Z_{0,1}}\right)\frac{\partial V}{\partial t} = -(v_{0,0}Z_{0,0})\frac{\partial^2 \rho}{\partial t^2}$$

$$-2\omega Cd_x(v_{0,0}Z_{0,0})\frac{\partial \rho}{\partial t}. \quad (88)$$

5. EQUIVALENT FOR MULTIDIMENSIONAL STRUCTURES

Equation (88) is the general differential equation for the potential $V(x, z, t)$ along a two-dimensional anisotropic line with a forcing function.

Before proceeding with the specialization of Eq. (88) to lossy, isotropic and one-dimensional situations, it is well to examine the form and characteristics of Eq. (88) further. To accomplish this we apply an old theorem for triply periodic wave functions developed by F. Bloch. An earlier treatment of the one-dimensional Mathieu equation used in circuit analysis, originally suggested by Floquet, is also applicable.

Floquet's theorem for one-dimensional structures states that the fields at any particular cross section of a periodic structure can differ from those one period away in either direction by at most a complex constant. Consider a two-dimensional lattice in which the wave function or wave potential may be expressed as follows, according to Floquet's theorem:

$$V(x, z) = \Phi(x, z)e^{-\gamma_1 z - \gamma_2 x}, \tag{89}$$

where $\Phi(x, z)$ is taken to be periodic in x and z, which are orthogonal directions of propagation. γ_1 and γ_2 are the respective directional propagation constants. It is necessary and convenient to assume that the function $\Phi(x, z)$ is separable and expandable in a Fourier series as an infinity of space harmonics. Thus

$$\begin{aligned}V(x, z) &= \exp(-j\beta_1 z - j\beta_2 x) \sum_m \sum_n A_{mn} \exp\left(-jm\frac{2\pi z}{p_z} - jn\frac{2\pi x}{p_x}\right) \\ &= \sum_m A_m \exp(-j\beta_{zm} z) \sum_n B_n \exp(-j\beta_{xn} x),\end{aligned} \tag{90}$$

where the respective phase constants β_{zm} and β_{xn} are defined as

$$\beta_{zm} \triangleq \beta_1 + \frac{2\pi m}{p_z}$$

and

$$\beta_{xn} \triangleq \beta_2 + \frac{2\pi n}{p_x}. \tag{91}$$

p_x and p_z are the periods along the respective axes of propagation. Double differentiation of Eq. (90) with respect to the two directions results in

$$\frac{\partial^2 V}{\partial x^2} + \frac{\partial^2 V}{\partial z^2} = -\sum_n B_n \beta_{xn}^2 e^{-j\beta_{xn} x} \sum_m A_m e^{-j\beta_{zm} z} \\ - \sum_n B_n e^{-j\beta_{xn} x} \sum_m A_m \beta_{zm}^2 e^{-j\beta_{zm} z}. \tag{92}$$

Thus we may write a wavelike equation by selecting a particular space harmonic of the total wave function and using

$$\frac{\partial^2 V_{m,n}}{\partial x^2} + \frac{\partial^2 V_{m,n}}{\partial z^2} + (\beta_{xn}^2 + \beta_{zn}^2)V_{m,n} = 0, \qquad (93)$$

which is the telegraphist's equation in the direction of wave propagation. Note that one-half of the time dependence is associated with each direction. The β's may be considered as operators of the form $j\beta = j(\omega/v) = 1/v(\partial/\partial t)$. In the direction orthogonal to both x and z the Laplace equation is obtained:

$$\frac{\partial^2 V_{m,n}}{\partial y^2} - (\beta_{xn}^2 + \beta_{zm}^2)V_{m,n} = 0. \qquad (94)$$

Of course the procedure outlined above has its one-dimensional counterpart.

In comparing the above results with the equivalent circuit of Fig. 14, we see why only one element of capacitance is required. This can also be argued on physical grounds. This condition implies that

$$\frac{Z_{0,0}}{Z_{0,1}} = \frac{v_{0,1}}{v_{0,0}}. \qquad (95)$$

Thus Eq. (88) may now be written as

$$v_{0,1}^2 \frac{\partial^2 V(x,z,t)}{\partial x^2} + v_{0,0}^2 \frac{\partial^2 V(x,z,t)}{\partial z^2} - 2\frac{\partial^2 V(x,z,t)}{\partial t^2} - 4\omega C d_x \frac{\partial V(x,z,t)}{\partial t}$$
$$= -(v_{0,0}Z_{0,0})\left[\frac{\partial^2 \rho(x,z,t)}{\partial t^2} + 2\omega C d_x \frac{\partial \rho(x,z,t)}{\partial t}\right]. \qquad (96)$$

Several special cases are of interest:

A. *Lossless Anisotropic Line*

$$v_{0,1}^2 \frac{\partial^2 V(x,z,t)}{\partial x^2} + v_{0,0}^2 \frac{\partial^2 V(x,z,t)}{\partial z^2} - 2\frac{\partial^2 V(x,z,t)}{\partial t^2} = -v_{0,0}Z_{0,0}\frac{\partial^2 \rho(x,z,t)}{\partial t^2}. \qquad (97)$$

B. *Lossy Isotropic Line*

$$\frac{\partial^2 V(x,z,t)}{\partial x^2} + \frac{\partial^2 V(x,z,t)}{\partial z^2} - \frac{2}{v_{0,0}^2}\frac{\partial^2 V(x,z,t)}{\partial t^2} - \frac{4\omega C d_z}{v_{0,0}^2}\frac{\partial V(x,z,t)}{\partial t}$$
$$= -\frac{Z_{0,0}}{v_{0,0}}\left[\frac{\partial^2 \rho(x,z,t)}{\partial t^2} + 2\omega C d_z \frac{\partial \rho(x,z,t)}{\partial t}\right]. \qquad (98)$$

5. EQUIVALENT FOR MULTIDIMENSIONAL STRUCTURES

C. Lossless Isotropic Line

$$\frac{\partial^2 V(x,z,t)}{\partial x^2} + \frac{\partial^2 V(x,z,t)}{\partial z^2} - \frac{2}{v_{0,0}^2}\frac{\partial^2 V(x,z,t)}{\partial t^2} = -\frac{Z_{0,0}}{v_{0,0}}\frac{\partial^2 \rho(x,z,t)}{\partial t^2}. \quad (99)$$

D. One-Dimensional Lossy Line

$$\frac{\partial^2 V(z,t)}{\partial z^2} - \frac{1}{v_0^2}\frac{\partial^2 V(z,t)}{\partial t^2} - \frac{2\omega C d_z}{v_0^2}\frac{\partial V(z,t)}{\partial t}$$
$$= -\frac{Z_0}{v_0}\left[\frac{\partial^2 \rho(z,t)}{\partial t^2} + 2\omega C d_z \frac{\partial \rho(z,t)}{\partial t}\right]. \quad (100)$$

E. One-Dimensional Lossless Line

$$\frac{\partial^2 V(z,t)}{\partial z^2} - \frac{1}{v_0^2}\frac{\partial^2 V(z,t)}{\partial t^2} = -\frac{Z_0}{v_0}\frac{\partial^2 \rho(z,t)}{\partial t^2}. \quad (101)$$

The above transmission-line equations have been developed for structures located a constant height above an electron beam below which is a ground plane. The beam is considered constrained to move in the z-direction. Although the mathematical development was made in the cartesian coordinate system, transformation to other systems is easily made. For example, in a cylindrically symmetric system characteristic of O-type devices we need only treat the x-coordinate as follows:

$$dx = a\, d\varphi,$$

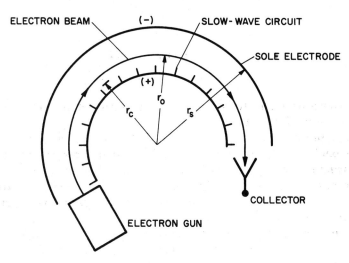

FIG. 16. E-type configuration. r_c = radius of the slow-wave circuit; r_0 = radius of the center-of-the-beam electron; r_s = radius of the sole electrode.

where a is the mean radius of the cylindrical structure and φ is the angle about the axis.

In devices employing interaction in a circular arc, e.g., E-type devices, as illustrated in Fig. 16, the following transformation is appropriate:

$$dz = a\, d\theta.$$

The coupling function ψ, which measures the fraction of flux emanating from the beam which terminates on the structure, introduced in Eq. 47, may be calculated for the particular geometrical system of interest from Ramo's[14] induced current theorem.

6 Backward-Wave Equivalent Circuits

Heretofore in this chapter we have considered equivalent circuits for structures propagating forward waves, i.e., waves with phase and group velocities in the same direction. Such modes are said to have positive dispersion. Another interesting set of space harmonic modes belonging to all periodic structures are those exhibiting negative dispersion; i.e., the phase and group velocities are oppositely directed. This relationship is clear from the following expression relating phase and group velocities:

$$v_g = \frac{v_p}{1 - \dfrac{\omega}{v_p}\dfrac{dv_p}{d\omega}},$$

where

$v_g \triangleq$ wave group velocity, and

$v_p \triangleq$ space harmonic phase velocity.

These structure characteristics are necessary to obtain voltage-tunable backward-wave interaction.

The application and examination of Floquet's theorem for one-dimensional wave propagation in periodic structures reveals that all such structures possess many space harmonic field components which exhibit negative dispersion. An ω-β diagram for a periodic structure is shown in Fig. 17, wherein forward and backward space harmonic field components are identified.

It is interesting to consider the development of an equivalent circuit for the backward-wave modes similar to that used for forward-wave modes (Rowe[15]). The forward-wave equivalent circuit has an inductive series element whose phase shift per section increases with frequency.

6. BACKWARD-WAVE EQUIVALENT CIRCUITS

In the backward-wave case we desire a decreasing phase shift per section with frequency, and hence an equivalent circuit can be obtained by simply reversing the positions of the series inductance and shunt capacitance as illustrated in Fig. 18. The backward-wave equivalent circuit exhibits a decreasing phase shift per section with frequency and a phase velocity which increases with frequency as

$$v_0 = -\omega^2 \sqrt{L_e C_e},$$

which is consistent with experiment.

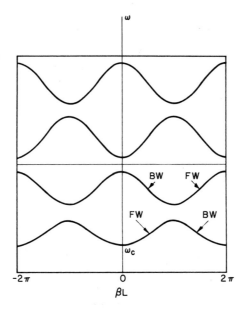

FIG. 17. ω-β diagram for a periodic propagating structure.

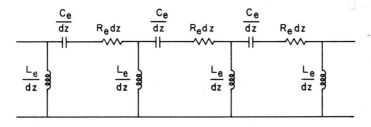

FIG. 18. Equivalent circuit for backward-wave devices. L_e (henrys · meters), C_e (farads · meters), R_e (ohms/meter); $v_0 = -\omega^2 \sqrt{L_e C_e}$; $Z_0 = \sqrt{Z/Y}$.

The telegraphist's equations for this backward-wave circuit are

$$\frac{dV(z,t)}{dz} - ZI(z,t) = 0 \qquad (102)$$

and

$$\frac{dI(z,t)}{dz} - YV(z,t) = 0, \qquad (103)$$

where

$$Z \triangleq R + 1/j\omega C_e, \text{ and}$$

$$Y \triangleq 1/j\omega L_e.$$

Combining Eqs. (102) and (103) and introducing the propagation constant $\gamma^2 = ZY$ gives

$$\frac{d^2V(z,t)}{dz^2} - \gamma^2 V(z,t) = 0. \qquad (104)$$

For small values of the series loss, the attenuation and phase constants are

$$\alpha = -\frac{R_e}{2}\sqrt{\frac{C_e}{L_e}}$$

and

$$\beta = \frac{1}{-\omega\sqrt{L_e C_e}}.$$

While the form suggested in Fig. 18 for the backward-wave equivalent circuit is intuitively satisfying and possesses the proper phase shift and phase velocity characteristics, it cannot be realized in terms of a smooth circuit (distributed constant) but only in terms of lumped parameters, to give the desired oppositely directed phase and group velocities. This form of the equivalent circuit also leads to difficulties in attaching physical significance to the various circuit elements.

Another approach to the development of a satisfactory circuit equation comes as a result of examining the fundamental interaction phenomena when a charged beam interacts with a backward-wave space harmonic. The space harmonic phase velocity is taken in the same direction as the drifting beam motion and hence the direction of travel of the initial modulation. Since the wave group velocity is oppositely directed, the electromagnetic energy flow is in the opposite direction to the beam travel.

An interesting philosophical speculation is, which is backwards, the beam travel or the energy flow? Recall that the circuit impedance may be defined in terms of the electric field at the beam position and the stored energy as $E^2/2\beta^2 W v_g$. Thus backward energy flow may be associated with the negative impedance, since v_g is assumed negative. Thus we may write the backward-wave circuit equation by changing the sign of the impedance in the forward-wave equation. The two forms are written together below for a one-dimensional circuit.

$$\frac{\partial^2 V(z,t)}{\partial t^2} - v_0^2 \frac{\partial^2 V(z,t)}{\partial z^2} + 2\omega Cd \frac{\partial V(z,t)}{\partial t}$$
$$= \pm v_0 Z_0 \left[\frac{\partial^2 \rho(z,t)}{\partial t^2} + 2\omega Cd \frac{\partial \rho(z,t)}{\partial t} \right], \quad (105)$$

where in the case of the double signs the upper refers to the forward wave and the lower to the backward wave.

7 Equivalent Circuits with Spatially Varying Line Parameters

In Chapter XIII phase-focusing techniques are considered as means of improving the electronic interaction efficiency of both M- and O-type amplifiers and oscillators. Phase "locking" of the charge bunches may be accomplished in a variety of ways: e.g., application of dc gradients, sudden phase shifts, and gradual variation of the phase shift per section or phase velocity of a periodic propagating circuit. It can be shown that such variations in the circuit cause an impedance variation, generally decreasing, with distance and hence the previous transmission-line equivalents with spatially independent parameters no longer apply. We are therefore obliged to consider the equivalent circuit form when there are spatially varying parameters (Meeker[9]).

The one-dimensional equivalent circuit to be considered, including the impressed currents due to the presence of the beam, is shown in Fig. 19. The fundamental Kirchhoff equations written for the assumed circuit are

$$V(z_{i-1}, t) - V(z_i, t) = R_e(z_{i-1}) \Delta z I(z_i, t) + L_e(z_{i-1}) \Delta z \frac{\partial I(z_i, t)}{\partial t} \quad (106)$$

and

$$I(z_i, t) - I(z_{i+1}, t) = \frac{\partial q(z_i, t)}{\partial t}. \quad (107)$$

The total charge brought on the capacitance per section is composed of that due to the line currents and that due to the beam-induced currents in the line. The charge on the capacitance is written as

$$Q(z_i, t) = C_e(z_i) \Delta z \, V(z_i, t) \tag{108}$$

and

$$Q(z_i, t) = q(z_i, t) + \rho(z_i, t) \Delta z. \tag{109}$$

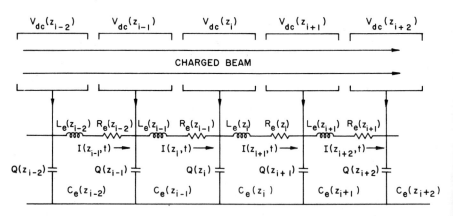

FIG. 19. One-dimensional equivalent transmission line including spatially dependent line parameters.

Equations (106)–(109) are combined using a procedure similar to that of the previous section. The following results

$$\frac{\partial^2 V(z,t)}{\partial z^2} - \frac{1}{L_e(z)} \frac{dL_e(z)}{dz} \frac{\partial V(z,t)}{\partial z} - C_e(z) L_e(z) \frac{\partial^2 V(z,t)}{\partial t^2} - R_e(z) C_e(z) \frac{\partial V(z,t)}{\partial t}$$

$$= -R_e(z) \frac{\partial \rho(z,t)}{\partial t} - L_e(z) \frac{\partial^2 \rho(z,t)}{\partial t^2}$$

$$- I(z,t) R_e(z) \left[\frac{1}{R_e(z)} \frac{dR_e(z)}{dz} - \frac{1}{L_e(z)} \frac{dL_e(z)}{dz} \right]. \tag{110}$$

The elimination of the $I(z, t)$ terms from Eq. (110) is not accomplished without difficulty. Direct elimination may proceed with the aid of Eq. (106) and a third-order differential equation is obtained. If one assumes that the resistance and inductance vary as the same function of z, a much simpler form is obtained. Fortunately this is a reasonable choice from a physical standpoint. Thus we see that

$$\frac{1}{R_e(z)} \frac{dR_e(z)}{dz} - \frac{1}{L_e(z)} \frac{dL_e(z)}{dz} \equiv 0.$$

7. SPATIALLY VARYING LINE PARAMETERS

Under this assumption the differential equation for the voltage along the line is

$$\frac{\partial^2 V(z,t)}{\partial z^2} - L_e(z)C_e(z)\frac{\partial^2 V(z,t)}{\partial t^2} - \frac{\partial}{\partial z}\ln L_e(z)\frac{\partial V(z,t)}{\partial z} - R_e(z)C_e(z)\frac{\partial V(z,t)}{\partial t}$$

$$= -L_e(z)\frac{\partial^2 \rho(z,t)}{\partial t^2} - R_e(z)\frac{\partial \rho(z,t)}{\partial t}. \quad (111)$$

It is convenient to introduce transmission-line parameters which are identical to uniform-line usage except that the parameters are now a function of the distance z. Thus,

$Z_0(z) \triangleq \sqrt{L_e(z)/C_e(z)}$, the characteristic line impedance, assuming a low-loss structure, now a function of distance, ohms;

$v_0(z) \triangleq 1/\sqrt{L_e(z)C_e(z)}$, the characteristic phase velocity of the line, varying with distance, meters/sec;

$2\omega Cd(z) = R_e(z)/L_e(z)$, relating $d(z)$, the loss parameter used in traveling-wave tubes, dimensionless;

$d(z) \triangleq 0.01836 l(z)/C$, the loss factor;

$l(z) =$ the series loss expressed in dB per undisturbed wavelength along the line;

$C(z) =$ the gain parameter defined by $C^3 \triangleq I_0 Z_0(z)/4V_0$;

$I_0 =$ the dc stream current, amps;

$V_0 =$ the dc stream voltage, at entrance to slow-wave structure, volts;

$u_0 = \sqrt{2\eta V_0}$, the dc stream velocity, at entrance, meters/sec;

$\eta = q/m$, the charge-to-mass ratio for the electron, coulombs/kg;

$\omega =$ the angular frequency of the wave impressed on the slow-wave structure, radians/sec.

Substitution of these newly defined circuit parameters in Eq. (111) yields

$$\frac{\partial^2 V(z,t)}{\partial z^2} - \frac{1}{v_0^2(z)}\frac{\partial^2 V(z,t)}{\partial t^2} - \frac{\partial}{\partial z}\ln\left(\frac{Z_0(z)}{v_0(z)}\right)\frac{\partial V(z,t)}{\partial z}$$

$$- \frac{2\omega Cd(z)}{v_0^2(z)}\frac{\partial V(z,t)}{\partial t} = -\frac{Z_0(z)}{v_0(z)}\left[\frac{\partial^2 \rho(z,t)}{\partial t^2} + 2\omega Cd(z)\frac{d\rho(z,t)}{\partial t}\right]. \quad (112)$$

References

1. Bloch, F., Über die Quantenmechanik der Elektronen in Kristallgittern. *Z. Physik* **52**, 555-600 (1928).
2. Bloom, S., and Peter, R. W., Transmission line analog of a modulated electron beam. *RCA Rev.* **15**, 95-112 (1954).
3. Brillouin, L., *Wave Propagation in Periodic Structures*. McGraw-Hill, New York, 1946.
4. Chu, C. M., Propagation of waves in helical wave guides. *J. Appl. Phys.* **29**, No. 1, 88-99 (1958).
5. Chu, L. J., and Jackson, J. D., Field theory of traveling wave tubes. *Proc. IRE* **36**, No. 7, 853-863 (1948).
6. Doehler, O., Epsztein, B., and Arnaud, J., Nouveaux types de lignes pour tubes hyperfréquences. *Proc. 1st Intern. Congr. Hyperfrequencies, Travaux du Congrès, Vol. 1, Paris, 1956* pp. 499-508.
7. Kino, G. S., and Paik, S. F., Circuit theory of coupled transmission systems. *J. Appl. Phys.* **33**, No. 10, 3002-3009 (1962).
8. Maxwell, J. C., *Electricity and Magnetism*, Vols. 1 and 2. Oxford Univ. Press, London and New York, 1873.
9. Meeker, J. G., "Phase Focusing in Linear-Beam Devices." Univ. of Michigan Electron Phys. Lab. Tech. Rept. No. 49 (August 1961).
10. Mourier, G., Circuits à structure périodique à deux et trois dimensions applications possibles aux tubes à ondes progressives. *Proc. 1st Intern. Congr. Hyperfrequencies, Travaux du Congrès, Vol. 1, Paris, 1956* pp. 493-498.
11. Paik, S. F., "A Study of Plasma Interaction with Traveling Waves." Stanford Electron. Labs. Tech. Rept. No. 408-1 (October 1961).
12. Palluel, P., and Arnaud, J., Results on delay lines for high-power traveling-wave tubes. *Proc. IEE (London)* **B-105**, 727-729 (1958).
13. Pierce, J. R., *Traveling Wave Tubes*. Van Nostrand, Princeton, N. J., 1950.
14. Ramo, S. I., Currents induced by electron motion. *Proc. IRE* **27**, No. 9, 584-586 (1939).
15. Rowe, J. E., Analysis of nonlinear O-type backward-wave oscillators. *Proc. Symp. Electronic Waveguides, Brooklyn Polytech. Inst., 1958* Vol. 8, pp. 315-339. Wiley (Interscience), New York, 1958.
16. Schelkunoff, S. A., Conversion of Maxwell's equations into generalized telegraphist's equations. *Bell System Tech. J.* **34**, No. 5, 995-1043 (1955).
    Schelkunoff, S. A., Generalized telegraphist's equations for waveguides. *Bell System Tech. J.* **31**, 784-801 (1952).
17. Sensiper, S., "Electromagnetic Wave Propagation on Helical Conductors." Sc.D. Thesis, Mass. Inst. Technol., Cambridge, Mass., 1951; also in abbreviated form, *Proc. IRE* **43**, 149-161 (1955).
18. Suhl, H., A proposal for a ferromagnetic amplifier in the microwave range. *Phys. Rev.* **106**, 384-385 (1957).
19. Tien, P. K., Traveling-wave tube helix impedance. *Proc. IRE* **41**, No. 11, 1617-1624 (1953).
20. Tien, P. K., and Suhl, H., A traveling-wave ferromagnetic amplifier. *Proc. IRE* **46**, No. 4, 700-706 (1958).
21. Thomson, Sir William (Lord Kelvin), *Mathematical and Physical Papers*, Vol. 2, p. 79. Cambridge Univ. Press, London and New York, 1884.

CHAPTER

IV | Space-Charge-Field Expressions

1 Introduction

We come now to one of the most intractable portions of the beam-electromagnetic wave interaction problem: the development of space-charge models and field expressions. A physically realizable electron or ion beam to be injected into an interaction region may have a uniform charge density, uniform current density, or uniform velocity over its cross section; or in fact none of them may be uniform. In reality we know that there is a continuous distribution of velocities (Maxwellian) characteristic of the entering beam.

Focusing methods may be electrostatic, magnetostatic, or both; and beam shapes may be rectangular, solid cylindrical, or hollow cylindrical. Both magnetically confined flow (infinite-B field) and finite field flow (Brillouin focusing) will be treated for axially symmetric beams when the interaction problem is analyzed in succeeding chapters. The problem is to develop a physically reasonable and consistent manner of evaluating the Coulomb forces for physically realizable beams.

The development of Eulerian space-charge-field expressions for one- and two-dimensional single-velocity flow analyses proceeds in a straightforward manner if Poisson's equation is solved for the potential on the axis or at the midplane of the beam and then this is expanded in the form of a power series for positions significantly away from the axis or plane of symmetry. Poisson's equation can be solved analytically only for a few special geometries and beam configurations. Several authors in the field of electron optics [1,2] have utilized such an approach in the development of paraxial-ray equations.

The two cases of particular interest to the development of the non-linear interaction theory are (1) planar symmetric beams; and (2) axially symmetric beams. Physical models used in such elementary treatments of the problem are illustrated in Fig. 1. The field due to the space charge itself may be easily found for small-dimension constant-charge-density beams by direct integration of Poisson's equation, since $\rho(x, y, z)$ or

70   IV. SPACE-CHARGE-FIELD EXPRESSIONS

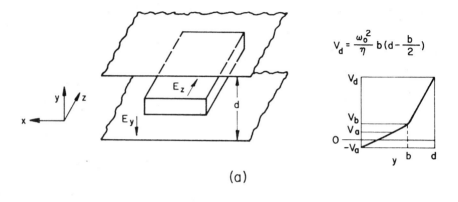

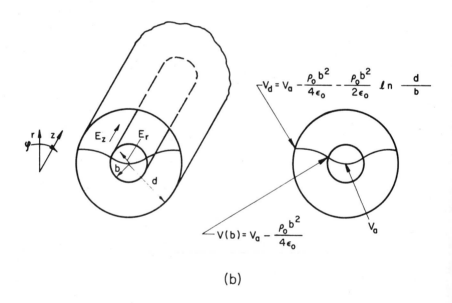

Fig. 1. Simple single-velocity flow space-charge-field models: (a) planar symmetry; (b) axial symmetry.

1. INTRODUCTION

$\rho(r, \varphi, z)$ is a constant. The potential distribution in such systems is given by the following:

(A) An axially symmetric system with the potential independent of the z-position and $\rho = \rho_0$, constant over the cross section (Brillouin flow):

$$V(r) = V_a - \frac{\rho_0 b^2}{4\epsilon_0} - \frac{\rho_0 b^2}{2\epsilon_0} \ln \frac{r}{b}, \qquad r \geqslant b \tag{1}$$

where b is the beam radius and V_a is the potential at $r = 0$.

(B) Rectangular beam between parallel plates (Brillouin flow):

$$V(y) = \frac{\omega_c^2}{2\eta} y^2 \quad \text{(within the stream)}$$

$$V(y) = \text{(a linear variation outside the stream)} \tag{2}$$

and is illustrated in Fig. 1 for the case of no variation in the z-direction.

The above method is applicable only to small-cross-section low-density beams and is thus limited to simple flow problems in view of the usual assumption on the charge density distribution.

In nonlinear interaction problems we have already encountered the problem of electron crossover (Chapter I), which leads to a multivelocity flow. The unmanageability of the Eulerian equations for multivelocity flow occasioned the use of a Lagrangian system as one means of solving the problems. The basis of the Lagrangian or particle method is the characterization of all charge particles in the flow by one or more parameters, such as entrance time and velocity, etc., and the subsequent tracking of individual particles or "charge groups" through the field system. For reasons which will become apparent later in this chapter, the "charge groups" referred to above will be considered as constituted of numerous electrons or ions distributed over a finite volume rather than as individual elementary charge elements. The essence is to obtain charge groups with nonzero dimension. The utility and convenience of the Lagrangian method is balanced by the problem of developing appropriate space-charge-field expressions for the analysis. The methods to be developed involve the calculation of Coulomb forces between discrete charge groups, as opposed to a more elegant and satisfying treatment which would give an analytic expression for the potential and hence fields from a continuous charge distribution.

Since the active beam-wave devices to be considered are of both the rectangular and the axial symmetry varieties, both configurations are analyzed. Under the most general conditions three-dimensional effects must be considered, although some physical systems are adequately described using two- and one-dimensional formulations. The principal analytical method to be utilized in this chapter involves the solution of Poisson's equation using Green's function techniques, although some other, equivalent methods will be mentioned to illustrate concepts and add clarity. The solution of potential problems using the Green's function is thoroughly outlined by Smythe[4] and Panofsky-Phillips[5] and their work may be used directly. Both uniform and bunched beams are analyzed, since both are considered in subsequent chapters.

2 Green's Function Method for Potential Problems

In the analysis of beam-wave devices the flow of charged beams of particular shapes in regions containing composite electrostatic, magnetostatic and radio-frequency fields must be considered. The detailed motion of the charged particles within the beam depends not only upon the impressed fields but also on the fields between elements of charge, i.e., Coulomb fields. For a particular beam shape and electrode configuration surrounding the beam, Poisson's equation must be solved within the

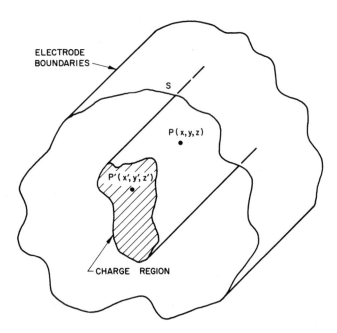

Fig. 2. Potential problem for an arbitrary geometry.

2. GREEN'S FUNCTION METHOD FOR POTENTIAL PROBLEMS

beam and Laplace's equation in the charge-free region between the beam and the electrodes.

The solution of such potential problems as illustrated in Fig. 2 is facilitated by the use of Green's theorem. From Poisson's equation we may write the potential at an arbitrary point $P(x, y, z)$ within the electrode boundaries due to elementary charge elements located at the points $P'(x', y', z')$ as

$$V(P) = \frac{1}{4\pi\epsilon_0} \int_{\text{vol}} \frac{\rho(P') \, d\tau(P')}{|\mathbf{r}(P, P')|}, \qquad (3)$$

where $\rho(P')$ represents the distribution of charge within the volume and $\mathbf{r}(P, P')$ is the radius vector between the charge location and the point in question. Of course, knowledge of $\rho(P')$ allows immediate calculation of the potential function. The class of potential problems depicted in Fig. 2 can be conveniently solved by generating the potential function solution from the Green's function solution for an elementary charge unit (point, rod, ring, or disk) within the boundaries, when the boundaries are at zero potential.

Green's theorem is

$$\int_\tau (V\nabla^2 G - G\nabla^2 V) \, d\tau = \int_S (V\nabla G - G\nabla V) \cdot d\mathbf{S}, \qquad (4)$$

where V is the desired potential function and G is the Green's function discussed above, both being nonsingular throughout τ bounded by S. The Green's function may be written, then, as the sum of that due to the point-charge potential function plus that due to the charge induced on the boundary S.

$$G(\mathbf{r}) = \frac{1}{4\pi\epsilon_0 r} + \psi_S. \qquad (5)$$

Since the boundaries are at zero potential, substituting Eq. (5) into Green's theorem gives the potential solution as

$$V(P) = -\epsilon_0 \left[\int G\nabla^2 V \, d\tau + \int V_S \nabla G \cdot d\mathbf{S} \right]$$

$$= -\epsilon_0 \int G\nabla^2 V \, d\tau. \qquad (6)$$

Equation 6 will be applied to particular electrode and beam configurations in the following sections of this chapter.

3 Potential Function for the Cartesian Coordinate System

The geometry to be considered for the rectangular beam is that characteristic of various crossed electric and magnetic field devices in which a rectangular strip beam flows between two parallel plates, one being an rf structure and the other a high-conductivity planar surface. For the purpose of solving the potential problem, the rf structure will be replaced by a smooth, perfectly conducting sheet. The geometry to be considered is shown in Fig. 3. The validity of this operation is examined

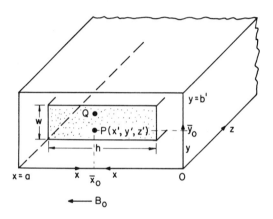

FIG. 3. Strip beam geometry for space-charge-field calculation.

in a later section. The space-charge field is thus found by a superposition of the fields produced inside the boundaries $x = a$, $y = b'$ by individual charges arranged according to the beam model illustrated.

A rigorous treatment of the problem begins with the retarded potential, and potential expressions valid for relativistic particle velocities are obtained conveniently by the use of Green's dyadic. We will consider chiefly nonrelativistic particle velocities; however, a one-dimensional relativistic treatment will be given. For particle velocities small compared to the velocity of light, i.e., $(v/c)^2 \ll 1$, the particle self-magnetic field may be neglected and only the electrostatic potential need be considered. All theorems and methods common to electrostatic problems are then applicable.

In Fig. 3, the potential distribution within the boundaries due to a distributed space charge is written as

$$V(x, y, z) = \iiint_\tau \rho(x'y', z', t) G(x, y, z, x', y', z') \, dx' \, dy' \, dz', \tag{7}$$

3. CARTESIAN COORDINATE SYSTEM

where τ indicates the volume of the region occupied by the charge, $\rho(x', y', z', t)$ is the space-charge density at a point (x', y', z'), and the Green's function $G(x, y, z, x', y', z')$ is a function of the exact geometry and measures the potential produced at a point (x, y, z) due to a unit charge placed at (x', y', z') when both points are within the boundaries $x = a$, $y = b'$.

A unit charge Q is located at a point $P(x', y', z')$ in the interaction region as illustrated in Fig. 3 with the potentials of the boundary walls taken as zero; i.e., $V = 0$ at $x = 0$, a and $V = 0$ at $y = 0$, b'. The potential at any point within the walls (x, y, z) due to this unit charge at $P(x', y', z')$ is given by the solution of Laplace's equation

$$\nabla^2 G(x, y, z) = 0$$

or
$$\nabla^2 V(x, y, z) = 0 \tag{8}$$

except at the point P. The potential function must be symmetrical about the z-coordinate of the source at $P(x', y', z')$ and must be zero on the boundaries. It must also vanish at a large distance, z, from P. The point charge at P must have a finite dimension, ϵ, in order to maintain the potential function everywhere bounded.

It is straightforward to expand the potential function in terms of harmonics of the form

$$V(x, y, z) = X(x)Y(y)Z(z) \tag{9}$$

and apply the boundary conditions outlined above. The general solution for the potential when the region is infinitely extended in z and $-z$ is given by

$$V(x, y, z) = \sum_{k=1}^{\infty}\sum_{m=1}^{\infty} A_{mk} \sin\frac{k\pi x}{a} \sin\frac{m\pi y}{b'} \exp\left[-\frac{\pi(m^2 a^2 + k^2 b'^2)^{\frac{1}{2}} |z - z'|}{ab'}\right]. \tag{10}$$

The constants A_{mk} of Eq. (10) are found by placing $\partial V/\partial z = 0$ everywhere in the $z = z'$ plane except at $P(x', y', z')$, where the unit charge is located. Only one-half of the flux emanating from the unit charge at z' goes towards a plane at z; thus an integration over the ΔA area occupied by the charge at P and the application of Gauss' theorem yields

$$\iint_{\Delta A} \left(\frac{\partial V}{\partial z}\right)\bigg|_{z'} dx\, dy = \frac{|Q|}{2\epsilon_0}. \tag{11}$$

Thus A_{mk} may be evaluated and the potential function is expressed as

$$V(x, y, z) = \frac{2Q}{\pi\epsilon_0} \sum_{k=1}^{\infty} \sum_{m=1}^{\infty} (m^2a^2 + k^2b'^2)^{-\frac{1}{2}} \exp\left[-\frac{\pi(m^2a^2 + k^2b'^2)^{\frac{1}{2}}}{ab'}|z - z'|\right]$$

$$\cdot \sin\frac{m\pi y}{b'} \sin\frac{m\pi y'}{b'} \sin\frac{k\pi x}{a} \sin\frac{k\pi x'}{a}. \tag{12}$$

For comparison purposes the potential function $V(x, y, z)$ is given for the similar geometry except that the region is of finite extent in z and $-z$. It is easily shown that

$$V(x, y, z) = \frac{4Q}{ab'\epsilon_0} \sum_{k=1}^{\infty} \sum_{m=1}^{\infty} \frac{\sinh A_{mk}(c - z') \sinh A_{mk}z}{A_{mk} \sinh A_{mk}c}$$

$$\cdot \sin\frac{k\pi x'}{a} \sin\frac{k\pi x}{a} \sin\frac{m\pi y'}{b'} \sin\frac{m\pi y}{b'}, \tag{13}$$

where

$$A_{mk} = \frac{\pi(m^2a^2 + k^2b'^2)^{\frac{1}{2}}}{ab'}.$$

The rectangular box extends from $z = 0$ to c and the above considers $z > z'$. For $z < z'$ simply interchange z and z'. Equation (12) is sufficiently general for the space-charge potential since the interaction region is generally long compared to all the transverse dimensions.

The unit charge used in the derivation may be related to the beam charge density by appropriate integration over the volume of the interaction region. The charge Q is replaced by $-|\rho_0| dx_0 dy_0 dz_0$ after integration over dx_0, dy_0, and dz_0, which denote the initial charge coordinates. The unit charge positions x', y', and z' are written as functions of x_0, p_0, Φ_0, and q, normalized variables to be defined in Chapter VIII. $p_0 = y_0/w$ indicates the initial normalized y-position, Φ_0 the initial phase position of the charge relative to the rf wave, and $q = D\omega z/\bar{u}_0$ the normalized z-coordinate. The space-charge potential is now written as

$$V_{sc} = -\frac{2|\rho_0|wh\bar{u}_0}{\pi\epsilon_0\omega} \sum_{k=1}^{\infty} \sum_{m=1}^{\infty} \int_0^{2\pi} \int_{(\frac{1}{s}-\frac{1}{2})}^{(\frac{1}{s}+\frac{1}{2})} \int_{(\frac{1}{\delta}-\frac{1}{2})}^{(\frac{1}{\delta}+\frac{1}{2})}$$

$$\cdot (m^2a^2 + k^2b'^2)^{-\frac{1}{2}} \sin\frac{m\pi y}{b'} \sin\frac{m\pi y'}{b'} \sin\frac{k\pi x}{a} \sin\frac{k\pi x'}{a}$$

$$\cdot \exp\left[-\frac{\pi(m^2a^2 + k^2b'^2)^{\frac{1}{2}}|z - z'|}{ab'}\right]\left(\frac{u_{zi}}{u_z}\right) dq_0 \, dp_0 \, d\Phi_0, \tag{14}$$

where

$\bar{u}_0 \triangleq$ average z-directed beam velocity,
$u_{zi} \triangleq$ initial normalized z-directed velocity of a charge group,
$u_z \triangleq$ normalized z-directed velocity of a charge group,
$\bar{y}_0 \triangleq$ mean beam position in the y-direction,
$\bar{x}_0 \triangleq$ mean beam position in the x-direction,
$\Phi_0 \triangleq$ entrance phase position of a charge group,
$p_0 \triangleq (y_0/w)$,
$p \triangleq (y/w)$,
$s \triangleq (w/\bar{y}_0)$,
$r \triangleq (\bar{y}_0/b')$,
$g_0 \triangleq (x_0/h)$,
$g \triangleq (x/h)$,
$\bar{\alpha} \triangleq (x_0/a)$, and
$\delta \triangleq (h/\bar{x}_0)$.

The triple integration indicated in Eq. (14) extends over the transverse plane of the beam and over the entire entering charge, as indicated by the initial charge phase positions covering one complete cycle of the rf wave.

Equation (14), for the space-charge potential in the interaction region bounded by a rectangular wall configuration which is infinite in extent in z and $-z$, is a three-dimensional expression where the unit charges are finite-size "points" sufficient to keep the potential and field functions everywhere bounded. It is to be used later in a study of lateral beam spreading in crossed-field devices employing singly periodic structures and also has particular application in crossed-field devices employing doubly or biperiodic propagating structures. Another, not so common application is in linear-beam devices (O-type) employing strip beams and planar structures. The difficulty here is one of beam focusing, particularly with regard to lateral spreading and "curling" of the beam edges.

The appropriate space-charge-field components for the three directions are obtained by systematic differentiation of Eq. (14). The following results appear:

$$E_{sc-x} = -\frac{\partial V_{sc}}{\partial x} = \frac{2|\rho_0|wh\bar{u}_0}{\pi\epsilon_0\omega} \sum_{k=1}^{\infty}\sum_{m=1}^{\infty}$$

$$\cdot \int_0^{2\pi} \int_{(\frac{1}{s}-\frac{1}{2})}^{(\frac{1}{s}+\frac{1}{2})} \int_{(\frac{1}{\delta}-\frac{1}{2})}^{(\frac{1}{\delta}+\frac{1}{2})} (m^2a^2 + k^2b'^2)^{-\frac{1}{2}} \left(\frac{k\pi}{a}\right) \sin\frac{m\pi y}{b'} \sin\frac{m\pi y'}{b'}$$

$$\cdot \cos\frac{k\pi x}{a} \sin\frac{k\pi x'}{a} \left(\frac{u_{zi}}{u_z}\right) \exp\left[-\frac{\pi(m^2a^2 + k^2b'^2)^{\frac{1}{2}}|z-z'|}{ab'}\right]$$

$$\cdot dg_0\, dp_0\, d\Phi_0, \tag{15}$$

$$E_{sc-y} = -\frac{\partial V_{sc}}{\partial y} = \frac{2|\rho_0|wh\bar{u}_0}{\pi\epsilon_0\omega}\sum_{k=1}^{\infty}\sum_{m=1}^{\infty}\int_0^{2\pi}$$

$$\cdot \int_{(\frac{1}{s}-\frac{1}{2})}^{(\frac{1}{s}+\frac{1}{2})}\int_{(\frac{1}{\delta}-\frac{1}{2})}^{(\frac{1}{\delta}+\frac{1}{2})}(m^2a^2+k^2b'^2)^{-\frac{1}{2}}\left(\frac{m\pi}{b'}\right)\cos\frac{m\pi y}{b'}\sin\frac{m\pi y'}{b'}\sin\frac{k\pi x}{a}$$

$$\cdot \sin\frac{k\pi x'}{a}\left(\frac{u_{zi}}{u_z}\right)\exp\left[-\frac{\pi(m^2a^2+k^2b'^2)^{\frac{1}{2}}|z-z'|}{ab'}\right]dg_0\,dp_0\,d\Phi_0, \tag{16}$$

and finally, in the direction of primary flow

$$E_{sc-z} = -\frac{\partial V_{sc}}{\partial z} = -\frac{2|\rho_0|wh\bar{u}_0}{\pi\epsilon_0\omega}\left(\frac{\pi}{ab'}\right)\sum_{k=1}^{\infty}\sum_{m=1}^{\infty}$$

$$\cdot \int_0^{2\pi}\int_{(\frac{1}{s}-\frac{1}{2})}^{(\frac{1}{s}+\frac{1}{2})}\int_{(\frac{1}{\delta}-\frac{1}{2})}^{(\frac{1}{\delta}+\frac{1}{2})}\sin\frac{m\pi y}{b'}\sin\frac{m\pi y'}{b'}\sin\frac{k\pi x}{a}\sin\frac{k\pi x'}{a}\left(\frac{u_{zi}}{u_z}\right)$$

$$\cdot \exp\left[-\frac{\pi(m^2a^2+k^2b'^2)^{\frac{1}{2}}|z-z'|}{ab'}\right]\operatorname{sgn}(z-z')\,dg_0\,dp_0\,d\Phi_0, \tag{17}$$

where

$$\operatorname{sgn}(z-z') = 1 \quad \text{for} \quad z > z'$$

and

$$\phantom{\operatorname{sgn}(z-z')} = -1 \quad \text{for} \quad z < z'.$$

The sign of the term $(z-z')$ as obtained above gives the direction of force on the unit charge.

Examination of Eqs. (15)–(17) for the component space-charge fields reveals that they are intractable in the analytic sense due to the very large number of terms required in the summations in order to secure convergence. As will be evident in many sections of this book, the high-speed computer can render valuable assistance. Although such summations and integrations as are contained in the above equations are completed in a few minutes on today's computers (in seconds on tomorrow's), their repeated evaluation at several hundred z-displacement planes requires inordinately long times and large subsidies. It is thus well to explore possible simplifications and shortened numerical methods.

It is convenient to define space-charge-field weighting functions and integrals of weighting functions from Eqs. (15)–(17) for the fields.

3. CARTESIAN COORDINATE SYSTEM

Interchanging integration and summation gives the following space-charge-field weighting functions:

$$F_{3-x} = \sum_{k=1}^{\infty} \sum_{m=1}^{\infty} k \left[m^2 \left(\frac{a^2}{b'^2} \right) + k^2 \right]^{-\frac{1}{2}} \sin m\pi \frac{\beta y}{\beta b'} \sin m\pi \frac{\beta y'}{\beta b'} \cos k\pi \frac{\beta x}{\beta a}$$

$$\cdot \sin k\pi \frac{\beta x'}{\beta a} \exp \left\{ -\frac{\pi}{\beta a} \left[m^2 \left(\frac{a^2}{b'^2} \right) + k^2 \right]^{\frac{1}{2}} | \Phi - \Phi' | \right\}, \quad (18)$$

$$F_{3-y} = \sum_{k=1}^{\infty} \sum_{m=1}^{\infty} m \left[m^2 + \left(\frac{b'^2}{a^2} \right) k^2 \right]^{-\frac{1}{2}} \cos m\pi \frac{\beta y}{\beta b'} \sin m\pi \frac{\beta y'}{\beta b'} \sin k\pi \frac{\beta x}{\beta a}$$

$$\cdot \sin k\pi \frac{\beta x'}{\beta a} \exp \left\{ -\frac{\pi}{\beta a} \left[m^2 \left(\frac{a^2}{b'^2} \right) + k^2 \right]^{\frac{1}{2}} | \Phi - \Phi' | \right\}, \quad (19)$$

and

$$F_{3-z} = \sum_{k=1}^{\infty} \sum_{m=1}^{\infty} \sin m\pi \frac{\beta y}{\beta b'} \sin m\pi \frac{\beta y'}{\beta b'} \sin k\pi \frac{\beta x}{\beta a} \sin k\pi \frac{\beta x'}{\beta a}$$

$$\cdot \exp \left\{ -\frac{\pi}{\beta a} \left[m^2 \left(\frac{a^2}{b'^2} \right) + k^2 \right]^{\frac{1}{2}} | \Phi - \Phi' | \right\}, \quad (20)$$

where $| z - z' |$ has been replaced by $| \Phi - \Phi' |$ in accordance with the procedure outlined in Chapters V and VIII. The space-charge-field expressions are thus written as

$$E_{sc-x} = \frac{2 | \rho_0 | wh\bar{u}_0}{\pi \epsilon_0 \omega} \left(\frac{\pi}{ab'} \right) \int_0^{2\pi} \int_{(\frac{1}{s}-\frac{1}{2})}^{(\frac{1}{s}+\frac{1}{2})} \int_{(\frac{1}{\delta}-\frac{1}{2})}^{(\frac{1}{\delta}+\frac{1}{2})} F_{3-x} \left(\frac{u_{zi}}{u_z} \right)$$

$$\cdot dg_0 \, dp_0 \, d\Phi_0, \quad (21)$$

$$E_{sc-y} = \frac{2 | \rho_0 | wh\bar{u}_0}{\pi \epsilon_0 \omega} \left(\frac{\pi}{ab'} \right) \int_0^{2\pi} \int_{(\frac{1}{s}-\frac{1}{2})}^{(\frac{1}{s}+\frac{1}{2})} \int_{(\frac{1}{\delta}-\frac{1}{2})}^{(\frac{1}{\delta}+\frac{1}{2})} F_{3-y} \left(\frac{u_{zi}}{u_z} \right)$$

$$\cdot dg_0 \, dp_0 \, d\Phi_0, \quad (22)$$

and

$$E_{sc-z} = -\frac{2 | \rho_0 | wh\bar{u}_0}{\pi \epsilon_0 \omega} \left(\frac{\pi}{ab'} \right) \int_0^{2\pi} \int_{(\frac{1}{s}-\frac{1}{2})}^{(\frac{1}{s}+\frac{1}{2})} \int_{(\frac{1}{\delta}-\frac{1}{2})}^{(\frac{1}{\delta}+\frac{1}{2})} F_{3-z} \left(\frac{u_{zi}}{u_z} \right)$$

$$\cdot \text{sgn}(\Phi - \Phi') \, dg_0 \, dp_0 \, d\Phi_0. \quad (23)$$

The space-charge-field weighting functions defined in Eqs. (18)–(20) have been computed for typical values of $\beta b'$ and of a/b'. The computations were made by dividing the interaction region into fifteen rectangles as illustrated in Fig. 4; the beam may span part or all of the rectangles.

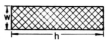

FIG. 4. Beam and interaction space configuration for computing three-dimensional space-charge-field weighting functions.

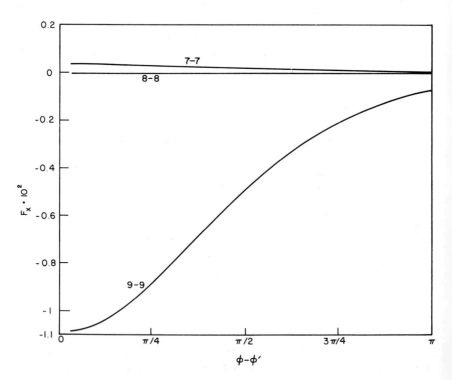

FIG. 5. x-directed space-charge-field weighting function. ($\beta b' = 2$, $\beta a = 8$.)

4. TWO-DIMENSIONAL RECTANGULAR SYSTEM

The weighting functions F_x, F_y, and F_z are shown in Figs. 5, 6, and 7 for $\beta b' = 2$ and $b'/a = 0.25$. As noted on the figures, F_y and F_z do not vary over the beam cross section.

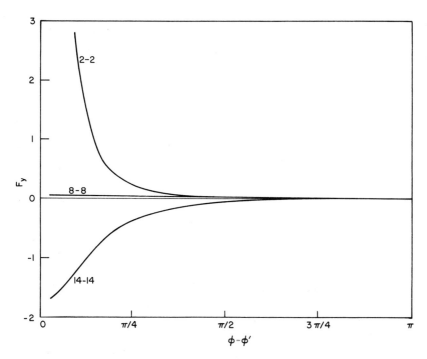

FIG. 6. y-directed space-charge-field weighting function. ($\beta b' = 2$, $\beta a = 8$.)

4 Potential Function for a Two-Dimensional Rectangular System

In the previous section the general three-dimensional potential problem for rectangular beams was solved considering the charges as elementary volume elements. In devices with singly periodic rf structures the circuit width in the x-direction is usually less than $\lambda_0/2$ and the beam width somewhat less than that in order for the rf voltage to be constant over the beam cross section. Under these conditions the calculation of the space-charge potential and fields is simplified, since variations with the x-coordinate may be neglected. In order to account for end effects a finite beam width, h, will be considered. Figure 3, illustrating the beam cross section, is applicable and Eqs. (21)–(23) for the space-charge field may be simplified.

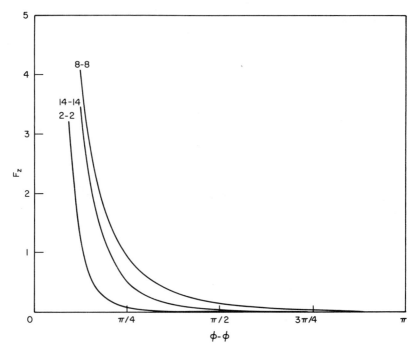

FIG. 7. z-directed space-charge-field weighting function. ($\beta b' = 2$, $\beta a = 8$.)

First consider the exponential factor $(m^2 a^2 + k^2 b'^2)^{\frac{1}{2}}$ and expand this using the binomial theorem:

$$(m^2 a^2 + k^2 b'^2)^{\frac{1}{2}} = (ma + kb') \left\{ 1 - \frac{makb'}{(ma + kb')^2} - \frac{1}{2}\left[\frac{makb'}{(ma + kb')^2}\right]^2 \right.$$
$$\left. + \frac{1}{2}\left[\frac{makb'}{(ma + kb')^2}\right]^3 + \cdots \right\}. \quad (24)$$

Under many practical conditions the first term in the expansion is sufficient. The exponential is thus simplified and correction factors C_f and C_f' are introduced when a particular set of parameters dictates a need.

$$E_{sc-y} = \frac{2|\rho_0| w \bar{u}_0}{\pi \epsilon_0 \omega} C_f \int_0^{2\pi} \int_{\frac{\bar{y}_0}{w} - \frac{1}{2}}^{\frac{\bar{y}_0}{w} + \frac{1}{2}} \left(\frac{u_{zi}}{u_z}\right) \sum_{k=1}^{\infty} \sum_{m=1}^{\infty} \left(\frac{m\pi}{b'}\right)$$
$$\cdot (m^2 a^2 + k^2 b'^2)^{-\frac{1}{2}} \cos \frac{m\pi y}{b'} \sin \frac{m\pi y'}{b'} \left(\frac{2a}{k\pi}\right) \sin \frac{k\pi x}{a} \sin \frac{k\pi h}{2a}$$
$$\cdot \sin \frac{k\pi}{2} \exp\left[-\frac{\pi(ma + kb')|z - z'|}{ab'}\right] dp_0 \, d\Phi_0 \quad (25)$$

4. TWO-DIMENSIONAL RECTANGULAR SYSTEM

and

$$E_{sc-z} = -\frac{2|\rho_0|w\bar{u}_0}{\pi\epsilon_0\omega}\left(\frac{\pi}{ab'}\right)C_f'\int_0^{2\pi}\int_{\bar{y}_0-\frac{1}{w}-\frac{1}{2}}^{\bar{y}_0+\frac{1}{w}+\frac{1}{2}}\left(\frac{u_{zi}}{u_z}\right)\text{sgn}(z-z')$$

$$\cdot\left[\sum_{m=1}^{\infty}\sin\frac{m\pi y}{b'}\sin\frac{m\pi y'}{b'}\exp\left(-\frac{m\pi|z-z'|}{b'}\right)\right]\left[\sum_{k=1}^{\infty}\sin^2\frac{k\pi}{2}\left(\frac{2a}{k\pi}\right)\right.$$

$$\left.\cdot\sin\frac{k\pi h}{2a}\exp\left(-\frac{k\pi|z-z'|}{a}\right)\right]dp_0\,d\Phi_0, \quad (26)$$

where the integration over the x-dimension of the beam has been eliminated by restricting our consideration to the midplane $x = a/2$ of the beam.

The infinite series of Eqs. (25) and (26) may be summed and the space-charge-field components for this special two-dimensional case written as

$$E_{sc-y} = \frac{2|\rho_0|w\bar{u}_0}{\pi\epsilon_0 b'\omega}F_{2-y} \quad (27)$$

and

$$E_{sc-z} = \frac{2|\rho_0|w\bar{u}_0}{\pi\epsilon_0 b'\omega}F_{2-z}, \quad (28)$$

where F_{2-y} and F_{2-z} are space-charge integrals defined as follows:

$$F_{2-y} = C_f\int_0^{2\pi}\int_{\left(\frac{1}{s}-\frac{1}{2}\right)}^{\left(\frac{1}{s}+\frac{1}{2}\right)}[(1+e^{-2|\xi|})\cos Y - 2e^{-|\xi|}\cos Y']\chi\,dp_0\,d\Phi_0 \quad (29)$$

and

$$F_{2-z} = C_f'\int_0^{2\pi}\int_{\left(\frac{1}{s}-\frac{1}{2}\right)}^{\left(\frac{1}{s}+\frac{1}{2}\right)}\sin Y(1-e^{-2|\xi|})\chi\,\text{sgn}(\xi)\,dp_0\,d\Phi_0. \quad (30)$$

The following parameters have been defined:

$$\xi \triangleq (\pi/b')(z-z'),$$
$$Y \triangleq (\pi y/b') = \pi rsp,$$
$$Y' \triangleq \pi rsp',$$
$$s \triangleq (w/\bar{y}_0),$$
$$r \triangleq (\bar{y}_0/b'),$$
$$p \triangleq (y/w)$$

and
$$\chi = \frac{u_{zi}/u_z \sin Y' e^{-|\xi|} F(h_a, |b_a\xi|)}{[(1 + e^{-2|\xi|}) - 2e^{-|\xi|}\cos(Y + Y')][(1 + e^{-2|\xi|}) - 2e^{-|\xi|}\cos(Y - Y')]}$$
$$h_a = h/a,$$
$$b_a = b'/a.$$

The function $F(h_a, |b_a\xi|)$ is a space-charge-field weighting function for the magnetic field direction (x-direction) and arises due to the considered finite width of the electron beam. This function is shown in Fig. 8 and is written as

$$F(h_a, |b_a\xi|) = \tan^{-1}\left[\frac{\sin(\pi h_a)/2}{e^{b_a|\xi|} - \cos(\pi h_a)/2}\right]$$
$$+ \tfrac{1}{2}\tan^{-1}\left[\frac{2\sin(\pi h_a)/2\,[e^{b_a|\xi|} + \cos(\pi h_a)/2]}{[e^{b_a|\xi|} + \cos(\pi h_a)/2]^2 - \sin^2(\pi h_a)/2}\right]. \quad (31)$$

Both h_a and b_a are generally less than 1.

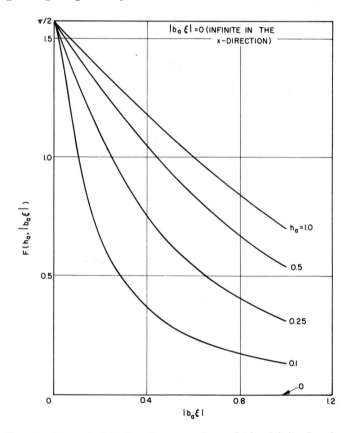

FIG. 8. Magnetic field direction space-charge-field weighting function.

4. TWO-DIMENSIONAL RECTANGULAR SYSTEM

The space-charge integrals $F_{2-y}(Y, \xi)$ and $F_{2-z}(Y, \xi)$ given in Eqs. (29) and (30) may now be computed for any particular division of the stream. There, of course, arises a question as to the optimum number of stream divisions for a given thickness. It is shown later that each layer can effectively represent approximately 4–5% of the interaction thickness. To indicate the nature and the range of effectiveness of the space-charge forces, the field integrals $F_{2-y, j-k}$ and $F_{2-z, j-k}$ are shown in Figs. 9–12 for a two-layer beam. Physically speaking, the integrals $F_{2-y, j-k}$ and $F_{2-z, j-k}$

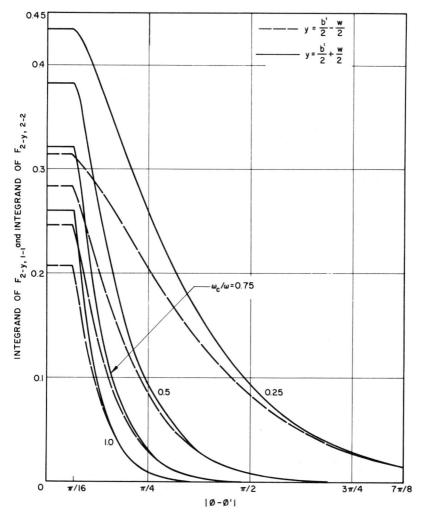

FIG. 9. Variation of E_{sc-y} with phase. ($D = 0.1$, $r = 0.75$, $s = 0.067$, $\omega_p = \omega_c$.)

represent the y- and z-directed space-charge fields appearing at electrons in the jth layer due to electrons in the kth layer. It is noted immediately that the range of effectiveness of the forces decreases as ω_c/ω increases, indicating that for very large magnetic field values fewer charge groups need be considered in order to compute accurately the space-charge forces. Since the computation time approximately quadruples for a doubling of charge groups this can amount to a considerable savings.

The abscissa of Figs. 9–12 is related to $z - z'$ and thus is a measure

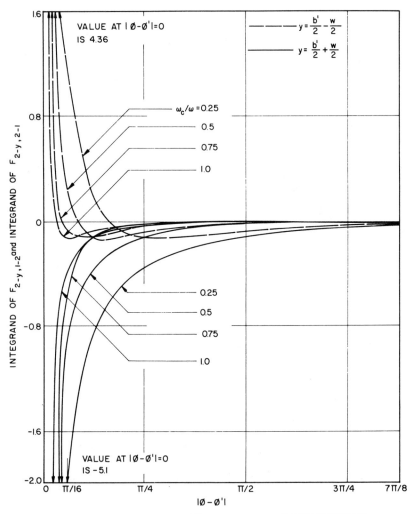

Fig. 10. Variation of E_{sc-y} with phase. ($D = 0.1$, $r = 0.75$, $s = 0.067$, $\omega_p = \omega_c$.)

4. TWO-DIMENSIONAL RECTANGULAR SYSTEM

of the separation of charge groups. The reason for considering the charge group to have a finite dimension is apparent when it is noted that $F_{2-y,1-1}$, $F_{2-y,2-2}$, $F_{2-z,1-1}$, $F_{2-z,2-2}$, $F_{2-z,1-2}$, and $F_{2-z,2-1}$ all tend to infinity as $\Phi - \Phi'$ or $z - z'$ approaches zero. To avoid this mathematical difficulty we consider only the macroscopic behavior of charge groups and assume that the integrals are constant for $|\Phi - \Phi'|$ or $|z - z'|$ less than the initial inter-charge group spacing. The value to be used in computation is that corresponding to $|\Phi - \Phi'| = 2\pi/m$, m being the

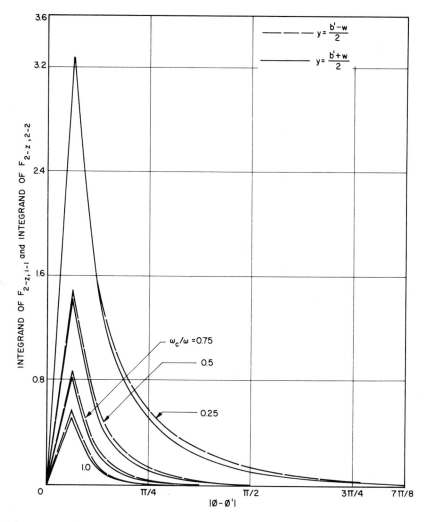

Fig. 11. Variation of E_{sc-z} with phase. ($D = 0.1$, $r = 0.75$, $s = 0.067$, $\omega_p = \omega_c$.)

number of charge groups. The actual values of y and y' for the charge group in question are used. In the case of the z-directed fields it is assumed that the field increases linearly in the region from $|\Phi - \Phi'| = 0$ to $2\pi/m$. To be certain, the method of handling the space-charge-field integrals for $|\Phi - \Phi'| \approx 0$ does affect the computations and the very accurate method of using a large number of charge groups is costly of computation time. The method outlined above has been found to be quite satisfactory in yielding reasonably accurate results.

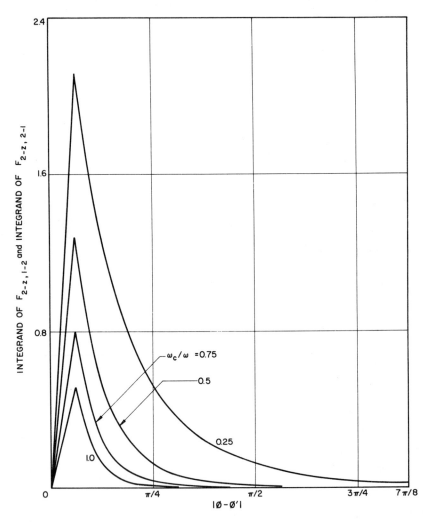

FIG. 12. Variation of E_{sc-z} with phase. ($D = 0.1$, $r = 0.75$, $s = 0.067$, $\omega_p = \omega_c$.)

5 Space-Charge Fields for Rods of Charge

An alternate approach to the development of space-charge-field expressions for a two-dimensional rectangular interaction problem is to consider that the beam is made up of rods of charge, either infinite or finite in length in the x-direction, as illustrated in Fig. 13. The field

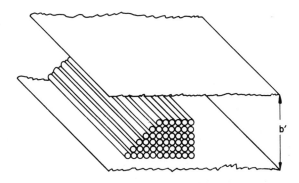

FIG. 13. Rod of charge model for space-charge-field calculation.

equations have been worked out by Smythe[4] and were used by Sedin[6] in his treatment of the nonlinear crossed-field problem. In view of the fact that the expressions for E_{sc-y} and E_{sc-z} are the same as those obtained in previous sections, the details of the derivation procedure are left to the reader.

By either Green's method or the method of images and an appropriate conformal transformation, the potential function for the configuration of Fig. 13 may be written as

$$V(y, z) = \frac{Q_l}{2\pi\epsilon_0} R_e \left\{ \ln \frac{\exp\left[\frac{\pi}{b'} | z - z' | + j\frac{\pi}{b'} y\right] - \exp\left[-j\frac{\pi}{b'} y'\right]}{\exp\left[\frac{\pi}{b'} | z - z' | + j\frac{\pi}{b'} y\right] - \exp\left[j\frac{\pi}{b'} y'\right]} \right\}, \quad (32)$$

where Q_l denotes the magnitude of the line charge.

6 Replacement of Rf Structure by an Impedance Sheet

In the previous sections of this chapter the space-charge fields were calculated from electrostatics after the rf structure was replaced by a perfectly conducting sheet. Question arises as to the validity of this procedure, since the structure is more accurately characterized as an impedance sheet. The replacement of the circuit by a perfectly conducting sheet was justified on the basis of neglecting the fringing fields between the circuit and bottom plate in view of the short-range nature of the space-charge forces.

That assumption will be evaluated by replacing the slow-wave structure by an impedance sheet of the following form at $y = b'$ and then proceeding with the Green's function technique in a manner similar to that outlined previously:

$$\frac{E_z(y=b')}{E_y(y=b')} \triangleq \frac{-\omega\epsilon_0}{\gamma} Z, \qquad (33)$$

where

$Z \triangleq$ the sheet impedance and

$\gamma \triangleq$ field propagation constant assuming a variation of the form $e^{-j\gamma z}$.

Satisfying the other boundary conditions, that the potential be zero along $x = 0$, $x = a$ and $y = 0$, gives for the new potential function

$$V_{sc}(x, y, z) = \sum_{n_k} \sum_{k=1}^{\infty} A_{nk} \sin y \left[n_k^2 - \left(\frac{k\pi}{a}\right)^2 \right]^{\frac{1}{2}} \sin \frac{k\pi x}{a} e^{-n_k|z-z'|}, \qquad (34)$$

where n_k is chosen to satisfy the boundary condition of Eq. (34) at $y = b'$.

Assume that the factor $\bar{n}_k = [n_k^2 - (k\pi/a)^2]^{\frac{1}{2}}$ defines an orthogonal set of values for all n_k and that the field is zero everywhere in the interaction region except in the neighborhood ϵ of the unit charge, Q, as previously. Then the space-charge potential becomes

$$V_{sc}(x, y, z) = -\frac{|Q|}{\epsilon_0} \sum_{n_k} \sum_{k=0}^{\infty} \frac{\sin \bar{n}_k y' \sin \frac{k\pi x'}{a}}{\left(\frac{ab'n_k}{2}\right)\left[1 - \frac{\sin 2\bar{n}_k b'}{2\bar{n}_k b'}\right]}$$

$$\cdot \sin \bar{n}_k y \sin \frac{k\pi x}{a} e^{-n_k|z-z'|}. \qquad (35)$$

The characteristic equation for n_k is obtained by calculating E_{sc-y} and E_{sc-z} from Eq. (35) and then substituting into Eq. (33), allowing the n, k field component. The result is

$$\frac{\bar{n}_k}{n_k} \cot \bar{n}_k b' = \frac{-n_k}{\omega\epsilon_0 X_{n,k}}, \qquad (36)$$

where $Z_{n,k} \triangleq jX_{n,k}$. Equation (36) is transcendental and its solution is seen in Fig. 14.

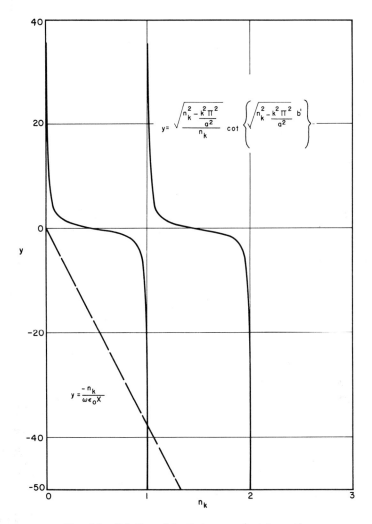

FIG. 14. Solution of the $\bar{n}_k$ transcendental equation.

The large negative slope for $-n_k/\omega\epsilon_0 X$ indicated in Fig. 14 shows that n_k may be written approximately as

$$n_k \approx \left[\left(\frac{n\pi}{b'}\right)^2 + \left(\frac{k\pi}{a}\right)^2\right]^{\frac{1}{2}} \tag{37}$$

and substitution of Eq. (37) into Eq. (35) results in the same form as given by Eq. (12). Thus the original assumption is justified. Such a proof is independent of the actual geometry of the interaction space configuration and thus is valid for axially symmetric systems with solid or hollow beam shapes.

7 Space-Charge Potentials for Cylindrical Systems

Many of the beam-wave interaction systems to be studied in later chapters may be classified as axially symmetric systems, as for instance the klystron, the traveling-wave tube, and the beam-plasma system. In most cases there is symmetry about the cylindrical axis and a two-dimensional system results. However in more exotic systems such as the class of devices utilizing biperiodic guiding structures, triply periodic wave functions must be employed. This necessitates a similar consideration of the space-charge potential and thus the development corresponding to Section 3 is made for cylindrical systems. Beam shapes employed in this configuration are restricted to solid cylindrical and hollow ring. In the general three-dimensional case, the charge is considered to occupy a nonzero volume so as to maintain the boundedness of both the potential and the field functions.

The model to be used is illustrated in Fig. 15. The Green's function method is utilized as developed by Smythe[4]. Wilson[7] has applied this method to a number of cases with axial symmetry. The beam is assumed

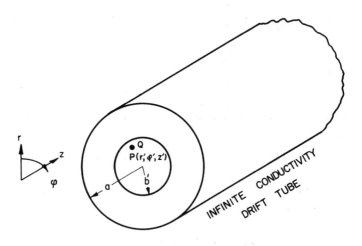

FIG. 15. Cylindrical space-charge model for calculation of space-charge potential.

7. SPACE-CHARGE POTENTIALS FOR CYLINDRICAL SYSTEMS

to be contained in an infinite-conductivity drift tube which replaces the circuit, and a three-dimensional Green's function is used which has the properties that

$$G(r = a) = 0, \qquad (38a)$$

$$G(|z| \to \infty) \to 0, \qquad (38b)$$

which say that the drift tube is held at zero potential and that the potential function is bounded about z'. The unit charge is located at $P(r', \varphi', z')$ and represents a discontinuity in the field for a region ϵ in the plane $z = z'$. The Green's function $G(r, \varphi, z, r', \varphi', z')$ denotes the potential distribution everywhere within the drift tube due to the unit charge, Q.

Except for the region ϵ around P, the potential function satisfies Laplace's equation; i.e.,

$$\nabla^2 V(r, \varphi, z) = 0. \qquad (39)$$

Laplace's equation may be solved using circular harmonics. The space-charge potential is obtained as

$$V_{sc}(r, \varphi, z) = \sum_{l=1}^{\infty} \sum_{s=0}^{\infty} A_{ls} e^{-\mu_{ls}|z-z'|} J_s(\mu_{ls} r) \cos s(\varphi - \varphi'), \qquad (40)$$

for the system illustrated in Fig. 15 and extended to $\pm\infty$ in z. The modified Bessel function $Y_s(\mu_{ls} r)$ does not appear in Eq. (39) because the potential must remain bounded on the system axis.

The argument μ_{ls} is chosen so that $J_s(\mu_{ls} a) = 0$. The factor A_{ls} may be calculated from the fact that $\partial V/\partial z = 0$ everywhere except for $z = z'$. Thus

$$\left.\frac{\partial V}{\partial z}\right|_{z=z'} = -\sum_{l=1}^{\infty} \sum_{s=0}^{\infty} \mu_{ls} A_{ls} J_s(\mu_{ls} r) \cos s(\varphi - \varphi') \qquad (41)$$

and therefore

$$A_{ls} = \frac{-(2 - \delta_s^0)}{\pi \mu_{ls} [a J_{s+1}(\mu_{ls} a)]^2} \iint \left(\frac{\partial V}{\partial z}\right)_{z=z'} r J_s(\mu_{ls} r) \cos s(\varphi - \varphi') \, dr \, d\varphi, \qquad (42)$$

where

$$\delta_s^0 = 1 \quad \text{for} \quad s = 0$$
$$= 0 \quad \text{for} \quad s \neq 0.$$

The double integral of Eq. (42) is evaluated by noting that at $z = 0$ the area in which $(\partial V/\partial z)|_{z=z'} \neq 0$ is sufficiently small that $J_s(\mu_{ls} r) \approx J_s(\mu_{ls} r')$

and $\cos s(\varphi - \varphi') \approx 1$. Carrying out this procedure and introducing Gauss' theorem to relate flux and charge, we obtain the space-charge potential as

$$V_{sc}(r, \varphi, z) = -\iiint \frac{\rho r' \, d\varphi' \, dr' \, dz'}{2\pi\epsilon_0 a^2} \sum_{l=1}^{\infty} \sum_{s=0}^{\infty} (2 - \delta_s^0) e^{-\mu_{ls}|z-z'|}$$

$$\cdot \cos s(\varphi - \varphi') \frac{J_s(\mu_{ls}r')J_s(\mu_{ls}r)}{\mu_{ls}[J_{s+1}(\mu_{ls}a)]^2} \cdot \qquad (43)$$

The charge at any point r, φ, z may be related to the entering charge by the continuity relationship

$$\rho r \, dr \, d\varphi \, dz = \rho_0 r_0 \, dr_0 \, d\varphi_0 \, dz_0 \qquad (44)$$

and $\rho_0 = I_0/\pi b'^2 u_0$.

Before proceeding with the derivation of the space-charge-field components it is convenient to introduce normalized variables into Eq. (43). Define

$$\nu_{ls} \triangleq \left(\frac{u_0}{C\omega}\right) \mu_{ls}, \qquad (45a)$$

$$y \triangleq \frac{C\omega}{u_0} z, \qquad (45b)$$

$$x \triangleq \frac{C\omega}{u_0} r, \qquad (45c)$$

$$x_0 \triangleq \frac{C\omega}{u_0} r_0, \quad (r_0 = \text{mean radius of the charge ring at } z = 0) \qquad (45d)$$

$$x_a \triangleq \frac{C\omega}{u_0} a, \quad (x_a = \text{normalized circuit or drift tube radius}) \qquad (45e)$$

$$x_{b'} \triangleq \frac{C\omega}{u_0} b', \quad (x_{b'} = \text{normalized outer stream radius}) \qquad (45f)$$

and

$$\Phi_0 \triangleq \frac{\omega}{u_0} z_0. \qquad (45g)$$

The space-charge potential is then

$$V_{sc}(x, \varphi, y) = -\left(\frac{1}{C}\right)\left(\frac{u_0}{\omega}\right)^2 \frac{\omega_p^2}{|\eta|} \int_{x_0} \int_{\varphi_0} \int_{\Phi_0} \frac{x_0' \, d\varphi_0' \, dx_0' \, d\Phi_0'}{2\pi x_a^2} \sum_{l=1}^{\infty} \sum_{s=0}^{\infty}$$

$$\cdot (2 - \delta_s^0) e^{-\nu_{ls}|y-y'|} \cos s(\varphi - \varphi') \frac{J_s(\nu_{ls}x')J_s(\nu_{ls}x)}{\nu_{ls}[J_{s+1}(\nu_{ls}x_a)^2]}, \qquad (46)$$

where the radian plasma frequency $\omega_p^2 \triangleq |\eta| I_0/\pi\epsilon b'^2 u_0$ has also been introduced.

8. RING OF CHARGE IN AXIALLY SYMMETRIC SYSTEM

The component space-charge fields are then obtained by appropriate differentiation of Eq. 46. The results are

$$E_{sc-z} = -\frac{\partial V_{sc}}{\partial z} = -\left(\frac{u_0}{\omega}\right)\frac{\omega_p^2}{|\eta|}\int_{x_0}\int_{\varphi_0}\int_{\Phi_0}\frac{x_0'\,d\varphi_0'\,dx_0'\,d\Phi_0'}{2\pi x_a^2}$$
$$\cdot F_{3-z}\,\text{sgn}(y-y'), \tag{47a}$$

where

$$F_{3-z} \triangleq \sum_{l=1}^{\infty}\sum_{s=0}^{\infty}(2-\delta_s^0)\,e^{-\nu_{ls}|y-y'|}\cos s(\varphi-\varphi')$$
$$\cdot \frac{J_s(\nu_{ls}x')J_s(\nu_{ls}x)}{[J_{s+1}(\nu_{ls}x_a)]^2}; \tag{47b}$$

$$E_{sc-r} = -\frac{\partial V_{sc}}{\partial r} = -\left(\frac{u_0}{\omega}\right)\frac{\omega_p^2}{|\eta|}\int_{x_0}\int_{\varphi_0}\int_{\Phi_0}\frac{x_0'\,d\varphi_0'\,dx_0'\,d\Phi_0'}{2\pi x_a^2}F_{3-r}, \tag{48a}$$

where

$$F_{3-r} \triangleq \sum_{l=1}^{\infty}\sum_{s=0}^{\infty}(2-\delta_s^0)\,e^{-\nu_{ls}|y-y'|}\frac{J_s(\nu_{ls}x')[J_{s+1}(\nu_{ls}x)-J_{s-1}(\nu_{ls}x)]}{2[J_{s+1}(\nu_{ls}x_a)]^2}, \tag{48b}$$

and

$$E_{sc-\varphi} = -\frac{1}{r}\frac{\partial V_{sc}}{\partial \varphi} = -\left(\frac{u_0}{\omega}\right)\frac{\omega_p^2}{|\eta|}\int_{x_0}\int_{\varphi_0}\int_{\Phi_0}\frac{x_0'\,d\varphi_0'\,dx_0'\,d\Phi_0'}{2\pi x_a^2}F_{3-\varphi}, \tag{49a}$$

where

$$F_{3-\varphi} \triangleq \sum_{l=1}^{\infty}\sum_{s=0}^{\infty}s(2-\delta_s^0)\,e^{-\nu_{ls}|y-y'|}\sin s(\varphi-\varphi')\frac{J_s(\nu_{ls}x')J_s(\nu_{ls}x)}{(\nu_{ls}x)[J_{s+1}(\nu_{ls}x_a)]^2}. \tag{49b}$$

The problem now is to calculate the various space-charge weighting functions for the coordinate directions as defined by Eqs. (47b), (48b) and (49b). This problem is directly similar to the three-dimensional rectangular case treated in Section 3 and one must again resort to the high-speed digital computer. First the beam is divided into rings of charge and then these rings are further divided into annular segments. The degree of subdivision is somewhat critical, depending upon the value of ω_p^2.

8 Potential Function for a Ring of Charge in an Axially Symmetric System

If one is interested in neither the angular spreading of the beam nor the effects of biperiodic circuits on the beam dynamics, a considerable simplification can be made in the development of the space-charge potential and field expressions. For cylindrical symmetry the potential is independent of φ and thus the Green's function for a delta function ring of charge in an infinite-conductivity cylinder may be used in lieu of that for an elementary volume of charge. By dividing the total beam into a number of concentric annular rings of charge, two-dimensional space-charge effects, i.e., radial motion and spreading, may be studied. This is illustrated in Fig. 16.

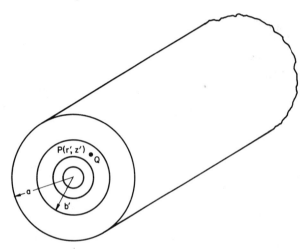

FIG. 16. Rings of charge in a drift tube: two-dimensional space-charge model.

The Green's function for the ring of charge in a drift tube is bounded in the same manner as that for the volume element treated in the previous section; i.e., $G(r = a) = 0$ and $G(|z| \to \infty) \to 0$. Solution of $\nabla^2 G(r, z) = 0$ gives the Green's function as

$$G(r, z) = \sum_{l=1}^{\infty} \alpha_l(r') e^{-\mu_l |z-z'|} J_0(\mu_l r), \qquad (50)$$

where μ_l is determined by the successive zeros of $J_0(\mu_l a)$. These are tabulated in Appendix A.

Again the field is everywhere zero in the interaction region except at the plane $z = z'$, where the ring charge is located. At this position $\partial G/\partial z$ is equal to $-1/\epsilon_0$ times the surface-charge density represented by the ring. The surface-charge density is given by $2\pi r' \rho' \, dr' \, dz' \, \delta(r - r')$, where $\delta(r - r')$ is the Dirac delta function defined by

$$2\pi \int_0^a \delta(r - r') r \, dr = 1. \qquad (51)$$

8. RING OF CHARGE IN AXIALLY SYMMETRIC SYSTEM

Following a procedure similar to that outlined in the previous section and defining $\rho_0 = I_0/\pi b'^2 u_0$ gives the space-charge potential as

$$V_{sc}(r, z) = -\iint \frac{I_0 r_0' \, dr_0' \, dz_0'}{u_0 \pi b'^2 a^2 \epsilon_0} \sum_{l=1}^{\infty} \frac{J_0(\mu_l r') J_0(\mu_l r)}{\mu_l [J_1(\mu_l a)]^2} e^{-\mu_l |z-z'|}. \tag{52}$$

Equation (52) could also have been obtained directly by appropriately simplifying Eq. (43) of Section 7 in view of the angular symmetry.

By a procedure parallel to that used in Section 7, the space-charge-field components for this two-dimensional ring charge model are obtained as

$$E_{sc-z} = -\frac{\partial V_{sc}}{\partial z} = \frac{-C^2 \omega I_0}{u_0^2 \pi \epsilon_0 x_a^2 x_{b'}^2} \int_0^{x_{b'}} x_0' \, dx_0' \int_0^{2\pi} d\Phi_0' F_{2-z} \, \text{sgn}(y - y'), \tag{53a}$$

where

$$F_{2-z} \triangleq \sum_{l=1}^{\infty} \frac{J_0(\nu_l x') J_0(\nu_l x)}{[J_1(\nu_l x_a)]^2} e^{-\nu_l |y-y'|} \tag{53b}$$

and

$$E_{sc-r} = -\frac{\partial V_{sc}}{\partial r} = \frac{-C^2 \omega I_0}{u_0^2 \pi \epsilon_0 x_a^2 x_{b'}^2} \int_0^{x_{b'}} x_0' \, dx_0' \int_0^{2\pi} d\Phi_0' F_{2-r}, \tag{54a}$$

where

$$F_{2-r} \triangleq \sum_{l=1}^{\infty} \frac{J_0(\nu_l x') J_1(\nu_l x)}{[J_1(\nu_l x_a)]^2} e^{-\nu_l |y-y'|}, \tag{54b}$$

and

$$\nu_l = \frac{u_0}{C\omega} \mu_l.$$

Integration of the space-charge weighting functions over the beam diameter and the entering phase positions of the charge groups is readily accomplished once they have been found. As in all space-charge problems the chief difficulty and time consumer is the calculation of the various weighting functions.

The weighting functions defined by Eqs. (53b) and (54b) are shown in Figs. 17–20 for a solid beam which has been subdivided into seven annular rings of charge. The short-range nature of the space-charge forces is evident from these graphs, as from those in the previous section.

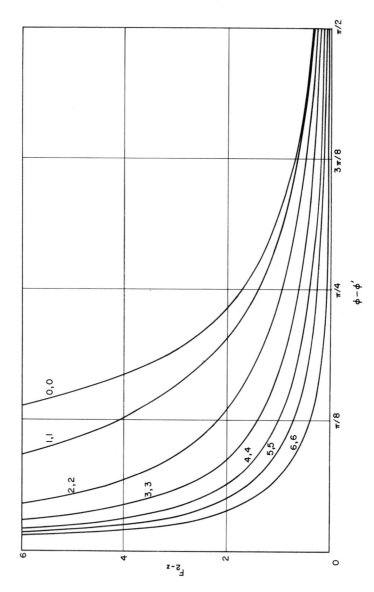

FIG. 17. Axial space-charge-field weighting function forces between particles in the same ring. ($b'/a = 0.7$.)

8. RING OF CHARGE IN AXIALLY SYMMETRIC SYSTEM

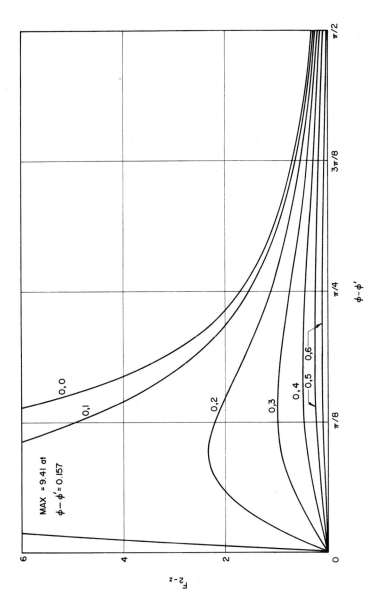

FIG. 18. Axial space-charge-field weighting function forces between particles in different rings. ($b'/a = 0.7$.)

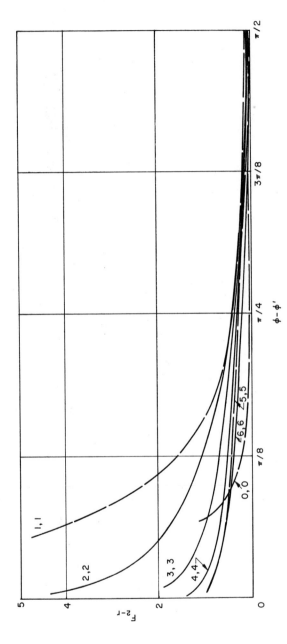

FIG. 19. Radial space-charge-field weighting function forces between particles in the same ring. ($b'/a = 0.7$.)

8. RING OF CHARGE IN AXIALLY SYMMETRIC SYSTEM

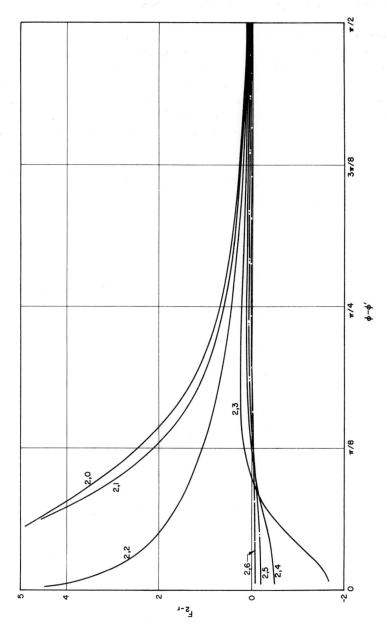

FIG. 20. Radial space-charge-field weighting function forces between particles in different rings. ($b'/a = 0.7$.)

9 Potential Functions for Hollow Beams

Many cylindrical interaction systems employ hollow electron beams or ion beams in order to obtain high beam perveance and high interaction efficiency. The space-charge potential and field functions derived in Sections 7 and 8 were quite general, since the only boundary condition imposed at $r = 0$ was that the potential function remain finite, thus eliminating any dependence on $Y(\mu_{ls}r)$. Thus it should be clear that the resulting space-charge-field expressions can be used in the three- or two-dimensional treatment of hollow beams.

In analysis of such hollow beams they are presumed to be made up of concentric rings of charge between $r = b_i'$ and $r = b_0'$ (see Fig. 21)

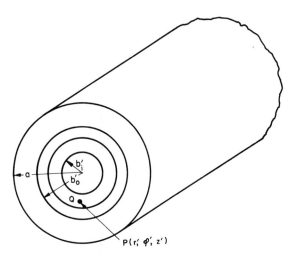

Fig. 21. Space-charge model for hollow cylindrical beams.

and to be either finite or infinite in extent in the z- and $-z$-directions. The calculations are made by injecting the charge distributed among the rings as illustrated in Fig. 21. Integration now extends from $r = b_i'$ to $r = b_0'$. For finite B field flow, the electrons can not only change rings but also change the inner and outer boundaries.

Another interesting potential problem which is left to the student is the geometry of Fig. 21 with a metal center conductor.

10 One-Dimensional Disk Space-Charge Model

Under large magnetic field conditions (confined flow) the operation of linear-beam devices like the klystron and traveling-wave tube may be satisfactorily described in terms of a one-dimensional theory. In such devices the radial electron motion is inhibited and hence one need not account for radial variations in the space-charge potential, providing that the beam diameter is not too large. The appropriateness and range of validity of the one-dimensional model will be examined in detail in succeeding chapters, related to specific classes of devices.

An appropriate one-dimensional space-charge model may be developed directly from the Green's function results presented in Section 8. In this model the beam charge for a cylindrical beam of radius b' in an infinite drift tube of radius a is taken to be compressed into an infinite array of "delta-function disks" arranged in a periodic fashion along the axis of the cylindrical system. The disks are perpendicular to the axis, with a radius less than that of the drift tube, and arranged as illustrated in Fig. 22. Such a space-charge model was used by Tien et al.[8] in their

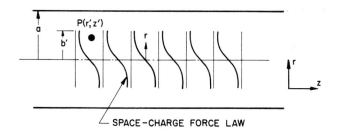

FIG. 22. One-dimensional delta-function disk space-charge model.

studies of the traveling-wave amplifier and by Webber[9] in his klystron calculations.

The formal procedure is to use the Green's function for a point charge in an infinite cylinder given in Eq. (43) of Section 7 and then integrate it into a ring, a disk and finally into an infinite array of disks. In this one-dimensional problem we are only interested in axial forces on charge groups; thus only the z-component of space-charge field E_{sc-z} or $-(\partial G/\partial z)$ need be calculated.

In order to facilitate comparison with the expression used by Tien, et al., a slightly different normalization is used than in Sections 7 and 8; i.e., the argument of the Bessel functions is taken as $\mu_i(r/a)$ rather than

μ_l. This produces no essential differences in the results and affects only the detailed numerical procedures. The Green's function written in this manner is then

$$G(r, \varphi, z, r', \varphi', z') = \frac{1}{2\pi\epsilon_0 a} \sum_{l=1}^{\infty} \sum_{s=0}^{\infty} (2 - \delta_s^0) \exp - \frac{\mu_{ls} |z - z'|}{a}$$

$$\cdot \cos s(\varphi - \varphi') \frac{J_s\left(\mu_{ls} \frac{r'}{a}\right) J_s\left(\mu_{ls} \frac{r}{a}\right)}{\mu_{ls}[J_{s+1}(\mu_{ls})]^2}. \tag{55}$$

Assuming the one-dimensional disk charge model illustrated in Fig. 22, where the charge disks have a radius equal to the beam radius, dictates the following form for the space-charge density distribution (the charge is assumed to be distributed uniformly over the disk):

$$\rho(r, z, z') = \frac{Q}{\pi b'^2} \delta(z - z') \quad \text{for} \quad 0 < r < b'$$

$$= 0 \quad \text{for} \quad r > b'. \tag{56}$$

Substituting into the z-component derivative of Eq. (55) and carrying out the indicated operations gives the following expression for the space-charge weighting function:

$$F_{1-z} = \sum_{l=1}^{\infty} \exp\left[-\frac{\mu_{l0}}{a} |z - z'|\right] \left[\frac{J_1(\mu_{l0}(b'/a))}{\mu_{l0} J_1(\mu_{l0})}\right]^2 \mathrm{sgn}(z - z') \tag{57}$$

where

$$\mathrm{sgn}(z - z') = 1 \quad \text{for} \quad z > z'$$
$$= -1 \quad \text{for} \quad z < z'.$$

The weighting function F_{1-z} is related to the $B(z)$ used by Tien et al., by

$$F_{1-z} \triangleq 2\pi\epsilon_0 b'^2 B(z). \tag{58}$$

In this case μ_{l0} is determined by the zeros of $J_0(\mu_{l0})$.

The space-charge field is then obtained by appropriately integrating the above force law over the infinite array pictured in Fig. 22. The result is

$$E_{sc-z}(z - z') = \frac{1}{2\pi\epsilon_0 b'^2} \int_{-\infty}^{\infty} \sum_{l=1}^{\infty} \rho(z', t) \exp\left[-\frac{\mu_{l0}}{a} |z - z'|\right] \left[\frac{J_1\left(\mu_{l0} \frac{b'}{a}\right)}{\mu_{l0} J_1(\mu_{l0})}\right]^2$$

$$\cdot \mathrm{sgn}(z - z') \, dz'. \tag{59}$$

10. ONE-DIMENSIONAL DISK SPACE-CHARGE MODEL

The problem has thus again been reduced to the tedious task of calculating the space-charge weighting function F_{1-z}, although happily in this case only one of them is required. The numerics have been carried out and are shown in Fig. 23.

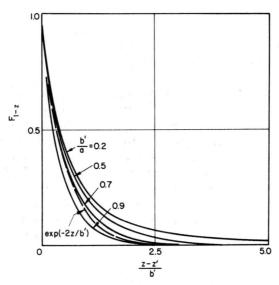

FIG. 23. Space-charge-field weighting function for delta-function disk one-dimensional model.[8]

The weighting function plots of Fig. 23 indicate a near exponential force law as would be expected from elementary considerations. Two useful facts appear from the results, namely that, (1) independent of the normalized beam diameter, b'/a, all curves have the same behavior for $(z - z')/b'$ near zero and have the same limit as $(z - z')/b' \to 0$; and (2) the space-charge repulsion force falls off rapidly along the axis and in fact charge disks are not significantly affected by other charge disks more than one or two cycles away.

As a result the family of curves shown in Fig. 23 can be effectively replaced by an exponential approximation to the weighting function. The exponential approximation given by

$$F_{1-z} = \exp\left[-\frac{2|z-z'|}{b'}\right] \text{sgn}(z - z') \tag{60}$$

is also shown in Fig. 23. The space-charge-field expression is then

$$E_{sc-z} = \frac{1}{2\pi\epsilon_0 b'^2} \int_{-\infty}^{\infty} \rho(z', t) \exp\left[-\frac{2|z-z'|}{b'}\right] \text{sgn}(z - z') \, dz'. \tag{61}$$

The order and significance of this approximation are discussed further in a later section.

In his treatment of this problem Tien[8] transforms from a space distribution to the equivalent time distribution in calculating space-charge fields.

11 Harmonic Method for Calculating the One-Dimensional Space-Charge Field

In previous sections of this chapter space-charge potentials and fields were derived using the Green's functions, where at a given time the contributions of force on a charge group are added for a particular charge distribution in the beam. In all cases the nonrelativistic assumption was made, so that the scalar potential rather than the retarded potential could be used. A relativistic treatment is given in Chapter V, Section 7. An alternate approach to obtaining the one-dimensional space-charge-field expression, which will be shown to be equivalent to the expression obtained from the Green's function, is to expand the beam space-charge density in a Fourier time series, assuming a particular spatial distribution:

$$\rho(z, t) = \sum_{n=1}^{\infty} \rho_n(z, t) e^{-j\omega n t}, \qquad (62)$$

where n extends over all of the harmonics of ω in the beam. Such a model has been used by Poulter[10] and Rowe[11,12] in their studies of the traveling-wave tube.

In calculating forces due to space charge it is not sufficient to include one or a few harmonics: all must be considered. The beam is assumed to be drifting with the average velocity $u_0 \ll c$ in an ideal drift tube as previously. Thus the time distribution of Eq. (62) can be converted to a spatial distribution, since we know the charge positions z_0 at some time t_0 previous to t.

As in previous one-dimensional treatments, we assume that the beam is in confined flow and that the average charge distribution is ion-neutralized. Of course the crux of the analysis lies with the primordial assumption on the space-charge density distribution. Suppose that the rf wave impressed on a beam-structure configuration bunches the beam so that the space-charge-density is constant in amplitude and varies sinusoidally with distance. We now remove the propagating structure, replace it with the drift tube, and calculate the steady-state space-charge potential for this distribution. Thus the space-charge density is assumed to have the form

$$\rho(z) = \rho_0 e^{-j\beta z}. \tag{63}$$

The axial electric field is calculated due to each harmonic of the charge distribution

$$\rho_n(z, z') = \rho_{n0} \exp\left[-\frac{j\omega n(z - z')}{u_0}\right], \tag{64}$$

and then summed over all harmonics. The choice of velocity for the propagating charge waves of Eq. (64) or the β of Eq. (63) is somewhat open to question. The β used is $\beta_e(1 + Cb)$, which will be justified later.

The problem is then to solve Poisson's equation in the beam for the assumed charge variation and match the solutions to those for Laplace's equation in the region between the beam and the drift tube. Hence the appropriate electrostatic equations are

$$\nabla \times \mathbf{E} = 0, \tag{65}$$

$$\mathbf{E} = -\nabla V, \tag{66}$$

and

$$\nabla^2 V_{sc}(r, z) = -\frac{\rho(z)}{\pi \epsilon b'^2} = -A\rho_0 e^{-j\beta z}. \tag{67}$$

Poisson's equation (67) may be solved by any of several methods, including the product solution method. The solution may be written directly as

$$V_{sc}(r)e^{-j\beta z} = \left[\frac{A\rho_0}{\beta^2} + BI_0(\beta r)\right]e^{-j\beta z}, \tag{68}$$

where β is determined by the impressed wave and $I_0(\beta r)$ is the zero-order modified Bessel function of the first kind. The axial electric field is then

$$E_{sc-z}(r, z) = j\beta\left[\frac{A\rho_0}{\beta^2} + BI_0(\beta r)\right]e^{-j\beta z}. \tag{69}$$

The constants A and B of Eq. (69) must be determined through the following boundary conditions:

$$E_r{}^i(b') = E_r{}^0(b'), \tag{70}$$

$$E_z{}^i(b') = E_z{}^0(b'), \tag{71}$$

and

$$E_z{}^0(a) = 0, \tag{72}$$

where Eqs. (70) and (71) express continuity of the electric field across the beam boundary and Eq. (72) assumes that the electric field is zero at the drift-tube wall (infinite conductivity). The conditions expressed by Eqs. (70)–(72) define an eigenvalue problem, since there are an infinite number of space-charge modes which are superposed to satisfy these conditions. The eigenvalues are the infinite set of propagation constants for the modes. Applying these to Eq. (69) yields

$$E_{sc-z}(r, z) = \frac{j\rho_0 A}{\beta} \left\{ 1 - \beta b' \frac{I_0(\beta r)}{I_0(\beta a)} \left[I_1(\beta b')K_0(\beta a) + I_0(\beta a)K_1(\beta b') \right] \right\} e^{-j\beta z}. \quad (73)$$

The desired space-charge field for the one-dimensional problem is then obtained by evaluating Eq. (73) on the cylindrical axis:

$$E_{sc-z} = \frac{j\rho_0 A}{\beta} R^2 e^{-j\beta z}, \quad (74)$$

where

$$R^2 \triangleq 1 - \frac{\beta b'}{I_{0a}} (I_{1b'} K_{0a} + I_{0a} K_{1b'}),$$

$$I_{1b'} \equiv I_1(\beta b'),$$

and

$$K_{0a} \equiv K_0(\beta a).$$

It is convenient here to introduce the radian plasma frequency

$$\omega_p^2 \triangleq \frac{|\eta||I_0|}{\pi \epsilon_0 b'^2 u_0}$$

and define an effective radian plasma frequency

$$\omega_{qn}^2 \triangleq \omega_p^2 R_n^2,$$

where

$$R_n^2 \triangleq 1 - \frac{n\beta b'}{I_{0na}} (I_{1nb'} K_{0na} + I_{0na} K_{1nb'}).$$

The quantity R_n is called a plasma frequency reduction factor for axial symmetry and is shown in Fig. 24.

Introduction of the above definitions into Eq. (74) gives

$$E_{sc-z} = -\sum_{n=1}^{\infty} \rho_n \exp\left[-j\left(n\Phi + \frac{\pi}{2}\right)\right] \left(\frac{\omega_{qn}}{\omega}\right)^2 \frac{\omega}{(|I_0|/u_0^2)|\eta| n(1 + Cb)}, \quad (75)$$

where the instantaneous phase of the rf wave causing the bunching has been introduced. The phase constant β for the wave-impressed bunching is approximated by

$$\beta \approx \left(\frac{\omega}{u_0}\right)\left(\frac{u_0}{v_0}\right) = \beta_e(1 + Cb). \tag{76}$$

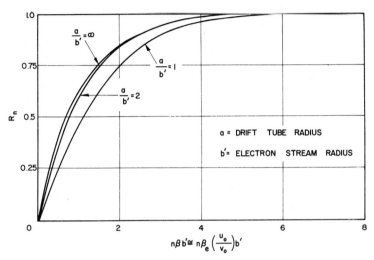

Fig. 24. Electron plasma frequency reduction factor for axial symmetry.

The space-charge density is now expanded in a Fourier series in Φ (see Chapter V) and the summation of Eq. (75) becomes

$$\sum_{n=1}^{\infty} \frac{\rho_n}{n} \exp\left[-j\left(n\Phi + \frac{\pi}{2}\right)\right]\left(\frac{\omega_{qn}}{\omega}\right)^2$$

$$= \frac{-|I_0|}{u_0} \sum_{n=1}^{\infty} \left(\frac{1}{n}\right)\left(\frac{\omega_p}{\omega}\right)^2 R_n^2 \exp\left[-j\left(n\Phi + \frac{\pi}{2}\right)\right]$$

$$\cdot \left\{\frac{1}{\pi}\int_0^{2\pi} \frac{e^{jn\Phi(y,\Phi_0')}\, d\Phi_0'}{1 + 2Cu(y,\Phi_0')}\right\}. \tag{77}$$

The expansion of the space-charge-field components into a Fourier time (phase) series at a particular displacement plane assumes that the rf wave growth during any one cycle is small. As shown in the previous section the space-charge-field pattern for two adjacent cycles will be very nearly alike, but ones further away may be different. Fortunately the effect does not extend far.

We have transformed a distribution in time to one in space and it is argued that the distribution in space for constant time is very nearly the same as the distribution in time for a small interval in z or y, providing there is no appreciable change in amplitude and that the influence of the space charge does not extend more than two or three cycles in either direction. Also, since the beam bunch is traveling at the average beam velocity, the distribution in velocities (ac) around this value does not have a significant effect even for large values.

Since the integrand of Eq. (77) is uniformly convergent, the order of summation and integration is switched and the right-hand side becomes

$$\frac{-2|I_0|}{u_0}\left(\frac{\omega_p}{\omega}\right)^2 \int_0^{2\pi} \left\{\sum_{n=1}^{\infty} \exp\left[-jn(\Phi-\Phi')-j\frac{\pi}{2}\right]\frac{R_n^2}{2\pi n}\right\} \frac{d\Phi_0'}{[1+2Cu(y,\Phi_0')]}. \tag{78}$$

Taking the real part of Eq. (77), with the aid of Eq. (78), yields

$$\text{Re}\left\{\sum_{n=1}^{\infty}\frac{\rho_n}{n}\exp\left[-j\left(n\Phi+\frac{\pi}{2}\right)\right]\left(\frac{\omega_{qn}}{\omega}\right)^2\right\} = \frac{2|I_0|}{u_0}\left(\frac{\omega_p}{\omega}\right)^2$$

$$\int_0^{2\pi}\left[\sum_{n=1}^{\infty}\frac{\sin n(\Phi-\Phi')R_n^2}{2\pi n}\right]\frac{d\Phi_0'}{(1+2Cu(y,\Phi_0'))}. \tag{79}$$

The space-charge-field expression is thus

$$E_{sc-z} = \frac{-2\omega u_0}{|\eta|(1+Cb)}\left(\frac{\omega_p}{\omega}\right)^2 \int_0^{2\pi}\frac{F_{1-z}(\Phi-\Phi')\,d\Phi_0'}{[1+2Cu(y,\Phi_0')]}, \tag{80}$$

where

$$F_{1-z}(\Phi-\Phi') \triangleq \sum_{n=1}^{\infty}\frac{\sin n(\Phi-\Phi')R_n^2}{2\pi n}, \tag{81}$$

and is the axial space-charge weighting function. Those familiar with the Gibbs phenomena will recognize the difficulty in performing the summation in Eq. (81) when $(\Phi-\Phi') \to 0$.

The proper evaluation of Eq. (81) begins with the recognition that $(1-R_n)$ varies nearly exponentially with $n\beta b'$ and hence $\ln(1-R_n)$ versus $n\beta b'$ will be a straight line with a slope determined by the value of a/b'. Designate this slope as $f(a/b')$ and define

$$B \triangleq \beta b'. \tag{82}$$

The weighting function is then written as

$$F_{1-z}(\Phi-\Phi') = \sum_{n=1}^{\infty}\frac{\sin n(\Phi-\Phi')}{2\pi n}[1-e^{-nBf(a/b')}]^2. \tag{83}$$

11. HARMONIC METHOD

The summation now proceeds satisfactorily for all but the first term after the quadratic factor is expanded, since the convergence for this term is not uniform. This difficulty is avoided if one writes

$$\lim_{N\to\infty} \sum_{n=1}^{N} \frac{\sin n(\Phi - \Phi')}{2\pi n}, \tag{84}$$

which clearly has a finite limit as $(\Phi - \Phi') \to 0$. Consider the following general term of Eq. (83)

$$\sum_{n=1}^{\infty} \frac{\sin n(\Phi - \Phi')}{n} e^{-n\alpha} \quad \text{for} \quad \alpha > 0.$$

The relation

$$\sum_{n=1}^{\infty} \frac{Z^n}{n} = \ln \frac{1}{1-Z} \quad \text{for} \quad |Z| < 1,$$

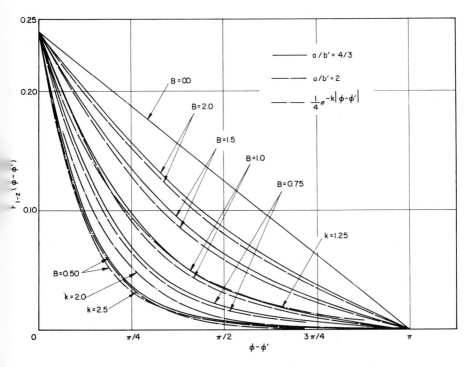

FIG. 25. One-dimensional space-charge-field weighting function.

where Z may be complex and the locus of $|Z| = 1$ is the circle of convergence, is used to rewrite the general term of the weighting function as

$$\sum_{n=1}^{\infty} \frac{\sin n(\Phi - \Phi')}{n} e^{-n\alpha} = \tan^{-1} \frac{\sin(\Phi - \Phi')}{e^{\alpha} - \cos(\Phi - \Phi')}. \tag{85}$$

Fortunately Eq. (85) is valid on the circle of convergence, i.e., $\alpha = 0$, and thus Eq. (83) may now be written as follows over the range $0-2\pi$:

$$F_{1-z}(\Phi - \Phi') = \frac{1}{2\pi} \left[\frac{\pi - (\Phi - \Phi')}{2} - 2 \tan^{-1} \frac{\sin(\Phi - \Phi')}{e^{Bf(a/b')} - \cos(\Phi - \Phi')} \right.$$

$$\left. + \tan^{-1} \frac{\sin(\Phi - \Phi')}{e^{2Bf(a/b')} - \cos(\Phi - \Phi')} \right]. \tag{86}$$

Equation (86) for the space-charge-field weighting function is easily evaluated as a function of B and a/b' and is shown in Fig. 25.

12 Equivalence of the Green's Function and Harmonic Methods for the One-Dimensional Problem

An interesting question arises as to the equivalence of the Green's function method and the Fourier method of calculating the one-dimensional space-charge-field weighting function. If the assumption as to the form of the space-charge distribution is correct and if the time distribution is properly converted into a spatial distribution, then the weighting functions should be approximately the same. The two weighting functions can in fact be transformed into one another after much detailed manipulation, provided that the integrands are uniformly convergent so that the order of integration and summation may be interchanged.

A simpler numerical procedure may also be used to show the equivalence. First it is noted that both Rowe (see Fig. 25) and Tien (see Fig. 23) have found that the form and details of the weighting function are relatively independent of a/b'; hence we will choose simply $a/b' = 2$ for the comparison. The difference in scale factor between the two expressions is adjusted and the Green's function and harmonic method weighting functions are plotted in Fig. 26 along with an exponential approximation used by Tien. The equivalence of the weighting functions is noted, thus justifying the assumptions made in the harmonic method. The same equivalence holds true for other values of a/b'.

12. EQUIVALENCE OF THE GREEN'S FUNCTION AND HARMONIC METHODS

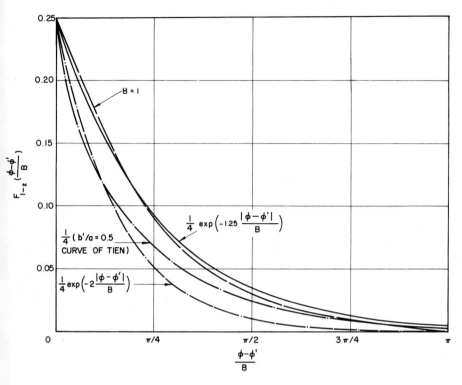

FIG. 26. Comparison of harmonic and Green's function one-dimensional weighting functions.

The space-charge weighting function obtained from the disk model may be approximated by an exponential function

$$e^{-k|\Phi-\Phi'|},$$

where the transformation from distance to phase is made by using

$$\frac{|z-z'|}{b'} = \frac{u_0}{\omega b'}|\Phi - \Phi'|$$

$$= \frac{|\Phi - \Phi'|}{\beta b'} = \frac{|\Phi - \Phi'|}{B}.$$

The constant k varies from 1 to 5, and in particular its value depends upon the ratio of the stream-to-helix diameters. On the basis of the approximate form the relationship between k and B is given by

$$Bk = 2.$$

If the exact form is used then the relationship is

$$Bk = 1.25.$$

Further evidence of the equivalence is shown in Chapter VI, where efficiency calculations for the traveling-wave amplifier made using both methods are virtually the same.

13 Space-Charge Fields for Specialized Configurations

In the previous sections of this chapter the Green's functions were calculated for elementary volume element charges both between parallel plates (infinite and finite) and within a cylindrical drift tube. Various degenerate cases were examined such as rods of charge, rings of charge and one-dimensional charged disks. After the Green's functions are obtained the appropriate space-charge potential is calculated from

$$V_{sc}(\mathbf{r}, \mathbf{r}') = \int G(\mathbf{r}, \mathbf{r}') \rho(\mathbf{r}, \mathbf{r}') \, d\mathbf{r},$$

where we have simply integrated the product of the charge density and the Green's function to obtain the potential in the given geometry. $\mathbf{r}$ denotes a generalized position vector in the configuration of interest.

In this section we will summarize the results of similar calculations for certain special cases of interest. Since the method again follows from Smythe[4] and some of these have also been investigated by Wilson,[7] the details are omitted.

a. Charged Disk, Infinite in Extent

The simplest space-charge model would be to consider a truly one-dimensional problem so that variations can occur only along the z-direction and so that the system is infinite in extent in the transverse plane, giving rise to no boundary conditions. Such a system is fairly well approximated by some confined flow systems.

We may calculate the space-charge field from the Green's function for charge disks, infinite in transverse dimensions, located along the z-coordinate direction. For low velocities the electrostatic field equations

13. SPACE-CHARGE FIELDS FOR SPECIALIZED CONFIGURATIONS

are used and it is assumed that the charge per unit area is σ coulombs/meter² distributed uniformly over the disk. Then

$$G(z, z') = -\frac{1}{2\epsilon_0} |z - z'|$$

and the z-component electric field is given by

$$E_{sc-z}(z, z') = \frac{Q}{2\epsilon_0 A} \operatorname{sgn}(z - z'),$$

where A represents an area on the disk, i.e., πr^2 or xy.

b. Rectangular Bunches in a Drift Tube

In the analysis of linear-beam devices various space-charge distributions must be investigated in studying the various bunching and interaction processes. The previous sections of this chapter have been concerned with several uniform beam states for both rectangular and cylindrical geometries. In succeeding chapters it will be seen that some beam-wave analyses can be carried out analytically without the aid of high-speed digital computers when the beam is composed of ideal rectangular bunches. In Chapter XIV the power required to create such an ideal beam is evaluated. The starting point is again the generalized Green's function given in Eq. (55), of Section 10. The directional space-charge weighting functions and various component space-charge fields are computed from the appropriate directional derivatives and the application of Gauss' theorem.

Considering rectangular ideal bunches as illustrated in Fig. 27, we assume angular symmetry and no dependence of the space-charge

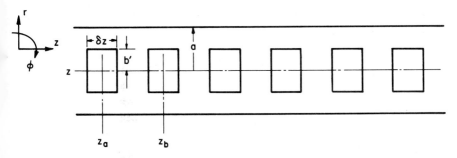

Fig. 27. Uniform rectangular bunches in a hollow drift tube.

density on radius. Thus we may write the space-charge density, for the model illustrated, as follows

$$\rho(z) = \frac{Q}{\pi b'^2 \delta z} \qquad 0 < r' < b'$$
$$\left(z_a - \frac{\delta z}{2}\right) < z' < \left(z_a + \frac{\delta z}{2}\right)$$
$$= 0 \qquad \text{for all other } z. \tag{87}$$

In view of the invariance with radius and angle about the axis, the only space-charge-field component is in the z-direction and hence only $-\partial G/\partial z$ need be calculated. Invoking these conditions on Eq. (55) of Section 10 and calculating the force exerted on one bunch by another bunch gives the following space-charge weighting function:

$$F_{1-z} = \sum_{l=1}^{\infty} \left[\frac{J_1\left(\mu_l \frac{b'}{a}\right)}{\left(\frac{\mu_l b'}{a}\right) J_1(\mu_l)}\right] \begin{pmatrix} g_i & \text{for} & |z_b - z_a| < \delta z \\ g_0 & \text{for} & |z_b - z_a| > \delta z \end{pmatrix}, \tag{88}$$

where

$$g_i \triangleq 2 \exp\left[-\frac{\mu_l \delta z}{2a}\right] \left(\frac{a}{\mu_l \delta z}\right)^2 \left\{\cosh \frac{\mu_l \delta z}{2a} \left(1 - \cosh \frac{\mu_l}{a} |z_b - z_a|\right)\right.$$

$$+ \sinh \frac{\mu_l \delta z}{2a} \left(1 - \exp\left[-\frac{\mu_l}{a} |z_b - z_a|\right]\right)$$

$$\left. + \sinh \frac{\mu_l}{a} |z_b - z_a|\right\} \operatorname{sgn}(z_b - z_a)$$

and

$$g_0 \triangleq 4 \left(\frac{a}{\mu_l \delta z}\right)^2 \sinh^2 \frac{\mu_l \delta z}{2a} \exp\left[-\frac{\mu_l}{a} |z_b - z_a|\right] \operatorname{sgn}(z_b - z_a).$$

This weighting function is shown in Fig. 28 for various bunch dimensions.

c. Space-Charge Field within an Ideal Delta Function Bunch

In evaluating the performance of prebunched-beam devices with regard to efficiency it is necessary to calculate the power required to bunch the beam. This requires a knowledge of the space-charge potential and electric field within the bunch. Whereas in part (*b*) of this section the force between rectangular bunches distributed in z was found, here we must calculate the field within one ideal bunch.

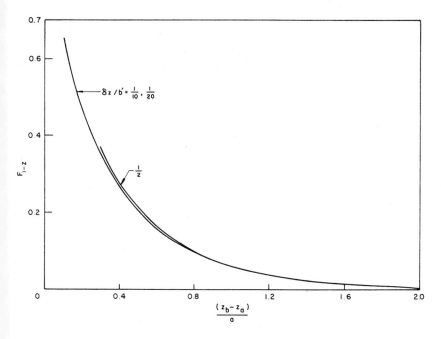

FIG. 28. Rectangular bunch space-charge-field weighting function. $(b'/a = 0.5.)$

The problem is thus to calculate the appropriate Green's function for a thin rectangular bunch of radius b' and thickness δz inside a drift tube of radius a. The Green's function is obtained again from the general form of Eq. (55), Section 10, by appropriate integrations over the angular and radial directions. The result is

$$G(z, z', r) = \frac{1}{\pi \epsilon_0 b'} \sum_{l=1}^{\infty} \frac{1}{(\mu_l)^2} \left[\frac{J_0\left(\mu_l \frac{r}{a}\right) J_1\left(\mu_l \frac{b'}{a}\right)}{[J_1(\mu_l)]^2} \right] \exp -\frac{\mu_l}{a} |z - z'|, \tag{89}$$

where again μ_l is found from $J_0(\mu_l) = 0$. The exponential factor of Eq. (89) points out the fact that the range of effectiveness of the space-charge forces is less than the distance between bunches. The normalized length is a/μ_l.

As previously, we obtain the space-charge potential within the bunch from the integral of G with the charge density as given by

$$\rho(z) = \rho_0 \qquad -\frac{\delta z}{2} \leq z \leq \frac{\delta z}{2}$$

$$= 0 \qquad \text{for all other } z. \tag{90}$$

The total bunch charge is taken as Q coulombs and thus $\rho_0 = Q/\delta z$ coulombs/meter. Since the bunch is created by an rf wave of frequency ω and period T, $Q = TI_0$. The space-charge or bunch potential is thus given by

$$V_{sc}(r, z', z) = \frac{I_0 \lambda_0}{\pi b'} \left(\frac{\mu_0}{\epsilon_0}\right)^{\frac{1}{2}} \sum_{l=1}^{\infty} \frac{1}{(\mu_l)^2} \left(\frac{2a}{\mu_l \delta z}\right)$$

$$\cdot \left[\frac{J_0\left(\mu_l \frac{r}{a}\right) J_1\left(\mu_l \frac{b'}{a}\right)}{[J_1(\mu_l)]^2}\right] \left\{1 - \exp -\frac{\mu_l \delta z}{2a}\right.$$

$$\left. \cdot \cosh\left[\left(\frac{\mu_l \delta z}{2a}\right)\left(\frac{z - z'}{a}\right)\right]\right\}, \tag{91}$$

where λ_0 is the rf wave free-space wavelength and I_0 is the average beam current. The bunch length δz may also be conveniently written in terms of the time width of the bunch relative to the rf period as $\delta z/\lambda_0 = \tau/T$,

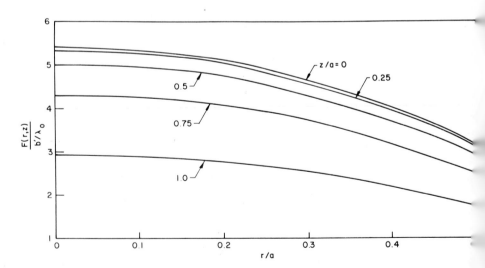

Fig. 29. Bunch potential weighting function. ($b'/a = 0.5$, $\delta z/\lambda_0 = 0.1$, $a/\lambda_0 = 0.05$.)

when $(u_0/c)^2 \ll 1$. τ/T is the fractional duration of the current pulse. The space-charge field within this delta function bunch is given by

$$E_{sc-z}(z, z', r) = -\frac{\partial V_{sc-z}}{\partial z} = \frac{I_0 \lambda_0}{\pi b' \delta z} \left(\frac{\mu_0}{\epsilon_0}\right)^{\frac{1}{2}} \sum_{l=1}^{\infty} \frac{2}{(\mu_l)^2}$$

$$\cdot \left[\frac{J_0\left(\mu_l \frac{r}{a}\right) J_1\left(\mu_l \frac{b'}{a}\right)}{[J_1(\mu_l)]^2}\right] e^{-\mu_l \delta z/2a} \sinh\left(\frac{\mu_l(z-z')}{a}\right). \quad (92)$$

These expressions for potential and electric field are valid everywhere within the bunch and are used in Chapter XIV in considering the subject of prebunching. The maximum field, of course, occurs at the bunch edge. The weighting function of Eq. (91) for the potential is shown in Fig. 29 for a typical set of parameters.

REFERENCES

1. Maloff, I. G., and Epstein, D. W., *Electron Optics in Television*. McGraw-Hill, New York, 1938.
2. Myers, L. M., *Electron Optics: Theoretical and Practical*. Van Nostrand, Princeton, New Jersey, 1939.
3. Sturrock, P. A., *Static and Dynamic Electron Optics*. Cambridge Univ. Press, London and New York, 1955.
4. Smythe, W. R., *Static and Dynamic Electricity*. McGraw-Hill, New York, 1939.
5. Panofsky, W. K. H., and Phillips, M., *Classical Electricity and Magnetism*. Addison-Wesley, Reading, Massachusetts, 1955.
6. Sedin, J. W., "A Large-Signal Analysis of Beam-Type Crossed-Field Traveling-Wave Tubes." Hughes Res. Labs. Tech. Memo. No. 520 (July 1958).
7. Wilson, R. N., "Large-Signal Space-Charge Theory of Klystron Bunching." Microwave Lab. Tech. Rept. No. 750, Stanford Univ. (September 1960).
8. Tien, P. K., Walker, L. R., and Wolontis, V. M., A large-signal theory of traveling-wave amplifiers." *Proc. IRE* **43**, No. 3, 260-277 (1955).
9. Webber, S. E., Large-signal analyses of the multicavity klystron. *IRE Trans. Electron Devices* **5**, 98-108 (1958).
10. Poulter, H. C., "Large Signal Theory of the Traveling-Wave Tube." Electron. Res. Lab. Tech. Rept. No. 73, Stanford Univ. (January 1954).
11. Rowe, J. E., A large-signal analysis of the traveling-wave amplifier: Theory and general results. *IRE Trans. Electron Devices* **3**, 39-57 (1956).
12. Rowe, J. E., One-dimensional traveling-wave tube analyses and the effect of radial electric field variations. *IRE Trans. Electron Devices* **7**, 16-22 (1960).

CHAPTER

V | Klystron Analysis

1 Introduction

In this chapter the first application of the general theory developed in the previous chapters is made to the linear-beam tube known as the klystron amplifier. It will be considered first not because it is any more important than other devices to be analyzed but only because it is the simplest device to analyze. This simplicity arises from the fact that the important processes of velocity modulation of the beam, subsequent conversion of the velocity modulation to density modulation, and the excitation of the output circuit all occur in distinctly separate regions of the device and thus may be analyzed in a sequential manner. This separability of functions aids materially in developing an understanding of modulation and bunching phenomena that will be useful in considering other, more complicated devices.

Although the linear analysis of the klystron will not be treated in detail in this book, many of its results will be used as a basis of comparison and means of evaluating nonlinear effects; hence the reader should be aware of the various small-signal theories.

The first linear analysis of the device was a ballistic one given by Webster[1] which is valid up to the point of first "overtaking," or electron crossover in the beam. Further ballistic analyses have been given by Warnecke and Guenard[2]. Another approach, using the space-charge-wave concept, was developed by Hahn[3] and Ramo[4] and later treated in infinite detail in unpublished notes by Feenberg[5,6]. The latter contribution was to include in the summation not only the first-order space-charge mode used by Ramo but also higher-order modes. A later analysis by Zitelli[7] among other things extends the Feenberg treatment to account for relativistic effects. Mihran[8] has compared the results of the above methods and pointed out some errors and limitations of application.

The simplest klystron configuration was illustrated in Fig. 1 of Chapter I, where only two cavities are utilized. A time-varying voltage (sinusoidal, triangular, etc.) is applied to the first cavity to velocity modulate the entering beam. The beam then travels through the drift tube, where the velocity modulation is converted to density modulation, and then the bunched beam interacts with the fields in the output gap. Also illustrated in Fig. 1 of Chapter I is the multicavity klystron amplifier, in which several cavities have been inserted between the buncher (input) and catcher (output) cavities. These are added in order to enhance the beam bunching for excitation of the output cavity. In general, it is not necessary to limit the bunching and catching circuits to cavities; indeed they may be propagating structures for increased operating bandwidth. The analysis developed in this chapter allows for either possibility.

The nonlinear theory of klystron amplifiers has received much attention with analysis methods extending from a large-amplitude perturbation treatment (Paschke[9]) to a general Lagrangian analysis. Early treatments of the nonlinear problem by Feenberg[6], Doehler-Kleen[10] and others were seriously limited by assumptions relating to the neglect of space-charge forces and/or the neglect of electron overtaking (crossover). More recently Lagrangian calculations have been made by Webber[11,12] and Meeker[13] which include the effects of both space charge and electron overtaking. Solymar[14,15,16] has made extensive large-signal klystron calculations including space-charge effects which, however, are strictly valid only up to the point of crossover. The subject of bunching in long-transit-angle cavities has also been investigated by Turner[17] using both ballistic and space-charge-wave analyses. Two-dimensional effects in confined-flow klystrons were studied by Wilson[18].

The nonlinear analysis developed in succeeding sections of this chapter is a Lagrangian approach and is generalized to include the effects of several dimensions, finite beam size and various debunching effects. As an aid to the student of the subject the analysis is developed in the order of increasing complexity, beginning with a relatively simple one-dimensional treatment.

2 One-Dimensional Klystron Analysis

It is easily shown that the ballistic treatment rather than the space-charge-wave theory is most appropriate to the general Lagrangian treatment of the nonlinear klystron problem. In this analysis a number of representative charge groups are traced through the input cavity, drift and output cavity regions, accounting for the presence of modulating fields, finite gap transit times and space-charge forces between charge groups. Thus the problem is to add up all the forces on the electrons in the beam at each displacement plane beyond the input plane of the initial modulating gap. If we assume nonrelativistic mechanics the force equation for the ballistic model may be written as

$$\frac{d\mathbf{v}}{dt} = -|\eta|[(\mathbf{E}_c + \mathbf{E}_{sc}) + \mathbf{v} \times \mathbf{B}], \tag{1}$$

where $\eta \triangleq e/m$, the charge-to-mass ratio for the electrons, $\mathbf{v}$ is the velocity vector, $\mathbf{E}_{sc}$ is the space-charge field and $\mathbf{E}_c$ is the circuit field. For a purely modulating cavity or gap in which there is no spatially varying field, a time-varying field produces a velocity or density modulation of the electrons and $\mathbf{E}_c$ disappears from the equation, leaving (for the region beyond the gap)

$$\frac{d\mathbf{v}}{dt} = -|\eta|[\mathbf{E}_{sc} + \mathbf{v} \times \mathbf{B}]. \tag{2}$$

However, in klystrons with traveling-wave resonators $\mathbf{E}_c$ must be retained. It will be seen later that the distinction between such a klystron and a traveling-wave amplifier is slight. Restricting the analysis to one dimension, i.e., neglecting radial and angular variations, simplifies Eq. (2) to (static magnetic field in the z-direction)

$$\frac{dv_z}{dt} = \frac{d^2z}{dt^2} = -|\eta|E_{sc-z}, \tag{3}$$

where z is the axial dimension.

To facilitate the analysis and the later comparison with the traveling-wave-amplifier equations the same variable normalizations will be used in the two theories. The distance variable is normalized to the stream wavelength as follows:

$$X \triangleq \frac{\alpha}{2} \frac{\omega z}{u_0} = \frac{\alpha \pi z}{\lambda_s} = \pi \alpha N_s, \tag{4}$$

where $\lambda_s \triangleq (u_0/f)$, the stream wavelength and N_s is the number of stream wavelengths in a given axial distance. u_0 is the initial average velocity of the electron beam determined by $u_0^2 = 2|\eta|V_0$. The symbol X is used to conform to the bunching parameter of Webster[1]. The α parameter of Eq. (4) is the depth of modulation index of klystron theory defined by

$$\alpha \triangleq \beta \frac{V_g}{V_0}, \tag{5}$$

where β is the gap coefficient (to be evaluated under initial conditions) and V_g/V_0 is the normalized modulation voltage as shown in Fig. 1 of Chapter I.

Since discrete charge groups are considered, their entrance into the interaction region is conveniently described in terms of their entrance phase positions relative to one cycle of the modulating rf signal $V_g e^{j\omega t}$. An entrance phase variable is defined by

$$\Phi_0 \triangleq \frac{\omega z_0}{u_0} = \omega t_0, \qquad (6)$$

where z_0 is an initial position coordinate and t_0 is the entrance time. The representative electron "charge groups" are injected in some fashion over a 2π phase region of the modulating wave form. In the event that the entering beam is not density modulated then the initial phases are taken as

$$\Phi_{0,j} = \frac{2\pi j}{m} \qquad j = 0, 1, 2, ..., m, \qquad (7)$$

where $m \triangleq$ the number of charge groups and j is simply a recording index. It should be mentioned, however, that any arbitrary distribution of Φ_0's can be used.

After injection the charge-group velocity and phase position are modulated and it is convenient to keep track of their movement in terms of X and Φ, a new phase variable. The new phase variable Φ is defined with reference to Φ_0 and the unmodulated electron flight line determined by u_0:

$$\Phi \triangleq \frac{\omega z}{u_0} - \omega t. \qquad (8)$$

The definition of this variable and the electron flight-line diagram are illustrated in Fig. 1. This diagram is contrasted with Fig. 6 of Chapter 2 for traveling-wave tubes. It remains only to define the velocity of the charge groups as follows:

$$u_t(X, \Phi_0) = \frac{2u_0}{\alpha\omega} \frac{dX}{dt} \triangleq u_0[1 + \alpha u(X, \Phi_0)], \qquad (9)$$

where $\alpha u(X, \Phi_{0,j})$ represents the normalized rf velocity of the jth charge group.

Before substituting the newly defined variables into the Newton force equation it is well to find a relation between the new dependent variables. X and $\Phi_{0,j}$ are the independent variables of the problem.

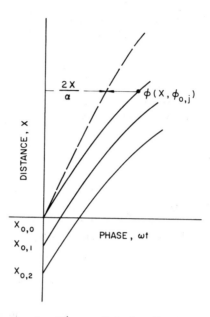

Fig. 1. Klystron flight-line diagram.

Solving Eq. (8) for z, converting to X through Eq. (4), taking the time derivative of the result, and equating to Eq. (9) gives the velocity-phase equation:

$$\frac{\partial \Phi(X, \Phi_0)}{\partial X} = \frac{2}{\alpha}\left[\frac{1}{1+\alpha u(0, \Phi_0)} - \frac{1}{1+\alpha u(X, \Phi_0)}\right]. \tag{10}$$

The partial derivative used in Eq. (10) requires some explanation. As mentioned earlier X and Φ_0 are the independent variables, X being continuous whereas Φ_0 will be treated in a discrete sense. Derivatives with respect to X occur and to obtain the effect of the total *beam* we sum over all $\Phi_{0,j}$. Equation (10) relates the dependent variables in velocity and phase position as a function of X and Φ_0.

The conversion of the left-hand side of the Newton force equation, Eq. (3), is accomplished as follows:

$$\frac{dv_X}{dt} = \frac{\partial v_X}{\partial X}\frac{dX}{dt} = \frac{\alpha^2}{2}u_0\omega[1+\alpha u(X,\Phi_0)]\frac{\partial u(X,\Phi_0)}{\partial X} \tag{11}$$

and thus Eq. (3) is rewritten as

$$[1+\alpha u(X,\Phi_0)]\frac{\partial u(X,\Phi_0)}{\partial X} = -\frac{2|\eta|}{\alpha^2 u_0 \omega}E_{sc-x}(X,\Phi). \tag{12}$$

2. ONE-DIMENSIONAL KLYSTRON ANALYSIS

It now remains only to incorporate the one-dimensional space-charge-field expression from Chapter IV into Eq. (12). The Lagrangian variables have already been introduced into the space-charge-field equations.

It was shown in Chapter IV that the space-charge-field expression for disk electrons and the expression obtained by the harmonic method are identical (Section IV.12); hence either may be used in the klystron study. Webber[11] has used the disk model expression and Meeker[13] has utilized the harmonic formulation. The equivalence can be shown by identical klystron calculations. The space-charge-field expression obtained by the harmonic method appropriate to the klystron problem is (electrons)

$$E_{sc-X} = -\frac{2\omega u_0}{|\eta|}\left(\frac{\omega_p}{\omega}\right)^2 \int_0^{2\pi} \frac{F_{1-x}(\Phi - \Phi')\,d\Phi_0'}{[1 + \alpha u(X, \Phi_0')]}, \quad (13)$$

where

$$F_{1-x}(\Phi - \Phi') = \sum_{n=1}^{\infty} \frac{\sin n(\Phi - \Phi') R_n^2}{2\pi n}. \quad (14)$$

Substitution of Eq. (13) into Eq. (12) yields the following final expression for the force equation:

$$[1 + \alpha u(X, \Phi_0)]\frac{\partial u(X, \Phi_0)}{\partial X} = 4\left(\frac{\omega_p}{\omega\alpha}\right)^2 \int_0^{2\pi} \frac{F_{1-x}(\Phi - \Phi')\,d\Phi_0'}{[1 + \alpha u(X, \Phi_0')]}. \quad (15)$$

Equation (10), relating variables, and Eq. (15), the force equation, are now to be solved simultaneously subject to the prescription of initial conditions and the selection of operating parameters. The space-charge-field weighting function given by Eq. (14) weights the influence of one charge group on another as determined by the range parameter $B = \beta_e b' = \gamma b'$. This weighting function is shown in Fig. 25 of Chapter IV for a representative range of B. It is shown later that the one-dimensional assumption is not strictly valid for $B > 1$. The other parameter to be selected is the beam modulation index α, along with the normalized radian plasma frequency, which measures the beam charge density and hence the strength of the space-charge forces. Fortunately these occur together as $(\omega_p/\omega\alpha)^2$, a coefficient of the space-charge-force integral, so that only this composite parameter need be specified.

The disk electron space-charge-field term used by Tien and Webber is written as

$$E_{sc-X} = -\frac{u_0 \omega_p^2}{2\omega\eta}\int_{-\infty}^{\infty} \exp\left(-\frac{2}{B}|\Phi(\Phi_0 + \theta, X) - \Phi(\Phi_0, X)|[1 + \alpha u(\Phi_0 + \theta, X)]\right)$$
$$\cdot \operatorname{sgn}[\Phi(\Phi_0 + \theta, X) - \Phi(\Phi_0, X)]\,d\theta, \quad (16)$$

where the distribution in space has been transformed to a distribution in time or phase. Thus Tien makes the same assumption as Poulter and Rowe did, namely that since the space-charge forces are short-range ones the distributions in space and time are the same providing that the velocities do not vary markedly over one period (or range of effectiveness).

The transformation from the spatial distribution to the time distribution is facilitated by noting that because of different initial phase positions Φ_0 and Φ_0' two electrons will reach a given displacement plane X at different times t and t' given by

$$t - t' = \frac{1}{\omega}\left[\omega t - \frac{\omega}{u_0}z - \left(\omega t' - \frac{\omega}{u_0}z\right)\right] = \frac{1}{\omega}[\Phi(X, \Phi_0') - \Phi(X, \Phi_0)]. \quad (17)$$

Multiply the time difference of Eq. (17) by the velocity as given by Eq. (9) and obtain the spatial separation,

$$(z' - z)\mid_t = \frac{u_0}{\omega}[\Phi(X, \Phi_0') - \Phi(X, \Phi_0)][1 + \alpha u(X, \Phi_0)]. \quad (18)$$

Clearly the above constitutes only the first term of the Taylor expansion of $t - t'$ as given by

$$(z' - z)\bigg|_t = \frac{dz}{dt}\bigg|_t (t - t') + \frac{1}{2}\frac{d^2z}{dt^2}\bigg|_t (t - t')^2 + \cdots. \quad (19)$$

It is probable that this assumption is not severely limiting even in very strong interaction systems. The variable θ used by Tien in Eq. (16), or $\Phi_0 + \theta$, serves only to indicate the disk of charge which is creating the force or field on the disk at Φ_0. Thus Tien's $\mid \Phi(\Phi_0 + \theta, X) - \Phi(\Phi_0, X)\mid$ corresponds directly to Rowe's $\mid \Phi(X, \Phi_0) - \Phi(X, \Phi_0')\mid$. As pointed out in Chapter IV the sgn $(\Phi - \Phi')$ indicates the correct direction of the force since it reverses as $\Phi - \Phi'$ passes through $n\pi$. Substituting Eq. (16) into Eq. (12) yields the final alternate force equation:

$$[1 + \alpha u_j(X, \Phi_0)]\frac{\partial u_j(X, \Phi_0)}{\partial X} = \frac{4\pi}{m}\left(\frac{\omega_p}{\omega\alpha}\right)^2 \sum_{\substack{n=0 \\ n \neq j}}^{m} \exp\left\{-\frac{2}{B}\mid \Phi_n(\Phi_0 + \theta, X)\right.$$

$$\left. - \Phi_j(\Phi_0, X)\mid [1 + \alpha u_n(\Phi_0 + \theta, X)]\right\} \operatorname{sgn}[\Phi_n(\Phi_0 + \theta, X) - \Phi_j(\Phi_0, X)], \quad (20)$$

where the integration has been replaced by a summation over the entering charge as $d\theta \approx (2\pi \Delta m/m)$.

In addition to obtaining velocity and phase information on the charge groups from the simultaneous solution of Eqs. (10) and (15) or Eqs. (10)

2. ONE-DIMENSIONAL KLYSTRON ANALYSIS

and (20) subject to initial conditions, one can obtain information on the rf current amplitudes in the beam from a knowledge of $\Phi(X, \Phi_0)$ and $u(X, \Phi_0)$. To do this we make use of the continuity equation, which states the conservation of entering charge in terms of beam-dependent variables. The charge entering the device, $\rho(0, t_0)\, dz(0, t_0)$, due to all charge groups whose initial positions lie between z_0 and $z_0 + dz_0$ must equal the charge $\rho(z, t)\, dz(z, t)$ at some later displacement plane. This is expressed as

$$\rho(z, t) = \rho(0, t_0) \frac{dz(0, t_0)}{dz(z, t)}. \tag{21}$$

The distance increments of Eq. (21) may be written in terms of the velocity

$$dz(z, t) = u_t(z, t)\, dt \tag{22a}$$

and

$$dz(0, t_0) = u_t(0, t_0)\, dt_0. \tag{22b}$$

Thus Eq. (21) is rewritten as

$$\rho(z, t) = \rho(0, t_0) \frac{u_t(0, t_0)}{u_t(z, t)} \left| \frac{d\Phi_0}{d\Phi} \right|, \tag{23}$$

where the absolute value is taken in order to insure inclusion of all current components. Equation (23) is rewritten in terms of the newly defined variables as follows:

$$\rho(X, \Phi) = \rho(0, \Phi_0) \frac{[1 + \alpha u(0, \Phi_0)]}{[1 + \alpha u(X, \Phi_0)]} \left| \frac{d\Phi_0}{d\Phi} \right|, \tag{24}$$

where $\rho(0, \Phi_0)$ represents the entering charge and $1 + \alpha u(0, \Phi_0)$ measures the initial velocity modulation. In the special case of an unmodulated beam

$$\rho(0, \Phi_0) \equiv \rho_0 \tag{25a}$$

and

$$1 + \alpha u(0, \Phi_0) \equiv 1. \tag{25b}$$

The total current $i_t(X, \Phi)$ may conveniently be expanded in terms of a Fourier series in the variable $\Phi(X, \Phi_0)$ as follows:

$$i_t(X, \Phi_0) = \frac{1}{2\pi}\int_0^{2\pi} i_t(X, \Phi)\, d\Phi + \sum_{n=1}^{\infty} \left\{ \frac{\cos n\Phi}{\pi} \int_0^{2\pi} i_t(X, \Phi) \cos n\Phi\, d\Phi \right.$$
$$\left. + \frac{\sin n\Phi}{\pi} \int_0^{2\pi} i_t(X, \Phi) \sin n\Phi\, d\Phi \right\}. \tag{26}$$

However, the total current may also be written in terms of ρ as

$$i_t(X, \Phi) = \rho(0, \Phi_0) u_0 [1 + \alpha u(0, \Phi_0)] \left| \frac{d\Phi_0}{d\Phi} \right|. \tag{27}$$

When this is combined with Eq. (26), we obtain

$$i_t(X, \Phi) = \frac{u_0}{2\pi} \int_0^{2\pi} \rho(0, \Phi_0')[1 + \alpha u(0, \Phi_0')] \, d\Phi_0'$$

$$+ \sum_{n=1}^{\infty} \frac{u_0 \cos n\Phi}{\pi} \int_0^{2\pi} \rho(0, \Phi_0')[1 + \alpha u(0, \Phi_0')] \cos n\Phi \, d\Phi_0'$$

$$+ \sum_{n=1}^{\infty} \frac{u_0 \sin n\Phi}{\pi} \int_0^{2\pi} \rho(0, \Phi_0')[1 + \alpha u(0, \Phi_0')] \sin n\Phi \, d\Phi_0'. \quad (28)$$

Equation (28) may be used to calculate the ratio of the fundamental or any harmonic current amplitude to the dc as shown in Eqs. (29) and (30).

$$\left| \frac{i_n}{I_0} \right| = \frac{\left\{ \left(\frac{1}{\pi} \int_0^{2\pi} \rho(0, \Phi_0')[1 + \alpha u(0, \Phi_0')] \cos n\Phi(X, \Phi_0') \, d\Phi_0' \right)^2 + \left(\frac{1}{\pi} \int_0^{2\pi} \rho(0, \Phi_0')[1 + \alpha u(0, \Phi_0')] \sin n\Phi(X, \Phi_0') \, d\Phi_0' \right)^2 \right\}^{\frac{1}{2}}}{\frac{1}{2\pi} \int_0^{2\pi} \rho(0, \Phi_0')[1 + \alpha u(0, \Phi_0')] \, d\Phi_0'}. \quad (29)$$

Consistent with the discussion of Section 1, the beam entering the drift tube may have emanated from either a cavity modulation region or a distributed interaction region.

The initial conditions required in the solution of Eqs. (10) and (15) or Eqs. (10) and (20) simply specify the state of the entering beam; i.e.,

$$\left. \begin{array}{l} \Phi(0, \Phi_0) \\ \rho(0, \Phi_0) \\ 1 + \alpha u(0, \Phi_0) \end{array} \right\} \quad \text{must be specified.} \quad (30)$$

For an unbunched entering beam with a sinusoidal voltage modulation the conditions are

$$\rho(0, \Phi_0) = \rho_0$$

$$\Phi(0, \Phi_{0j}) = \frac{2\pi j}{m} \quad j = 0, 1, 2, ..., m$$

$$1 + \alpha u(0, \Phi_0) = (1 + \alpha \sin \Phi_{0,j})^{\frac{1}{2}}. \quad (31)$$

The system is complete and may now be solved, unfortunately not by hand, but on a computer.

In the case of a long-transit-angle modulation gap, the gap coupling coefficient, β, is not unity since the electrons do not experience a constant field during their gap transit. If the gap transit angle is $\theta \triangleq (\omega d/u_0)$, where d is the gap distance, and a sinusoidal voltage is applied, then averaging over the mean electron gap transit time yields

$$\frac{u_0}{d} \int_{t-[d/(2u_0)]}^{t+[d/(2u_0)]} V_g \sin \omega t \, dt = V_g \sin \omega t \, \frac{\sin \theta/2}{(\theta/2)} \quad (32)$$

and thus

$$\beta \triangleq \frac{\sin \theta/2}{(\theta/2)}. \quad (33)$$

In calculating the various harmonic current amplitudes in the beam according to Eq. (29) by a Fourier analysis of the velocity-phase information, one must be careful to utilize sufficient charge groups. It has been shown that at least 32 representative charge groups are required for $QC < 0.5$ in order to insure an accurate calculation of i_n/I_0 (n up to 5) in the klystron and of rf voltage level in the traveling-wave amplifier. If one is interested in peaks beyond the first of i_n/I_0 then the question as to the sufficiency of the number of charge groups must again be examined.

A representative plot of i_1/I_0 versus X for several peaks and different numbers of charge groups is shown in Fig. 2 when $\alpha = 0.4$ and $\omega_p/\omega = 0$. It is apparent that a minimum of 32 charge groups is required and that for high accuracy at large X, 64 are needed.

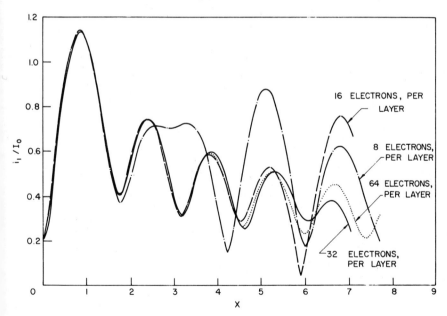

FIG. 2. Klystron fundamental current versus X and number of charge groups ($\alpha = 0.4$, $\omega_p/\omega = 0$).

3 One-Dimensional Klystron Results

Examination of the working equations for the klystron reveals that in the absence of space-charge forces the velocity variable $u(X, \Phi_0)$ is constant and hence the electron trajectories are purely ballistic, of the type studied by Webster[1]. For finite space-charge forces as measured by

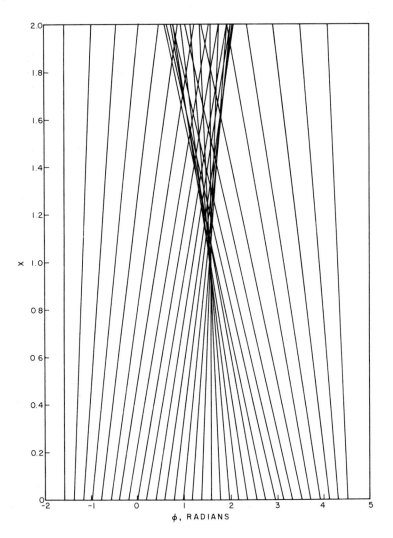

Fig. 3. Klystron zero-space-charge ballistic trajectories ($\alpha = 0.2$, $\omega_p/\omega = 0$, $B = 1$, $a/b' = 2$).

$(\omega_p/\omega\alpha)^2 \neq 0$ and very large α, i.e., depth of modulation, the trajectory character is again ballistic and there are several crossovers. Mihran[8] and others have shown that when $\alpha \ll 1$ and $(\omega_p/\omega\alpha)^2 \neq 0$ the klystron behaves according to space-charge-wave predictions.

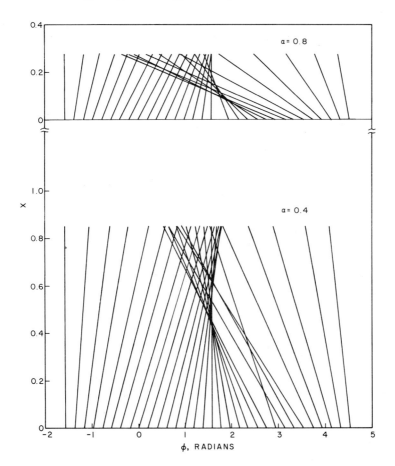

FIG. 4. Klystron zero-space-charge ballistic trajectories ($\alpha = 0.4, 0.8$, $\omega_p/\omega = 0$, $B = 1$, $a/b' = 2$).

The ballistic zero-space-charge trajectories are shown in Figs. 3 and 4 as a function of the modulation index α. As expected all electrons behave as though the others are absent and for very small modulation indices the first crossover point occurs at $X/\lambda_q = 1$ for the unmodulated electron or charge group. Others slightly adjacent to this electron cross somewhat further down the drift region and the most strongly perturbed or modulated particles experience their crossover much further down the drift region. Note that as the modulation index increases the crossover point occurs closer to the modulating grid and the spatial defocusing occurs much more rapidly. The crossover point as a function of space charge and modulation index is shown in Fig. 5.

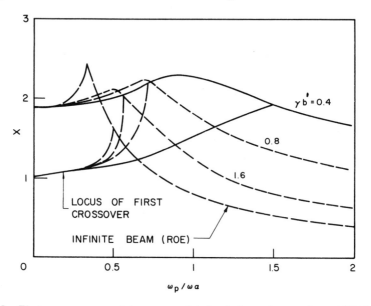

FIG. 5. Electron crossover point versus modulation index and space charge (Webber[11]).

The normalized fundamental and harmonic current amplitudes for these cases are shown in Fig. 6. The ballistic predictions are confirmed, as $|i_1/I_0|$ is a maximum at $X = 1.84$ for $\alpha \ll 1$. Note that the amplitudes decrease as α is increased, again in conformity with zero-space-charge ballistic predictions. As α increases the length for maximum current amplitude is markedly different from the value of $\lambda_q/4$ which is predicted by the Hahn-Ramo space-charge-wave theory. Mihran's[8] experimental results also confirm this fact. The small-amplitude ballistic theory of Webster[1] predicts that the maximum fundamental current occurs when $J_1(X)$ has a first maximum. This occurs for $X = 1.84$ and has a value of $2J_1(1.84) = 1.16$. This ballistic limit is indicated on the curves of Fig. 6 and it is seen that the nonlinear calculation of i_1/I_0 approaches this limit asymptotically.

If small space-charge forces are introduced through B and $(\omega_p/\omega\alpha)^2 \neq 0$, then electron crossings are prevented for small modulation indices as shown in Fig. 7. It should be mentioned that identical results

3. ONE-DIMENSIONAL KLYSTRON RESULTS

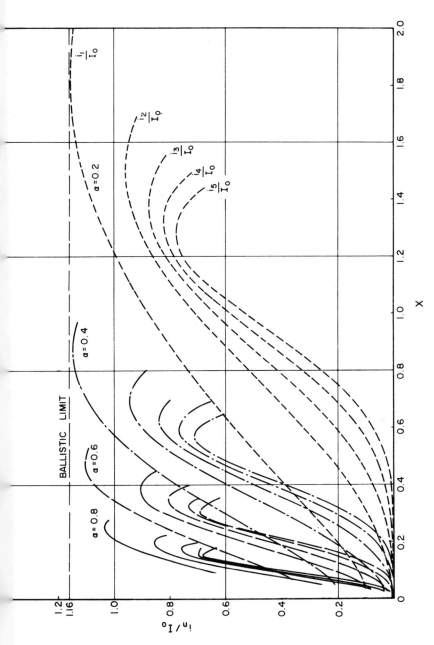

FIG. 6. Fundamental and harmonic-current amplitudes in a drift region ($\omega_p/\omega = 0$, $B = 1$, $a/b' = 2$).

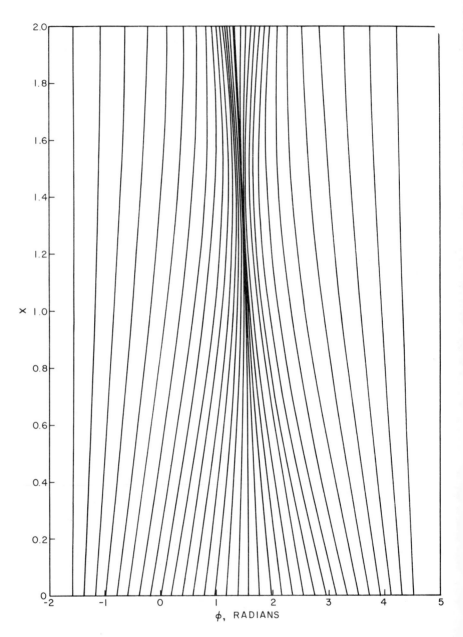

Fig. 7. Klystron space-charge trajectories ($\alpha = 0.2$, $\omega_p/\omega = 0.1251$, $B = 1$, $a/b' = 2$).

3. ONE-DIMENSIONAL KLYSTRON RESULTS

are obtained using either pair of working equations, again verifying the equivalence of the space-charge expressions. Note the effect of space-charge repulsion preventing crossovers. As α is increased, however, crossover again occurs as shown in Fig. 8. The current amplitudes for

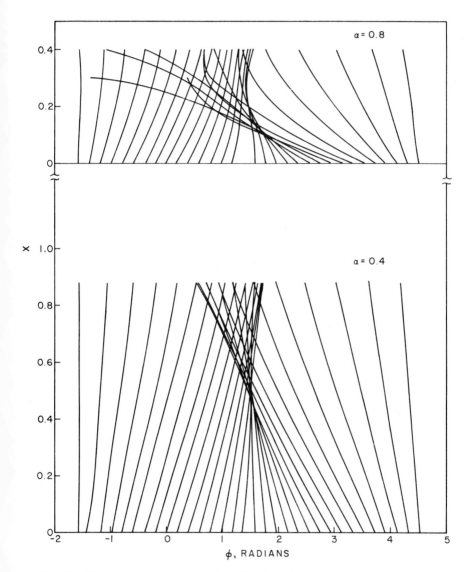

FIG. 8. Klystron space-charge trajectories ($\alpha = 0.4, 0.8$, $\omega_p/\omega = 0.1251$, $B = 1$, $a/b' = 2$).

136 V. KLYSTRON ANALYSIS

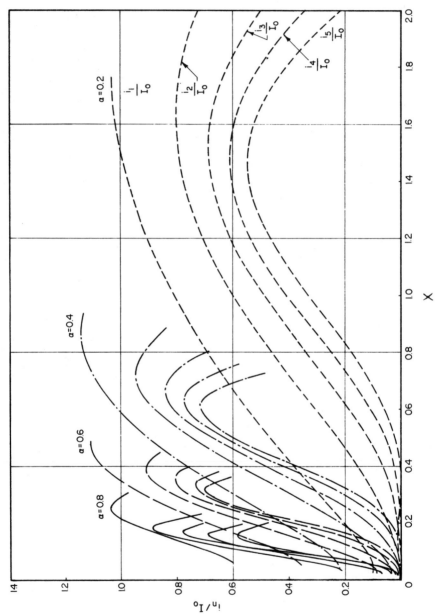

FIG. 9. Fundamental and harmonic-current amplitudes in the drift region ($\omega_p/\omega = 0.1251$, $B = 1$, $a/b' = 2$).

both a moderate and a high value of ω_p/ω are presented in Figs. 9 and 10.

For low levels of modulation, $\alpha \ll 1$, the space-charge-wave behavior is followed but for larger α the return to a ballistics behavior is noted. For small α the current amplitudes fall short of the $\omega_p/\omega = 0$ results due to the space-charge spreading, whereas as α is increased the ballistic limit is once again approached. In all these figures note that the harmonic currents reach a maximum slightly before the fundamental and thus two slightly separated cavities resonant to different harmonics could be utilized to obtain harmonic powers.

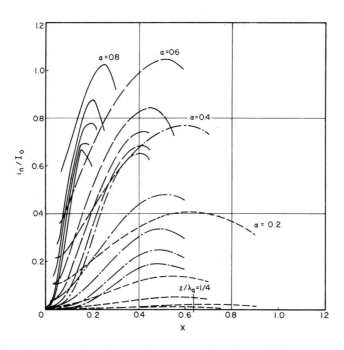

FIG. 10. Fundamental and harmonic-current amplitudes in a drift region ($\omega_p/\omega = 0.41$, $B = 1$, $a/b' = 2$).

Webber[11] has made extensive calculations on the two-cavity klystron using the above theory and these are summarized here. Unfortunately there are many parameters, $\omega_p/\omega\alpha$, B, initial bunching $\rho(0, t_0)$ and initial velocity modulation $1 + \alpha u(0, \Phi_0)$, which must be investigated systematically to gain a complete understanding even of the seemingly simple two-cavity device. Whereas this creates joy in the hearts of the computer people it brings nightmares to the klystron engineer. An approximate plasma frequency reduction factor,

$$R = \frac{\omega_q}{\omega_p} = \frac{1}{\sqrt{1 + \left(\frac{2}{B}\right)^2}}, \tag{34}$$

may be used to show compositely the maximum fundamental-current amplitude and the position at which it occurs, i.e., the value of the bunching parameter X, versus the degree of space charge as measured by the amplitude factor $\omega_p/\omega\alpha$. This is shown in Fig. 11, including the

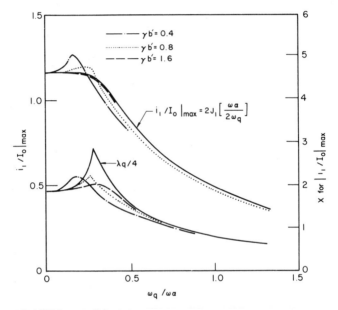

FIG. 11. Maximum fundamental current and bunching parameter at maximum current versus space-charge parameter ($\omega_q/\omega\alpha$) (Webber[11]).

effect of a variation in the beam diameter as measured by the range parameter B. Note that the ballistic expression

$$\frac{i_1}{I_0} = 2J_1\left(\frac{\omega\alpha}{2\omega_q}\right) \tag{35}$$

is quite accurate for a considerable range of $\omega_q/\omega\alpha$ and B. We also see that the length and current amplitude predictions confirm the existence of ballistic and space-charge-wave regions as cited in reference to Figs. 3–10.

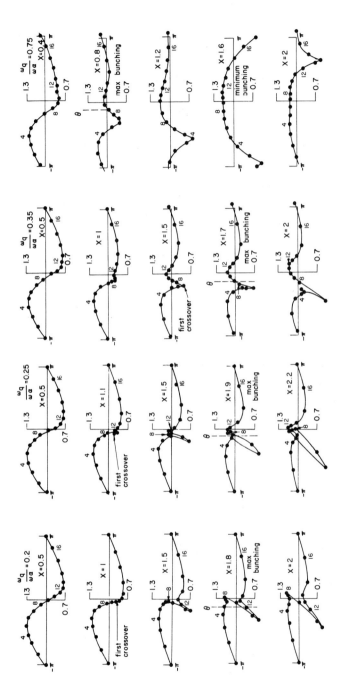

Fig. 12. Phase diagrams for various space-charge conditions (velocity versus phase position). θ = phase of fundamental current, $\alpha = 0.6$, $\gamma b' = 0.8$ (Webber[11]).

The velocity distribution from which the fundamental and harmonic current amplitudes are calculated are shown in Fig. 12 for particular values of $B = \gamma b'$ and α as a function of distance X. The position of first trajectory crossover is clear and occurs further down the drift tube as $\omega_q/\omega\alpha$ is increased due to the increased debunching of the space-charge forces. As the degree of bunching increases a vortex point develops in the velocity-phase curves and of course the space-charge forces are intense within the bunch.

Webber[11] has estimated the efficiency of such a two-cavity klystron using

$$\eta = \frac{1}{2}\frac{i_1}{I_0}\left(\frac{V_g}{V_0}\right)_{\text{eff.}}\eta_c, \qquad (36)$$

where $\eta_c \triangleq$ the circuit efficiency,

$$\left(\frac{V_g}{V_0}\right)_{\text{eff.}} \triangleq 1 - 0.6\alpha, \qquad (37a)$$

the effective voltage swing at the output gap as calculated by Warnecke and Guenard[2] and

$$\frac{i_1}{I_0} \approx 2J_1\left(\frac{\omega\alpha}{2\omega_q}\right). \qquad (37b)$$

The circuit efficiency was taken as

$$\eta_c = 1 - \frac{1}{R_u V_0^{\frac{5}{2}} \hat{P}}, \qquad (38)$$

where

$R_u =$ unloaded output cavity impedance, and

$\hat{P} =$ beam microperveance.

The resulting expression is (for $\omega_q/\omega = 0.1\hat{P}^{\frac{1}{2}}$)

$$\eta = J_1\left(\frac{\alpha}{0.2\hat{P}^{\frac{1}{2}}}\right)(1 - 0.6\alpha)\left(1 - \frac{1}{R_u V_0^{\frac{5}{2}} \hat{P}}\right) \qquad (39)$$

and is plotted in Fig. 13. This is an approximate calculation and of course the absolute answers will vary directly with i_1/I_0 and other parameter values. However, the trend indicating highest efficiency for low perveance with high voltage and for high perveance at low voltage should be correct.

The treatment of multicavity klystrons illustrated in Fig. 1 of Chapter I may proceed using the same equations and dividing the device into a series of two-cavity devices with the bunched and modulated output of the nth two-cavity section being the input to the $(n+1)$st section. Of course a further complication develops as the modulation form to succeeding cavities and the phase angle between modulation and current also enter. Again there is joy in "computerville" and despair in the "klystron factory" as we have expanded the parameter space by many dimensions. A general treatment of the nth cavity (including the output cavity) is thus tedious and time-consuming.

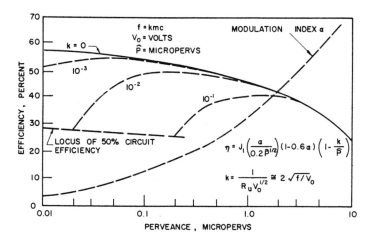

FIG. 13. Efficiency of the two-cavity klystron versus beam perveance (Webber[11]).

Some specialized and idealized calculations on the multicavity klystron problem have been made by Webber[12] to give some insight into the general problem. Since his studies are based on a number of idealizations the absolute results are questionable although the general pattern (qualitative behavior) is probably correct. Fortunately one of the questions raised above as to the optimum phase between the cavity voltage and the bunch has been answered by the experimental work of Dehn and Branch as shown by Webber[12] (see Fig. 14). The optimum phase angle for maximum power output is seen to be 90°; i.e., the current lags the voltage by $\pi/2$ for large-signal conditions. Assuming that it is always true leads to a simplification and reduction in the necessary calculations. The calculated performance as a function of the cavity phase angle also confirms this value of angle (Fig. 15).

The choice of appropriate input conditions is complex and many different types should be investigated. Since the space-charge wave and ballistics theories are easily understood it is well to use these results as input conditions. Under conditions of low modulation and finite space-charge forces the space-charge-wave theory was seen to apply; thus at the output of such a two-cavity device the beam would be density modulated with $i_1/I_0 \approx 1.16$ and the velocity modulation is zero. With these as initial conditions to the second cavity the current ratios of Fig. 16 are found. Clearly the bunching has been enhanced and for $z/\lambda_q < 0.15$ the final bunching is not critically dependent on the degree of

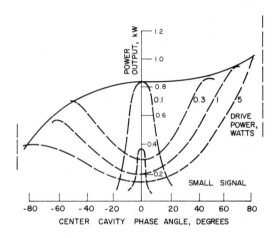

Fig. 14. Measured power output versus phase of center-cavity voltage with rf drive power as a parameter (Webber[12]).

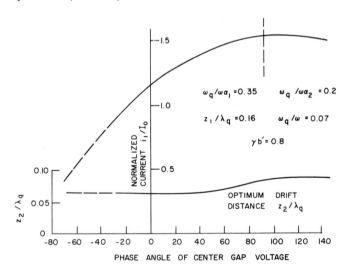

Fig. 15. Normalized current versus phase of center-gap voltage (Webber[12]).

prebunching. For $z/\lambda_q > 0.15$ however there is a criticality. Notice that curves of constant $\omega_q/\omega\alpha_2$ indicate that z/λ_q does not vary rapidly with the degree of prebunching. If instead the output beam state obtained from the two-cavity calculations is used as input to the second cavity, the

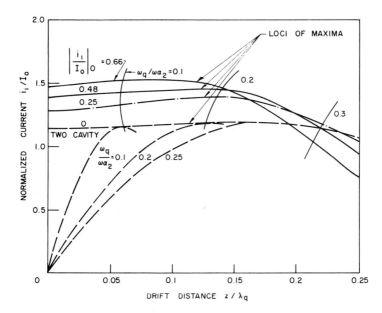

FIG. 16. Maximum current versus distance for various conditions of modulation. Small perveance, small-signal prebunching (density only) (Webber[12]).

results of Fig. 17 are obtained for a particular value of ω_q/ω. The results are seen to be quite similar, indicating that the velocity spread has not been particularly deleterious, at least for the case studied. Results for larger values of ω_q/ω tend to be of the same type and bear out the conclusions reached above. Other than sinusoidal velocity modulation at the second gap, however, could produce quite different results.

The enhanced bunching provided by the addition of an intermediate cavity should lead to greater efficiency for multicavity klystrons relative to two-cavity devices. If the effect of space charge is neglected the approximate efficiency of a three-cavity device can be calculated easily from the change in kinetic energy experienced by the electrons as they traverse the output gap. Such results are shown in Fig. 18; the enhancement is clear. The absolute numbers in Fig. 18 should not be taken seriously, however, as space charge and velocity spread effects have been neglected in the output gap.

The reader can now see how the complexity of the problem develops as more and more cavities are added and is probably already discouraged with the number of selections to be investigated.

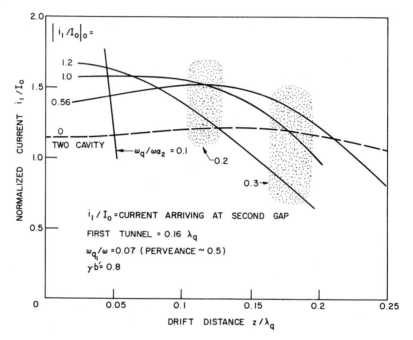

FIG. 17. Maximum current versus length of second tunnel for various conditions of modulation index at first and second cavity, $\omega_q/\omega = 0.07$ (Webber[12]).

4 Two-Dimensional Klystron Analysis

After considering the results of the one-dimensional analysis one immediately wonders how important radial motion and radial debunching are with regard to the predicted degree of bunching and efficiency of the klystron, two-cavity or multicavity. The next step in the generalization of the analysis is to include the effects of radial space-charge forces but to assume angular symmetry and that the beam is in confined flow, i.e., the beam diameter is smooth.

The analysis proceeds in the same manner as previously except that two components (radial and axial) of the vector force equation must be used. The space-charge field to be used is that for rings of charge as

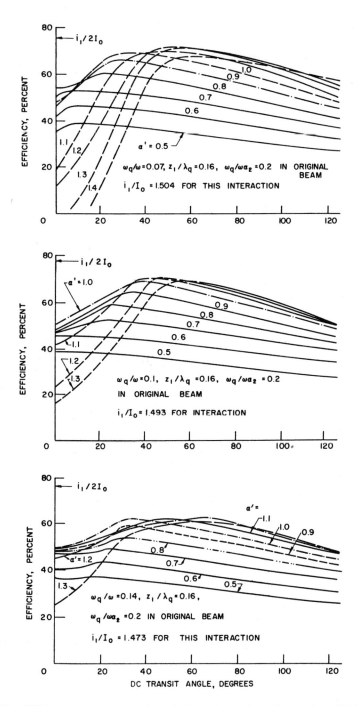

FIG. 18. Efficiency versus gap transit angle for various values of gap voltage (Webber[12]).

derived in Section 8. The definition of the independent space and phase variables follows the previous treatment:

$$X = y \triangleq \frac{\alpha \omega z}{2u_0}, \tag{40a}$$

$$x \triangleq \frac{\alpha \omega r}{2u_0}, \tag{40b}$$

$$x_0 \triangleq \frac{\alpha \omega r_0}{2u_0} \tag{40c}$$

and

$$\Phi_0 \triangleq \frac{\omega z_0}{u_0} = \omega t_0. \tag{40d}$$

The phase variable for the displaced electrons is again written as

$$\Phi(z, t) = \frac{\omega z}{u_0} - \omega t. \tag{41}$$

The electron velocity has both axial and radial components and is defined by

$$u_z = \frac{2u_0}{\alpha \omega} \frac{dX}{dt} \triangleq u_0[1 + \alpha u_y(X, \Phi_0, x_0)] \tag{42}$$

and

$$u_r = \frac{2u_0}{\alpha \omega} \frac{dx}{dt} \triangleq \alpha u_0 u_x(X, \Phi_0, x_0). \tag{43}$$

The independent variable x_0 indicates the normalized radial position at which the electron enters the device. Equations (41) and (42) are combined to give the following velocity-phase relation

$$\frac{\partial \Phi(X, \Phi_0, x_0)}{\partial X} = \frac{2}{\alpha} \left[\frac{1}{1 + \alpha u_y(0, \Phi_0, x_0)} - \frac{1}{1 + \alpha u_y(X, \Phi_0, x_0)} \right], \tag{44}$$

where $u_y(0, \Phi_0, x_0)$ represents the initial velocity modulation in the z-direction. If $u_y(0, \Phi_0, x_0) \equiv 0$, then Eq. (44) reduces to Eq. (10).

The x-position of the charge at any displacement plane is treated as a dependent variable of the system and is rewritten as follows:

$$x(X, \Phi_0, x_0) = x_0 + \int_0^t \frac{dx}{dt} dt = x_0 + \alpha \int_0^X \frac{u_x(X, \Phi_0, x_0)}{1 + \alpha u_y(X, \Phi_0, x_0)} dX. \tag{45}$$

4. TWO-DIMENSIONAL KLYSTRON ANALYSIS

The force equations for the two directions, radial and axial, are

$$\frac{d^2r}{dt^2} = -|\eta| E_{sc-r} \tag{46}$$

and

$$\frac{d^2z}{dt^2} = -|\eta| E_{sc-z}. \tag{47}$$

Introducing the new reduced variables into Eqs. (46) and (47) and simplifying yields the following:

$$\frac{\partial u_x}{\partial X}(1+\alpha u_y) = -\frac{2|\eta|}{\alpha^2 u_0 \omega} E_{sc-x} \tag{48}$$

and

$$\frac{\partial u_y}{\partial X}(1+\alpha u_y) = -\frac{2|\eta|}{\alpha^2 u_0 \omega} E_{sc-X}. \tag{49}$$

The space-charge-field expressions are obtained directly from Eqs. (53) and (54) of Chapter IV. These are

$$E_{sc-x} = -\int_0^{x_{b'}/x_a} \frac{x_0' \, dx_0'}{x_a \, x_a} \int_0^{2\pi} \frac{\alpha^2 \omega I_0}{4 u_0^2 \pi \epsilon_0 x_{b'}^2} \sum_{l=1}^{\infty} \exp{-\frac{\nu_l}{x_a} |X - X'|}$$

$$\cdot \frac{J_0\left(\nu_l \frac{x'}{x_a}\right) J_1\left(\nu_l \frac{x}{x_a}\right)}{[J_1(\nu_l)]^2} \, d\Phi_0'$$

and

$$E_{sc-X} = -\int_0^{x_{b'}/x_a} \frac{x_0' \, dx_0'}{x_a \, x_a} \int_0^{2\pi} \frac{\alpha^2 \omega I_0}{4 u_0^2 \pi \epsilon_0 x_{b'}^2} \sum_{l=1}^{\infty} \exp{-\frac{\nu_l}{x_a} |X - X'|}$$

$$\cdot \frac{J_0\left(\nu_l \frac{x'}{x_a}\right) J_0\left(\nu_l \frac{x}{x_a}\right)}{[J_1(\nu_l)]^2} \operatorname{sgn}(X - X') \, d\Phi_0',$$

where

$$\nu_l \triangleq \frac{2\mu_l u_0}{\omega \alpha},$$

$$x_a = \frac{\alpha \omega a}{2 u_0},$$

and

$$x_{b'} = \frac{\alpha \omega b'}{2 u_0}.$$

The final form of the force equations is

$$\frac{\partial u_x}{\partial X}(1 + \alpha u_y) = 2\left(\frac{\omega_p}{\omega\alpha}\right)^2 \int_0^{x_{b'}/x_a} \frac{x_0'\,dx_0'}{x_a\,x_a} \int_0^{2\pi}$$

$$\cdot \sum_{l=1}^{\infty} \exp\left[-\frac{\nu_l}{2x_a}\alpha\,|\Phi - \Phi'|\right] \frac{J_0\left(\nu_l \dfrac{x'}{x_a}\right) J_1\left(\nu_l \dfrac{x}{x_a}\right)}{[J_1(\nu_l)]^2}\,d\Phi_0' \quad (50)$$

and

$$\frac{\partial u_y}{\partial X}(1 + \alpha u_y) = 2\left(\frac{\omega_p}{\omega\alpha}\right)^2 \int_0^{x_{b'}/x_a} \frac{x_0'\,dx_0'}{x_a\,x_a} \int_0^{2\pi} \operatorname{sgn}(X - X')$$

$$\cdot \sum_{l=1}^{\infty} \exp\left[-\frac{\nu_l}{2x_a}\alpha\,|\Phi - \Phi'|\right] \frac{J_0\left(\nu_l \dfrac{x'}{x_a}\right) J_0\left(\nu_l \dfrac{x}{x_a}\right)}{[J_1(\nu_l)]^2}\,d\Phi_0', \quad (51)$$

where

$$\omega_p^2 \triangleq \frac{|\eta|I_0}{\pi\epsilon_0 b'^2 u_0}.$$

The working equations are Eqs. (44), (45), (50), and (51) for the two-dimensional confined-flow klystron. The independent variables are

$$X, \quad x_0 \quad \text{and} \quad \Phi_0.$$

The dependent variables are

$$x, \quad \Phi(X, \Phi_0), \quad u_y(X, \Phi_0, x_0) \quad \text{and} \quad u_x(X, \Phi_0, x_0).$$

The input conditions are similar to those used in the one-dimensional analysis.

(a) The entering charge per ring is distributed over one cycle of the modulating wave as

$$\Phi_{0,j} = \frac{2\pi j}{m} \quad j = 0, 1, 2, ..., m.$$

Various rings are specified by values of x_0. Of course, both solid and hollow beams are included.

(b) Initial velocities are specified by both

$$u_x = u_x(0, \Phi_0, x_0)$$

and

$$u_y = u_y(0, \Phi_0, x_0).$$

4. TWO-DIMENSIONAL KLYSTRON ANALYSIS

(c) Operating parameters are

$$B = \gamma b', \qquad \omega_p/\omega\alpha, \qquad a/b'.$$

The fundamental and harmonic current amplitudes in the beam are again calculated from the continuity equation as previously. This conservation of charge equation is written as

$$\rho r \, dr \, dz = \rho_0 r_0 \, dr_0 \, dz_0 \tag{52}$$

and a linear charge density at any displacement plane is written as

$$\sigma = -2\pi \int_0^{b'} \rho r \, dr. \tag{53}$$

Combining Eqs. (52) and (53) yields

$$\sigma = -2\pi \int_0^{b'} \rho_0 r_0 \left| \frac{\partial z_0}{\partial z} \right| dr_0, \tag{54}$$

where

$$\rho_0 \triangleq \frac{I_0}{\pi b'^2 u_0}. \tag{55}$$

The electron phase displacement is introduced through

$$\left| \frac{\partial z_0}{\partial z} \right| = \frac{1}{1 + \alpha u_y} \left| \frac{\partial \Phi_0}{\partial \Phi} \right| \tag{56}$$

and Eq. (54) becomes, after introduction of normalized variables,

$$\sigma = \frac{-2I_0}{u_0 x_{b'}^2} \int_0^{x_{b'}} \left| \frac{\partial \Phi_0}{\partial \Phi} \right| \frac{x_0}{1 + \alpha u_y} dx_0. \tag{57}$$

The amplitudes of the various beam harmonic currents may be calculated in much the same manner as in the one-dimensional theory. The linear charge density in any annular ring is written as follows in terms of normalized variables:

$$d\sigma = -2\pi \rho(z, t, r) r \, dr = -2\pi \rho(0, t_0, r_0) \frac{u_t(0, t)}{u_t(z, t)} \left| \frac{\partial \Phi_0}{\partial \Phi} \right| r_0 \, dr_0. \tag{58}$$

Since there is no net current in the radial direction the incremental current is

$$dI(X, \Phi, x_0) = -2\pi \rho(0, \Phi_0, x_0) u_0 [1 + \alpha u_y(0, \Phi_0, x_0)] \left| \frac{\partial \Phi_0}{\partial \Phi} \right| \left(\frac{2u_0}{\omega\alpha} \right)^2 x_0 \, dx_0. \tag{59}$$

Further simplification and integration yields

$$I(X, \Phi) = -\frac{8\pi\rho_0 u_0^3}{\omega^2 \alpha^2} \int_0^{x_{b'}} [1 + \alpha u_y(0, \Phi_0', x_0')] \left| \frac{\partial \Phi_0'}{\partial \Phi} \right| x_0' \, dx_0'. \tag{60}$$

After expansion in a Fourier series and various simplifications, the harmonic current amplitudes are expressed as

$$\left| \frac{i_n(X)}{I_0} \right| = \frac{\left(\left\{ \frac{1}{\pi} \int_0^{2\pi} \int_0^{x_{b'}} [1 + \alpha u_y(0, \Phi_0', x_0')] x_0' \cos n\, \Phi(X, \Phi_0', x_0') \, dx_0' \, d\Phi_0' \right\}^2 + \left\{ \frac{1}{\pi} \int_0^{2\pi} \int_0^{x_{b'}} [1 + \alpha u_y(0, \Phi_0', x_0')] x_0' \sin n\, \Phi(X, \Phi_0', x_0') \, dx_0' \, d\Phi_0' \right\}^2 \right)^{\frac{1}{2}}}{\frac{1}{2\pi} \int_0^{2\pi} \int_0^{x_{b'}} [1 + \alpha u_y(0, \Phi_0', x_0')] x_0' \, dx_0' \, d\Phi_0'}.$$

$$\tag{61}$$

When only one layer of charge is utilized the above expression reduces to the previously derived one-dimensional expression.

5 Three-Dimensional Klystron Interaction

The two-dimensional analysis of Section 4 is somewhat restricted in that it assumes a confined flow so that the beam boundary is smooth and angular motion of electrons about the axis is neglected. We wish now to generalize the klystron drift-space analysis further to account for interaction in a finite focusing field and motion about the axis as well as radially. The system is more complex but the analysis may be developed along the same lines as used previously.

The normalized distance and initial phase variables of Eqs. (40) are again used. A new phase variable

$$\Phi(z, t, \varphi) \triangleq \omega \left(\frac{z}{u_0} - t \right) \tag{62}$$

is defined where φ is the angle about the cylindrical axis. The dependent velocity variables are defined as

$$\frac{2u_0}{\alpha\omega} \frac{dX}{dt} \triangleq u_0[1 + \alpha u_y(X, x_0, \Phi_0, \varphi_0)], \tag{63}$$

$$\frac{2u_0}{\alpha\omega} \frac{dx}{dt} \triangleq u_0[\alpha u_x(X, x_0, \Phi_0, \varphi_0)], \tag{64}$$

5. THREE-DIMENSIONAL KLYSTRON INTERACTION

and

$$\frac{d\varphi}{dt} \triangleq u_0 \left[\frac{\alpha}{r} u_\varphi(X, x_0, \Phi_0, \varphi_0) \right]. \tag{65}$$

The development of the relation between dependent variables is made in a manner similar to the previous treatments, with the result that

$$\frac{\partial \Phi(X, x_0, \Phi_0, \varphi_0)}{\partial X} \doteq \frac{2}{\alpha} \left[\frac{1}{1 + \alpha u_y(0, x_0, \Phi_0, \varphi_0)} - \frac{1}{1 + \alpha u_y(X, x_0, \Phi_0, \varphi_0)} \right]. \tag{66}$$

The three components of the vector force equation for low velocities are written as follows including the effect of a *finite* axial focusing field:

$$\frac{d^2 r}{dt^2} - r \left(\frac{d\varphi}{dt} \right)^2 = -|\eta| \left[E_{sc-r} + E_{o-r} + B_z r \frac{d\varphi}{dt} \right], \tag{67}$$

$$\frac{1}{r} \frac{d}{dt} \left(r^2 \frac{d\varphi}{dt} \right) = -|\eta| \left[E_{sc-\varphi} - B_z \frac{dr}{dt} \right], \tag{68}$$

and

$$\frac{d^2 z}{dt^2} = -|\eta| [E_{sc-z}], \tag{69}$$

where the magnetic field is assumed to be entirely axially directed over the interaction region and E_{o-r} is the radial field due to the dc space charge. At $z = 0$, $d\varphi/dt$ may have a specified value. If it is assumed that no flux threads the cathode then we may write

$$\frac{d\varphi}{dt} = \frac{\eta B_z}{2}$$

from Busch's theorem for the special case of Brillouin flow.

The space-charge-field components are obtained from Chapter IV, where the field is computed for elementary volume element charges, in Section IV.7. These field components are

$$E_{sc-X} = -\left(\frac{u_0}{\omega} \right) \frac{\omega_p^2}{|\eta|} \iiint \frac{1}{2\pi x_a^2} \sum_{l=1}^{\infty} \sum_{s=0}^{\infty} \mathrm{sgn}(X - X')(2 - \delta_s^0)$$

$$\cdot e^{-\nu_{ls}|X-X'|} \cos s(\varphi - \varphi') \frac{J_s(\nu_{ls} x') J_s(\nu_{ls} x)}{[J_{s+1}(\nu_{ls} x_a)]^2} x_0' \, d\varphi_0' \, dx_0' \, d\Phi_0', \tag{70}$$

$$E_{sc-x} = -\left(\frac{u_0}{\omega} \right) \frac{\omega_p^2}{|\eta|} \iiint \frac{1}{2\pi x_a^2} \sum_{l=1}^{\infty} \sum_{s=0}^{\infty} (2 - \delta_s^0) \, e^{-\nu_{ls}|X-X'|}$$

$$\cdot \cos s(\varphi - \varphi') \frac{J_s(\nu_{ls} x')[J_{s+1}(\nu_{ls} x) - J_{s-1}(\nu_{ls} x)]}{2[J_{s+1}(\nu_{ls} x_a)]^2} x_0' \, d\varphi_0' \, dx_0' \, d\Phi_0' \tag{71}$$

and

$$E_{sc-\varphi} = -\left(\frac{u_0}{\omega}\right)\frac{\omega_p^2}{|\eta|}\iiint \frac{1}{2\pi x_a^2}\sum_{l=1}^{\infty}\sum_{s=0}^{\infty}s(2-\delta_s^0)$$
$$\cdot e^{-\nu_{ls}|X-X'|}\sin s(\varphi-\varphi')\frac{J_s(\nu_{ls}x')J_s(\nu_{ls}x)}{(\nu_{ls}x)[J_{s+1}(\nu_{ls}x_a)]^2}x_0'\,d\varphi_0'\,dx_0'\,d\Phi_0'. \quad (72)$$

The electric field due to the dc space charge in the radial force equation [Eq. (67)] may be written as follows for a solid cylindrical electron beam:

$$E_{0-r} = -\frac{\omega_p^2}{\eta}\frac{r_0^2}{2r}.$$

The above is rewritten in terms of the normalized radial coordinate as

$$E_{0-x} = -\frac{\omega_p^2}{\eta}\left(\frac{u_0}{\alpha\omega}\right)\frac{x_0^2}{x}. \quad (73)$$

Equations (63)–(65) for the velocity components, and Eqs. (70)–(73) for the space-charge-field components are substituted in a now familiar manner into the force equations, Eqs. (67)–(69), and the following form of the force equations is obtained.

$$\frac{\partial u_x}{\partial X}(1+\alpha u_y) = \frac{\alpha}{x}u_\varphi^2 - \left(\frac{2\omega_c}{\omega\alpha}\right)u_\varphi + \left(\frac{\omega_p}{\omega}\right)^2\frac{2}{\alpha^3}\frac{x_0^2}{x} + 4\left(\frac{\omega_p}{\omega\alpha}\right)^2\int_0^{2\pi}\int_0^{x_b'}\int_0^{2\pi}\frac{1}{4\pi x_a^2}$$
$$\cdot \sum_{l=1}^{\infty}\sum_{s=0}^{\infty}(2-\delta_s^0)\,e^{-\nu_{ls}|X-X'|}\cos s(\varphi-\varphi')$$
$$\cdot \frac{J_s(\nu_{ls}x')[J_{s+1}(\nu_{ls}x)-J_{s-1}(\nu_{ls}x)]}{2[J_{s+1}(\nu_{ls}x_a)]^2}\,d\Phi_0'\,x_0'\,dx_0'\,d\varphi_0', \quad (74)$$

$$\frac{\partial u_y}{\partial X}(1+\alpha u_y) = 4\left(\frac{\omega_p}{\omega\alpha}\right)^2\int_0^{2\pi}\int_0^{x_b'}\int_0^{2\pi}\frac{1}{4\pi x_a^2}\sum_{l=1}^{\infty}\sum_{s=0}^{\infty}$$
$$\cdot \text{sgn}(X-X')(2-\delta_s^0)\,e^{-\nu_{ls}|X-X'|}\cos s(\varphi-\varphi')$$
$$\times \frac{J_s(\nu_{ls}x')J_s(\nu_{ls}x)}{[J_{s+1}(\nu_{ls}x_a)]^2}\,d\Phi_0'\,x_0'\,dx_0'\,d\varphi_0' \quad (75)$$

$$\frac{\partial u_\varphi}{\partial X}(1+\alpha u_y) = \left(\frac{2\omega_c}{\omega\alpha}\right)u_x - \frac{\alpha}{x}u_x u_\varphi + 4\left(\frac{\omega_p}{\omega\alpha}\right)^2\int_0^{2\pi}\int_0^{x_b'}\int_0^{2\pi}\frac{1}{4\pi x_a^2}$$
$$\cdot \sum_{l=1}^{\infty}\sum_{s=0}^{\infty}(2-\delta_s^0)\,e^{-\nu_{ls}|X-X'|}\sin s(\varphi-\varphi')$$
$$\times \frac{J_s(\nu_{ls}x')J_s(\nu_{ls}x)}{(\nu_{ls}x)[J_{s+1}(\nu_{ls}x_a)]^2}\,d\Phi_0'\,x_0'\,dx_0'\,d\varphi_0', \quad (76)$$

5. THREE-DIMENSIONAL KLYSTRON INTERACTION

where the parameter ω_c/ω measures the strength of the axial focusing magnetic field which may be a function of distance. The velocity-phase relationship is again

$$\frac{\partial \Phi(X, x_0, \Phi_0, \varphi_0)}{\partial X} = \frac{2}{\alpha} \left[\frac{1}{1 + \alpha u_y(0, x_0, \Phi_0, \varphi_0)} - \frac{1}{1 + \alpha u_y(X, x_0, \Phi_0, \varphi_0)} \right], \tag{77}$$

where $u_y(0, x_0, \Phi_0, \varphi_0)$ measures the input modulation. In this three-dimensional drift-space analysis both the variables x and φ are treated as dependent variables, as x was in the two-dimensional theory. The x-position variable is written as

$$x(X, x_0, \Phi_0, \varphi_0) = x_0 + \alpha \int_0^X \frac{u_x(X, x_0, \Phi_0, \varphi_0)}{[1 + \alpha u_y(X, x_0, \Phi_0, \varphi_0)]} \, dX. \tag{78}$$

The angular position variable φ may be treated in a directly similar manner:

$$\varphi = \varphi_0 + \int_0^t \frac{d\varphi}{dt} \, dt$$

$$= \varphi_0 + \int_0^z \frac{d\varphi}{dt} \frac{dz}{u_0[1 + \alpha u_y]}.$$

Introducing Eq. (65) for $d\varphi/dt$ into the above yields

$$\varphi(X, x_0, \Phi_0, \varphi_0) = \varphi_0 + \alpha \int_0^X \frac{u_\varphi}{x} \frac{dX}{[1 + \alpha u_y(X, x_0, \Phi_0, \varphi_0)]}. \tag{79}$$

The working equations to be solved simultaneously for this three-dimensional drift region are Eqs. (73)–(79). The independent variables are

$$X, \quad x_0, \quad \Phi_0 \quad \text{and} \quad \varphi_0.$$

The dependent variables are

$$x, \quad \varphi(\varphi_0), \quad \Phi(X, \Phi_0), \quad u_y(X, \Phi_0, x_0, \varphi_0), \quad u_x(X, \Phi_0, x_0, \varphi_0)$$

and

$$u_\varphi(X, \Phi_0, x_0, \varphi_0).$$

The input conditions are similar to the one- and two-dimensional analyses:

(a) The entering charge per ring is distributed over one cycle of the modulating wave as

$$\Phi_{0,j} = \frac{2\pi j}{m} \quad j = 0, 1, 2, ..., m$$

and various rings are specified by values of x_0. Each ring is segmented into elements $\varphi_{0,j} = 2\pi i/k$ where $i = 0, 1, 2, ..., k$.

(b) Initial velocities are specified by

$$u_x = u_x(0, \Phi_0, x_0, \varphi_0)$$
$$u_y = u_y(0, \Phi_0, x_0, \varphi_0)$$
$$u_\varphi = u_\varphi(0, \Phi_0, x_0, \varphi_0) = \frac{r_0}{\alpha u_0}\left(\frac{d\varphi}{dt}\right)\bigg|_{X=0}.$$

For shielded Brillouin flow

$$\frac{d\varphi}{dt} = \frac{\eta B_z}{2}$$

and then

$$u_\varphi(0, x_0, \Phi_0, \varphi_0) = \frac{\omega_c}{\omega}\left(\frac{x_0}{\alpha^2}\right).$$

(c) Operating parameters are

$$\gamma b', \quad \frac{\omega_p}{\omega\alpha}, \quad \frac{\omega_c}{\omega\alpha}, \quad \frac{a}{b'} \quad \text{and} \quad \alpha.$$

The calculation of fundamental and harmonic current amplitudes can be obtained from the three-dimensional Lagrangian continuity equation which states

$$\rho' r' \, dr' \, d\varphi' \, dz' = \rho_0' r_0' \, dr_0' \, d\varphi_0' \, dz_0'. \tag{80}$$

The linear charge density at any displacement plane is

$$\sigma = -\int_0^{b'} \frac{r \, dr}{a} \rho', \tag{81}$$

where ρ' is the volume charge density. Combining Eqs. (80) and (81) gives the following:

$$\sigma = -\frac{1}{a}\int_0^{b'} \frac{\partial \Phi_0'}{\partial \Phi} \frac{\partial \varphi_0'}{\partial \varphi} \frac{I_0 r_0' \, dr_0'}{\pi b'^2 u_0} \frac{1}{(1 + \alpha u_y)}. \tag{82}$$

The final force equations, Eqs. (74)–(76), contain the three-dimensional space-charge-force weighting functions and thus involve a great deal of computing time due to the doubly infinite sum over the various Bessel functions. An interesting and useful simplification of this system is to assume that the space-charge fields are axially symmetric even though the electrons move about the axis. The force equations then simplify to

$$\frac{\partial u_x}{\partial X}(1+\alpha u_y) = \frac{\alpha}{x}u_\varphi^2 - \left(\frac{2\omega_c}{\omega\alpha}\right)u_\varphi + \left(\frac{\omega_p}{\omega}\right)^2 \frac{2}{\alpha^3}\frac{x_0^2}{x} + 2\left(\frac{\omega_p}{\omega\alpha}\right)^2 \int_0^{x_{b'}/x_a} \frac{x_0'\,dx_0'}{x_a\,x_a} \int_0^{2\pi}$$

$$\cdot \sum_{l=1}^{\infty} \exp\left(-\frac{\nu_l\alpha}{2x_a}|\Phi - \Phi'|\right) \frac{J_0\!\left(\nu_l\frac{x'}{x_a}\right) J_1\!\left(\nu_l\frac{x}{x_a}\right)}{[J_1(\nu_l)]^2}\,d\Phi_0', \qquad (83)$$

$$\frac{\partial u_y}{\partial X}(1+\alpha u_y) = 2\left(\frac{\omega_p}{\omega\alpha}\right)^2 \int_0^{x_{b'}/x_a} \frac{x_0'\,dx_0'}{x_a\,x_a} \int_0^{2\pi} \mathrm{sgn}(X - X')$$

$$\cdot \sum_{l=1}^{\infty} \exp\left(-\frac{\nu_l\alpha}{2x_a}|\Phi - \Phi'|\right) \frac{J_0\!\left(\nu_l\frac{x'}{x_a}\right) J_1\!\left(\nu_l\frac{x}{x_a}\right)}{[J_1(\nu_l)]^2}\,d\Phi_0' \qquad (84)$$

and

$$\frac{\partial u_\varphi}{\partial X}(1+\alpha u_y) = \left(\frac{2\omega_c}{\omega\alpha}\right)u_x - \frac{\alpha}{x}u_x u_\varphi. \qquad (85)$$

The above system permits a study of beam interaction and stability in a finite magnetic focusing field without undue complications. Radio-frequency current amplitudes and velocity-phase information calculated using this model are presented in the following section and are compared with the infinite and zero magnetic field cases. For an injected Brillouin beam from a shielded cathode, the first two terms (Lorentz force) on the right-hand side of Eq. (83) exactly balance the third (space-charge force) at the input plane ($X = 0$).

6 Radial and Angular Effects in Klystrons

Radial and angular space-charge fields and motion about the axis of symmetry (finite magnetic field effects) were included in the equations of the previous section. A numerical solution of the three-dimensional equations would be quite time-consuming (even on a computer) due to the doubly infinite series of Bessel functions in the space-charge weighting functions. It is generally expected that angular space-charge-field variations would not be important so that the simpler equations given at the end of Section 5 may be utilized. Electron motion about the axis is still considered as indicated by the presence of terms proportional to ω_c/ω.

The tunnel (drift tube) diameter is specified by the parameter $x_a = \alpha\gamma a/2$ and hence for $\alpha = 0.2$ and $x_a = 0.15$ corresponds to

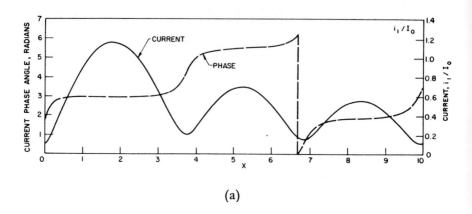

(a)

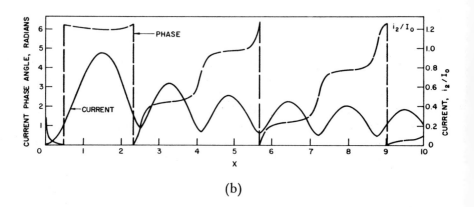

(b)

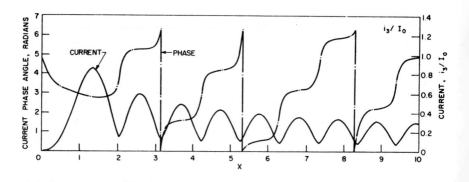

(c)

6. RADIAL AND ANGULAR EFFECTS IN KLYSTRONS

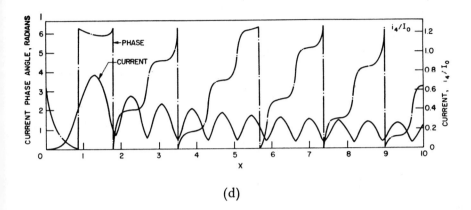

(d)

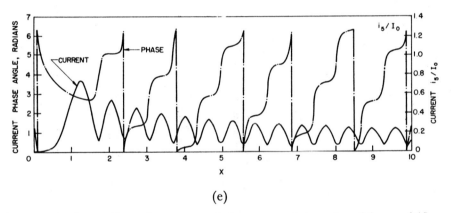

(e)

FIG. 19. Stream harmonic currents and phases versus distance ($\alpha = 0.2$, $x_a = 0.15$, $x_{b'} = 0.105$, $\omega_p/\omega = \omega_c/\omega = 0$). (a) i_1/I_0; (b) i_2/I_0; (c) i_3/I_0; (d) i_4/I_0; (e) i_5/I_0.

$\gamma a = 1.5$. The stream diameter is correspondingly given by $x_{b'} = \alpha \gamma b'/2$ and for $\gamma b' \leqslant 1$ three annular rings and 32 charge groups per ring are sufficient for high accuracy. This of course depends somewhat upon the value of the stream plasma frequency, ω_p. For $\omega_p > 0.3$ it is desirable to utilize 64 or more charge groups per layer. When $\omega_c/\omega > 0$ the optimum integration increment ΔX depends upon ω_c/ω. It has been found empirically that ΔX should be selected in accordance with the following:

$$\Delta X < \frac{\alpha \pi}{120 \left(\dfrac{\omega_c}{\omega}\right)},$$

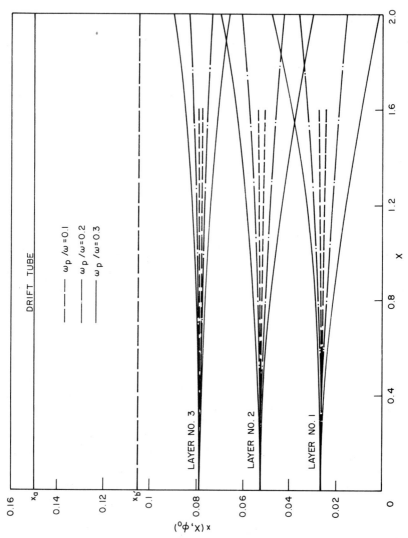

Fig. 20. Electron trajectories in the normalized r-z plane ($\alpha = 0.2$, $x_a = 0.15$, $x_{b'} = 0.105$, $\omega_c/\omega = 0$).

6. RADIAL AND ANGULAR EFFECTS IN KLYSTRONS 159

which arises from the requirement that the forward integration step must be small compared with the cyclotron wavelength. Solutions may be obtained again for streams with arbitrary initial velocity and current distributions. Two of the important special cases of interest are confined flow ($\omega_c/\omega \to \infty$) and Brillouin flow.

Prior to considering radial and angular effects it is interesting to examine the drift-region harmonic currents and phases (calculated from the arctangent of the ratio of terms under the radical in Eq. (61) over several stream wavelengths). A typical calculation is shown in Fig. 19, where the first five harmonics of the rf beam current and their corresponding phases are shown for a particular case of interest. These calculations were made using three stream layers and 128 charge groups

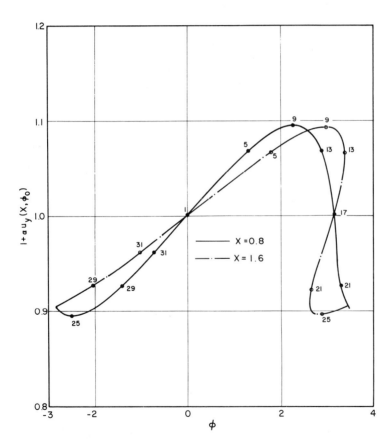

FIG. 21. Drift-space interaction ($\alpha = 0.2$, $x_a = 0.15$, $\omega_c/\omega = 0$, $\omega_p/\omega = 0.1$). (a) Axial velocity versus

per layer. In this particular case, the successive peaks of each harmonic current component continually decrease and there is an abrupt change of phase of π radians each time the harmonic current goes through a minimum.

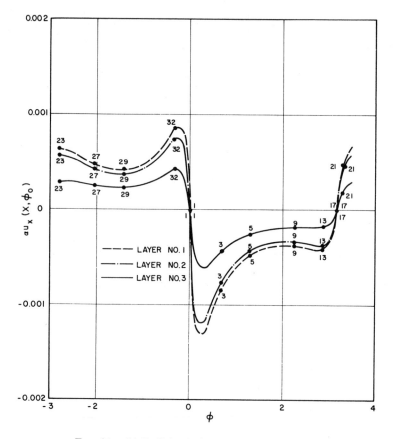

FIG. 21. (b) Radial velocity versus phase ($X = 0.8$).

For a case in which the magnetic field is zero, i.e., $\omega_c/\omega = 0$ but finite ω_p/ω, the stream gradually expands as illustrated in Fig. 20, where the profile of each layer is shown for typical values of ω_p/ω encountered in experimental klystrons. The drift-tube position is shown and for this range of parameters there is no interception in this two-cavity case. The various directional velocities are shown in Fig. 21 for zero magnetic field.

The important parameters to be evaluated in this multidimensional interaction process are the magnetic field, ω_c/ω; the radian plasma frequency, ω_p/ω; and the ratio of stream and tunnel diameters, $x_{b'}/x_a = \gamma b'/\gamma a$. The dependence of the fundamental rf current amplitude on each of these is shown in Figs. 22 and 23 when $\alpha = 0.2$. The rf harmonic-current amplitudes are calculated from Eq. (61), involving an integration over the stream cross section.

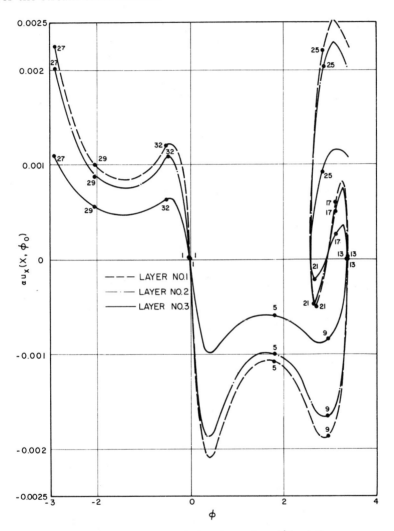

FIG. 21. (c) Radial velocity versus phase ($\dot{X} = 1.6$).

Since the stream has been divided into annular rings for computation purposes, it is useful to examine the individual layer currents and their arithmetic average. These are also shown in Figs. 22 and 23. It is interesting to note that even though the layer values of i_1/I_0 may differ significantly the integrated value and the simple arithmetic average are virtually the same, particularly for confined flow. The dependence of the fundamental rf current amplitude on the relative stream diameter,

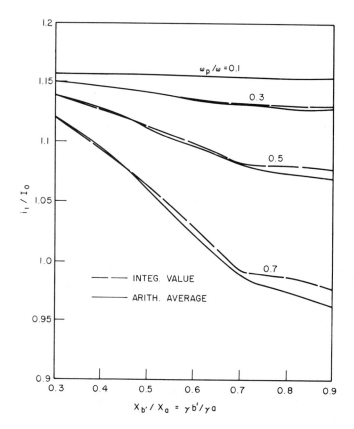

FIG. 22. Dependence of fundamental current amplitude on ω_p/ω and $x_{b'}$ at zero magnetic field ($\alpha = 0.2$, $x_a = 0.15$, $\omega_c/\omega = 0$). (a) Integrated and average values of i_1/I_0.

$\gamma b'/\gamma a$, is not marked, particularly at large magnetic fields. The significant beam expansion at large values of $\gamma b'$ accounts for the drop-off of i_1/I_0 when ω_p/ω is large. The effects of magnetic field and electron plasma frequency on the beam profile are shown in Fig. 24, indicating the considerable focusing action provided by the magnetic field.

6. RADIAL AND ANGULAR EFFECTS IN KLYSTRONS 163

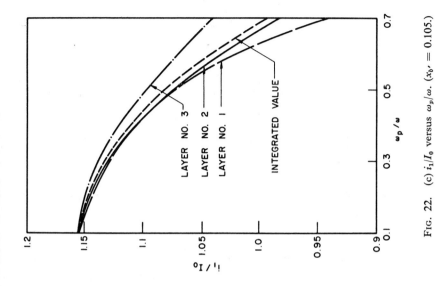

Fig. 22. (c) i_1/I_0 versus ω_p/ω. ($x_{b'} = 0.105$.)

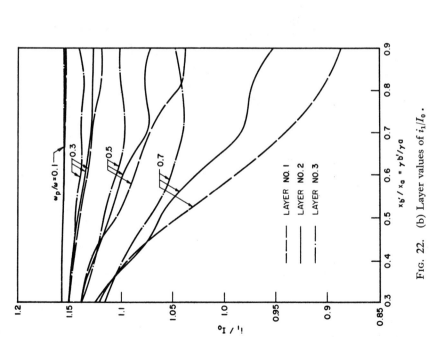

Fig. 22. (b) Layer values of i_1/I_0.

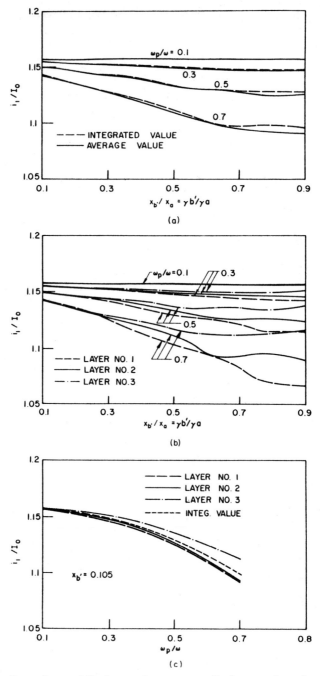

Fig. 23. Dependence of fundamental current amplitude on ω_p/ω and $x_{b'}$ at finite magnetic field ($\alpha = 0.2$, $x_a = 0.15$, $\omega_c/\omega = 0.4$). (a) Integrated and average values of $i_1 I_0$; (b) layer values of i_1/I_0; (c) i_1/I_0 versus ω_p/ω.

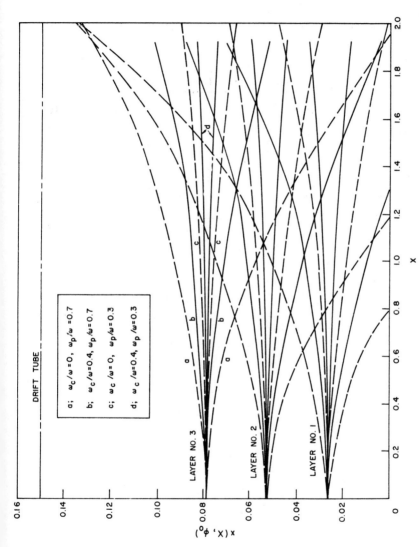

Fig. 24. Stream profile dependence on magnetic field and electron plasma frequency ($\alpha = 0.2$, $x_a = 0.15$, $x_{b'} = 0.105$).

For a finite magnetic field, $\omega_c/\omega > 0$ and space-charge forces included, i.e., $\omega_p/\omega \neq 0$, there is electron motion about the axis and u_φ is finite. The beam profile for a particular drift-space case is shown in Fig. 25. The electrons in the optimum rf phase (relative to the velocity-modulation voltage) are slowed the most and thus achieve the highest radial and angular velocities. The stream expansion is evident as rf bunching develops.

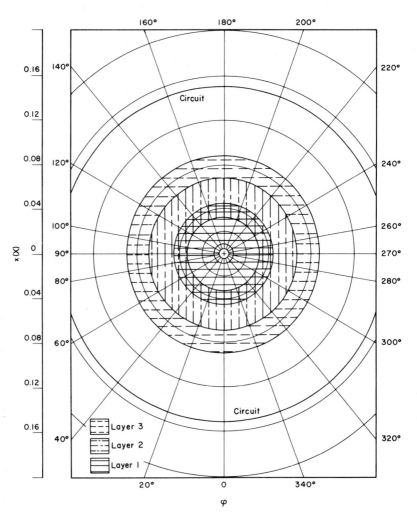

FIG. 25. Velocities and trajectories for a klystron interaction ($\alpha = 0.2$, $x_a = 0.15$, $x_{b'} = 0.105$, $\omega_c/\omega = 0.4$, $\omega_p/\omega = 0.5$, $X = 1.76$).

In the event that the entering stream has a significant velocity spread about the average velocity then the initial velocities must be assigned in accordance with the distribution function. Two particular distributions of interest are shown in Fig. 26. In some instances (e.g., the potential

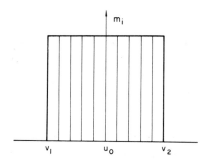

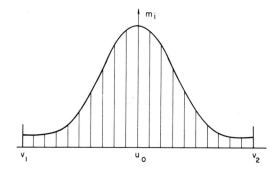

FIG. 26. Typical velocity distributions. m is the number of charge groups per velocity class.

minimum region of a diode) a half-Maxwellian distribution is appropriate. The division of a given distribution into a number of velocity classes is treated as follows. Recall that the total charge group velocity is given by

$$u_t(X, \Phi_{0j}) = u_0[1 + \alpha u(X, \Phi_{0j})] \tag{86}$$

and hence an initial distribution of velocities is incorporated by simply specifying $u(0, \Phi_{0j})$. The velocity perturbation $u(X, \Phi_{0j})$ is conveniently written as

$$u(X, \Phi_{0j}) = \frac{S}{\alpha}\left(\frac{n - \frac{1}{2}}{N} - \frac{1}{2}\right), \tag{87}$$

where $S = (v_2 - v_1)/u_0 = \Delta v/u_0$, and $N =$ the number of velocity classes with $n = 1, 2, ..., N$. Note that S is a measure of the percentage velocity spread about u_0. In view of the above definition Eq. (86) may be written as

$$u_t(X, \Phi_{0j}) = u_0 \left[1 + S\left(\frac{n - \frac{1}{2}}{N}\right) - \frac{1}{2}\right]. \tag{88}$$

When nonsquare velocity distributions are treated, the number of charge groups per velocity class must be additionally specified.

7 Relativistic Klystron Analysis

The push toward higher and higher power levels for klystrons eventually resulted in voltage levels where relativistic corrections to the equations of motion are required. A relativistic one-dimensional analysis of the klystron or drift-space interaction is made starting with the time-dependent Lagrangian for a charged particle in an electromagnetic field:

$$L(t) = [-mc^2 \sqrt{1 - (v/c)^2} - eV + e\mathbf{v} \cdot \mathbf{A}], \tag{89}$$

where v is the electron velocity relative to the laboratory system.

The generalized momentum is given by

$$p_i = \frac{\partial L(t)}{\partial v_i} \tag{90}$$

and the action integral

$$I = \int_{t_1}^{t_2} L(t)\, dt \tag{91}$$

is a scalar invariant. Applying Hamilton's variational principle to the action integral such that

$$\delta I = \delta \int_{t_1}^{t_2} L(t)\, dt = 0 \tag{92}$$

leads to the Euler-Lagrangian equations

$$\frac{d}{dt}\left(\frac{m_0 v^s}{\sqrt{1 - k_e^2}}\right) = eE^s + \frac{e}{c} v^r H^{sr}, \tag{93}$$

where H^{sr} is the electromagnetic field tensor (the term $(e/c)v^r H^{sr}$ is zero for a one-dimensional confined flow) and m_0 is the rest mass of the electron. In the following a one-dimensional system is assumed. Thus we have the momentum equation

$$f = \frac{dp}{dt} = m\frac{dv}{dt} + v\frac{dm}{dt} \tag{94}$$

and the time rate of change of mass is

$$\frac{dm}{dt} = \frac{m_0 v \dfrac{dv}{dt}}{c^2(1-k_e^2)^{3/2}} \tag{95}$$

where

$$k_e \triangleq u_0/c.$$

Combining Eqs. (94) and (95) yields

$$\frac{dv}{dt} = -|\eta|(1-k_e^2)^{3/2} E_{sc}, \tag{96}$$

where $\eta_0 = e/m_0$, the charge-to-mass ratio. Since only a one-dimensional analysis is being considered then only the z component of space-charge field is needed.

The space-charge field in the axial direction, E_{sc-z}, must now be calculated from the four-dimensional generalization of Poisson's equation, the inhomogeneous wave equation.

$$\Box^2 V_{sc} = -\rho/\epsilon. \tag{97}$$

The D'Alembertian operator is invariant to a proper transformation and the derivation of a suitable expression can proceed directly from the previous theory. The harmonic method is used because of its convenience. Assume that

$$\Box^2 V_{sc} = \frac{-\rho(z)}{\pi \epsilon b'^2} = -A\rho e^{j(\omega t - \beta z)}, \tag{98}$$

where b' represents the electron stream radius. Equation (98) applies inside the stream, whereas between the stream and surrounding drift tube the source term is zero; hence

$$\Box^2 V = 0. \tag{99}$$

The solution of Eq. (98) is determined as

$$V(r)e^{j(\omega t - \beta z)} = \left[\frac{A\rho}{\beta^2(1-k_e^2)} + BI_0(\beta r \sqrt{1-k_e^2})\right] e^{j(\omega t - \beta z)}. \tag{100}$$

Equation (100) may be verified directly by substituting directly into Eq. (98). The magnetic vector potential must be included as follows:

$$\mathbf{E}_{sc} = -\left(\nabla V_{sc} + \frac{\partial \mathbf{A}}{\partial t}\right).$$

The space-charge field may be obtained from Eq. (100) as

$$E_{sc-z} = j\beta(1 - k_e^2)\left[\frac{A\rho}{\beta^2(1 - k_e^2)} + BI_0(\beta r \sqrt{1 - k_e^2})\right] \cdot e^{j(\omega t - \beta z)}. \quad (101)$$

In the charge-free region between the stream and the drift tube E_{sc-z} has a Bessel function dependence. The continuity of radial and longitudinal components of electric field at the drift-tube wall defines an eigenvalue problem whose solution gives the space-charge-wave propagation constants (eigenvalues of the system). The resultant expression for E_{sc-z} is

$$E_{sc-z} = \frac{j\rho A}{\beta}\left\{1 - \beta b' \sqrt{1 - k_e^2} \frac{I_0(\beta r \sqrt{1 - k_e^2})}{I_0(\beta a \sqrt{1 - k_e^2})}\right.$$

$$\cdot [I_1(\beta b' \sqrt{1 - k_e^2}) K_0(\beta b' \sqrt{1 - k_e^2})$$

$$\left. + I_0(\beta a \sqrt{1 - k_e^2}) K_1(\beta b' \sqrt{1 - k_e^2})]\right\} e^{j(\omega t - \beta z)}. \quad (102)$$

The space-charge field is evaluated at $r = 0$ and the time dependence is suppressed:

$$E_{sc-z} = \frac{jA\rho}{\beta} R^2 e^{-j\beta z}, \quad (103)$$

where

$$R^2 \triangleq 1 - \frac{B_r}{I_{0a}}(I_{1b'}K_{0a} + I_{0a}K_{1b'}),$$

$$I_{1b'} \triangleq I_1(\beta b' \sqrt{1 - k_e^2}),$$

and

$$B_r \triangleq \beta b' \sqrt{1 - k_e^2} = B \sqrt{1 - k_e^2}.$$

The I and K functions are the modified Bessel functions, first and second kind.

The derivation of the final form for E_{sc} proceeds directly along the lines used in Chapter IV, Section 11. The electron plasma frequency may be defined in terms of either the laboratory system particle mass or the relativistic mass, i.e.,

$$\omega_p^2 \triangleq \frac{|I_0| |\eta_0|}{\pi \epsilon b'^2 u_0}.$$

7. RELATIVISTIC KLYSTRON ANALYSIS

The resultant relativistic expression for E_{sc-X} in the klystron case is

$$E_{sc-X} = -\frac{2\omega u_0}{|\eta_0|}\left(\frac{\omega_p}{\omega}\right)^2 \int_0^{2\pi} \frac{F_{1-X}(\Phi - \Phi')\,d\Phi_0'}{1 + \alpha u(X, \Phi_0')}. \quad (104)$$

The equation stating the relationship between the dependent variables remains the same as in the nonrelativistic case. The force equation upon introduction of the space-charge field given by Eq. (104) and the normalized one-dimensional variables yields the following result.

$$\frac{\partial u(X, \Phi_0)}{\partial X}[1 + \alpha u(X, \Phi_0)] = 4(1 - k_e^2)^{\frac{3}{2}}\left(\frac{\omega_p}{\omega\alpha}\right)^2$$

$$\cdot \int_0^{2\pi} \frac{F_{1-X}(\Phi - \Phi')\,d\Phi_0'}{[1 + \alpha u(X, \Phi_0')]}. \quad (105)$$

Some representative calculations of the beam rf current harmonic structure are illustrated in Fig. 27, where the harmonic-to-dc-current ratio is plotted versus n for a specific value of $\omega_p/\omega\alpha$. As a basis of comparison the original space-chargeless theory of Webster may be used to calculate the harmonic content of the stream. Webster's procedure was to use a

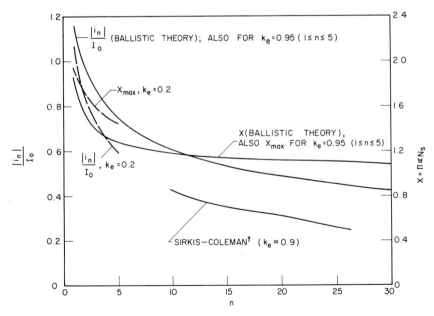

FIG. 27. Klystron harmonic-current amplitudes and optimum drift distances. $(\omega_p/\omega\alpha) = 0.63$. †M. D. Sirkis and P. D. Coleman, *J. Appl. Phys.* **28**, 944 (1957).

ballistic model and to neglect space-charge forces. If one assumes an ideal velocity-modulation gap then the harmonic-to-dc-current ratio in the stream is given by

$$\frac{|i_n|}{I_0} = 2J_n(nX)|_{\max}, \tag{106}$$

where $J_n(nX)|_{\max}$ is the maximum of the J_n Bessel function; this of course also gives information on the distance from the modulating gap to where the nth harmonic current reaches a maximum.

The maximum current ratio and value of X at which it occurs are also given in Fig. 27. It is seen that for low values of k_e the nonlinear relativistic current ratio is less than that for the ballistic calculations. However, independent of $\omega_p/\omega\alpha$, for high values of k_e the harmonic-to-dc-current ratios are just those given by the ballistic theory. The comparison of the optimum drift distances indicates the same thing. This is to be expected from a consideration of Eq. (105), since at high relativistic velocities the right-hand side approaches zero independent of $\omega_p/\omega\alpha$ and hence space-charge forces are not important. As has been shown by Webber[11], it is probably possible to find combinations of parameters which will give current ratios greater than those predicted by ballistic theory. However, even in these cases very high energy beams would behave ballistically. Even larger harmonic-to-dc-current ratios are achievable through the use of adequately prebunched electron beams. Again it is apparent that high-energy beams would have great utility in millimeter-wave generators and amplifiers.

In order to give some comparison with actual operating relativistic beams, some of the results of Sirkis and Coleman are also shown in Fig. 27, where the computed harmonic-to-dc-current amplitudes are given for a rebatron-harmodotron system where k_e is approximately 0.9. These experimental data do not represent the best attainable. The current amplitudes were calculated by Fourier analysis at a distance of 10 cm from the exit aperture of the accelerating cavity. These are not maximum obtainable current ratios and are not meant to correlate directly with the calculations presented here.

8 Voltage Stepping in Klystrons

The output power and efficiency of klystrons are basically limited by the extent of the rf velocity spread in the beam at the output cavity. Generally in very efficient high-power klystrons the amplitude of the rf voltage at the output cavity is such that the slowest electrons are just stopped or

8. VOLTAGE STEPPING IN KLYSTRONS

even possibly turned back. One possible means for increasing the efficiency of any klystron is to step up the dc voltage across or before the output cavity.[19,20] This permits an adjustment of the output cavity impedance to increase R/Q so as to increase the rf voltage amplitude before electrons are stopped and hence a greater efficiency is achievable. Such action may in fact decrease the bandwidth. The analysis of such voltage-stepped beams is outlined below.

Since the rf modulation on a beam is conserved across a dc velocity jump we note that

$$W = u_t^2 - 2\eta V_0 \tag{107}$$

is conserved. The following energy equation may be written in a region where the beam is unmodulated (Region 1).

$$u_{01}^2 - 2\eta V_{01} = 0 = \frac{1}{m}\sum_{i=1}^{m} u_{01i}^2 - 2\eta V_{01}, \tag{108}$$

where

$u_{01} \triangleq$ the initial dc velocity,

$m \triangleq$ the number of charge groups considered, and

$u_t \triangleq$ actual total electron velocity.

Following the nonlinear theory the total charge-group velocity is written as

$$u_t(X, \Phi_{0i}) \triangleq [1 + \alpha u(X, \Phi_{0i})], \tag{109}$$

where X and Φ_{0i} are as defined previously. At any displacement plane removed from an unmodulated region

$$\frac{1}{m}\sum_{i=1}^{m} u_{t1i}^2 - 2\eta V_{01} = -2\eta V_{rf1}, \tag{110}$$

where V_{rf} indicates the rf energy content of the beam at some particular plane. At some greater displacement plane, where the dc voltage has been changed,

$$\frac{1}{m}\sum_{i=1}^{m} u_{t2i}^2 - 2\eta V_{02} = 2\eta V_{rf2}. \tag{111}$$

Since W is invariant across such a gap then

$$V_{rf1} = V_{rf2} \tag{112}$$

and

$$2\eta(V_{01} - V_{02}) = \frac{1}{m}\sum_{i=1}^{m}(u_{t1i}^2 - u_{t2i}^2) \tag{113}$$

or

$$\sum_{i=1}^{m}(u_{t1i}^2 - u_{01}^2) = \sum_{i=1}^{m}(u_{t2i}^2 - u_{02}^2). \tag{114}$$

Assume that the deviation in the squared velocity over the sum may be replaced by that over each charge, so that

$$u_{t1i}^2 - u_{01}^2 = u_{t2i}^2 - u_{02}^2. \tag{115}$$

Equation (115) is rearranged to give

$$\left(\frac{u_{t2i}}{u_{02}}\right)^2 = 1 + \left(\frac{V_{01}}{V_{02}}\right)\left[\left(\frac{u_{t1i}}{u_{01}}\right)^2 - 1\right]. \tag{116}$$

In terms of normalized velocities Eq. (116) is

$$(1 + \alpha u_{2i})^2 - 1 = \frac{V_{01}}{V_{02}}\left[(1 + \alpha u_{1i})^2 - 1\right]. \tag{117}$$

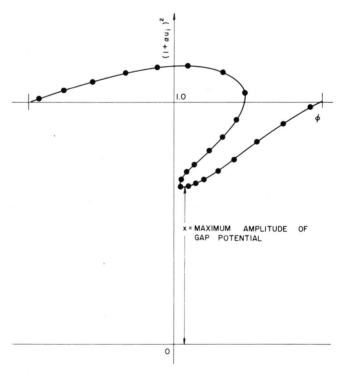

FIG. 28. Velocity-phase plot for a klystron with a velocity step.

8. VOLTAGE STEPPING IN KLYSTRONS

It is clear that the percentage velocity spread in the output beam is reduced by a dc velocity jump. Before calculating the effect on efficiency the effect of the voltage jump on the plasma frequency is examined. The current is an invariant across the gap so that

$$I_{01} = I_{02}$$

and

$$\rho_{01} u_{01} = \rho_{02} u_{02} .$$

The normalized radian plasma frequency is

$$\left(\frac{\omega_p}{\omega}\right)^2 = \frac{|\eta| I_0}{\pi \epsilon b'^2 u_0 \omega^2}$$

so that

$$\left(\frac{\omega_p}{\omega}\right)_2^2 = \frac{u_{01}}{u_{02}} \left(\frac{\omega_p}{\omega}\right)_1^2 = \left(\frac{V_{01}}{V_{02}}\right)^{\frac{1}{2}} \left(\frac{\omega_p}{\omega}\right)_1^2 .$$

The efficiency is calculated using a simple graphical procedure as illustrated in Fig. 28, since a dc voltage jump is simply a shift in the origin of the velocity-phase curve. The efficiency versus the degree of voltage change is shown in Fig. 29, where a considerable improvement in efficiency is noted for voltage ratios up to 2. If the voltage step is placed

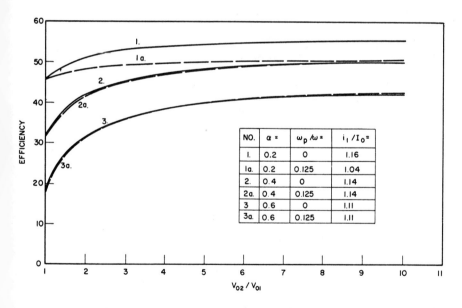

NO.	α =	ω_p/ω =	i_1/I_0 =
1.	0.2	0	1.16
1a.	0.2	0.125	1.04
2.	0.4	0	1.14
2a.	0.4	0.125	1.14
3	0.6	0	1.11
3a.	0.6	0.125	1.11

FIG. 29. Efficiency versus V_{02}/V_{01} for a klystron.

across the output gap then the effect both on the transit angle and on the rf coupling factor must be considered. Hefni[19] has carried out numerous experiments on this effect and realized a significant enhancement in the efficiency. The greatest improvement in experimental efficiency was realized when the bias was applied across the penultimate cavity.

REFERENCES

A. Small-Signal Theory

1. Webster, D. L., Cathode ray bunching. *J. Appl. Phys.* **10**, 501-508 (1939).
2. Warnecke, R., and Guenard, P., "Tubes Electroniques à Commande par Modulation de Vitesse." pp. 260-277. Gauthier-Villars, Paris, 1951.
3. Hahn, W. C., Small signal theory of velocity modulated electron beams. *Gen. Elec. Rev.* **42**, 258-270 (1939).
4. Ramo, S. I., The electronic-wave theory of velocity modulated tubes. *Proc. IRE* **27**, 757-763 (1939).
5. Feenberg, E., Elementary treatment of longitudinal debunching in a velocity modulation system. *J. Appl. Phys.* **17**, 852-855 (1946).
6. Feenberg, E., "Notes on Velocity Modulation." Sperry Gyroscope Co. Lab. Rept. No. 5221-1043 (September 1945).
7. Zitelli, L., "Space-Charge Effects in Gridless Klystrons." Microwave Lab. Stanford Univ. Rept. No. 149 (October 1951).
8. Mihran, T. G., The effect of space charge on bunching in a two-cavity klystron. *IRE Trans. Electron Devices* **6**, No. 1, 54-64 (1959).

B. Nonlinear Theory

9. Paschke, F., On the nonlinear behavior of electron-beam devices. *RCA Rev.* **18**, 221-242 (1957).
10. Doehler, O., and Kleen, W., Phenomenes non lineaires dans les tubes à propagation d'onde. *Ann. Radioelec.* **3**, 124-143 (1948).
11. Webber, S. E., Ballistic analysis of a two-cavity finite beam klystron. *IRE Trans. Electron Devices* **5**, No. 2, 98-109 (1958).
12. Webber, S. E., Large signal analysis of the multicavity klystron. *IRE Trans. Electron Devices* **5**, No. 4, 306-316 (1958).
13. Meeker, J. G., "Phase Focusing in Linear-Beam Devices." Electron Phys. Lab. Univ. of Michigan Tech. Rept. No. 49, Chapter 6 (August 1961).
14. Solymar, Exact solution of the one-dimensional bunching problem. *J. Electron. Control* **10**, No. 3, 165-181 (1961).
15. Solymar, L., Extension of the one-dimensional (klystron) solution to finite gaps. *J. Electron. Control* **10**, No. 5, 361-385 (1961).
16. Solymar, L., Large signal calculations of the admittance of an electron beam traversing a high frequency gap. *J. Electron. Control* **12**, No. 4, 313-319 (1962).
17. Turner, C. W., "Electron Bunching in Long Transit Angle Cavities." Stanford Univ. Microwave Lab. Sci. Rept. No. 32 (July 1961).
18. Wilson, R. N., "Large-Signal Space-Charge Theory of Klystron Bunching." Stanford Univ. Microwave Lab. Tech. Rept. No. 750 (September 1960).
19. Hefni, I., The variable-drift biased-gap klystron. *Proc. IRE* **52**, No. 1, 102 (1964).
20. Rowe, J. E., Efficiency improvement by voltage stepping in klystrons. *Proc. IRE* **52**, No. 3, 328-329 (1964).

CHAPTER

VI | Traveling-Wave Amplifier Analysis

1 Introduction

In the cavity-drift space-cavity interaction studied in the last chapter the modulating fields extended over only a short distance of travel of the electron beam. Thus the separation of the velocity modulation from the density modulation region is discrete and the operation is relatively easily understood. The bandwidth of such devices is determined primarily by the cavity bandwidth, which usually is extremely narrow. In an effort to obtain a broadband interaction the possibility is considered of an infinite chain of stagger-tuned cavities, in the limit an electromagnetic wave propagating structure. Two examples of such structures are illustrated in Fig. 1, where a is a periodically loaded resonant waveguide structure and b is a periodic nonresonant helical waveguide. Both structures support the propagation of a time-varying space-varying electromagnetic wave. If an electron beam, of the same type as used in the klystron, is arranged to interact with the field of the traveling wave then such a device is called a traveling-wave tube. The first experiments carried out by Kompfner were on a tube using a helical propagating structure. As in the klystron, some focusing system, such as a magnet system producing an axial field, is required to obtain transmission of the beam.

Since the invention of the traveling-wave amplifier in 1943 the science and technology of such devices has progressed markedly, with the result that today there exist tubes covering the frequency range from 50 Mc to 500 Gc at power levels from milliwatts to megawatts. The objective of this chapter is to study the nonlinear interaction process in detail so as to determine the gain and efficiency characteristics as a function of various operating parameters. Many factors such as structure impedance, beam perveance, space-charge forces, radial variations, and finite focusing fields are to be considered. In the following sections all these effects are treated mathematically for the basic traveling-wave amplifier. In Chapter XIII various means for improving the performance are enunciated. As in the klystron case we begin with the simplest case and progress to generalize this analysis in succeeding sections.

The first work on the nonlinear Lagrangian analysis (zero space charge) of the traveling-wave amplifier was contained in a memorandum by A. T. Nordsieck[1]; this was later published in its original form in 1953.

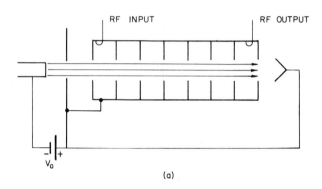

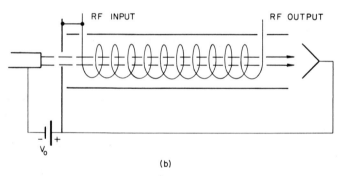

FIG. 1. (a) Resonant loaded, and (b) nonresonant helical, waveguide TWA structures.

Earlier studies by Doehler and Kleen[2] (1948), Brillouin[3] and Slater[4] approached the problem in a different manner but their work served to point out the complexity of the problem. Many limiting assumptions such as the neglect of space-charge forces and crossover effects restricted their validity and application.

Following the lead of Nordsieck later work by Wang[5] (1952), Poulter[6] (1954), Rowe[7,8] (1955), Tien et al.[9] (1955) and Tien[10] (1956) served to outline the problem and present solutions for the one-dimensional

2. MATHEMATICAL ANALYSIS OF THE ONE-DIMENSIONAL TWA

case. Vainshtein[11] in Russia has studied the one-dimensional TWA using the same basic method. Different but yet equivalent methods were used by the above authors. These differences and their equivalence are discussed in detail in Section 10. Since the numerous detail aspects and generalizations of the problem have been considered by several authors, the reader is referred to the bibliography at the end of this chapter for a complete listing. Because of its great utility and wide range of application the Lagrangian method is developed and discussed in detail in this book.

2 Mathematical Analysis of the One-Dimensional TWA

a. General

The basic equations of the analysis are the Lorentz force equation, the circuit voltage equation, and some form of the conservation of charge relationship. Again we call on the generalized results of Chapters III and IV for the necessary equations. A Lagrangian formulation as outlined in Chapter II and developed in detail in Chapter V is the basic approach utilized.

b. Circuit Equation

It is more convenient to utilize the transmission-line equivalent circuits of Chapter III than to solve Maxwell's equations for the fields of each individual propagating structure. The equivalence of the two methods has been shown and thus one can proceed without the need of further justification. Of course, if the predicted results agree well with experiments then this provides some additional justification.

The appropriate circuit equation is Eq. (100) of Chapter III for the one-dimensional lossy transmission line.

$$\frac{\partial^2 V(z,t)}{\partial z^2} - \frac{1}{v_0^2}\frac{\partial^2 V(z,t)}{\partial t^2} - \frac{2\omega Cd}{v_0^2}\frac{\partial V(z,t)}{\partial t}$$
$$= -\frac{Z_0}{v_0}\left[\frac{\partial^2 \rho(z,t)}{\partial t^2} + 2\omega Cd \frac{\partial \rho(z,t)}{\partial t}\right], \qquad (1)$$

where, for the reader's convenience, we repeat the definitions:

$v_0 \triangleq \dfrac{1}{\sqrt{ZY}}$, the characteristic phase velocity of the one-dimensional line;

$Z_0 = \sqrt{(Z/Y)}$, the characteristic impedance of the one-dimensional line;

$$C^3 \triangleq \frac{Z_0 I_0}{4V_0},\qquad \text{the beam-circuit coupling parameter;}$$

$$d \triangleq \frac{R}{2\omega LC},\qquad \text{the circuit loss parameter (axial); and}$$

$V(z,t),\ \rho(z,t) \triangleq$ the *rf* voltage on the structure and the linear charge density in the beam respectively.

The addition of this equation describing the variation of the rf voltage along the circuit when driven by the beam constitutes the generalization and increased complexity of the present system over that for the klystron of the last chapter.

c. *Lorentz Force Equation*

The general form of the Lorentz equation is given by

$$\frac{d\mathbf{v}}{dt} = -|\eta|[\mathbf{E} + \mathbf{v} \times \mathbf{B}]. \tag{2}$$

For the case at hand the above equation is specialized to one-dimensional axial (z) motion and it is assumed that $(v/c)^2 \ll 1$ so that nonrelativistic mechanics may be assumed and the beam self-magnetic field may be neglected. Incidentally, this field provides a self-focusing force and will be evaluated later. Under these assumptions, separating the electric field into circuit and space-charge-field components results in the following force equation form:

$$\frac{d^2 z}{dt^2} = |\eta|\left[\frac{\partial V_c(z,t)}{\partial z} + \frac{\partial V_{sc}(z,t)}{\partial z}\right]. \tag{3}$$

The solutions of the above force equation may be expressed either as

$$z = f(z_0, t), \tag{4a}$$

or its inverse

$$z_0 = G(z, t), \tag{4b}$$

where z_0 denotes the charge group position at $t = 0$. Note that z_0 is a multivalued function of z.

d. *Conservation of Charge*

The entering beam charge may be bunched or distributed uniformly in time. Since charge must be conserved we note that a particular amount of charge $\rho(0,0)$ entering the device over a short distance dz_0 must appear at some new displacement plane at a later time. Mathematically

$$\rho(z,t)\,dz = \rho(0,0)\,dz_0, \tag{5}$$

2. MATHEMATICAL ANALYSIS OF THE ONE-DIMENSIONAL TWA

the entering charge is related to the dc beam current by $\rho(0, 0) = I_0/u_0$, and thus Eq. (5) becomes

$$\rho(z, t) = \frac{I_0}{u_0} \left| \frac{dz_0}{dz} \right|_t . \tag{6}$$

The absolute value sign of Eq. (6) results from the fact that electrons cross over and thus z_0 is no longer a single-valued function of z; the sum of all branches of this multivalued function must be accounted for. Here it is seen why z_0 is used as the independent variable rather than z since z is a single-valued function of z_0. Suppose that a bunched beam is injected; then the conservation equation is written as

$$\rho(z, t)\, dz = \rho(0, t)\, dz(0, t). \tag{7}$$

Separating the time dependence out and noting that the charge spacing is inversely proportional to the same time function, we find

$$\rho(0, t) = \rho(0, 0) f(t) \tag{8}$$

and

$$dz(0, t) = \frac{dz(0, 0)}{f(t)}, \tag{9}$$

with the result that

$$\rho(z, t)\, dz = \rho(0, 0)\, dz_0 , \tag{10}$$

as previously. In such a Lagrangian formulation the charge density $\rho(z, t)$ is interpreted as the particle charge times the number density of electrons.

e. Introduction of Normalized Variables and Functions

The independent variables of the system are taken as the displacement z and the initial charge position z_0. As in the klystron analysis it is convenient to introduce a distance normalization. The two characteristic velocities or wavelengths of the system are those associated with the electron stream and the unperturbed wave on the structure. There is no particular advantage in one over the other since they differ by the factor $1 + Cb$; hence the following normalization is chosen:

$$y \triangleq \frac{C\omega z}{u_0} = 2\pi C N_s , \tag{11}$$

where it is noticed that $y = X$ of the previous chapter if $\alpha \triangleq 2C$.

In lieu of specifying $z_{0,j}$ we may denote the time $t_{0,j}$ at which a particular charge group enters the interaction region. This is conveniently transformed to an entering phase variable

$$\Phi_{0,j} = \omega t_{0,j}, \tag{12}$$

which denotes the entering phase position of a charge group relative to one cycle of the rf wave at the input, i.e., $y = 0$. We now proceed to the definition of appropriate dependent variables relating to the electron velocity, phase position and the voltage along the rf structure.

Since the rf electron velocity represents a departure from its initial average velocity u_0 it is convenient to define variables in a reference system traveling at u_0. The particle velocity in terms of y and $\Phi_{0,j}$ is written as

$$u_t(y, \Phi_0) = \frac{u_0}{C\omega} \frac{dy}{dt} \triangleq u_0[1 + 2Cu(y, \Phi_{0,j})], \tag{13}$$

where $2Cu_0 u(y, \Phi_{0,j})$ denotes the rf velocity of a particular charge group at a given displacement plane and will have an average value (negative) when particle kinetic energy is given to the rf wave.

Referred to a hypothetical rf wave traveling at the velocity u_0 the actual rf wave on the structure will experience a phase shift (lag) due to the beam loading as energy is given to the wave. We define $\theta(y)$ as this phase lag and then the charge group phases at any value of y, $\Phi(y, \Phi_{0,j})$, denote phase positions relative to the wave at that y value as illustrated in Fig. 6 of Chapter II for both sets of independent variables. Thus we see that any displacement plane the following relation evolves:

$$\Phi_{0,j} + \frac{y_i}{C} - \theta(y_i) = \Phi_{0,j} + \omega t + \Phi(y_i, \Phi_{0,j})$$

or

$$\theta(y) \triangleq \frac{y}{C} - \omega t - \Phi(y, \Phi_0). \tag{14}$$

Also, from Fig. 6 of Chapter II, the circuit phase velocity and actual wave phase velocity are

$$v_0(y) \triangleq \frac{u_0}{1 + Cb(y)} \tag{15a}$$

and

$$v(y) \triangleq \frac{u_0}{1 - C\dfrac{d\theta(y)}{dy}}. \tag{15b}$$

2. MATHEMATICAL ANALYSIS OF THE ONE-DIMENSIONAL TWA

In general all of the above quantities may be functions of distance and are treated as such in a later chapter.

The final dependent variable to be defined relates to the circuit voltage. From the chapter on circuit fields and equivalent circuits we note that a wave-type solution exists and that the voltage can be defined as the product of two slowly varying functions, one of distance and the other of time. Physically this is justified by assuming that all of the structure impedance occurs at the fundamental frequency so that even though $\rho(z, t)$ is rich in harmonics only ρ_1 of the beam produces a voltage on the circuit. The voltage is thus defined as

$$V(z, t) = \text{Re}\left[V(z) \exp j\left(\omega t - \int_0^z \beta(z)\, dz\right)\right]$$
$$= \text{Re}[V(z) \exp -j\Phi], \tag{16}$$

where

$$\Phi \triangleq -\omega t + \int_0^z \beta(z)\, dz.$$

$\beta(z)$ denotes the possibility of the circuit phase constant changing with distance and it is noted that the variable Φ again appears although not in a unique sense. Equation (16) is now written in terms of the new variables as

$$V(y, \Phi) \triangleq \text{Re}\left[\frac{Z_0 I_0}{C} A(y) e^{-j\Phi}\right], \tag{17}$$

where $A(y) \triangleq$ the normalized voltage amplitude along the circuit.

Upon introduction of the normalized dependent and independent variables defined in this section, the circuit, force and continuity equations yield the following:

Circuit Equation

$$-C\left(\frac{\omega}{1+Cb}\right)^2 Z_0 I_0$$

$$\cdot\left[\frac{d^2 A(y)}{dy^2} - A(y)\left\{\left(\frac{1}{C} - \frac{d\theta(y)}{dy}\right)^2 - \left(\frac{1+Cb}{C}\right)^2\right\}\right] \cos \Phi(y, \Phi_0)$$

$$+ \left[\left(\frac{1}{C} - \frac{d\theta(y)}{dy}\right)\left(-2\frac{dA(y)}{dy}\right) + A(y)\frac{d^2\theta(y)}{dy^2} - \frac{2d}{C}(1+Cb)^2 A(y)\right] \sin \Phi(y, \Phi_0)$$

$$= v_0 Z_0 \left[\frac{\partial^2 \rho_1}{\partial t^2} + 2\omega Cd\, \frac{\partial \rho_1}{\partial t}\right], \tag{18}$$

where

$b \triangleq$ a velocity parameter such that $u_0/v_0 = 1 + Cb$,

$C^3 \triangleq (Z_0 I_0 / 4 V_0)$, and

$\rho_1 \triangleq$ fundamental component of linear charge density driving the circuit, consistent with the previous assumption.

Force Equation

$$\frac{d^2 y}{dt^2} = |\eta| \left\{ \frac{Z_0 I_0 \omega}{u_0} \left[\frac{dA(y)}{dy} \cos \Phi(y, \Phi_0) \right. \right.$$
$$\left. \left. - A(y) \sin \Phi(y, \Phi_0) \left(\frac{1}{C} - \frac{d\theta(y)}{dy} \right) \right] - E_{sc-z}(y, \Phi) \right\}. \quad (19)$$

Conservation of Charge Equation. In order to change variables in this equation it is convenient to rewrite Eq. (6) as

$$\rho(z, t) = \frac{I_0}{u_0} \left| \frac{\partial z_0/\partial t}{\partial z/\partial t} \right|_t \quad (20)$$

and note that in the framework of the new variables

$$\frac{\partial z}{\partial t} = u_t(y, \Phi_0) = u_0 [1 + 2Cu(y, \Phi_0)], \quad (21a)$$

$$\frac{\partial z_0}{\partial t} = \frac{\partial z_0}{\partial \Phi_0} \frac{\partial \Phi_0}{\partial \Phi} \frac{\partial \Phi}{\partial t} = -u_0 \frac{\partial \Phi_0}{\partial \Phi} \quad (21b)$$

and hence Eq. (20) becomes

$$\rho(y, \Phi) = \frac{I_0}{u_0} \left| \frac{\partial \Phi_0}{\partial \Phi} \right| \frac{1}{1 + 2Cu(y, \Phi_0)}, \quad (22)$$

where the absolute value sign of Eq. (22) is taken in the same sense as in Eq. (6).

f. Final Formulation of the Generalized Amplifier Equations

The similarity of Eqs. (18), (19), and (22) written in Lagrangian variables to their predecessors in (z, t) coordinates is apparent and need not be discussed further. It remains now to eliminate the charge density and acceleration from these equations and introduce a space-charge-field expression to obtain their final form.

Circuit Equation. It is well known that the beam charge density $\rho(y, \Phi)$ is rich in harmonics (see Chapter *V*) and thus it is convenient to

2. MATHEMATICAL ANALYSIS OF THE ONE-DIMENSIONAL TWA

expand this into a Fourier series in the phase variable Φ. We have assumed that only ρ_1 excites the circuit since Z_0 is all at the fundamental frequency. Harmonic voltages may be accounted for by retaining ρ_n; however, one must then know Z_{0n} and what the phase relationship is between the various circuit voltage components.

The Fourier expansion is written as

$$\rho(z, t) = \sum_{n=1}^{\infty} [A_n \sin(-n\Phi) + B_n \cos(-n\Phi)], \tag{23}$$

where

$$(-n\Phi) = n\omega t - \int_0^z n\beta(z)\, dz.$$

The coefficients A_n and B_n of Eq. (23) are

$$A_n = \frac{1}{\pi} \int_0^{2\pi} \rho_n(z, \Phi) \sin(-n\Phi)\, d\Phi \tag{24a}$$

and

$$B_n = \frac{1}{\pi} \int_0^{2\pi} \rho_n(z, \Phi) \cos(-n\Phi)\, d\Phi. \tag{24b}$$

The Lagrangian variable continuity equation, Eq. (22), is used to write Eq. (23) in final form.

$$\rho(y, \Phi) = \operatorname{Re}\left\{ \frac{I_0}{u_0 \pi} \sum_{n=1}^{\infty} e^{-jn\Phi} \left[\int_0^{2\pi} \frac{\cos n\Phi(y, \Phi_0')\, d\Phi_0'}{1 + 2Cu(y, \Phi_0')} \right. \right.$$
$$\left.\left. + j \int_0^{2\pi} \frac{\sin n\Phi(y, \Phi_0')\, d\Phi_0'}{1 + 2Cu(y, \Phi_0')} \right] \right\}, \tag{25}$$

where the prime simply denotes the variable of integration.

In the case at hand we are interested in ρ_1 and hence $n = 1$ in Eq. (25); the derivatives of Eq. (18) are then expressed as

$$\rho_1 = \rho_{1c} \cos \Phi + \rho_{1s} \sin \Phi, \tag{26a}$$

$$\frac{\partial \rho_1}{\partial t} = \frac{\partial \rho_1}{\partial \Phi} \frac{\partial \Phi}{\partial t} \tag{26b}$$

and

$$\frac{\partial^2 \rho_1}{\partial t^2} = \frac{\partial \rho_1}{\partial \Phi} \frac{\partial^2 \Phi}{\partial t^2} + \left(\frac{\partial \Phi}{\partial t}\right)^2 \frac{\partial^2 \rho_1}{\partial \Phi^2}. \tag{26c}$$

The right-hand side of the circuit equation is now handled by operating on Eq. (25) according to Eqs. (26) and then substituting into Eq. (18).

Since the coefficients of sin Φ and cos Φ on each side of the equal sign are independent of Φ and the sine and cosine are orthogonal, we equate coefficients and evolve the following two circuit equations:

$$\frac{d^2 A(y)}{dy^2} - A(y)\left[\left(\frac{1}{C} - \frac{d\theta(y)}{dy}\right)^2 - \left(\frac{1+Cb}{C}\right)^2\right]$$
$$= -\left(\frac{1+Cb}{\pi C}\right)\left[\int_0^{2\pi} \frac{\cos \Phi(y, \Phi_0') \, d\Phi_0'}{1 + 2Cu(y, \Phi_0')} + 2Cd \int_0^{2\pi} \frac{\sin \Phi(y, \Phi_0') \, d\Phi_0'}{1 + 2Cu(y, \Phi_0')}\right] \quad (27)$$

and

$$A(y)\left[\frac{d^2\theta(y)}{dy^2} - \frac{2d}{C}(1+Cb)^2\right] + 2\frac{dA(y)}{dy}\left(\frac{d\theta(y)}{dy} - \frac{1}{C}\right)$$
$$= -\left(\frac{1+Cb}{\pi C}\right)\left[\int_0^{2\pi} \frac{\sin \Phi(y, \Phi_0') \, d\Phi_0'}{1 + 2Cu(y, \Phi_0')} - 2Cd \int_0^{2\pi} \frac{\cos \Phi(y, \Phi_0') \, d\Phi_0'}{1 + 2Cu(y, \Phi_0')}\right]. \quad (28)$$

The left-hand sides of Eqs. (27) and (28) represent the homogeneous portions of the equation and the respective right-hand sides the inhomogeneous parts due to the fact that the circuit is being driven by the beam.

Force Equation. The force equation Eq. (19) contains acceleration, which must be eliminated for ease of solution. Expansion of dv/dt and substitution of the velocity variable proceed according to

$$\frac{dv}{dt} = 2Cu_0 \frac{\partial u}{\partial y} \frac{dy}{dt} = 2C^2 u_0 \omega [1 + 2Cu(y, \Phi_0)] \frac{\partial u(y, \Phi_0)}{\partial y}. \quad (29)$$

The space-charge-field expression is obtained from Chapter IV using either the harmonic- or disk-model results.

The one-dimensional space-charge-field expression obtained in Chapter IV by the harmonic method as applied to traveling-wave tubes is (no initial velocity modulation)

$$E_{sc-z}(y, \Phi) = \frac{-2\omega u_0}{|\eta|(1+Cb)}\left(\frac{\omega_p}{\omega}\right)^2 \int_0^{2\pi} \frac{F_{1-z}(\Phi - \Phi') \, d\Phi_0'}{1 + 2Cu(y, \Phi_0')}. \quad (30)$$

Substituting the above into the right-hand side of Eq. (19) and using Eq. (29) gives the following result:

$$[1 + 2Cu(y, \Phi_0)] \frac{\partial u(y, \Phi_0)}{\partial y} = -A(y)\left[1 - C\frac{d\theta(y)}{dy}\right] \sin \Phi(y, \Phi_0)$$
$$+ C\frac{dA(y)}{dy} \cos \Phi(y, \Phi_0) + \frac{1}{(1+Cb)}\left(\frac{\omega_p}{\omega C}\right)^2 \int_0^{2\pi} \frac{F_{1-z}(\Phi - \Phi_0') \, d\Phi_0'}{1 + 2Cu(y, \Phi_0')}. \quad (31)$$

Equation (31) relates the rate of change of charge group velocity to the circuit field, as represented by the first two terms on the right-hand side,

2. MATHEMATICAL ANALYSIS OF THE ONE-DIMENSIONAL TWA

and to the space-charge field, given by the last term on the right. If the alternate (and equivalent) disk-electron expression for the space-charge field is used the following form for the force equation is obtained:

$$[1 + 2Cu(y, \Phi_0)] \frac{\partial u(y, \Phi_0)}{\partial y}$$

$$= - A(y) \left[1 - C \frac{d\theta(y)}{dy}\right] \sin \Phi(y, \Phi_0)$$

$$+ C \frac{dA(y)}{dy} \cos \Phi(y, \Phi_0) + \frac{1}{(1 + Cb)} \left(\frac{\omega_p}{\omega C}\right)^2$$

$$\cdot \int_{-\infty}^{\infty} \exp\left[-\frac{2}{B} | \Phi(y, \Phi_0 + \theta) - \Phi(y, \Phi_0)| [1 + 2Cu(y, \Phi_0)]\right] d\Phi$$

$$\cdot \text{sgn}[\Phi(y, \Phi_0 + \theta) - \Phi(y, \Phi_0)]. \qquad (32)$$

Even though their appearance belies it at first glance, Eqs. (31) and (32) are equivalent. In the next section it is shown that the results obtained using the two expressions are virtually identical.

Relation Between Variables. The velocity-phase relationship may be developed from the velocity and phase variable definitions as in the case of the klystron except that $\theta(y)$ is now included in the definition of $\Phi(y, \Phi_0)$. The resulting expression is (including an initial velocity $[1 + 2Cu(0, \Phi_0)]$)

$$\frac{\partial \Phi(y, \Phi_0)}{\partial y} + \frac{d\theta(y)}{dy} = \frac{1}{C}\left[\frac{1}{1 + 2Cu(0, \Phi_0)} - \frac{1}{1 + 2Cu(y, \Phi_0)}\right]. \qquad (33)$$

Before proceeding further it is interesting to compare the traveling-wave amplifier working equations, Eqs. (27), (28), (31) or (32) and (33), with the klystron equations of Chapter V. Since in the klystron there is no traveling circuit field Eqs. (27) and (28) are eliminated along with the first two terms on the right-hand side of Eq. (31) or (32) and the $d\theta(y)/dy$ of Eq. (33). Then if we replace y by $X = \pi\alpha N_s$ and $2C$ by α we obtain exactly the klystron or drift-space equations.

g. Input Boundary Conditions

In proceeding to solve the above system of equations for the TWA one might treat the problem as a boundary-value problem and wish to specify conditions at $y = 0$ and $y = y_L$. This leads to some difficulty since the rf conditions are not known (they are being determined) at the output and hence an iterative solution would be required. The complexity of treatment as a boundary-value problem and the fact that the saturation length is wanted from the calculations lead us to a consideration of solving the problem as an initial-value problem. Recall that the rf voltage along the structure was specified as the product of two slowly varying functions, one of distance and one of phase. Now assume that the structure is everywhere terminated in its characteristic impedance and that no reflections occur from the output so that we are dealing only with forward-traveling waves. We may then solve the system as an initial-value problem, specifying values of the dependent variables and some of their derivatives at $y = 0$. The solution is then obtained by integrating along particle trajectories through the circuit region until saturation is reached. Then not only the saturated output, but the optimum device length is a result. This is equivalent to assuming that the beam-structure configuration extends to $y = \infty$ and is matched over the entire region.

In the initial-value framework we specify the following initial conditions.

A. Rf Signal.

(1) $A(0) \triangleq A_0$, the input signal level relative to CI_0V_0. (ψ_0 indicates the db level of A_0 relative to CI_0V_0.)

(2) $dA(y)/dy|_{y=0}$, the rate of change of the rf signal level at the input. For a lossless circuit and an initially unbunched beam $dA(y)/dy|_{y=0} \equiv 0$ since the beam cannot affect the signal until after beam modulation occurs. In general, however, we may find the above condition in the following way.

Consider Eq. (28) and assume that $d^2\theta(0)/dy^2 = 0$ (this is verified below). Then

$$-2 \frac{d}{C} A(0)(1 + Cb)^2 + 2 \frac{dA(y)}{dy} \left(\frac{d\theta(y)}{dy} - \frac{1}{C} \right) = 0.$$

It is shown below in (4) that $d\theta(0)/dy = -b$ and thus

$$\frac{dA(0)}{dy} = -dA_0(1 + Cb).$$

In the case of an entering bunched beam, assuming zero circuit loss and that

$$A(y) \frac{d^2\theta(y)}{dy^2} \ll 2 \frac{dA(y)}{dy} \left(\frac{d\theta(y)}{dy} - \frac{1}{C} \right),$$

we have

$$2 \frac{dA(y)}{dy} \left(\frac{d\theta(y)}{dy} - \frac{1}{C} \right) = -\frac{(1 + Cb)}{\pi C} \int_0^{2\pi} \sin \Phi \, d\Phi_0' = -\frac{2(1 + Cb)}{C} \sin \alpha,$$

where α is the bunch injection phase angle. Again, if $d\theta(0)/dy = -b$

$$\frac{dA(0)}{dy} = \sin \alpha.$$

(3) $\theta(0) \equiv 0$ for all conditions since the rf signal is applied and the beam enters at $y = 0$.
(4) $d\theta(y)/dy \mid_{y=0}$. From Eq. (27) it is readily shown that

$$\frac{d\theta(0)}{dy} = -b$$

since the integral over the beam at the input is zero.

B. *Beam Input Conditions.*
(1) The electron velocity, i.e., $u_0[1 + 2Cu(0, \Phi_{0,j})]$, must be specified at the input. In the case of an entering unmodulated beam $1 + 2Cu(0, \Phi_{0,j}) \equiv 1$ for all j. Arbitrary modulations may be applied and accounted for by specifying the above function.
(2) Beam bunching is specified in terms of $\Phi(0, \Phi_{0,j}) \equiv \Phi_{0,j}$. In the case of an unbunched beam input the charge groups are injected uniformly distributed in phase over one cycle of the rf wave at $y = 0$; i.e.,

$$\Phi_{0,j} = \frac{2\pi j}{m} \qquad j = 0, 1, 2, ..., m.$$

C. *Parameter Specification.* The following normalized parameters appearing in the final working equations must be specified to obtain a solution.
(1) C, the gain or beam-circuit coupling parameter.
(2) d, the circuit-loss parameter which may be a function of z for devices with attenuators for stability.
(3) $b = (u_0 - v_0)/Cv_0$, the injection velocity parameter.
(4) $B = \gamma b'$, the space-charge force range parameter (stream diameter).
(5) ω_p/ω, the normalized plasma frequency.

Notice that in the nonlinear theory two space-charge parameters, one a range and the other an amplitude factor, must be specified whereas in the small-signal theory only one, QC, was required. Note that

$$QC = \frac{1}{4C^2}\left(\frac{R\omega_p/\omega}{1 + R\omega_p/\omega}\right)^2$$

and actually both the range and amplitude factors are incorporated in QC since $R\omega_p \triangleq \omega_q$, the effective plasma frequency which depends on the geometrical factor R. It is difficult to calculate ω_q for a given beam-circuit configuration and hence it is believed that the procedure used in the large-signal formulation is superior.

h. Gain, Efficiency and Current Calculation

The rf output is obtained in terms of $A(y)$, the normalized rf voltage amplitude along the structure. Since the initial value was A_0, a gain factor is simply computed from

$$\text{Gain}(y) = 20 \log \frac{A(y)}{A_0}. \tag{34}$$

The total rf power along the structure and hence the efficiency are obtainable from a calculation of the Poynting vector,

$$P = \tfrac{1}{2} \text{Re}[V^*I]. \tag{35}$$

The conjugate voltage along the structure is easily obtained from Eq. (17) and the current along the transmission-line equivalent circuit is calculated from one of the first-order circuit equations:

$$\frac{C\omega}{u_0} \frac{\partial V(y, \Phi)}{\partial y} + \frac{Z_0}{v_0} \frac{\partial I}{\partial \Phi} \frac{\partial \Phi}{\partial t} = 0. \tag{36}$$

Utilizing the definitions of Φ and V and performing the indicated operations gives the current expressed as

$$I(y, \Phi) = \frac{I_0}{(1+Cb)} \left[\left(\frac{1}{C} - \frac{d\theta(y)}{dy} \right) A(y) e^{-j\Phi} + j \frac{dA(y)}{dy} e^{-j\Phi} \right]. \tag{37}$$

Substitution into Eq. (35) gives the power at a particular y-plane as

$$P(y) = 2CI_0V_0A^2(y) \frac{[1 - C\, d\theta(y)/dy]}{(1+Cb)} \tag{38}$$

and since the beam power is I_0V_0, the efficiency is

$$\eta(y) = 2CA^2(y) \frac{[1 - C\, d\theta(y)/dy]}{(1+Cb)}. \tag{39}$$

The ratio of the last two factors of Eq. (39) is actually the ratio of the unperturbed circuit wave phase velocity to the actual wave phase

2. MATHEMATICAL ANALYSIS OF THE ONE-DIMENSIONAL TWA

velocity and is very nearly unity for all cases (θ is a negative number). On this basis the efficiency is given approximately by

$$\eta(y) \approx 2CA^2(y). \tag{40}$$

In the above it has been assumed that A_0 is small compared to the saturated output level.

The calculation of the fundamental and harmonic current amplitudes in the beam proceeds in a directly parallel manner to that for the klystron when $2C = \alpha$ and hence only the result is given here:

$$\left|\frac{i_n}{I_0}\right| = \frac{\left\{\left(\frac{1}{\pi}\int_0^{2\pi} \rho(0, \Phi_0')[1 + 2Cu(0, \Phi_0')]\cos n\Phi(y, \Phi_0')\, d\Phi_0'\right)^2 + \left(\frac{1}{\pi}\int_0^{2\pi} \rho(0, \Phi_0')[1 + 2Cu(y, \Phi_0')]\sin n\Phi(y, \Phi_0')\, d\Phi_0'\right)^2\right\}^{\frac{1}{2}}}{\frac{1}{2\pi}\int_0^{2\pi} \rho(0, \Phi_0')[1 + 2Cu(y, \Phi_0')]\, d\Phi_0'}. \tag{41}$$

i. Small-C Equations

The working equations derived in the preceding sections are general and valid for arbitrary values of the parameters such as C, ω_p/ω, d and B. It is worthwhile to investigate their simplification when $C \ll 1$ and/or $\omega_p/\omega = 0$.

If we assume $C \ll 1$ and $d = 0$ and neglect high-order terms the circuit equations become

and

$$\frac{d\theta(y)}{dy} + b = -\frac{1}{2\pi A(y)}\int_0^{2\pi} \cos \Phi(y, \Phi_0')\, d\Phi_0' \tag{42}$$

$$\frac{dA(y)}{dy} = \frac{1}{2\pi}\int_0^{2\pi} \sin \Phi(y, \Phi_0')\, d\Phi_0', \tag{43}$$

the equation relating variables (no initial velocity modulation)

$$\frac{\partial \Phi(y, \Phi_0)}{\partial y} + \frac{d\theta(y)}{dy} = 2u(y, \Phi_0) \tag{44}$$

and the force equation

$$\frac{\partial u(y, \Phi_0)}{\partial y} = -A(y)\sin \Phi(y, \Phi_0) + \text{(space-charge integral)}. \tag{45}$$

The above equations are Nordsieck's[1] original set if the space-charge integral is neglected. These are quite useful and their solution by analytic methods will be explored further in Chapter XIII.

j. Alternate System of Equations

It was mentioned earlier that alternate definitions of the Lagrangian variables y and Φ could have been used if it had been decided to ride with a coordinate system moving with v_0 rather than u_0. Suppose that the following alternate system of variables is defined:

$$y \triangleq \frac{C\omega z}{v_0} = 2\pi C N_g, \tag{46}$$

where $\lambda_s/\lambda_g = 1 + Cb$.

$$\Phi(z, t) \triangleq \omega \left(\frac{z}{v_0} - t\right), \tag{47}$$

$$v_i(y, \Phi_0) \triangleq v_0[1 + Cv(y, \Phi_0)] \tag{48}$$

and the normalized rf circuit voltage is separated into vector components

$$V(y, \Phi_0) \triangleq \frac{Z_0 I_0}{4C}[a_1(y) \cos \Phi(y, \Phi_0) - a_2(y) \sin \Phi(y, \Phi_0)], \tag{49}$$

where

$$4A(y) = [a_1^2(y) + a_2^2(y)]^{\frac{1}{2}},$$

and

$$\tan[-\theta(y) - by] = \frac{a_2(y)}{a_1(y)}.$$

The above separation of $A(y)$ adds nothing new to the analysis and it should be noted that Φ is defined differently from previously.

Pursuing the equations to their final form yields the following results:

Circuit Equations

$$\frac{C}{2}\frac{d^2 a_1(y)}{dy^2} - \frac{da_2(y)}{dy} - a_2(y)d = -\frac{2}{\pi}\Bigg[\int_0^{2\pi} \frac{\cos \Phi(y, \Phi_0') \, d\Phi_0'}{1 + Cv(y, \Phi_0')}$$
$$+ 2Cd \int_0^{2\pi} \frac{\sin \Phi(y, \Phi_0') \, d\Phi_0'}{1 + Cv(y, \Phi_0')}\Bigg] \tag{50}$$

and

$$\frac{C}{2}\frac{d^2 a_2(y)}{dy^2} + \frac{da_1(y)}{dy} + a_1(y)d = \frac{2}{\pi}\Bigg[\int_0^{2\pi} \frac{\sin \Phi(y, \Phi_0') \, d\Phi_0'}{1 + Cv(y, \Phi_0')}$$
$$+ 2Cd \int_0^{2\pi} \frac{\cos \Phi(y, \Phi_0') \, d\Phi_0'}{1 + Cv(y, \Phi_0')}\Bigg]. \tag{51}$$

Force Equation

$$\frac{\partial}{\partial y}[1 + Cv(y, \Phi_0)]^2 = -C(1+Cb)^2 \left\{ \left[a_2(y) - C\frac{da_1(y)}{dy} \right] \cos \Phi(y, \Phi_0) \right.$$

$$+ \left[a_1(y) + C\frac{da_2(y)}{dy} \right] \sin \Phi(y, \Phi_0) \right\}$$

$$+ 4C(1+Cb)\left(\frac{\omega_p}{\omega C}\right)^2 \int_0^{2\pi} \frac{F_{1-z}(\Phi - \Phi')\,d\Phi_0'}{1 + Cv(y, \Phi_0')}. \quad (52)$$

The relation between variables is (no initial velocity modulation)

$$\frac{\partial \Phi(y, \Phi_0)}{\partial y} = \frac{v(y, \Phi_0)}{1 + Cv(y, \Phi_0)}. \quad (53)$$

The above system has no advantage over the former and in fact from a numerical procedure's standpoint the former requires less computation time.

3 One-Dimensional Results

a. General

The problem is now to solve the TWA equations, (27), (28), (31) or (32) and (33), subject to the initial conditions and parameter selection as outlined in Section 2.g. The independent variables are y and Φ_0 with $\Phi(y, \Phi_0)$, $u(y, \Phi_0)$, $A(y)$ and $\theta(y)$ being the dependent variables. Clearly the data reduction problem is heightened over that for the klystron. In this section the results of extensive calculations for the case of the unmodulated and unbunched beam injection are given. Other special cases will be treated in succeeding chapters.

b. Voltage, Gain and Phase Shift

Typical normalized rf voltage variation as a function of distance is shown in Fig. 2 for finite space-charge conditions and a large-C value. In general two ranges of the velocity parameter b are shown, namely, that which gives maximum small-signal gain and that which gives maximum power output. Of course the more b increases, the greater the length required to reach maximum output. In order for small-signal conditions to prevail at the input a signal level A_0 corresponding to approximately 30 dB below CI_0V_0 (i.e. $\psi_0 = -30$) was selected for the calculations. Recall that A_0 is an arbitrary quantity. For such small input signals there will be no gain if $b > b_{x_1=0}$. This is, however, not true for large A_0 since

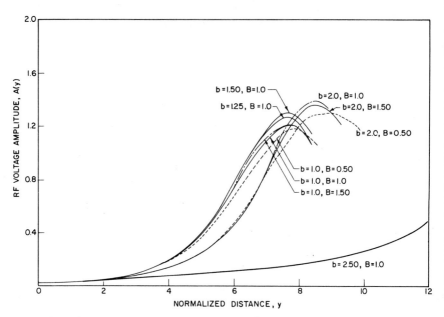

Fig. 2. Rf voltage versus distance ($C = 0.1$, $QC = 0.25$, $d = 0$, $\psi_0 = -30$).

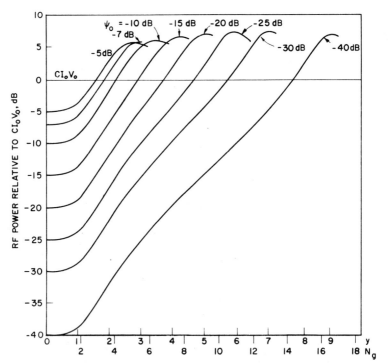

Fig. 3. Rf power relative to CI_0V_0 versus distance with ψ_0 as the parameter ($C = 0.1$, $QC = 0.125$, $d = 0$, $b = 1.5$).

3. ONE-DIMENSIONAL RESULTS

beating-wave amplification does exist. Solution beyond the first maximum in $A(y)$ results in an oscillatory curve and is generally not of great interest. Notice that the point of first electron overtaking indicated on some of the curves does not occur until well into the nonlinear region.

For a given value of b if A_0 is varied then the displacement plane at which saturation occurs increases for decreasing A_0 and decreases for increasing A_0 generally with little variation in the saturated output, as illustrated in Fig. 3. However, when a very large A_0 (drive signal) is

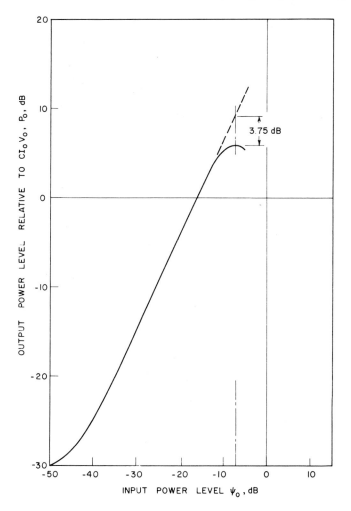

FIG. 4. Output power level relative to CI_0V_0, P_0 versus input power level below CI_0V_0, ψ_0, for fixed tube length ($C = 0.1$, $QC = 0.125$, $d = 0$, $b = 1.5$, $N_g = 5.5$).

applied then the signal immediately more than offsets the space-charge debunching and an increased output is obtained as evidenced in Fig. 3. When A_0 is large the efficiency expression of Eq. (40) should be computed on the basis of the beam energy converted to rf energy, so that the input rf energy must be subtracted from the output and a "conversion" or "interaction" efficiency defined at saturation as

$$\eta(\text{interaction}) \approx 2C(A_{\max}^2 - A_0^2). \tag{54}$$

From a plot such as shown in Fig. 3 a power output versus power input or "drive" curve may be constructed as shown in Fig. 4 at a particular displacement plane. Several things of interest are noted from this curve. The departure from a linear input-output relationship occurs usually around CI_0V_0 or slightly above and the saturated level is generally from 4–8 dB below the extrapolated linear output, depending upon the particular values of C and ω_p/ω. We also find that the saturated output varies from 7 dB above CI_0V_0 for small-C to 4 dB above CI_0V_0 for large-C.

The loading provided by the beam results in a phase lag of the actual rf wave whose velocity is $[1 - C(d\theta(y)/dy)]$ relative to the hypothetical

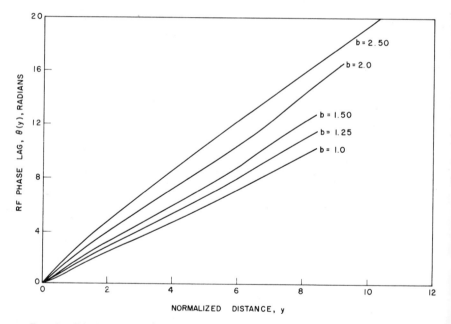

FIG. 5. Rf phase lag of the wave relative to the stream versus distance ($C = 0.1$, $QC = 0.25$, $d = 0$, $B = 1$, $\psi_0 = -30$).

wave velocity u_0. Thus $\theta(y)$ is a negative number and should increase negatively with y. In the linear region $\theta'(y)$ is nearly constant as predicted by the small-signal theory, the departure from this constant slope in the nonlinear region being a function of C, i.e., the strength of the interaction. These results are shown in Fig. 5. A composite curve of phase shift at saturation versus drive level is constructed and shown in Fig. 6. The

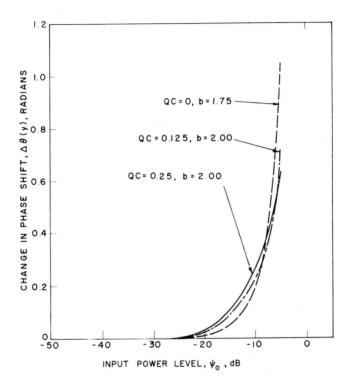

FIG. 6. Change in phase shift versus input drive level ($C = 0.05$, $N_g = 13$, $d = 0$, $B = 1$).

maximum phase shift is seen to be about π radians and decreases as one operates down from the saturated output into the linear region. Such curves may be constructed at various displacement planes of interest, e.g., somewhat prior to and somewhat beyond saturation.

c. *Electron Velocity-Phase and Flight-Line Diagrams*

As in the klystron or drift-space amplification case velocity and phase information is obtained and may be plotted in a variety of ways. For consideration of the variables $\Phi(y, \Phi_0)$ and $u(y, \Phi_0)$ the most useful and enlightening presentation is $1 + 2Cu(y, \Phi_0)$ at various y planes versus $\Phi(y, \Phi_0)$ and y versus $\Phi(y, \Phi_0)$ for particular Φ_0's. Actually the flight-line diagrams shown by the latter plots serve to show the details of bunch formation and crossover along the device and are thus useful in developing an understanding of the interaction. A fairly typical flight-line diagram is shown in Fig. 7 with the accelerating and decelerating field

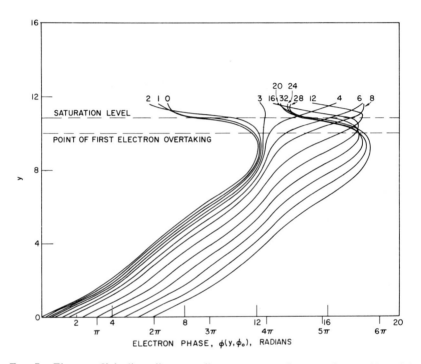

Fig. 7. Electron flight-line diagram: distance versus electron phase ($C = 0.1$, $QC = 0$, $d = 0$, $\psi_0 = -30$, $b = 2.0$).

regions indicated. The departure of the flight lines from straight lines results from the interaction: near saturation one sees the bunch forming in the decelerating field region and the antibunch forming in the accelerating field region. At saturation the antibunch is taking as much energy back from the circuit as the favorably phased bunch is delivering to it, resulting in no net energy transfer: by definition, saturation.

The velocity-phase information seems to have more significance since we can see the phase locations of the electrons relative to the wave along with their kinetic energy, all as a function of signal level along the rf structure. This information is shown in Figs. 8 and 9 for comparable

3. ONE-DIMENSIONAL RESULTS

cases with and without space charge. We also note the remaining kinetic energy of the electrons from these graphs and this information will be extremely useful in considering phase focusing and collector depression techniques to be outlined in later chapters. Initially (near the input) the velocity modulation is sinusoidal, but as the signal level increases the

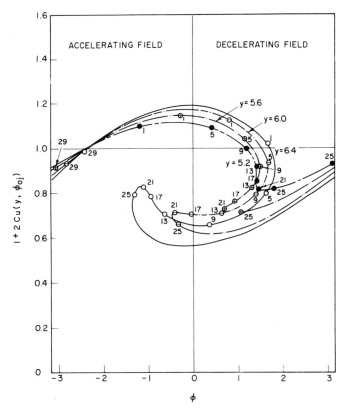

FIG. 8. Phase diagram of conventional TWA ($C = 0.1$, $QC = 0$, $b = 1$, $d = 0$, $\psi_0 = -30$ dB).

velocity-phase curves develop a vortex point around $\pi/2$ radians in a decelerating phase of the wave and these are the electrons contributing most to the enhancement of the circuit field. The irregularity and jaggedness of the curves in the finite-space-charge cases results from the fact that the weighting function is odd periodic and hence for $\Phi - \Phi' \approx 2n\pi$ there is an error, as pointed out earlier, due to the finite limited number of charge groups being treated. As ω_p/ω increases more and more charge groups become necessary for accuracy. For $\omega_p/\omega \leqslant 0.5$, 32 groups are sufficient whereas for $\omega_p/\omega > 0.5$, 64 or more are required. Unfortunately the computation time quadruples for a doubling of the number of charge groups. The additional time is spent in computing space-charge weighting functions and integrals.

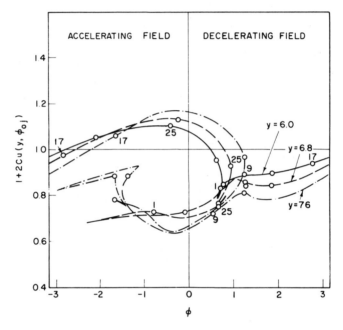

Fig. 9. Phase diagram of conventional TWA ($C = 0.1$, $d = 0$, $QC = 0.25$, $B = 1$, $a/b' = 2$, $\omega_p/\omega = 0.1827$, $b = 1$, $\psi_0 = -30$ dB).

d. Fundamental and Harmonic Current

In the klystron study it was extremely important to calculate the fundamental and harmonic current amplitudes at the output of the drift region since these represented the currents driving the output circuit. Multiple cavities and drift regions were beneficial in raising i_1/I_0 above the 1.16 value predicted by simple ballistic theory. Similarly in the case of the TWA we desire a strong interaction, high C and high P_μ so that i_1/I_0 will be large and the efficiency high.

As was discussed in Section 2.h, the normalized current amplitudes may be calculated from the velocity-phase information. It is also

3. ONE-DIMENSIONAL RESULTS

interesting to calculate the normalized charge density distribution in the stream as a function of signal level from

$$\frac{\rho(y, \Phi_0)}{\rho_0} = \frac{\rho(y, \Phi_0) u_0}{I_0} = \sum \left| \frac{\partial \Phi_0}{\partial \Phi} \right| \frac{1}{[1 + 2Cu(y, \Phi_0)]}. \quad (55)$$

Such a calculation of the charge density is shown in Fig. 10 for a typical

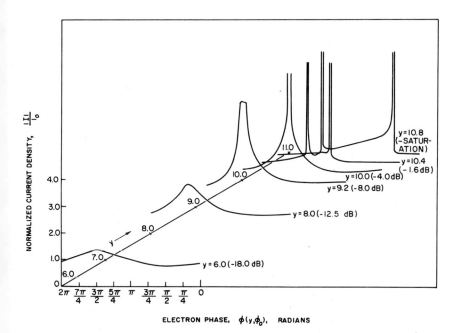

FIG. 10. Normalized linear current density in the stream versus electron phase ($C = 0.1$, $QC = 0$, $d = 0$, $\psi_0 = -30$, $b = 2$).

case. Note the formation of the bunch in the retarding field and at high signal levels its splitting into two or more bunches. The infinite peaks in the charge-density curves arise from the regions of zero slope in the Φ versus Φ_0 curves (overtaking). These infinite peaks do not indicate infinite current density, however, since charge is conserved. The initial bunch forms around $\pi/2$ and stays there until close to saturation, then splitting into two or more portions.

The expression for calculating the fundamental and harmonic current amplitudes is Eq. (41). For the case of an unmodulated, unbunched input beam the calculated values are shown in Fig. 11.

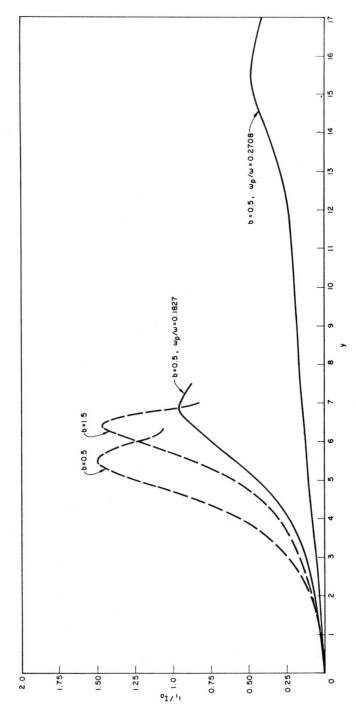

Fig. 11. Fundamental rf beam current versus distance ($C = 0.1$, $B = 1$, $d = 0$, $\psi_0 = -30$).

e. Saturation Efficiency and Length

The saturation efficiency and optimum length results from numerous calculations may be conveniently summarized in terms of their dependence on the velocity parameter b, the space-charge parameters

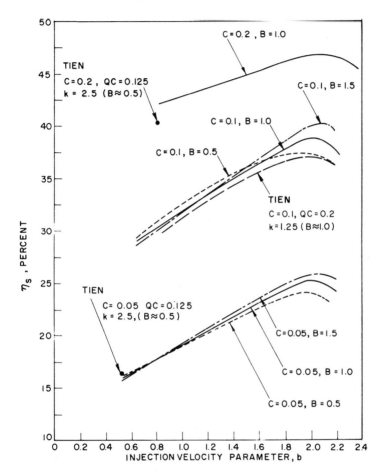

FIG. 12. Saturation efficiency versus b ($QC = 0.125$, $d = 0$).

ω_p/ω or QC and B, and the gain parameter C. For all calculations the circuit excitation A_0 is small and the rf structure is considered lossless. The effects of circuit loss will be evaluated in a later section. The aforementioned results are shown in Figs. 12–16. It is noted that a comparison is made with the results of Tien obtained by a different, although exactly equivalent, method. More will be said about this in

Section 10. The obvious significant results are that for highest efficiency a b value greater (higher voltage) than that for maximum low-level gain is required, and that the saturation efficiency increases linearly and rapidly up to $C \approx 0.12$ and thereafter increases at a much

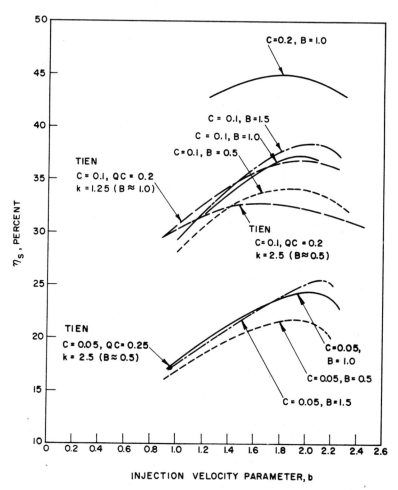

FIG. 13. Saturation efficiency versus b ($QC = 0.25$, $d = 0$).

slower rate. Interestingly enough, such a dependence has been determined experimentally by Cutler and reported in a definitive article[12] on TWA efficiency.

One of the most difficult experimental problems is that of obtaining *accurate* and *reproducible* efficiency data to compare with theoretical predictions. The experimental work of Cutler is summarized in Figs. 17 and 18, where operation at maximum low-level gain and maximum efficiency are depicted respectively. The results compare favorably with the one-dimensional theory when $B < 1$ for large C or $B < 2$ for small C, which indicates that for large B radial circuit field fall-off and radial space-charge forces are important.

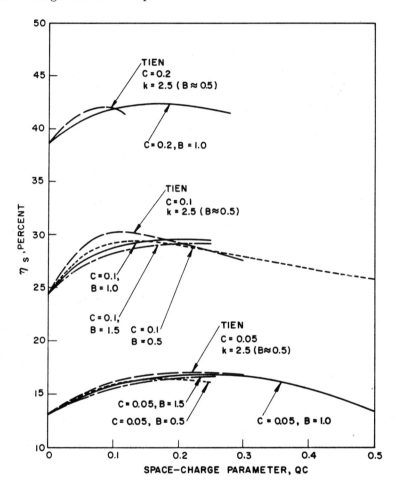

FIG. 14. Saturation efficiency versus space-charge parameter. b adjusted for maximum x_1. ($d = 0$.)

The dependence of efficiency on space-charge forces is not quite so simple, as we see that for operation near maximum low-level gain the efficiency is not critically dependent on QC and in fact for moderate values the efficiency is enhanced over that for $QC = 0$. On the other hand, operation at a voltage for maximum power output results in a decreased efficiency with increasing QC as a result of space-charge debunching in a manner expected.

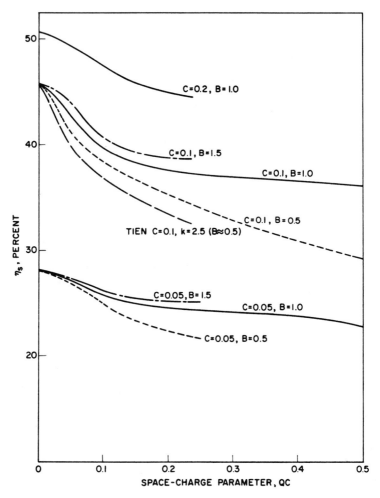

FIG. 15. Saturation efficiency versus space-charge parameter. b adjusted for maximum η_s. ($d = 0$.)

The space-charge range parameter $B = \gamma b'$ is also a measure of the beam diameter. The results of Fig. 15, indicating an increase in efficiency with increasing B, must be taken with caution. This one-

3. ONE-DIMENSIONAL RESULTS

dimensional theory does not account for circuit-field variations over the cross section nor for radial space-charge forces and hence is open to question for large B values. It is believed that these effects are not important until $B \geqslant 0.8$–1.0. This will be evaluated using a modified

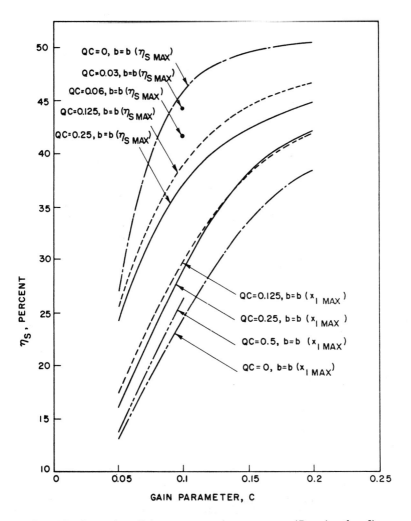

FIG. 16. Saturation efficiency versus gain parameter ($B = 1$, $d = 0$).

one-dimensional analysis in Section 4 and a general two-dimensional analysis in Section 5. Cutler's experimental results indicate an optimum value around 0.5–0.8.

The saturation length in terms of stream wavelengths is indicated in Fig. 19 for a wide variety of operating conditions. Fortunately and as expected, the optimum length is critically dependent only on C and not on the other operating parameters.

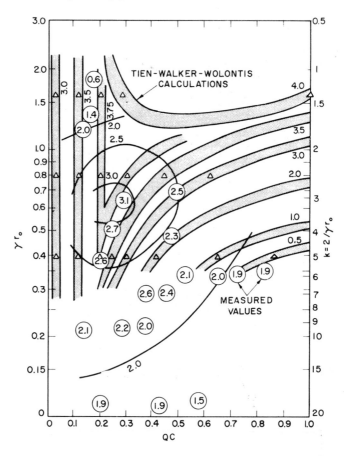

FIG. 17. Values of efficiency/C as a function of QC and γr_0 at the voltage giving maximum gain per unit length. The shaded contours and triangular points are from the computations of Tien et al.[9] The circled points are from the measurements and the line contours are estimated lines of constant efficiency. The most significant difference is for large beam radii, where the rf field varies over the beam radius in a way not accounted for in the computations (Cutler[12]).

f. Effect of Circuit Loss on Gain, Phase Shift, and Efficiency

It is intuitively obvious that the presence of circuit attenuation will result in a decreased power level at a given displacement compared to the zero-loss case. Unfortunately, significant loss in the form of an attenuator or circuit sever is necessary for stability in high gain tubes. The proper design of amount and placement of loss can result in no degradation in the saturation output as predicted by Cutler and

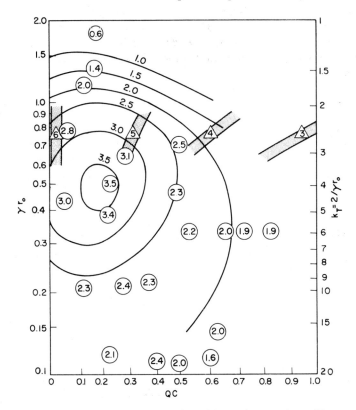

FIG. 18. Values of efficiency/C as a function of QC and γr_0 at elevated beam voltage. Raising the beam voltage has little effect at large QC and small γr_0, and less than expected anywhere. Again the triangular points are from Tien et al.[9] and the line contours are estimated from the measured data (Cutler[12]).

Brangaccio[13]. Basically one should avoid saturation in the attenuator region and have at least 20–25 dB gain beyond the end of the attenuator to develop the full saturation capability of the beam.

From the circuit model with loss shown in Chapter III the effect of loss on saturation may be evaluated. The reduction in saturation gain and change in phase shift for uniform loss extending from $y = 1.6$ ($CN_s \approx 0.3$) to saturation are shown in Fig. 20 for a particular set of operating parameters.

g. Severed-Circuit Calculations

In the previous section the effects of circuit attenuation, required for stability at high gain, were evaluated using the nonlinear one-dimensional theory. Many attenuator forms (d versus y) are possible and it is known that three guidelines should be considered in designing a good attenuator, i.e., one that doesn't limit the saturation output or efficiency. These are: (1) don't begin the attenuation too early or too fast; (2) have 20–25 dB free gain beyond the attenuator; and (3) avoid saturation within the attenuator.

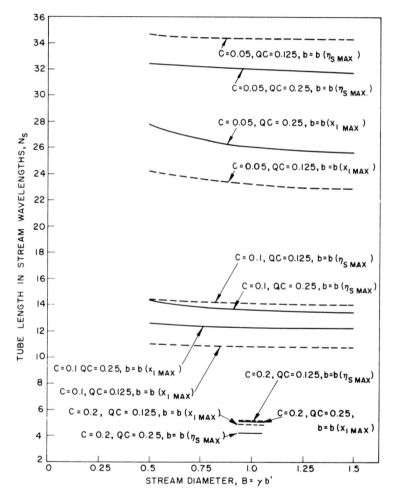

FIG. 19. Device length at saturation versus stream diameter ($\psi_0 = -30$, $d = 0$).

3. ONE-DIMENSIONAL RESULTS

Considering the attenuator for a moment, we note that passing to the limit of a highly concentrated short-length attenuator is equivalent to a circuit sever. The effects of circuit severs on traveling-wave tube performance are considered in this section using a one-dimensional theory.

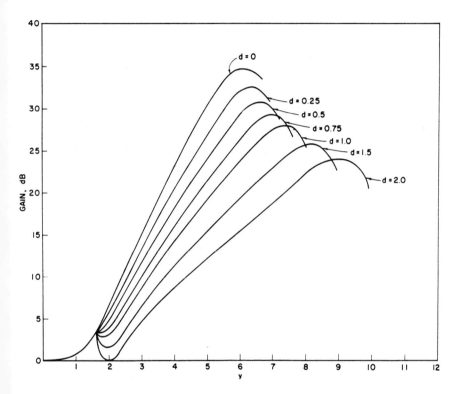

FIG. 20. Gain versus distance in a loss region ($C = 0.1$, $B = 1$, $b = 0.5$, $\omega_p/\omega = 0$, $\psi_0 = -30$).

The general case of a multiply severed amplifier in which the sever lengths may be of finite length is seen to be described by a set of equations which is a serial combination of the one-dimensional traveling-wave-amplifier system and the corresponding one-dimensional klystron system. If the first section of such a device is a traveling-wave amplifier section then those equations are used and the output beam conditions, with $A(y_1)$ and $\theta(y_1)$ placed equal to zero, are the input conditions to the sever

212   VI. TRAVELING-WAVE AMPLIFIER ANALYSIS

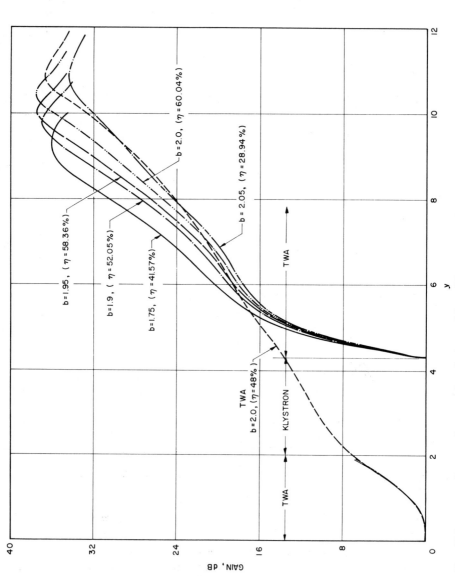

Fig. 21. Gain versus distance for a prebunched amplifier ($C = 0.1$, $\gamma a = 1.5$, $\gamma b' = 1.05$, $\psi_0 = -30$, $\omega_p/\omega = 0$).

3. ONE-DIMENSIONAL RESULTS

drift region. At the end of the sever drift region the modulated and bunched beam, along with $A(y_2) = 0$, $\theta(y_2) = 0$, $d\theta(y_2)/dy = -b$, constitutes the input to the next traveling-wave section and so on to the output. In moving from a circuit region through a drift region into a circuit region it is assumed that the only rf excitation of the new circuit section is that carried by the beam. The equations are sufficiently general to handle an additional rf circuit voltage, which is derived from the output of the previous circuit section appropriately attenuated and phase shifted.

The results for a severed-circuit device with particular operating parameters and various voltages after the sever are shown in Fig. 21.

Other severed-circuit calculations (Scott[14]) with operating parameters $C = 0.1$, $QC = 0.25$, $d = 0$, $B = 1$ indicate that the sever length is not particularly important, which does not agree with the above results. The most important effect seems to be the amount of small-signal gain beyond the sever. Both effects are seen to be of importance as expected, since too long a sever leads to significant defocusing of the stream due to space-charge forces and the gain of the output section must be sufficient to build the rf signal up from near zero to a level sufficient to contain the bunched beam. Clearly this is the same problem as that encountered in the prebunched beam studies of Chapter XIV. Scott's calculations were carried out using the same equations as the author's and his results are shown in Fig. 22 for $C = 0.1$, $QC = 0.25$, $d = 0$, $B = 1$. These are for very short sever lengths.

The obvious result of this calculation is that approximately 26 dB of gain is needed beyond the sever in order to achieve the saturation potential of the beam. The rf energy extracted from the beam and dissipated at the end of the first section has a negligible effect on the final result. Such a high-gain section could pose stability problems. As mentioned previously this high-gain requirement merely states that the rf circuit level must be built up to a value sufficient to maintain the bunch or a given value of i_1 in the beam where i_1 constitutes the current driving the circuit. These results are summarized in Fig. 23, where the effect on saturation output or efficiency is given along with some experimental results. Scott[14] has corrected the one-dimensional efficiency values by a factor of 0.8, appropriate to $B = \gamma b' = 1$ as predicted by Rowe[15]. Again it should be pointed out that these results are for very short sever lengths.

The net gain of a severed-circuit device is limited by both the sever and saturation effects. An approximate total reduction in gain due to these two effects is 11 dB, obtained from the addition of a 6-dB reduction in small-signal gain due to severing predicted by Pierce and an approximate 5-dB compression due to large-signal effects as calculated by both Rowe and Tien. Thus the net gain beyond the sever must be at least 11 dB below the small-signal gain beyond the sever. This is illustrated in Fig. 24.

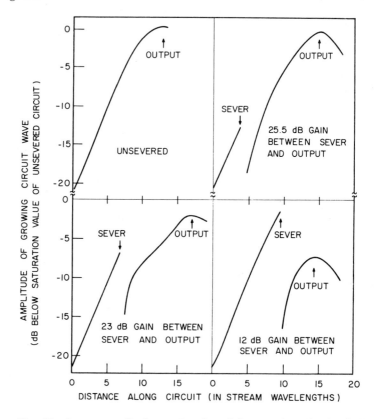

FIG. 22. Circuit wave amplitude as a function of distance along the circuit (Scott[14]).

Depressed collectors may, of course, be used on severed tubes as well as on unsevered tubes to enhance efficiency. Their effectiveness, for a given number of segments, in improving efficiency depends on the degree of velocity spread in the spent stream. For wide velocity spreads several segments of collector are needed to obtain a given improvement factor. This is seen in Chapter XV. If there is sufficient gain beyond the sever so as not to limit the efficiency, then a depressed collector of a given number of segments is as effective on a severed tube as on an unsevered tube.

Radial debunching and saturation effects in the beam also affect

3. ONE-DIMENSIONAL RESULTS

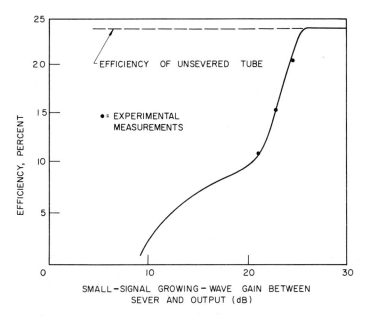

FIG. 23. Efficiency as a function of small-signal gain beyond the sever (Scott[14]).

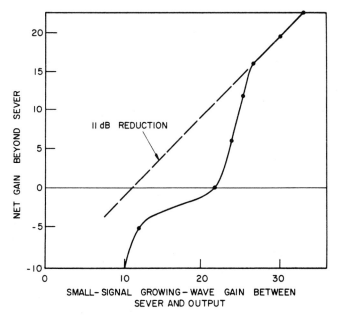

FIG. 24. Net saturation gain beyond sever as a function of small-signal gain beyond sever (Scott[14]).

sever performance; these are evaluated in Section 5 with a two-dimensional confined-flow theory.

h. Operation at High Drive and $b > b_{x_1=0}$

The traveling-wave amplifier is well known as a growing-wave device in which the signal along the rf structure increases exponentially until the nonlinear region is reached, thereafter growing at a lower rate. Examination of Pierce's small-signal theory indicates that if b is gradually increased beyond the maximum gain condition x_1 decreases and eventually there is a value of b, called $b_{x_1=0}$, beyond which x_1 is either extremely small in a lossy circuit or zero in a lossless circuit and the circuit waves cease to grow. It has been determined by Rowe[15] that there is now amplification due to a beating-wave phenomenon among constant-amplitude ($d = 0$) waves and that this is an efficient mode of operation. A device based on this principle of operation is called a Crestatron.

The operation can easily be illustrated through large-signal calculation results. The most efficient operation is obtained under conditions of high drive level. This is illustrated in Fig. 25, where the saturated efficiency is shown versus ψ_0 for b values both less than and greater than $b_{x_1=0}$. We see that high efficiency occurs for large ψ_0 at large b and as ψ_0 is

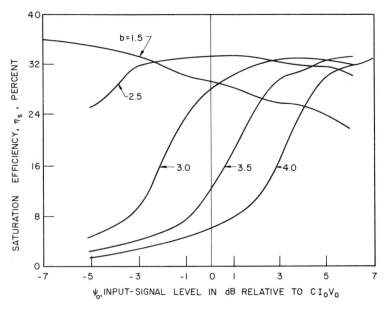

FIG. 25. Saturation efficiency versus drive level ($C = 0.1$, $QC = 0.125$, $B = 1$, $d = 0$, $b_{x_1=0} = 2.33$).

reduced for a constant b the output drops off rapidly. These predictions have been confirmed experimentally as shown in Fig. 26. In addition to the high efficiency characteristic of the Crestatron the optimum length is quite short, being

$$CN_s = \frac{\pi}{\sqrt{b - b_{x_1=0}}}.$$

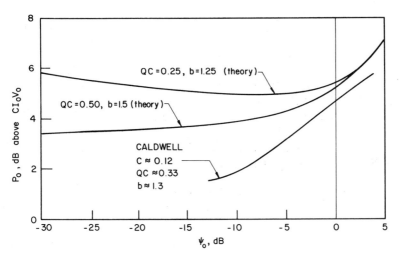

FIG. 26. Variation of saturation level with drive level ($C = 0.1$, $B = 1$, $d = 0$).

4 N-Beam TWA Analysis

It was seen in the last section that some discrepancy occurs between the one-dimensional theoretical efficiency predictions and the experimental results when B is large and there is a strong coupling to the circuit. These discrepancies have generally been attributed to radial variations of both the circuit and space-charge fields. Before solving the two-dimensional problem exactly, it is useful to modify the coupling to the beam as a function of radius to determine the circuit field effect. Rowe[16,17] has carried out detailed calculations by two different methods to evaluate the circuit and space-charge-field effects.

The first of these approximate methods[16] again assumes confined flow so that we have a smooth stream boundary and then incorporates a coupling and/or radial field variation function as follows. The potential is now defined by

$$V(b', y, \Phi) = \mathrm{Re}\left[\frac{Z_0 I_0}{C} A(y) f(B) e^{-j\Phi}\right], \tag{56}$$

where $f(B)$ represents an unknown radial function. The form of $f(B)$ will depend upon whether the stream is a thin hollow beam or a solid one. The following approximate functions obtained from the field equations are used:

$$f_h(B) = \frac{I_0(\gamma b')}{I_0(\gamma a)} \quad \text{for thin hollow beams} \tag{57a}$$

and

$$f_s(B) = \frac{[I_0^2(\gamma b') - I_1^2(\gamma b')]^{\frac{1}{2}}}{I_0(\gamma a)} \quad \text{for solid beams.} \tag{57b}$$

Since the circuit is located at $r = a$, the introduction of Eq. (56) into Eqs. (27) and (28) does not change the left-hand side since $f = 1$ at $r = a$. The driving terms, however, on the right-hand side are modified due to reduced coupling of the beam. The circuit field terms in the force equation are similarly modified. The equation relating dependent variables is unchanged since this is strictly a beam equation. The modified nonlinear equations are thus

$$\begin{Bmatrix} \text{Left-hand side} \\ \text{of Eq. (27)} \end{Bmatrix} = f(B) \begin{Bmatrix} \text{Right-hand side} \\ \text{of Eq. (27)} \end{Bmatrix}, \tag{58}$$

$$\begin{Bmatrix} \text{Left-hand side} \\ \text{of Eq. (28)} \end{Bmatrix} = f(B) \begin{Bmatrix} \text{Right-hand side} \\ \text{of Eq. (28)} \end{Bmatrix} \tag{59}$$

and

$$\begin{Bmatrix} \text{Left-hand side} \\ \text{of Eq. (31)} \end{Bmatrix} = f(B) \begin{Bmatrix} \text{Circuit-field} \\ \text{terms} \end{Bmatrix} + \begin{Bmatrix} \text{Space-charge-field} \\ \text{integral} \end{Bmatrix}. \tag{60}$$

Nonlinear calculations of efficiency as a function of B have been made using this method for a particular set of operating parameters, and are summarized in Fig. 27. The efficiency reduction as a function of B can be approximately written as

$$\frac{\eta_s(B)}{\eta_s|_{f=1}} \approx f^{\frac{1}{2}}(B) \quad \text{for} \quad b = b(x_{1_{\max}}) \tag{61a}$$

and

$$\frac{\eta_s(B)}{\eta_s|_{f=1}} \approx f^{\frac{1}{2}}(B) \quad \text{for} \quad b = n(\eta_{s_{\max}}). \tag{61b}$$

This approximate method of evaluating circuit field reduction effects on efficiency is in approximate agreement with Nordsieck's[1] two-beam approach for small-C equations neglecting space-charge effects.

An alternate approximate method [1,17] for handling the two-dimensional problem is to divide the stream into N annular streams, each having a different coupling to the slow-wave circuit. The space-charge field is calculated in the same manner as in the one-dimensional analysis, except that each stream is represented by a different space-charge-field weighting function $F_{1-2}(\Phi - \Phi')$. It is assumed that each stream carries the same

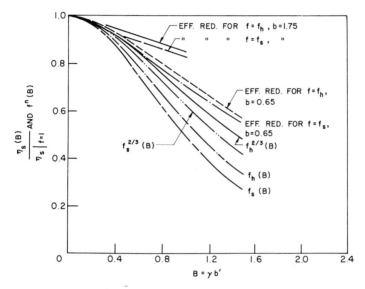

FIG. 27. Field reduction factor and efficiency reduction versus stream diameter ($C = 0.1$, $QC = 0.125$, $d = 0$, $a/b' = 2$).

current, I_0/N, and, since all streams are assumed to move forward at the same velocity, the space-charge density $\rho_0 = I_0/u_0$ is the same for all streams. The assumed stream model is shown in Fig. 28.

The current and space-charge density being the same for all streams, the individual areas are all equal and the radii at the stream boundaries are related in the following manner:

$$r_N = \sqrt{N}\, r_1. \tag{62}$$

Each of the N streams is treated exactly as was the entire stream in the one-dimensional analysis. The circuit field will be different at each stream and is indicated by E_{zi}, where $i = 1, 2, 3, ..., N$. Thus coupling varies with stream number and the equations may be modified directly in a manner similar to that of the last section.

Before writing out the modified equations it is well worth investigating the meaning of such parameters as C and QC in such an analysis. Each stream-circuit combination has a characteristic C, since the coupling factors are different for each; hence some problems arise in determining a value of C to be used in the reduced variables. A definition of C will be utilized for the N-stream tube which gives the same small-signal gain as that of a single-stream tube with the same value of C. Thus

$$C^3 \triangleq \frac{(\sum\limits_{i=1}^{N} Z_{0i})I_0/N}{4V_0}, \tag{63}$$

where Z_{0i} is the impedance seen by the ith stream as calculated from E_{zi}. If all Z_{0i} are equal, Eq. (63) reduces to the single-stream C.

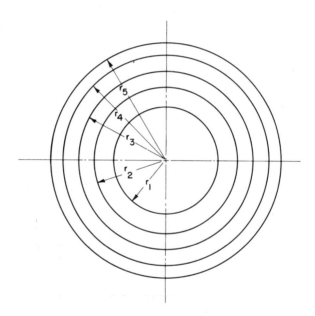

I_0 = TOTAL STREAM CURRENT

I_0/N = CURRENT PER STREAM

Fig. 28. N-stream model (I_0 = total stream current, I_0/N = current per stream).

The space-charge parameter QC will not be the same for each stream, since the plasma-frequency reduction factor is different for each stream. In view of the fact that QC is not a convenient parameter in this problem, the space-charge parameters $K = (\omega_p/\omega C)^2$ and $B_N = \gamma b_N'$ will be utilized. We have $(\omega_p/\omega)^2 = |I_0\eta|/\pi\epsilon b'^2 u_0\omega^2$, where I_0 = dc stream current and $b' = $ stream radius. The current density is the same for each stream: hence K is the same for all streams and is equal to the value of K for the single-stream one-dimensional case. Obviously B_N will be different for each stream.

The circuit coupling for the individual stream will be proportional to the ratio of radii of the circuit and the beam element considered. For the N-beam case the effect of the beam on the circuit will be weighted according to the circuit field acting on the stream. The ratio of the fields acting on the N streams is determined according to Eqs. (57).

The appropriately modified equations are as follows.

Circuit Equations

$$\begin{Bmatrix} \text{Left-hand side} \\ \text{of Eq. (27)} \end{Bmatrix} = -\frac{(1+Cb)}{\pi C} \sum_{i=1}^{N} \frac{f_i}{N} \left[\int_0^{2\pi} \frac{\cos \Phi_i(y,\Phi_0')\, d\Phi_0'}{1+2Cu_i(y,\Phi_0')} \right.$$

$$\left. + 2Cd \int_0^{2\pi} \frac{\sin \Phi_i(y,\Phi_0')\, d\Phi_0'}{1+2Cu_i(y,\Phi_0')} \right], \tag{64}$$

$$\begin{Bmatrix} \text{Left-hand side} \\ \text{of Eq. (28)} \end{Bmatrix} = -\frac{(1+Cb)}{\pi C} \sum_{i=1}^{N} \frac{f_i}{N} \left[\int_0^{2\pi} \frac{\sin \Phi_i(y,\Phi_0')\, d\Phi_0'}{1+2Cu_i(y,\Phi_0')} \right.$$

$$\left. - 2Cd \int_0^{2\pi} \frac{\cos \Phi_i(y,\Phi_0')\, d\Phi_0'}{1+2Cu_i(y,\Phi_0')} \right], \tag{65}$$

where $f_i = $ the weighting factor on the circuit field appearing at each stream as determined from Eqs. (57). Clearly $f_1 < f_2 < f_3 < \ldots f_N$, and the f_i's are related by

$$\sum_{i=1}^{N} f_i^2 = N. \tag{66}$$

The radial excursions of the electrons are ignored, consistent with the confined flow assumption, and thus only the axial force equation persists. It contains all of the dependent variables plus the space-charge-field expression. Thus

Force Equation

$$\frac{\partial u_i(y,\Phi_0)}{\partial y}[1+2Cu_i(y,\Phi_0)] = -f_i A(y)\left(1 - C\frac{d\theta(y)}{dy}\right)\sin \Phi_i(y,\Phi_0)$$

$$+ Cf_i \frac{dA(y)}{dy}\cos \Phi_i(y,\Phi_0)$$

$$+ \frac{K}{(1+Cb)}\int_0^{2\pi}\frac{F_{1-z}(\Phi-\Phi')_i\, d\Phi_0}{1+2Cu_i(y,\Phi_0)}$$

$$i = 1, 2, 3 \ldots N. \tag{67}$$

The appropriate value of QC may be calculated for each stream using K and the reduction factor R_N for that stream. An effective value of QC for the entire stream could be defined but, in view of the adequacy of the parameters K and B_N, this need not be done.

In the Lagrangian analysis individual charge groups are followed through the interaction region and at each position the circuit and space-charge-field forces on the charges are summed. Each individual stream is divided into representative charges and a relationship exists between the dependent variables describing the charge velocity and phase position with respect to the rf wave. Thus there will be a set of Φ's and u's for each of the N streams. Hence (no initial velocity modulation)

$$\frac{\partial \Phi_i(y, \Phi_0)}{\partial y} + \frac{d\theta(y)}{dy} = \frac{2u_i(y, \Phi_0)}{1 + 2Cu_i(y, \Phi_0)} \qquad i = 1, 2, 3 \ldots N. \qquad (68)$$

In the calculations to be presented $m = 32$; i.e., 32 charge groups per beam are taken so as to give the same accuracy as the one-dimensional calculations.

The weighting factors f_i on the electric field, as seen by each beam, are tabulated in Table I assuming $B = 1$ for the entire stream. The function

TABLE I

N-Beam Weighting Functions

$N = 1$	$f_1 = 1$	
$N = 2$	$f_1 = 0.7661$	$B_1 = 0.35$
	$f_2 = 1.1888$	$B_2 = 0.85$
$N = 3$	$f_1 = 0.7462$	$B_1 = 0.29$
	$f_2 = 0.9226$	$B_2 = 0.70$
	$f_3 = 1.2618$	$B_3 = 0.91$
$N = 4$	$f_1 = 0.7382$	$B_1 = 0.25$
	$f_2 = 0.8475$	$B_2 = 0.60$
	$f_3 = 1.0252$	$B_3 = 0.79$
	$f_4 = 1.2985$	$B_4 = 0.93$

$f(B)$ versus B is monotonically decreasing and can be synthesized quite accurately on a four-point basis. This leads to the conclusion that the N-stream results for $N > 4$ will not be appreciably different from the $N = 4$ ones. The input boundary conditions remain the same as in the single-stream case, and the gain and efficiency are computed in the same manner.

The results of numerous N-beam computations are summarized in Fig. 29 in terms of efficiency reduction due to radial variations. For $B = 1$ subdivision beyond $N = 2$ is not necessary and as expected the two-stream case saturates at a lower level and a longer length due to the reduced coupling of the inner ring of charge. The efficiency reduction calculated here is in general agreement with that of Fig. 27. It is seen,

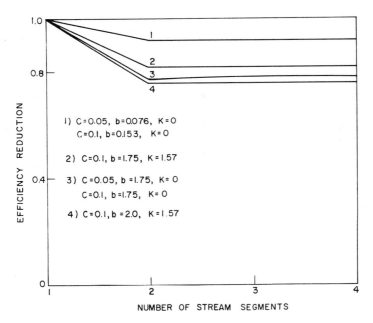

FIG. 29. Efficiency reduction versus degree of stream subdivision ($d = 0$, $B = 1$).

however, in Fig. 29 that the magnitude of the efficiency reduction is greater when b is adjusted for maximum efficiency than for maximum low-level gain, which is opposite to Fig. 27. The two-dimensional results will shed some light on this. The results of Fig. 29 are expected, since at low values of b the interaction is weak and the reduced coupling of the beam center is less important than in the strong interaction, maximum output operating condition.

The reduced coupling of the inner-stream segments to the circuit and its consequent saturation at a position beyond the point where the outer-stream element saturates is illustrated in the velocity-phase curves of Fig. 30. These curves are plotted at various positions and consequently signal levels along the rf structure, and indicate the state of modulation and bunching of the stream. The rf signal level is relative to saturation.

As expected the outer stream saturates first and delivers a maximum of energy to the circuit before the inner elements are significantly bunched. The fact that the inner elements saturate further along the device sometimes results in a broad saturation characteristic. The results for $N > 2$ are not much different from those for $N = 2$ when $B = 1$.

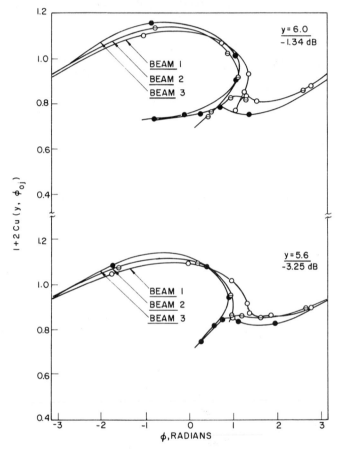

FIG. 30. Velocity versus phase, three-beam case ($C = 0.1$, $d = 0$, $K = 1.57$, $b = 0.65$, $y = 5.6$ and 6).

5 Two-Dimensional TWA Analysis

The radial circuit field variations and radial space-charge-field effects on efficiency and gain were evaluated in the previous section using a modified one-dimensional theory. In the present analysis we again assume confined flow so that the beam boundary is smooth; i.e., $\omega_c/\omega \gg 1$. Axial symmetry is again assumed and the general Lagrangian analysis methods are again utilized. This analysis follows that for the klystron except that circuit field effects are also considered.

The normalized independent space and phase Lagrangian variables are defined as

$$y \triangleq \frac{C\omega z}{u_0} = 2\pi C N_s, \tag{69}$$

$$x \triangleq \frac{C\omega r}{u_0}, \tag{70}$$

$$x_0 \triangleq \frac{C\omega r_0}{u_0}, \tag{71}$$

and

$$\Phi_0 \triangleq \frac{\omega z_0}{u_0} = \omega t_0. \tag{72}$$

A dependent phase variable which denotes the phase position of charge groups at a particular displacement plane relative to the rf wave is defined as

$$\Phi(z, t) = \frac{\omega z}{u_0} - \omega t - \theta_y. \tag{73}$$

The rf voltage along the circuit is defined as the product of slowly varying amplitude and phase functions

$$V(z, r, t) = V(y, x, \Phi) = \mathrm{Re}\left[\frac{Z_0 I_0}{C} A(y)\psi(x) e^{-j\Phi} e^{j\theta_x}\right], \tag{74}$$

where it is assumed that

$$\theta_x = \theta_x(x), \tag{75a}$$

$$\theta_y = \theta_y(y), \tag{75b}$$

and on the circuit itself, $r = a$,

$$\psi(a) = 1 \tag{76a}$$

and

$$\theta_x(a) = 0. \tag{76b}$$

The function $\psi(x)$ represents the radial variation of the circuit field.

The definition of dependent variables is completed with those for the electron velocities:

$$u_z = \frac{u_0}{C\omega}\frac{dy}{dt} \triangleq u_0[1 + 2Cu_y(y, x_0, \Phi_0)] \tag{77}$$

and

$$u_r = \frac{u_0}{C\omega}\frac{dx}{dt} \triangleq 2Cu_0u_x(y, x_0, \Phi_0). \tag{78}$$

The above definitions are combined using similar procedures to those outlined previously to obtain a generalized velocity-phase equation,

$$\frac{\partial\Phi(y, x_0, \Phi_0)}{\partial y} + \frac{d\theta_y(y)}{dy} = \frac{1}{C}\left[\frac{1}{1 + 2Cu_y(0, x_0, \Phi_0)} - \frac{1}{1 + 2Cu_y(y, x_0, \Phi_0)}\right], \tag{79}$$

where $u_y(0, x_0, \Phi_0)$ indicates an initial beam modulation. As previously, x is treated as a dependent variable and is written as

$$x(y, x_0, \Phi_0) = x_0 + \int_0^t \frac{dx}{dt}\,dt = x_0 + 2C\int_0^y \frac{u_x(y, x_0, \Phi_0)}{1 + 2Cu_y(y, x_0, \Phi_0)}\,dy. \tag{80}$$

In evaluating the radial function $\psi(x)$ we must solve the radial and axial wave equation,

$$\nabla^2 V(r, z, t) - \frac{1}{c^2}\frac{\partial^2 V(r, z, t)}{\partial t^2} = 0, \tag{81}$$

where rf space-charge effects are neglected. Substituting the appropriate derivatives of the potential defined by Eq. (74) in Eq. (81) and equating coefficients of $\sin(\Phi - \theta_y)$ and $\cos(\Phi - \theta_y)$ to zero yields the following two radial equations:

$$\psi(x)\frac{d^2A(y)}{dy^2} - A(y)\psi(x)\left(\frac{1}{C} - \frac{d\theta_y(y)}{dy}\right)^2 + A(y)\frac{d^2\psi(x)}{dx^2} - \left(\frac{d\theta_x(x)}{dx}\right)^2 A(y)\psi(x)$$
$$+ \frac{A(y)}{x}\frac{d\psi(x)}{dx} + A(y)\psi(x)\left(\frac{k_e}{C}\right)^2 = 0 \tag{82}$$

and

$$-2\psi(x)\left(\frac{1}{C} - \frac{d\theta_y(y)}{dy}\right)\frac{dA(y)}{dy} + A(y)\psi(x)\frac{d^2\theta_y(y)}{dy^2} + 2A(y)\frac{d\psi(x)}{dx}\frac{d\theta_x(x)}{dx}$$
$$+ A(y)\psi(x)\frac{d^2\theta_x(x)}{dx^2} + \frac{A(y)\psi(x)}{x}\frac{d\theta_x(x)}{dx} = 0, \tag{83}$$

where $k_e \triangleq u_0/c$. This procedure still leaves $\psi(x)$ as an undetermined function.

5. TWO-DIMENSIONAL TWA ANALYSIS

The conservation law for charge entering the interaction region is written as

$$\rho r \, dr \, dz = \rho_0 r_0 \, dr_0 \, dz_0. \tag{84}$$

A linear charge density σ is defined by

$$\sigma = -2\pi \int_0^{b'} \psi(r) |\rho_0| r_0 \left|\frac{\partial z_0}{\partial z}\right| dr_0, \tag{85}$$

where the entering dc charge density is taken as

$$\rho_0 = \frac{I_0}{\pi b'^2 u_0}.$$

The charge group phase displacement is introduced in Eq. (85) through

$$\left|\frac{\partial z_0}{\partial z}\right| = \left|\frac{\partial \Phi_0}{\partial \Phi}\right| \frac{1}{1 + 2Cu_y(y, x_0, \Phi_0)}$$

and thus

$$\sigma = -\frac{2|I_0|}{b'^2 u_0} \int_0^{b'} \left|\frac{\partial \Phi_0}{\partial \Phi}\right| \frac{\psi(r)r_0}{1 + 2Cu_y(y, x_0, \Phi_0)} dr_0. \tag{86}$$

Further specific normalized radii of the stream and circuit are

$$x_{b'} = \frac{C\omega b'}{u_0} = C\gamma b' \tag{87a}$$

and

$$x_a = \frac{C\omega a}{u_0} = C\gamma a \tag{87b}$$

and their introduction, along with Eq. (71), into Eq. (86) yields

$$\sigma = \frac{-2|I_0|}{x_{b'}^2 u_0} \int_0^{x_{b'}} \left|\frac{\partial \Phi_0}{\partial \Phi}\right| \frac{\psi(x)x_0}{1 + 2Cu_y(y, x_0, \Phi_0)} dx_0. \tag{88}$$

The development of the circuit equation requires the fundamental sine and cosine components of the linear charge density, which are

$$\sigma_{1s} = -\frac{\sin \Phi}{\pi} \left(\frac{2|I_0|}{u_0 x_{b'}^2}\right) \int_0^{x_{b'}} \int_0^{2\pi} \frac{\psi(x) \sin \Phi' \, x_0' \, dx_0' \, d\Phi_0'}{1 + 2Cu_y(y, x_0', \Phi_0')} \tag{89}$$

and

$$\sigma_{1c} = -\frac{\cos \Phi}{\pi} \left(\frac{2|I_0|}{u_0 x_{b'}^2}\right) \int_0^{x_{b'}} \int_0^{2\pi} \frac{\psi(x) \cos \Phi' \, x_0' \, dx_0' \, d\Phi_0'}{1 + 2Cu_y(y, x_0', \Phi_0')}. \tag{90}$$

The circuit equation for voltage variation along the equivalent transmission line located at $r = a$, including driving terms, is

$$\frac{\partial^2 V(z,t)}{\partial t^2} - v_0^2 \frac{\partial^2 V(z,t)}{\partial z^2} + 2\omega Cd \frac{\partial V(z,t)}{\partial t} = v_0 Z_0 \left[\frac{\partial^2 \sigma(z,t)}{\partial t^2} + 2\omega Cd \frac{\partial \sigma(z,t)}{\partial t} \right]. \quad (91)$$

Applying the circuit boundary conditions gives

$$V(r=a) = \text{Re}\left[\frac{Z_0 I_0}{C} A(z) e^{-j\Phi} \right]. \quad (92)$$

Substitute the appropriate derivatives of the potential into the left-hand side of Eq. (91) and the derivatives of σ_1 into the right-hand side and equate coefficients of sine and cosine to obtain the following two circuit equations:

$$\frac{d^2 A(y)}{dy^2} + A(y)\left[\frac{(1+Cb)^2}{C^2} - \left(\frac{1}{C} - \frac{d\theta_y(y)}{dy}\right)^2 \right]$$
$$= -\frac{2(1+Cb)}{\pi C x_{b'}^2} \left\{ \int_0^{x_{b'}} \left[\int_0^{2\pi} \frac{\psi(x) \cos \Phi' \, x_0' \, dx_0' \, d\Phi_0'}{1 + 2Cu_y(y, x_0', \Phi_0')} \right. \right.$$
$$\left. \left. + 2Cd \int_0^{2\pi} \frac{\psi(x) \sin \Phi' \, x_0' \, dx_0' \, d\Phi_0'}{1 + 2Cu_y(y, x_0', \Phi_0')} \right] \right\} \quad (93)$$

and

$$A(y)\left[\frac{d^2\theta_y(y)}{dy^2} - \frac{2d}{C}(1+Cb)^2 \right] + 2\frac{dA(y)}{dy}\left(\frac{d\theta_y(y)}{dy} - \frac{1}{C} \right)$$
$$= -\frac{2(1+Cb)}{\pi C x_{b'}^2} \left\{ \int_0^{x_{b'}} \left[\int_0^{2\pi} \frac{\psi(x) \sin \Phi' \, x_0' \, dx_0' \, d\Phi_0'}{1 + 2Cu_y(y, x_0', \Phi_0')} \right. \right.$$
$$\left. \left. - 2Cd \int_0^{2\pi} \frac{\psi(x) \cos \Phi' \, x_0' \, dx_0' \, d\Phi_0'}{1 + 2Cu_y(y, x_0', \Phi_0')} \right] \right\}. \quad (94)$$

Comparison of Eqs. (93) and (94) with the circuit equations of Section 2 indicates that the left-hand sides of both equations are identical in the two theories as expected. The right-hand sides are of course more general, since integration over the stream diameter is required. Reduction to the one-dimensional form is evident.

In formulating the Lorentz force equations we again assume low velocities, i.e., nonrelativistic mechanics, and thus that rf magnetic

5. TWO-DIMENSIONAL TWA ANALYSIS

fields are insignificant. In confined flow the following equations are written:

$$\frac{d^2z}{dt^2} = |\eta| \left[\frac{\partial V_c}{\partial z} + \frac{\partial V_{sc}}{\partial z} \right] = -|\eta| [E_{c-z} + E_{sc-z}] \quad (95a)$$

and

$$\frac{d^2r}{dt^2} = |\eta| \left[\frac{\partial V_c}{\partial r} + \frac{\partial V_{sc}}{\partial r} \right] = -|\eta| [E_{c-r} + E_{sc-r}]. \quad (95b)$$

Introducing the new dependent variables and expanding the differentials on the left-hand side gives

$$\frac{\partial u_y(y, x_0, \Phi_0)}{\partial y} [1 + 2Cu_y(y, x_0, \Phi_0)] = -\frac{|\eta|}{2Cu_0^2} \left[\frac{\partial V_c}{\partial y} + \frac{u_0}{C\omega} E_{sc-y} \right] \quad (96)$$

and

$$\frac{\partial u_x(y, x_0, \Phi_0)}{\partial y} [1 + 2Cu_y(y, x_0, \Phi_0)] = -\frac{|\eta|}{2Cu_0^2} \left[\frac{\partial V_c}{\partial x} - \frac{u_0}{C\omega} E_{sc-x} \right]. \quad (97)$$

The form is again familiar and it remains to introduce the space-charge-field expressions. In the two-dimensional case, assuming axial symmetry, the space-charge-field expressions were derived in Chapter IV for annular rings of charge. Equations (53a) and (54a) are substituted in Eqs. (96) and (97) above and the final force equations are obtained:

$$\frac{\partial u_y(y, x_0, \Phi_0)}{\partial y} [1 + 2Cu_y(y, x_0, \Phi_0)]$$
$$= -\left(1 - C\frac{d\theta_y(y)}{dy}\right) A(y)\psi(x) \sin(\Phi - \theta_x)$$
$$+ C\frac{dA(y)}{dy} \psi(x) \cos(\Phi - \theta_x) + \left(\frac{\omega_p}{\omega C}\right)^2 \frac{1}{2x_a^2}$$
$$\cdot \int_0^{x_{b'}} \left[\int_0^{2\pi} \sum_{l=1}^{\infty} e^{-\nu_l|y-y'|} \frac{J_0(\nu_j x) J_0(\nu_l x)}{[J_1(\nu_l x_a)]^2} \operatorname{sgn}(y - y') \, d\Phi_0' \right] x_0' \, dx_0' \quad (98)$$

and

$$\frac{\partial u_x(y, x_0, \Phi_0)}{\partial y} [1 + 2Cu_y(y, x_0, \Phi_0)]$$
$$= CA(y) \frac{d\psi(x)}{dx} \cos(\Phi - \theta_x)$$
$$+ CA(y)\psi(x) \frac{d\theta_x(x)}{dx} \sin(\Phi - \theta_x) + \left(\frac{\omega_p}{\omega C}\right)^2 \frac{1}{2x_a^2}$$
$$\cdot \int_0^{x_{b'}} \left[\int_0^{2\pi} \sum_{l=1}^{\infty} e^{-\nu_l|y-y'|} \frac{J_0(\nu_l x') J_1(\nu_l x)}{[J_1(\nu_l x_a)]^2} \, d\Phi_0' \right] x_0' \, dx_0'. \quad (99)$$

The independent variables are:

$$y, \quad x_0, \quad \text{and } \Phi_0.$$

The dependent variables are:

$$A(y) \quad \theta_y(y) \quad u_y(y, x_0, \Phi_0)$$
$$\psi(x) \quad \theta_x(x) \quad u_x(y, x_0, \Phi_0)$$
$$x \quad \Phi(y, x_0, \Phi_0).$$

The two-dimensional problem is solved in much the same way as the one-dimensional problem, i.e., as an initial-value problem wherein the dependent variables and their derivative values are specified.

A. *Rf Signal.*
(1) $A(0) \triangleq A_0$, relative to CI_0V_0
(2) $[dA(y)/dy] = -A_0(1 + Cb)d$
(3) $\theta_y(0) \equiv 0$
(4) $[d\theta_y(0)/dy] = -b$.

B. *Beam Input Conditions.*
(1) Input electron velocities

$$u_y(0, x_0, \Phi_0) \equiv 0 \quad \text{for an unmodulated beam}$$
$$u_x(0, x_0, \Phi_0) \equiv 0 \quad \text{for an unmodulated beam}$$

(2) Beam bunching

$$\Phi(0, \varphi_{0,j}) \equiv \Phi_{0,j}.$$

If the beam is unbunched, then

$$\Phi_{0,j} = \frac{2\pi j}{m} \quad j = 0, 1, 2, ..., m.$$

C. *Parameter Specification.*

$$C, \ d, \ b, \ \gamma b', \ \omega_p/\omega, \text{ and } a/b'.$$

The system of equations to be solved is Eqs. (79), (80), (82), (83), (93), (94), (98), and (99): a prodigious problem to say the least. These eight equations reduce to four in the one-dimensional case and are reduced to the klystron analysis by eliminating the circuit equations and the circuit field terms in the force equations and placing $b = 0$ and $\alpha = 2C$.

Before solving the above system to evaluate radial field effects on gain and efficiency we will proceed to generalize further the nonlinear interaction equations to allow for motion and field variations around the axis of symmetry.

6 Three-Dimensional O-TWA Analysis

The most general interaction configuration visualized for the O-TWA is one in which the rf circuit is periodic in both the axial and angular directions (biperiodic) and the electron beam is focused by a finite axial magnetic field, radial and angular motion of the electrons being allowed. An interesting special case of this analysis is that of a simply periodic circuit with the same beam conditions, i.e., radial and angular motions. The general analysis for this type of interaction system will utilize the general equivalent circuit of Section 5 of Chapter III and the space-charge-field expressions of Section 7 of Chapter IV.

The rf voltage of the system is again defined in terms of the product of slowly varying functions of distance and phase:

$$V(z, r, \varphi, t) = \mathrm{Re}\left[\frac{I_0}{2\pi a C Y_{0,0}} A(z, \varphi)\psi(r)e^{-j\Phi}e^{j\theta_x}\right], \tag{100}$$

where

$\Phi = \Phi(z, t, \varphi)$,

$\theta_x = \theta_x(r)$,

$\psi(a) = 1$,

$\theta_x(a) = 0$,

and

$Y_{0,0} = (C_{0,0}/L_{0,0})^{\frac{1}{2}}$, the characteristic admittance in the z-direction (axial).

The dependent Lagrangian variables are defined as previously used. They are:

$$y \triangleq \frac{C\omega}{u_0} z, \tag{101a}$$

$$x \triangleq \frac{C\omega}{u_0} r, \tag{101b}$$

$$x_0 \triangleq \frac{C\omega}{u_0} r_0 \tag{101c}$$

$$\Phi_0 \triangleq \frac{\omega}{u_0} z_0 = \omega t_0, \tag{101d}$$

$$\Phi(z, t, \varphi) \triangleq \omega \left(\frac{z}{u_0} - t\right) - \theta_c(z, \varphi), \tag{101e}$$

$$\frac{u_0}{C\omega} \frac{dy}{dt} \triangleq u_0[1 + 2Cu_y(y, x_0, \Phi_0, \varphi_0)], \tag{101f}$$

$$\frac{u_0}{C\omega} \frac{dx}{dt} \triangleq u_0[2Cu_x(y, x_0, \Phi_0, \varphi_0)] \tag{101g}$$

and

$$\frac{d\varphi}{dt} \triangleq u_0 \left[\frac{2C}{r} u_\varphi(y, x_0, \Phi_0, \varphi_0)\right]. \tag{101h}$$

The usual velocity-phase equation relating dependent variables is evolved by combining Eqs. (101e) and (101f). The result is

$$\frac{\partial \Phi(y, \Phi_0, \varphi_0)}{\partial y} + \frac{\partial \theta_c(y, \varphi)}{\partial y}$$
$$= \frac{1}{C} \left[\frac{1}{1 + 2Cu_y(0, x_0, \Phi_0, \varphi_0)} - \frac{1}{1 + 2Cu_y(y, x_0, \Phi_0, \varphi_0)}\right], \tag{102}$$

where again $u_y(0, x_0, \Phi_0, \varphi_0)$ indicates the initial velocity modulation of the beam. Recall that both the radial position x and the angular position φ are being treated as dependent variables and hence the following equations are evolved:

$$x(y, x_0, \Phi_0, \varphi_0) = x_0 + 2C \int_0^y \frac{u_x(y, x_0, \Phi_0, \varphi_0)}{1 + 2Cu_y(y, x_0, \Phi_0, \varphi_0)} dy \tag{103}$$

and

$$\varphi(y, x_0, \Phi_0, \varphi_0) = \varphi_0 + 2C \int_0^y \frac{u_\varphi(y, x_0, \Phi_0, \varphi_0)}{x} \frac{dy}{[1 + 2Cu_y(y, x_0, \Phi_0, \varphi_0)]}. \tag{104}$$

Within the circuit, i.e., $r < a$, the potential function must satisfy the following wave equation (neglecting space charge):

$$\nabla^2 V(r, z, \varphi) - \frac{1}{c^2} \frac{\partial^2 V}{\partial t^2} = 0. \tag{105}$$

Substitution of the appropriate derivatives of the potential function into Eq. (105) and independently setting the coefficients of $\sin(\Phi - \theta_x)$ and $\cos(\Phi - \theta_x)$ equal to zero yields the equations for the determination of the radial field weighting function, $\psi(x)$.

6. THREE-DIMENSIONAL O-TWA ANALYSIS

The two-dimensional equivalent circuit equation is given in Section 5 of Chapter III and may be conveniently separated into two equations after substituting the potential function of Eq. (100) and appropriate derivatives thereof. The driving function on the right-hand side is conveniently handled by the introduction of a planar charge density defined as

$$\sigma = -\int_0^{b'} \frac{\psi(r)\rho'}{a} r\, dr, \qquad (106)$$

where ρ' is the volume charge density and $\psi(r)$ represents the effective coupling coefficient. In a three-dimensional Lagrangian system the conservation of charge relationship is expressed as

$$\rho r\, dr\, d\varphi\, dz = \rho_0 r_0\, dr_0\, d\varphi_0\, dz_0 \qquad (107)$$

and since only the fundamental component, i.e., σ_1, is needed (it is again assumed that all of the circuit impedance is at the fundamental frequency) then σ is expanded in a Fourier series. Incorporation of the above along with the appropriate derivatives of the potential function yields the desired circuit equations.

The Lorentz force equations, for this geometry, may be written as follows, utilizing the previously indicated coordinate system and including the effect of an axial magnetic focusing field:

$$\frac{d^2 r}{dt^2} - r\left(\frac{d\varphi}{dt}\right)^2 = -|\eta|\left[E_{sc-r} + E_{o-r} - \frac{\partial V}{\partial r} + B_z r \frac{d\varphi}{dt}\right], \qquad (108)$$

$$\frac{1}{r}\frac{d}{dt}\left(r^2 \frac{d\varphi}{dt}\right) = -|\eta|\left[E_{sc-\varphi} - \frac{1}{r}\frac{\partial V}{\partial \varphi} - B_z \frac{dr}{dt}\right] \qquad (109)$$

and

$$\frac{d^2 z}{dt^2} = -|\eta|\left[E_{sc-z} - \frac{\partial V}{\partial z}\right], \qquad (110)$$

where the magnetic field is assumed to be entirely axially directed at the input to the interaction region and E_{o-r} is the radial field due to the dc space charge. At $z = 0$, $d\varphi/dt$ will have a specified value. If no flux threads the cathode then from Busch's theorem, $d\varphi/dt = \eta B_z/2$ for the special case of Brillouin flow. The various derivatives on the left-hand side of Eqs. (108)–(110) are formed through the use of Eqs. (101). The right-hand side terms are written with the aid of Eq. (100) and the space-charge-field expressions of Section 7 of Chapter IV and Section 5 of Chapter V.

7 Two-Dimensional Circuit, Three-Dimensional Flow

If the usual one-dimensional circuit equation modified to include radial field effects is assumed and yet electron motion in both the radial and angular directions is allowed then some simplification of the system of Section 6 is possible. The derivation of the working equations proceeds as outlined in previous sections of this chapter and hence only the results are quoted here.

Radial Field Equations

$$\psi \frac{d^2 A}{dy^2} - A\psi \left(\frac{1}{C} - \frac{d\theta_y}{dy}\right)^2 + A \frac{d^2\psi}{dx^2} - \left(\frac{d\theta_x}{dx}\right)^2 A\psi + \frac{A}{x}\frac{d\psi}{dx} + A\psi \left(\frac{k_e}{C}\right)^2 = 0 \tag{111}$$

$$-2\psi \left(\frac{1}{C} - \frac{d\theta_y}{dy}\right)\frac{dA}{dy} + A\psi \frac{d^2\theta_y}{dy^2} + 2A \frac{d\psi}{dx}\frac{d\theta_x}{dx} + A\psi \frac{d^2\theta_x}{dx^2} + \frac{A\psi}{x}\frac{d\theta_x}{dx} = 0. \tag{112}$$

Circuit Equations

$$\frac{d^2 A}{dy^2} + A\left[\frac{(1+Cb)^2}{C^2} - \left(\frac{1}{C} - \frac{d\theta_y}{dy}\right)^2\right]$$

$$= -\frac{2(1+Cb)}{\pi C x_{b'}^2} \left\{\int_0^{x_{b'}} \left[\int_0^{2\pi} \psi(x) \frac{\cos\Phi}{1+2Cu_y} d\Phi_0'\right.\right.$$

$$\left.\left. + 2Cd \int_0^{2\pi} \psi(x) \frac{\sin\Phi}{1+2Cu_y} d\Phi_0'\right] x_0' dx_0'\right\}, \tag{113}$$

$$A\left[\frac{d^2\theta_y}{dy^2} - \frac{2d}{C}(1+Cb)^2\right] + 2\frac{dA}{dy}\left(\frac{d\theta_y}{dy} - \frac{1}{C}\right)$$

$$= -\frac{2(1+Cb)}{\pi C x_{b'}^2} \left\{\int_0^{x_{b'}} \left[\int_0^{2\pi} \frac{\psi(x)\sin\Phi}{1+2Cu_y} d\Phi_0'\right.\right.$$

$$\left.\left. - 2Cd \int_0^{2\pi} \frac{\psi(x)\cos\Phi}{1+2Cu_y} d\Phi_0'\right] x_0' dx_0'\right\}. \tag{114}$$

Force Equations

$$\frac{\partial u_x}{\partial y}(1+2Cu_y) = AC \frac{d\psi}{dx} \cos(\Phi - \theta_x) + A\psi C \frac{d\theta_x}{dx} \sin(\Phi - \theta_x)$$

$$\cdot \frac{2C}{x} u_\varphi^2 - \left(\frac{\omega_c}{\omega C}\right) u_\varphi + \left(\frac{\omega_p}{\omega}\right)^2 \frac{1}{4C^3}\frac{x_0^2}{x} + \left(\frac{\omega_p}{\omega C}\right)^2 \int_0^{2\pi}\int_0^{x_{b'}}\int_0^{2\pi} \sum_{l=1}^{\infty}\sum_{s=0}^{\infty}(2-\delta_s^0)$$

$$\cdot e^{-\nu_{ls}|y-y'|} \cos s(\varphi - \varphi') \frac{J_s(\nu_{ls}x')[J_{s+1}(\nu_{ls}x) - J_{s-1}(\nu_{ls}x)]}{2[J_{s+1}(\nu_{ls}x_a)]^2} d\Phi_0' x_0' dx_0' d\varphi_0', \tag{115}$$

$$\frac{\partial u_y}{\partial y}(1+2Cu_y) = C\psi \frac{dA}{dy}\cos(\Phi-\theta_x) - \left(1 - C\frac{d\theta_y}{dy}\right) A\psi \sin(\Phi-\theta_x)$$

$$+ \left(\frac{\omega_p}{\omega C}\right)^2 \int_0^{2\pi}\int_0^{x_b'}\int_0^{2\pi} \sum_{l=1}^{\infty}\sum_{s=0}^{\infty} \text{sgn}(y-y')(2-\delta_s{}^0)$$

$$\cdot e^{-\nu_{ls}|y-y'|} \cos(\varphi-\varphi') \frac{J_s(\nu_{ls}x')J_s(\nu_{ls}x)}{[J_{s+1}(\nu_{ls}x_a)]^2} d\Phi_0' \, x_0' \, dx_0' \, d\varphi_0', \qquad (116)$$

$$\frac{\partial u_\varphi}{\partial y}(1+2Cu_y) = \left(\frac{\omega_c}{\omega C}\right)u_x - \frac{2C}{x}u_x u_\varphi + \left(\frac{\omega_p}{\omega C}\right)^2 \int_0^{2\pi}\int_0^{x_b'}\int_0^{2\pi}$$

$$\cdot \sum_{l=1}^{\infty}\sum_{s=0}^{\infty} (2-\delta_s{}^0) e^{-\nu_{ls}|y-y'|} \sin s(\varphi-\varphi') \frac{J_s(\nu_{ls}x')J_s(\nu_{ls}x)}{(\nu_{ls}x)[J_{s+1}(\nu_{ls}x_a)]^2} d\Phi_0' \, x_0' \, dx_0' \, d\varphi_0'.$$

(117)

A significant further simplification in the above system ensues when the space-charge potential is assumed axially symmetric so that the space-charge-field expressions used in Section 5 are appropriate. For shielded Brillouin flow $d\varphi/dt = \eta B_z/2$ so that

$$u_\varphi(0, x_0, \Phi_0, \varphi_0) = \frac{\omega_c}{\omega}\left(\frac{x_0}{4C^2}\right).$$

8 Effects of Transverse Variations on TWA Gain and Efficiency

a. Introduction

In the preceding sections of this chapter the general multidimensional theory of traveling-wave interaction, including radial and angular field effects (space-charge and circuit), is developed for beams focused in arbitrary magnetic field strengths. These general interaction equations are solved in basically the same way as the earlier derived one-dimensional equations, i.e., as an initial-value problem. As might be expected the computation time becomes quite excessive when all three components of the space-charge potential are computed, since this involves the calculation of weighting functions for electrons in various annular segments of the beam. The computation time associated with radial and angular circuit field variations is quite small compared to that associated with the space-charge potential.

A considerable simplification of the problem, and a resultant reduction in computation time, results when the space-charge potential is assumed to be axisymmetric. Motion about the axis is still allowable through terms of the type ω_c/ω in the force equations. In the computation results to be presented here this assumption has been made and we further confine ourselves to axisymmetric rf circuit fields. In finite ω_c/ω cases (motion about the axis) the longitudinal integration increment must be decreased in order to maintain accuracy; i.e., $\Delta y \ll$ cyclotron wavelength. Trajectory and beam velocity information is more sensitive to having a proper value of Δy than is the rf circuit voltage characteristic. Acceptable values of Δy are governed by the following relationships:

$$\Delta y < \frac{2\pi C}{120(\omega_c/\omega)} \quad \text{[for circuit variables]}$$

$$\Delta y < \frac{2\pi C}{600(\omega_c/\omega)} \quad \text{[for beam variables]}.$$

The following subsections are devoted to summaries of the findings and indicate the importance of transverse effects on rf output.

b. Stream Subdivision

Since angular field variations are not being considered, the stream subdivision involves a division into annular rings or layers of charge according to the value of $\gamma b'$ and then the division of each charge layer into representative electron charge groups as in the one-dimensional calculations. It should be recalled that the N-beam calculations indicate that $\gamma b' \leqslant 1$ for maximum efficiency and thus we will confine ourselves to that maximum. Since the stream is composed of annular rings of charge it is a simple matter to analyze hollow beams as illustrated in Fig. 31 where the normalized diameters are indicated.

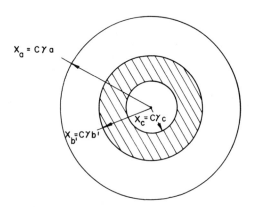

FIG. 31. Cross section of stream-circuit model for multidimensional calculations.

Typical results of calculations for various degrees of stream subdivision and number of charge groups per layer are shown in Fig. 32 for representative operating parameters. Note that both $\omega_c/\omega \neq 0$ and $\omega_p/\omega \neq 0$ have been specified; the electrons all enter the rf interaction region without radial velocity. It is concluded from these results that 3 annular rings of charge and 32 charge groups per layer are sufficient for the calculations. In all of these calculations γa has been taken as 1.5 since this is near optimum from a maximum bandwidth standpoint. In all remaining calculations 3 layers and 32 groups per layer are utilized.

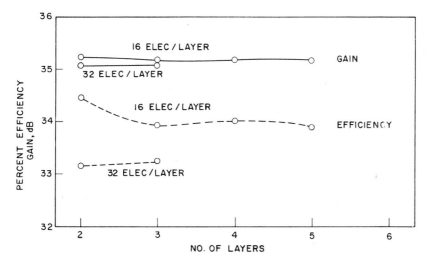

Fig. 32. Effect of stream subdivision on TWA gain and efficiency ($C = 0.1$, $b = 1$, $d = 0$, $\psi_0 = -30$, $x_a = 0.15$, $x_{b'} = 0.105$, $\omega_c/\omega = 0.4$, $\omega_p/\omega = 0.2$).

c. Stream Diameter

The principal departure of the multidimensional results from the one-dimensional predictions arises from the induced velocity slippage across the beam due to radial fields and rf circuit fields acting nonuniformly across the bunch. A result shown in detail later is that the excess stream velocity, as measured by $u_0/v_0 = 1 + Cb$, required for maximum power output is less when part of the stream energy is taken up in radial and angular motion.

The gain and efficiency for $\omega_p/\omega = 0$ are shown in Figs. 33 and 34 as a function of $\gamma b'/\gamma a$ for various magnetic field values and stream-circuit coupling parameter values. These results are compared with the one-dimensional results to indicate the importance of transverse circuit field variations. As mentioned earlier, one pronounced effect is

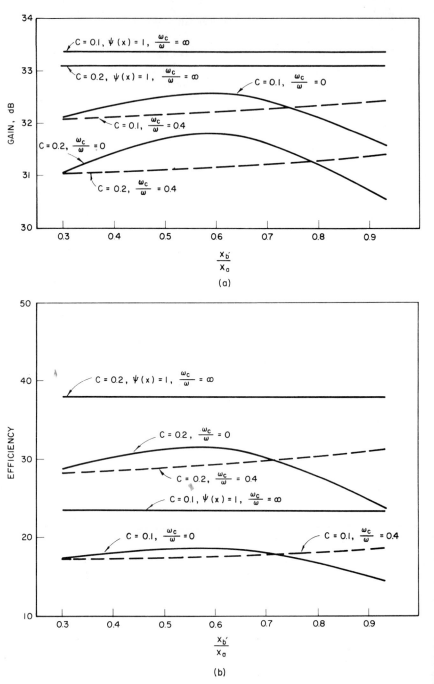

Fig. 33. Gain and efficiency versus stream diameter ($\omega_p/\omega = 0$, $b = 0$, $\psi_0 = -30$). (a) Gain; (b) efficiency.

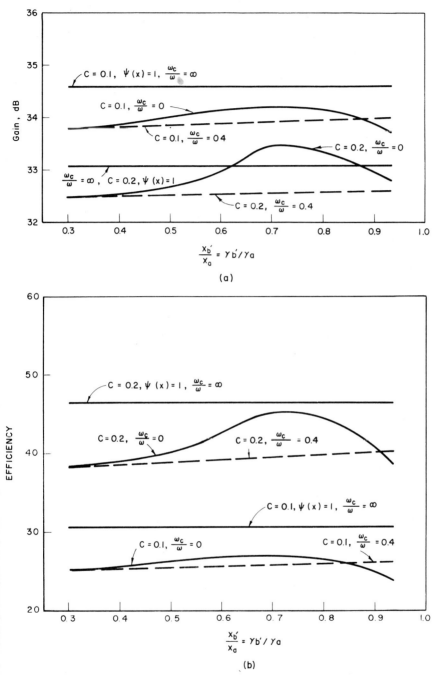

FIG. 34. Gain and efficiency versus stream diameter ($\omega_p/\omega = 0$, $b = 0.5$, $\psi_0 = -30$). (a) Gain; (b) efficiency.

that when radial variations are included the injection velocity required is considerably reduced compared to the one-dimensional predictions. This fact seems to agree well with experimental results in that the beam voltage for maximum power output is usually only a few percent above synchronism.

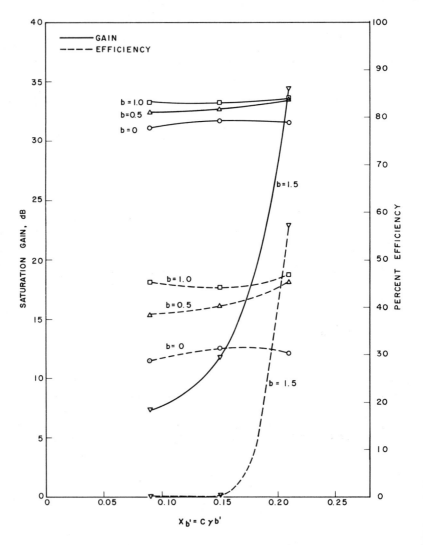

FIG. 35. Gain and efficiency versus $x_{b'}$ and b ($C = 0.2$, $\gamma a = 1.5$, $\omega_p/\omega = 0$, $\omega_c/\omega = 0$).

8. EFFECTS OF TRANSVERSE VARIATIONS

A striking result is that there is an optimum beam filling factor between 50 and 75 percent of the mean circuit diameter, depending on the beam injection velocity. This corresponds to an optimum $\gamma b'$ from 0.75 to 1.0 which agrees well with experimental characteristics. The rise in efficiency when $\omega_c/\omega \to 0$, of course, occurs as a result of beam expansion and the resultant stronger coupling to the rf circuit fields. This dependence on b is summarized for $C = 0.2$ in Fig. 35.

When space-charge forces are included there is an additional transverse effect corresponding to radial debunching and thus it is interesting to view again the question of optimum $\gamma b'$ when $\omega_p/\omega \neq 0$. Typical results are illustrated in Fig. 36 and it is apparent that the optimum is

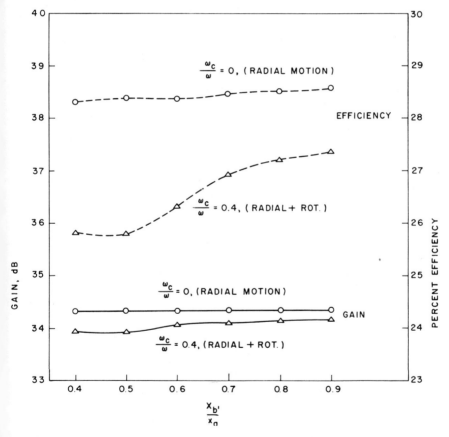

FIG. 36. Gain and efficiency versus stream diameter ($C = 0.1$, $b = 0.5$, $d = 0$, $x_a = 0.15$, $\psi_0 = -30$, $\omega_p/\omega = 0.1$).

less well defined but that strong coupling to the circuit is required for high efficiency. The effect of ω_p/ω on gain and efficiency at fixed $\gamma b'$ is shown in Fig. 37 for two different magnetic field conditions. Again as expected, one should design for low ω_p/ω to obtain maximum output and to realize maximum benefit from various phase-focusing techniques as detailed in Chapter XIII.

d. Stream Fundamental Current

The velocity and field variations across the stream give rise to a variation of current density across the stream. The fundamental and harmonic beam current amplitudes are calculated according to Eq. (61)

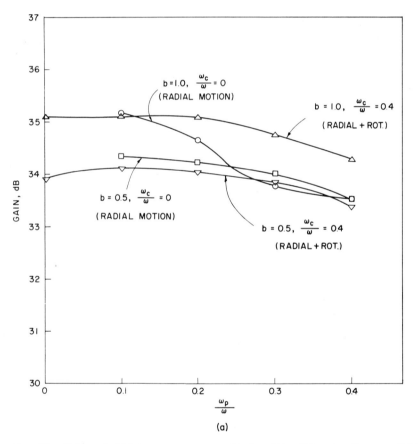

Fig. 37. Gain and efficiency versus ω_p/ω and ω_c/ω ($C = 0.1$, $d = 0$, $x_a = 0.15$, $x_b = 0.105$, $\psi_0 = -30$). (a) Gain; (b) efficiency.

of Chapter V by integrating over the entering charge and over the beam cross section at any displacement plane. Typical maximum values of the normalized fundamental current amplitude range between 1.1 and 1.4, depending upon ω_p/ω and ω_c/ω.

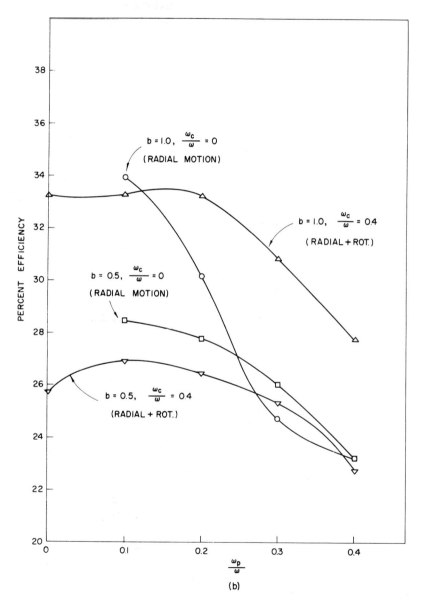

(b)

e. Rf Circuit Interception

When the axial magnetic field is weak or zero, the radial space-charge fields cause a beam expansion which results in tighter coupling to the rf field and eventually beam collection on the rf structure if the space-charge debunching is sufficiently strong. As expected, the interception on the circuit increases for increasing ω_p/ω at fixed ω_c/ω and decreasing ω_c/ω for fixed ω_p/ω. A typical set of results showing the effect of both ω_p/ω and ω_c/ω is given in Fig. 38, again for a specific input rf drive level. As

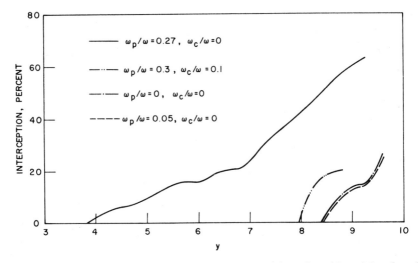

FIG. 38. Circuit interception versus distance and ω_p/ω ($C = 0.1$, $b = 1$, $x_a = 0.15$, $x_{b'} = 0.105$, $\psi_0 = -30$).

the applied rf signal is increased the saturation length of the device decreases and the interception occurs earlier. The amount of interception occurring is a result of the unbalance of the magnetic focusing field and the defocusing space-charge field. The amplitude of the rf field relative to the space-charge fields also influences the rate and amount of interception.

f. Rf Circuit Loss

We have seen in the previous one-dimensional calculations that rf attenuation does affect the performance of an amplifier if the attenuator is too close to the output or if saturation effects set in within the attenuator. It is instructive to evaluate radial effects on attenuator action. One of the important characteristics relates to velocity spread within the stream under the attenuator. Recall that the required excess beam

8. EFFECTS OF TRANSVERSE VARIATIONS

velocity is less when part of the energy goes into radial motion. Since the optimum interaction occurs for $0 \leqslant b \leqslant 1.0$ in the $d = 0$ case and it is known that the presence of loss slows the wave down, it is not surprising that the output for $d > 0$ is higher than that for $d = 0$ when b is large, say 1.5. Of course further increases in d eventually reduce the rf output level below the $d = 0$ level.

A number of similar results are compared with the one-dimensional ones on the basis of output power degradation versus d in Fig. 39. The

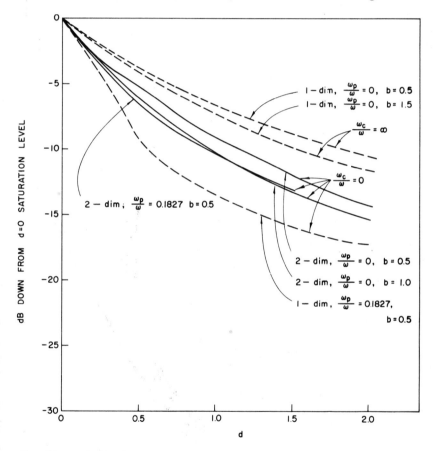

FIG. 39. Reduction in saturation power versus circuit loss ($C = 0.1$, $x_a = 0.15$, $x_{b'} = 0.105$, $\psi_0 = -30$).

additional compression of the output due to two-dimensional effects in the lossy region amounts to some 4 dB. It should be noted that the two-dimensional space-charge result lies above the corresponding one-dimensional prediction because $\omega_c/\omega = 0$ in the former case, allowing a beam expansion.

g. Trajectory and Velocity Characteristics

An excellent understanding of the bunching process and the energy exchange mechanism comes from the analysis of the electron trajectories

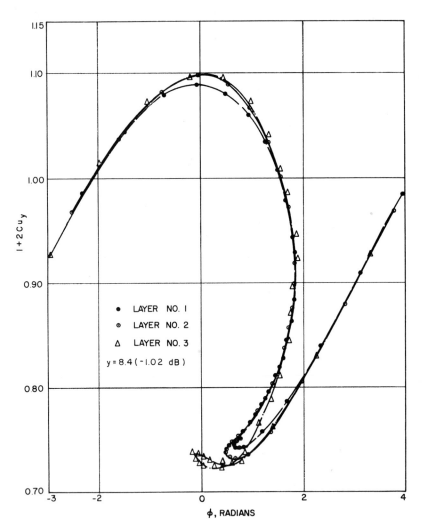

Fig. 40. Axial velocity versus phase ($C = 0.1$, $b = 1$, $d = 0$, $\psi_0 = -30$, $\omega_p/\omega = 0$, $x_a = 0.15$, $x_b = 0.105$, $\omega_c/\omega = 0$).

8. EFFECTS OF TRANSVERSE VARIATIONS

and the distribution of velocities with phase. Recall that the normalized directional velocities were defined as

$$\frac{1}{C\omega}\frac{dy}{dt} = 1 + 2Cu_y(y, x_0, \Phi_0, \varphi_0),$$

$$\frac{1}{C\omega}\frac{dx}{dt} = 2Cu_x(y, x_0, \Phi_0, \varphi_0)$$

and

$$\frac{x}{C\omega}\frac{d\varphi}{dt} = 2Cu_\varphi(y, x_0, \Phi_0, \varphi_0).$$

Thus it is apparent that the squares of the above relations are a measure of the longitudinal, radial and angular energies. A typical set of plots of each of the above is shown in Figs. 40–42 and the trajectory informa-

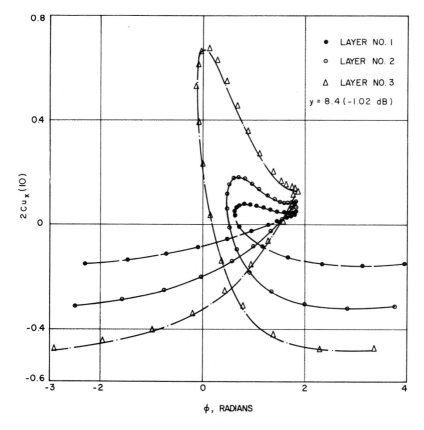

FIG. 41. Radial velocity versus phase, $y = 8.4$ ($C = 0.1$, $b = 1$, $d = 0$, $\psi_0 = -30$, $\omega_p/\omega = 0$, $x_a = 0.15$, $x_{b'} = 0.105$, $\omega_c/\omega = 0$).

tion, i.e., $x(y)$ versus y, is given in Fig. 43. Consideration of the velocity graphs indicates that the most favorably phased electrons, i.e., near $\Phi = \pi/2$ on the $1 + 2Cu_y$ graph, are the ones with the smallest radial and angular velocities as might be expected. The formation of the bunch is apparent, as is the defocusing due to nonuniform circuit fields and space-charge fields. Only the profile of each layer is shown in Fig. 43 to illustrate the degree of beam spread near the output plane. It is

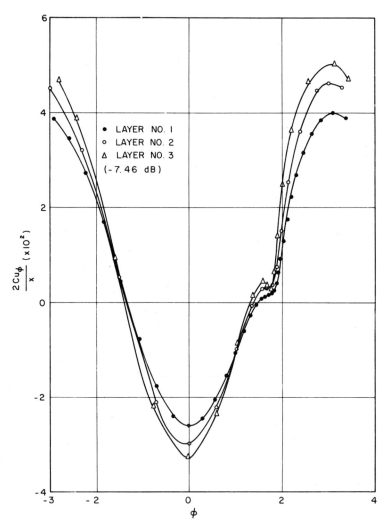

FIG. 42. Angular velocity versus phase, $y = 7.2$ ($C = 0.1$, $b = 1$, $d = 0$, $\psi_0 = -30$, $\omega_p/\omega = 0$, $x_a = 0.15$, $x_{b'} = 0.105$, $\omega_c/\omega = 0.8$).

apparent that some mechanism by which low axial energy electrons could be collected and removed from the interaction region would significantly improve the rf performance.

The generality of the multidimensional analysis is such as to permit the study and evaluation of many other facets of the interaction process including rf circuit severs, effect of rf on Brillouin focused beams, and periodic magnetic fields for focusing. As illustrated earlier, either solid or hollow beams are amenable to investigation simply by prescribing the initial charge and velocity distributions with radius.

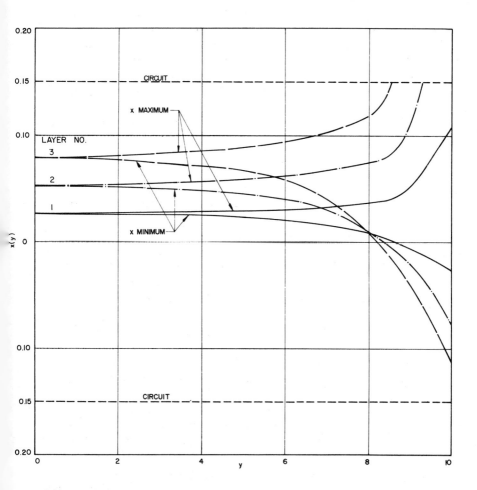

Fig. 43. Radial versus axial position ($C = 0.1$, $b = 1$, $d = 0$, $\psi_0 = -30$, $\omega_p/\omega = 0$, $x_a = 0.15$, $x_{b'} = 0.105$, $\omega_c/\omega = 0$).

9 Relativistic O-TWA

Many high-power traveling-wave amplifiers utilize beam voltages sufficiently high that relativistic corrections to the Lorentz force equation are required. Considering the basic mode of operation of beam-wave devices suggests that the gain should decrease as $u_0 \to c$ since in the limit an infinite force would be required to move or bunch the beam particles.[18,19] On the other hand the efficiency of a relativistic nonlinear beam-wave device would be expected to increase, since the velocity spread in the stream is not a significant fraction of the average velocity and hence it would have a less deleterious effect on bunching and the energy exchange process. Furthermore, for a given percentage decrease in velocity a relativistic beam gives up more energy than does a nonrelativistic beam. The Lorentz transformation will be written here as a linear transformation of the components of the vector $x^j(x^1, x^2, x^3, x^4 = ct)$ in a four-dimensional space. Implicit in the special theory is the invariance of certain proper quantities such as the velocity of light and the combined space-time interval. The Lorentz transformation is a proper one which satisfies the above principles. A fundamental postulate of the special theory is that the conservation of momentum for particles retains its classical form and hence is Lorentz covariant (independent of the choice of inertial frame). Thus it is seen that the particle mass cannot be invariant but must depend directly on the particle velocity in the particular frame of reference.

Current density and charge density are components of a four-vector and are simply different aspects of the same thing. The electronic charge, on the other hand, is a universal constant of relativity and thus in a Lagrangian analysis, such as is used in the nonlinear beam-wave interaction theory, a counting operation on the fundamental unit of charge is also an invariant.

Other useful consequences of the special theory are that both Maxwell's equations and the scalar and vector potentials lend themselves to a covariant description. It should be clear, then, that the formulation of the nonlinear equations for beam-wave interactions can proceed in the laboratory system with the major modifications to be made in the force equations and the fact that both the scalar-electric and magnetic-vector potentials must be considered along with a replacement of Poisson's equation by the inhomogeneous wave equation. Since the D'Alembertian operator is an invariant, the homogeneous equation governing electromagnetic wave propagation is in covariant form.

The circuit equations may be handled in any of several manners depending on the choice of definition of the potential and on the dimensionality of the problem. We assume a one-dimensional system in which the component of circuit electric field in the direction of electron travel is invariant to a proper transformation. Therefore we may choose the circuit electric field as given by $E_c = -\partial V_c/\partial z$, where V_c is the circuit potential. Thus the circuit equations used previously in the one-dimensional analysis are valid as they stand (Section 2). In a two-dimensional problem, wherein one takes all space harmonics of the total field into account including the complete energy storage circuit, each individual space-harmonic field component does not satisfy Maxwell's equations whereas the total field does. In this situation some space-harmonic components will be zero at $k_0 = 1$, i.e., $v_0/c = 1$, whereas others may not be. The exact solution of the two-dimensional problem plus boundary conditions will indicate this.

In the present analysis the optimum interaction condition occurs when the beam is synchronized with the wave phase velocity, and since the electron velocity is necessarily less than the velocity of light both k_e and k_0 are less than unity and there will be a finite (though small) axial electric field. Synchronism between the beam and the wave results in the rf wave appearing as a static wave and it in turn exerts long continued forces on the electrons. The helical waveguide satisfies these conditions in that the axial electric field becomes vanishingly small as $v_p \to c$ and in the limit a TEM wave exists. The presence of other space harmonics may be accounted for in the circuit equations by including space-harmonic terms as additional driving terms. A specific rf structure type would then have to be considered in order to specify analytically the form of the driving terms.

The force equation to be used is developed in a similar manner to that for the relativistic klystron (Section 7 of Chapter V) and is written as

$$\frac{dv}{dt} = -\eta_0(1 - k_e^2)^{\frac{3}{2}}(E_c + E_{sc}). \tag{118}$$

The space-charge field for the one-dimensional relativistic TWA is the same as for the one-dimensional relativistic klystron given in Section 7 of Chapter V. Thus the only equation in the one-dimensional analysis to be changed is the force equation:

$$\frac{\partial u(y, \Phi_0)}{\partial y}[1 + 2Cu(y, \Phi_0)] = (1 - k_e^2)^{\frac{3}{2}} \left\{ -A(y)\left(1 - C\frac{d\theta(y)}{dy}\right) \sin \Phi(y, \Phi_0) \right.$$

$$\left. + C\frac{dA(y)}{dy} \cos \Phi(y, \Phi_0) + \frac{1}{(1 + Cb)}\left(\frac{\omega_p}{\omega C}\right)^2 \int_0^{2\pi} \frac{F_{1-z}(\Phi - \Phi') \, d\Phi_0'}{1 + 2Cu(y, \Phi_0')} \right\}. \tag{119}$$

The relation between variables remains the same.

The calculation of efficiency for these relativistic interactions proceeds directly from the equivalence of mass and energy given by the Einstein relation $\Delta \mathscr{E} = \Delta mc^2$. The decrease in potential energy or increase in kinetic energy of a stream when accelerated through a large potential difference V_r is given by

$$eV_r = m_0 c^2 (p_e - 1) \tag{120}$$

and under the low-velocity assumption

$$eV_0 = \frac{m_0}{2} u_0^2. \tag{121}$$

Thus

$$\frac{V_r}{V_0} = \frac{2p_e^2}{p_e + 1}. \tag{122}$$

In making the relativistic calculations $C^3 = Z_0 I_0 / 4V_0$ was assumed constant and hence there must necessarily be an impedance change as V_r changes in order to maintain C constant. The impedance ratio is given by

$$\frac{Z_{0r}}{Z_0} = \frac{2p_e^2}{p_e + 1}. \tag{123}$$

The power output in general is given by

$$P = \frac{V_{rf}^2}{Z_{0r}}\bigg|_{\text{avg}} = 2CI_0 V_0 (A_{\max}^2 - A_0^2) \frac{(1 - C\, d\theta(y)/dy)}{(1 + Cb)} \frac{(p_e + 1)}{2p_e^2}, \tag{124}$$

where

$$A_{\max} = A(y)|_{\max},$$

and

$$A_0 = A(0).$$

The saturation efficiency is then

$$\text{Eff} = 2C(A_{\max}^2 - A_0^2)\frac{(1 - C\, d\theta(y)/dy)}{(1 + Cb)} \frac{(p_e + 1)^2}{4p_e^4}, \tag{125}$$

since the input power is $I_0 V_0 [2p_e^2/(p_e + 1)]$. Calculations on the spent beam indicate that the velocity spread in the beam is independent of the voltage.

It is important to note that as $k_e \to 1$ the gain parameter C becomes small and large interaction lengths would be required. It is thus very important to carry out beam bunching at low velocities and extract energy in a relativistic region.

A summary of efficiency calculations on relativistic TWA's is shown in Fig. 44 and is also compared with the asymptotic expression.

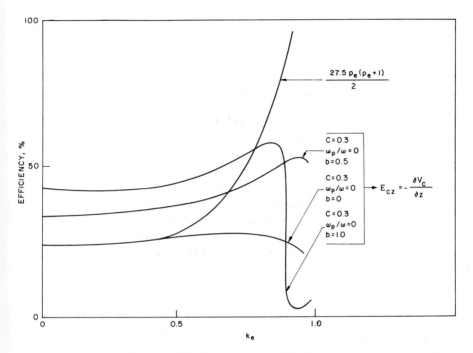

FIG. 44. Efficiency versus relativity factor.

10 Integral Equation Analysis

a. General

The basic analysis procedure utilized in this book for all types of devices involves the forward integration of the combined circuit and stream equations in Lagrangian form. In particular, the rf circuit is described in terms of an equivalent uniform transmission line and the second-order differential equation for voltage on the line is integrated numerically with the proper boundary conditions. Clearly this is not the only method of treating such a problem, since field analyses could be used (but not without great difficulty), and in fact the solution of the differential circuit equation can be obtained by other procedures. The numerical integration of the circuit differential equation does have the advantage of being direct and simple.

An alternate method which has been used both by Poulter[6] and by Tien[10] involves starting the numerical integration from the general solution of the differential equation written in semiclosed form. While formally correct, such a procedure is somewhat involved and does lead to conceptual difficulties. This integral equation treatment is developed and it is shown to be rigorously equivalent to the former method, both by analytical and numerical methods. Although the development and equivalence are applied to a one-dimensional system, the concept transfers directly to multidimensional systems.

b. Integral Equation for Circuit Voltage

An element of charge in a stream passing close to an rf circuit induces both forward traveling (left to right) and backward traveling (right to left) waves on the circuit. Generally, as a result of phase cancellation, the amplitude of the backward traveling wave at any position on the circuit is small and may be neglected. Separation of the total wave into these two components introduces no additional generality.

On a lossless equivalent uniform transmission line, the voltage satisfies the following second-order differential equation for a one-dimensional system:

$$\frac{\partial^2 V(z, t)}{\partial z^2} - \frac{1}{v_0^2} \frac{\partial^2 V(z, t)}{\partial t^2} = -\frac{Z_0}{v_0} \frac{\partial^2 \rho_1(z, t)}{\partial t^2}, \tag{126}$$

where $\rho_1(z, t)$ represents the fundamental component of charge density in the stream, which is the circuit forcing function. Since only $\rho_1(z, t)$ drives the circuit, the rf voltage on the circuit will also all be at the fundamental frequency.

The general solution of Eq. (126) is composed of a complementary function and a particular solution. The general solution is written as

$$V(z) = C_1 e^{-\Gamma_0 z} + C_2 e^{\Gamma_0 z} + e^{-\Gamma_0 z} \frac{\Gamma_0 v_0 Z_0}{2} \int_{z'}^{z} e^{\Gamma_0 z} \rho_1(z) \, dz$$
$$+ e^{\Gamma_0 z} \frac{\Gamma_0 v_0 Z_0}{2} \int_{z'}^{z} e^{-\Gamma_0 z} \rho_1(z) \, dz, \tag{127}$$

where $\Gamma_0 \triangleq$ the undisturbed circuit propagation constant and the $e^{-\Gamma_0 z}$ factor indicates a wave traveling in the positive z-direction. The time variation $\exp j\omega t$ is suppressed. The first two terms of Eq. (127) represent the unforced waves on the circuit and the last two terms indicate the voltage due to induced currents. The first and third terms represent a wave traveling to the right, and the second and fourth a wave moving to the left. Thus the total voltage at any displacement plane along the structure which extends from $z = 0$ to $z = D$ is composed of four parts: two due to the undisturbed wave and the other two due to the stream. The induced circuit voltage at z' traveling from left to right (forward) is due to all the charge elements to the left of z', and the induced voltage traveling from right to left (backward) is due to all the charge elements to the right of z'.

If it is assumed that the circuit is terminated in its characteristic impedance, then the amplitude of the backward wave is zero at $z = D$ and the forward-wave amplitude at $z = 0$ is that due to the applied input signal. Thus Eq. (127) can be rewritten as

$$V(z) = V_{in}e^{-\Gamma_0 z} + e^{-\Gamma_0 z}\frac{\Gamma_0 v_0 Z_0}{2}\int_0^z e^{\Gamma_0 z}\rho_1(z)\,dz$$
$$+ e^{\Gamma_0 z}\frac{\Gamma_0 v_0 Z_0}{2}\int_z^D e^{-\Gamma_0 z}\rho_1(z)\,dz. \qquad (128)$$

This equation forms the basis of the analyses of Poulter and Tien. The detailed treatment by the two authors from this point differs, although the end results are quite similar. The treatment by Tien is the one described here.

The Poulter-Tien integral solution for $V(z)$ is, in effect, a convolution integral of the spatial impulse response of the helix to the current induced in the circuit by current elements in the stream. This impulse response has two components, one of which can be interpreted as a forward wave, the other as a backward wave, but the two waves have the same propagation constant and have no separate physical existence.

From Eq. (128) the forward and backward wave voltages are written as

$$F(z) \triangleq V_{in}e^{-\Gamma_0 z} + e^{-\Gamma_0 z}\frac{\Gamma_0 v_0 Z_0}{2}\int_0^z e^{\Gamma_0 z}\rho_1(z)\,dz \qquad (129)$$

and

$$B(z) \triangleq e^{\Gamma_0 z}\frac{\Gamma_0 v_0 Z_0}{2}\int_z^D e^{-\Gamma_0 z}\rho_1(z)\,dz. \qquad (130)$$

The forward and backward components satisfy the following first-order differential equations:

$$\frac{\partial F(z, t)}{\partial z} + \frac{1}{v_0}\frac{\partial F(z, t)}{\partial t} = \frac{Z_0}{2}\frac{\partial \rho_1(z, t)}{\partial t} \qquad (131)$$

and

$$\frac{\partial B(z, t)}{\partial z} - \frac{1}{v_0}\frac{\partial B(z, t)}{\partial t} = -\frac{Z_0}{2}\frac{\partial \rho_1(z, t)}{\partial t}. \qquad (132)$$

Addition of the above two equations yields a relation between $F(z, t)$ and $B(z, t)$:

$$\frac{\partial F(z, t)}{\partial z} + \frac{1}{v_0}\frac{\partial F(z, t)}{\partial t} = -\frac{\partial B(z, t)}{\partial z} + \frac{1}{v_0}\frac{\partial B(z, t)}{\partial t}. \qquad (133)$$

Equations (131) and (132) are completely general and do not depend on the boundary conditions at $z = 0, D$. Thus

$$F = \frac{V + IZ_0}{2},$$

$$B = \frac{V - IZ_0}{2}.$$

The above equations constitute an alternate representation of the circuit voltage equation.

c. Traveling-Wave-Amplifier Equations

Both Poulter and Tien formulate the nonlinear TWA equations in a Lagrangian system using the normalizations originally introduced by Nordsieck. The normalized distance and phase independent variables are defined as in Eqs. (11) and (12):

$$y \triangleq \frac{C\omega}{u_0} z$$

and

$$\Phi_0 \triangleq \frac{\omega z_0}{u_0} = \omega t_0.$$

The dependent phase position variable and the charge group velocity are defined as

$$\Phi(y, \Phi_0) \triangleq \omega \left(\frac{z}{v_0} - t\right)$$

and

$$u_t(y, \Phi_0) \triangleq u_0[1 + Cw(y, \Phi_0)],$$

where $\Phi(y, \Phi_0)$ is defined in reference to the undisturbed circuit wave and the velocity variable $w(y, \Phi_0)$ refers to the initial average stream velocity, u_0. The usual velocity parameter $b = (u_0 - v_0)/Cv_0$ is used in the analysis.

The conservation of charge equation and the expansion of the stream space-charge density in a Fourier series proceeds as previously outlined and it is again assumed that only the fundamental component of space-charge density excites the circuit.

10. INTEGRAL EQUATION ANALYSIS

Circuit Equations. In view of the fact that the complementary functions of Eq. (127) are well out of synchronism with the beam, they may be neglected except in considering the boundary conditions. Since the particular solutions of Eqs. (131) and (132) are of interest, it is convenient to define the forward and backward amplitude functions in terms of components as

$$F(y, \Phi) \triangleq \frac{Z_0 I_0}{4C} [a_1(y) \cos \Phi - a_2(y) \sin \Phi] \qquad (134)$$

and

$$B(y, \Phi) \triangleq \frac{Z_0 I_0}{4C} [b_1(y) \cos \Phi - b_2(y) \sin \Phi]. \qquad (135)$$

The components are simply convolutions of the space charge with a cold forward and backward wave respectively. The separation of the total wave into components and these into further components enhances neither the generality nor the accuracy regardless of the degree of beam-circuit coupling.

The new definitions of F and B are now substituted into Eq. (133) and the backward-wave amplitude functions are found to be

$$b_1(y) = -\frac{C}{2(1 + Cb)} \frac{d}{dy} [a_2(y) + b_2(y)] \qquad (136)$$

and

$$b_2(y) = \frac{C}{2(1 + Cb)} \frac{d}{dy} [a_1(y) + b_1(y)]. \qquad (137)$$

These equations are combined to give

$$B(y, \Phi) = -\frac{Z_0 I_0}{4C} \frac{C}{2(1 + Cb)} \left[\frac{d[a_2(y) + b_2(y)]}{dy} \cos \Phi \right.$$
$$\left. + \frac{d[a_1(y) + b_1(y)]}{dy} \sin \Phi \right]. \qquad (138)$$

By the method of successive approximations the series solution of Eq. (138) may be written as

$$B(y, \Phi) = \frac{Z_0 I_0}{4C} \left\{ \frac{-C}{2(1 + Cb)} \left[\frac{da_1(y)}{dy} \sin \Phi + \frac{da_2(y)}{dy} \cos \Phi \right] \right.$$
$$\left. + \frac{C^2}{4(1 + Cb)^2} \left[-\frac{d^2 a_1(y)}{dy^2} \cos \Phi + \frac{d^2 a_2(y)}{dy^2} \sin \Phi \right] + \cdots \right\}. \qquad (139)$$

Under small-C conditions the C^2 term of Eq. (139) is generally negligible and hence may be neglected except near or beyond saturation. Of course the total backward-wave amplitude is composed of Eq. (139) plus the complementary part.

The final circuit equations are obtained by substituting Eq. (134) and the expression for $\rho_1(z, t)$ into Eq. (132). Two equations appear when the coefficients of sin Φ and cos Φ on each side are equated:

$$\frac{da_1(y)}{dy} = -\frac{2}{\pi}\int_0^{2\pi}\frac{\sin\Phi(y,\Phi_0')\,d\Phi_0'}{1+Cw(y,\Phi_0')} \tag{140}$$

and

$$\frac{da_2(y)}{dy} = -\frac{2}{\pi}\int_0^{2\pi}\frac{\cos\Phi(y,\Phi_0')\,d\Phi_0'}{1+Cw(y,\Phi_0')}. \tag{141}$$

The first-order equations above are a result of the separation of Eq. (126) into the two first-order equations (131) and (132).

Force Equation. The Newton force equation for a one-dimensional system was given in Eq. (3). Following a procedure paralleling Section 2 and using the velocity and phase definitions previously indicated gives the force equation in the following form:

$$2[1 + Cw(y,\Phi_0)]\frac{\partial w(y,\Phi_0)}{\partial y} = (1+Cb)[a_1(y)\sin\Phi + a_2(y)\cos\Phi]$$
$$-\frac{C}{2}\left[\frac{da_1(y)}{dy}\cos\Phi - \frac{da_2(y)}{dy}\sin\Phi\right]$$
$$+\frac{C^2}{4(1+Cb)}\left[\frac{d^2a_1(y)}{dy^2}\sin\Phi + \frac{d^2a_2(y)}{dy^2}\cos\Phi\right] - \frac{2\eta}{u_0\omega C^2}E_{sc-z}, \tag{142}$$

where all but E_{sc-z} have been included. Since various forms for the space-charge field were discussed in Chapters IV–VI, this discussion will not be repeated here.

Velocity-Phase Relation. A relation between the various normalized dependent variables is obtained after taking the appropriate derivatives of the velocity and phase variable (no initial velocity modulation):

$$\frac{\partial \Phi(y,\Phi_0)}{\partial y} - b = \frac{w(y,\Phi_0)}{1+Cw(y,\Phi_0)}. \tag{143}$$

The boundary conditions to be used in conjunction with Eqs. (140)–(143) are those of Section 2. g. The various solutions obtained by Poulter and Tien can be found in references 6 and 10.

d. Equivalence of Section 10.c Equations to Those of Section 2.j

It is convenient to carry out the comparison using the alternate equations derived in Section 2.j except that the different normalized distance variables will introduce a $1 + Cb$ factor. The definition of normalized voltage used there may be written as

$$V(y, \Phi) \triangleq \frac{Z_0 I_0}{4C} [A_1(y) \cos \Phi - A_2(y) \sin \Phi], \tag{144}$$

where

$$y \triangleq \frac{C\omega}{v_0} z = \frac{C\omega}{u_0} z(1 + Cb)$$

and

$$\Phi(y, \Phi_0) \triangleq \frac{y}{C} - \omega t.$$

The capital letters are used in Eq. (144) to avoid confusion with those of the previous section.

The rf voltage in Section 10.c is written as

$$\begin{aligned} V(y, \Phi) &= F(y, \Phi) + B(y, \Phi) \\ &= \frac{Z_0 I_0}{4C} \{[a_1(y) + b_1(y)] \cos \Phi - [a_2(y) + b_2(y)] \sin \Phi\}, \end{aligned} \tag{145}$$

where again $F(y, \Phi)$ and $B(y, \Phi)$ are respectively the voltages of the forward and backward waves on the structure. In that section $y \triangleq (C\omega/u_0)z$.

Comparison is greatly facilitated by defining

$$A_1(y) \triangleq a_1(y) + b_1(y)$$

and

$$A_2(y) \triangleq a_2(y) + b_2(y).$$

Substituting these into Eqs. (136) and (137) gives the backward-wave amplitudes

$$b_1(y) = -\frac{C}{2(1 + Cb)} \frac{dA_2(y)}{dy} \tag{146a}$$

and

$$b_2(y) = \frac{C}{2(1 + Cb)} \frac{dA_1(y)}{dy}. \tag{146b}$$

Then from above the forward-wave amplitudes are

$$a_1(y) = A_1(y) + \frac{C}{2(1+Cb)} \frac{dA_2(y)}{dy} \tag{147a}$$

and

$$a_2(y) = A_2(y) - \frac{C}{2(1+Cb)} \frac{dA_1(y)}{dy}. \tag{147b}$$

The next step is to differentiate Eqs. (147) with respect to y and substitute into Eqs. (140) and (141). The result is

$$\frac{dA_1(y)}{dy} + \frac{C}{2(1+Cb)} \frac{d^2A_2(y)}{dy^2} = -\frac{2}{\pi} \int_0^{2\pi} \frac{\sin \Phi(y, \Phi_0') \, d\Phi_0'}{1 + Cw(y, \Phi_0')} \tag{148}$$

and

$$\frac{dA_2(y)}{dy} - \frac{C}{2(1+Cb)} \frac{d^2A_1(y)}{dy^2} = -\frac{2}{\pi} \int_0^{2\pi} \frac{\cos \Phi(y, \Phi_0') \, d\Phi_0'}{1 + Cw(y, \Phi_0')}. \tag{149}$$

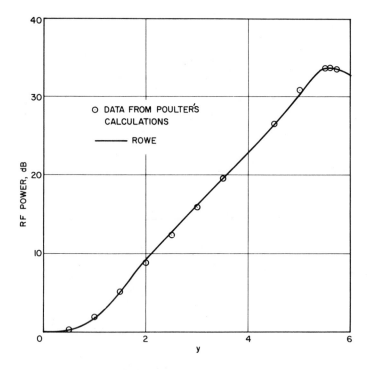

FIG. 45. Comparison of differential and integral equation solutions ($C = 0.2$, $d = 0$, $\omega_p/\omega = 0$, $\psi_0 = -30$, $b = 1.5$).

If $a_1(y)$ is replaced by $A_1(y)$ and $a_2(y)$ by $A_2(y)$ in Eqs. (50)–(52) of Section 2.j the above are seen to be identical (for zero loss) to the integral equations without the introduction of any approximations. The slight difference in the dependent variable equation is due simply to the different normalizations used in y and $\Phi(y, \Phi_0)$ and the minus sign on the right-hand side is due to a different sign convention used in I_0.

In both cases the equations are solved as an initial-value problem. Since small-signal conditions are assumed at the input, there is no backward wave at the input plane. The output of the circuit is terminated so that there is a reflection in the presence of the stream which exactly cancels the backward traveling wave produced by the modulated electron stream. The lack of synchronism of the backward traveling wave with the stream precludes any significant interaction between them even when C is large.

Poulter's integral equation method is somewhat different from Tien's, although the results are essentially the same.

Finally it is helpful to show that the numerical results obtained using the various methods are quite the same. A comparison of solutions using Rowe's method and Poulter's method is shown in Fig. 45.

Thus the *complete equivalence* of differential and integral equation methods has been proved, which was certainly expected in view of similar experiences from circuit analysis.

References

1. Nordsieck, A. T., Theory of the large-signal behavior of traveling-wave amplifiers. *Proc. IRE* **41**, No. 5, 630-637 (1953).
2. Doehler, O., and Kleen, W., Phénomènes non linéaires dans les tubes à propagation d'onde. *Ann. Radioelec.* **3**, 124-143 (1948).
3. Brillouin, L., The traveling-wave tube (Discussion of waves for large amplitudes). *J. Appl. Phys.* **20**, No. 12, 1196-1206 (1949).
4. Slater, J. C., *Microwave Electronics*. Van Nostrand, Princeton, N.J., 1950.
5. Wang, C. C., "Solution of Partial Integral-Differential Equations of Electron Dynamics Using Analogue Computers with Storage Devices," Project Cyclone, Symposium II on Simulation and Computing Techniques, Part 2. Reeves Instr. Corp, New York, 1952.
6. Poulter, H. C., "Large Signal Theory of the Traveling-Wave Tube." Stanford Univ. Tech. Rept. No. 73; ONR Contract No. N6ONR 251(07) (January 1954).
7. Rowe, J. E., "A Large-Signal Analysis of the Traveling-Wave Amplifier." Univ. of Michigan Electron Phys. Lab. Tech. Rept. No. 19 (April 1955).
8. Rowe, J. E., A large-signal analysis of the traveling-wave amplifier: Theory and general results. *IRE Trans. Electron Devices* **3**, 39-57 (1956).
9. Tien, P. K., Walker, L. R., and Wolontis, V. M., A large-signal theory of traveling-wave amplifiers. *Proc. IRE* **43**, No. 3, 260-277 (1955).

10. Tien, P. K., A large signal theory of traveling-wave amplifiers. *Bell System Tech. J.* **35**, 349-374 (1956).
11. Vainshtein, L. A., The non-linear theory of traveling-wave tubes: Part I, Equations and laws of conservation. *Radio Eng. Electron. (USSR) (Engl. Transl.)* **2**, No. 7, 92-108 (1957).
    Vainshtein, L. A., The non-linear theory of traveling-wave tubes: Part II, Numerical results. *Radio Eng. Electron (USSR) (Engl. Transl.)* **2**, No. 8, 110-139 (1957).
    Vainshtein, L. A. (with Filimonov, G. F.), The non-linear theory of the traveling-wave tube: Part III, Influence of the space-charge forces. *Radio Eng. Electron. (USSR) (Engl. Transl.)* **3**, No. 1, 116-123 (1958).
12. Cutler, C. C., The nature of power saturation in traveling-wave tubes. *Bell System Tech. J.* **35**, 841-876 (1956).
13. Brangaccio, D. J., and Cutler, C. C., Factors affecting traveling-wave tube power capacity. *IRE Trans. Electron Devices,* **3**, 9-24 (1953).
14. Scott, A. W., Why a circuit sever affects traveling-wave tube efficiency. *IRE Trans. Electron Devices* **9**, No. 1, 35-41 (1962).
15. Rowe, J. E., Theory of the Crestatron: A forward-wave amplifier. *Proc. IRE* **47**, No. 4, 536-545 (1959).
16. Rowe, J. E., One-dimensional traveling-wave tube analyses and the effect of radial electric field variations. *IRE Trans. Electron Devices* **7**, No. 1, 16-21 (1960).
17. Rowe, J. E., N-beam nonlinear traveling-wave amplifier analysis. *IRE Trans. Electron Devices* **8**, No. 4, 279-284 (1961).
18. Rowe, J. E., Relativistic beam-wave interactions. *Proc. IRE,* **50**, No. 2, 170-179 (1962).
19. Rowe, J. E., Relativistic beam-wave interactions. *Arch. Elek. Ubertragung* **18**, No. 2, 108 (1964).
20. Rowe, J. E., and Hok, G., When is a backward wave not a backward wave? *Proc. IRE* **44**, No. 8, 1060 (1956).

CHAPTER

VII | O-Type Backward-Wave Oscillators

1 Introduction

The O-type backward-wave oscillator or O-carcinotron is another member of the family of linear-beam devices which has found wide application in modern electronic systems. It is characterized by its moderate to low efficiency as compared to the O-type amplifier and its wide electronic tuning range as compared to other electronic oscillators. The beam electronics are identical with those for O-type amplifiers and klystrons; an axial focusing field is usually employed.

The basic operation of the backward-wave oscillator depends upon the interaction of the electron stream with a space harmonic field component whose phase and group velocities are oppositely directed. All periodic rf structures support such space-harmonic fields, as evidenced by an examination of Floquet's theorem.

The interaction process in the backward-wave device is essentially that of the forward-wave device in that an rf signal impressed on the circuit or produced by noise in the electron stream velocity-modulates the stream, which in turn produces a density modulation which induces an rf field onto the circuit in such a way as to add to the existing circuit field. The induced wave will then travel along the circuit with a phase velocity approximately equal to the electron velocity. The wave energy traveling opposite to the electron flow (negative group velocity) produces additional bunching, which in turn produces an increased energy on the circuit, resulting in regenerative amplification for extremely small stream currents and unstable regeneration or oscillation at higher stream currents, provided that other necessary oscillation conditions are satisfied.

Thus backward-wave devices, both oscillators and amplifiers, will be inherently less efficient than forward-wave amplifiers of the same type, since in the forward-wave amplifier the density modulation in the stream and the circuit field producing it travel in the same direction, whereas in the backward-wave device they travel in opposite directions, which results in the circuit field being strongest where the modulation is weakest and vice versa. It should be recalled that this backward traveling wave is not a reflected wave from the output transducer but is a property of the mode of operation of the circuit. Reflected waves from the output or from the attenuator in the forward-wave amplifier do not velocity-modulate the stream and lead to electronically tunable oscillation because the reflected-wave phase velocity is opposite to the electron velocity.

The circuit fields and the rf current in the stream as a function of distance were shown in Chapter I for both forward-wave amplifiers and backward-wave oscillators. In the forward-wave device it was seen that the circuit voltage grows at an exponential rate until the nonlinear regime is reached and finally the circuit voltage achieves a saturation value. As would be expected, the fundamental component of current in the stream reaches a maximum approximately a quarter circuit wavelength before the circuit voltage reaches its maximum. In the backward-wave device the circuit voltage falls off approximately as a cosine function and eventually goes through zero with an oscillation condition or reaches a nonzero minimum value in the case of a backward-wave amplifier. The fundamental current in the stream may reach a maximum before the end of the device, i.e., the length for infinite gain, at the critical length or it may still be increasing at the critical length, depending upon the level of oscillation.

External feedback paths applied to forward-wave amplifiers can produce unstable regenerative amplification, but it is usually characterized by mode instability which results in a discontinuous tuning curve. In both the amplifier and the oscillator the interaction is between the slow space-charge wave of the stream and the circuit wave. In fact, under usual conditions in the oscillator the fast space-charge wave is excited to only a negligible extent.

A considerable amount of theoretical work has been done on backward-wave oscillators using the modified linear forward-wave amplifier theory and neglecting the effects of space charge or large C or both. Some work by Grow and Watkins[3] has incorporated both of these effects into the linear theory and then made estimates of the efficiency assuming specific values for the ratio of fundamental rf current to the dc current in the stream. Sedin[20] reported on some nonlinear calculations pertaining to the backward-wave oscillator where he used Nordsieck's[17] equations, suitably modified and neglecting the effects of finite C, QC and circuit loss. Subsequently this author presented an analysis of the O-BWO including circuit loss and space-charge effects.[19]

2 Backward-Wave Circuits

In backward-wave interaction as well as in forward-wave interaction it is required that the circuit wave and the electron stream be in approximate synchronism. This requirement results in a voltage-tunable device in the case of backward-wave interaction, since the rf circuit must be dispersive in order to obtain oppositely directed phase and group velocities. The phase and group velocities of periodic structures are written as

$$v_p = \frac{\omega}{\beta} \qquad (1)$$

and

$$v_g = \frac{v_p}{1 - \dfrac{\omega}{v_p}\dfrac{dv_p}{d\omega}}. \qquad (2)$$

A negative group velocity requires that the circuit exhibit negative dispersion; i.e., the denominator must remain negative.

Any rf propagating circuit that satisfies the requirement of oppositely directed phase and group velocities for at least one space harmonic component of the total field can be used as a backward-wave device circuit. These space harmonics exist in all periodic structures, as indicated by the periodic nature of the boundary conditions. The fields can be expressed over all periods of the rf structure in terms of their values over one period through the use of Floquet's theorem.

It was shown in Chapter III that forward-wave circuits can be represented by an equivalent circuit in the form of an L-C transmission line (low-pass filter) when operating in the fundamental forward-wave mode. In this embodiment of the equivalent circuit the inductance and any series circuit loss are represented by the series elements and the capacitance by a shunt element. The equation for the voltage along the line as a function of distance and time is a linear second-order partial differential equation containing driving terms when an electron stream is present. This type of circuit has a phase shift per section which increases with frequency due to the series inductance. In the nondispersive region where it is usually operated the phase velocity is independent of frequency and is given by $v_0 = 1/\sqrt{LC}$.

Usually the differential equation for the circuit voltage in a backward-wave device is found by simply changing the sign of the circuit impedance to account for backward energy flow. The equivalent circuit may be drawn by interchanging the positions of the series inductance and the shunt capacitance to give a high-pass type of circuit with a phase shift per section characteristic which decreases with frequency. The phase velocity for this circuit is an increasing function of frequency and is given by $v_0 = -\omega^2 \sqrt{LC}$. A differential equation, of fourth order in time and second order in the distance, for the voltage along the line can be derived for this circuit. This equation can be reduced to a second-order equation if a time variation of $\exp j\omega t$ is introduced. The presence of an electron stream also adds driving terms to the circuit equation.

The forms of the two equivalent circuits are shown in Figs. 9 and 18 of Chapter III; the differential equation used both for the nonlinear forward-wave amplifier and for the backward-wave device is given below. In those terms preceded by a double sign the upper sign refers to the forward-wave device and the lower sign to the backward-wave device. The backward-wave device equations apply equally well to the oscillator and the amplifier.

$$\frac{\partial^2 V(z,t)}{\partial t^2} - v_0^2 \frac{\partial^2 V(z,t)}{\partial z^2} + 2\omega Cd \frac{\partial V(z,t)}{\partial t}$$
$$= \pm v_0 Z_0 \left[\frac{\partial^2 \rho(z,t)}{\partial t^2} + 2\omega Cd \frac{\partial \rho(z,t)}{\partial t} \right]. \tag{3}$$

While it is intuitively satisfying the backward-wave equivalent circuit cannot be realized in terms of a smooth circuit but only in terms of lumped parameters to give the desired, oppositely directed phase and group velocities. This form of the circuit also leads to difficulties in attaching physical significance to the circuit elements. It is generally most satisfying simply to change the sign of the circuit impedance.

3 Mathematical Analysis

In view of this simple change in the circuit equation for the O-BWO as compared to that for the O-FWA, the nonlinear Lagrangian analysis for the oscillator proceeds in a manner directly similar to that for the amplifier. Thus it is deemed necessary only to present the final equations. The one-dimensional system is used and initial velocity modulation is neglected.

Relation Between Variables

$$\frac{\partial \Phi(y, \Phi_0)}{\partial y} + \frac{d\theta(y)}{dy} = \frac{2u(y, \Phi_0)}{1 + 2Cu(y, \Phi_0)}. \tag{4}$$

3. MATHEMATICAL ANALYSIS

Circuit Equations

$$\frac{d^2 A(y)}{dy^2} - A(y)\left[\left(\frac{1}{C} - \frac{d\theta(y)}{dy}\right)^2 - \frac{(1+Cb)^2}{C^2}\right]$$
$$= \frac{(1+Cb)}{\pi C}\left[\int_0^{2\pi} \frac{\cos \Phi(y, \Phi_0')\, d\Phi_0'}{1 + 2Cu(y, \Phi_0')} + 2Cd \int_0^{2\pi} \frac{\sin \Phi(y, \Phi_0')\, d\Phi_0'}{1 + 2Cu(y, \Phi_0')}\right], \quad (5)$$

$$A(y)\left[\frac{d^2\theta(y)}{dy^2} - \frac{2d}{C}(1+Cb)^2\right] + 2\frac{dA(y)}{dy}\left(\frac{d\theta(y)}{dy} - \frac{1}{C}\right)$$
$$= \frac{(1+Cb)}{\pi C}\left[\int_0^{2\pi} \frac{\sin \Phi(y, \Phi_0')\, d\Phi_0'}{1 + 2Cu(y, \Phi_0')} - 2Cd \int_0^{2\pi} \frac{\cos \Phi(y, \Phi_0')\, d\Phi_0'}{1 + 2Cu(y, \Phi_0')}\right]. \quad (6)$$

Force Equation

$$[1 + 2Cu(y, \Phi_0)]\frac{\partial u(y, \Phi_0)}{\partial y} = -A(y)\left[1 - C\frac{d\theta(y)}{dy}\right]\sin \Phi(y, \Phi_0)$$
$$+ C\frac{dA(y)}{dy}\cos \Phi(y, \Phi_0) + \frac{1}{(1+Cb)}\left(\frac{\omega_p}{\omega C}\right)^2$$
$$\cdot \int_0^{2\pi} \frac{F_{1-z}(\Phi - \Phi_0')\, d\Phi_0'}{[1 + 2Cu(y, \Phi_0')]}. \quad (7)$$

The independent and dependent variables contained in the above equations are the same as defined previously and the basic assumptions concerning the circuit and beam characteristics are exactly the same as for the one-dimensional amplifier analysis.

The boundary conditions to be used in the nonlinear backward-wave device analysis are very similar to those used in solving the forward-wave amplifier equations. The rf structure is assumed to be matched to its characteristic impedance over its entire length and the problem is handled as an initial-value problem rather than a boundary-value problem. The dependent variables and their derivatives are specified at the input boundary and the equations are then integrated until a minimum in the rf voltage as a function of distance is reached.

Boundary Conditions

Circuit Variables

(1) $A(0) \equiv A_0$,
(2) $dA(y)/dy\,|_{y=0} = 0$ for a lossless circuit,
(3) $\theta(0) = 0$,
(4) $d\theta(y)/dy\,|_{y=0} = -b$, the apparent phase constant of the total wave at the input.

Beam Variables

(1) $u(0, \Phi_0) \equiv 0$ for an unmodulated entering beam,
(2) $\Phi(0, \Phi_{0,j}) \equiv \Phi_{0,j} = 2\pi j/m$, $j = 0, 1, 2, ..., m$.

Operating Parameters

(1) $C, \omega_p/\omega, B, d$.

4 Solution Procedure

The nonlinear backward-wave oscillator equations can be solved by assuming specific values of the injection velocity b and the amplitude (A_0) of the circuit voltage at $y = O$ for any set of C, QC and d. Then the equations are integrated numerically using the appropriate difference equation formulation consistent with accuracy and stability of the system until the circuit voltage goes through a minimum in the case of a backward-wave amplifier or goes through zero in the case of an oscillator. The number of representative electrons which have to be used in finding the solutions depends critically upon the value of the space-charge parameter, increasing as QC becomes quite large (around 0.5). As the collector end of the tube is approached, i.e., as the circuit field becomes very small, electrons begin to overtake one another and a maximum in the fundamental component of stream current is reached as in the forward-wave amplifier.

It is possible that in a backward-wave oscillator that is relatively short (CN_s approximately 0.3) the effect of crossing electron flight lines may not be as important as in the forward-wave amplifier where the typical circuit lengths are quite great, since crossing occurs first in a backward-wave oscillator very near the collector end where the circuit voltage is small and decreasing approximately as a cosine function. Overtaking of electrons does not result in an abrupt change in the slope of the voltage versus distance curve. Of course in the backward-wave amplifier where long structures would be used, overtaking of electrons, resulting in multivalued velocities, would be important as in the forward-wave amplifier. In the event that electron trajectory crossings are not of great importance, the Eulerian formulation of the equations could be used, and hence the hydrodynamical form of the continuity equation is appropriate. This would result in a somewhat simpler set of equations to be solved and the effects of space charge could be more easily included. Such an analysis has been made and overtaking does affect the start-oscillation conditions to a considerable extent.

4. SOLUTION PROCEDURE

Several trials are required in order to determine the combination of b and CN_s that leads to an oscillation condition. The sensitivity of the start-oscillation conditions to the value of A_0 at constant b for a particular set of C, QC and d is shown in Fig. 1, where oscillation occurred for A_0 equal to 0.68 and 0.685 but not for 0.67 and 0.69.

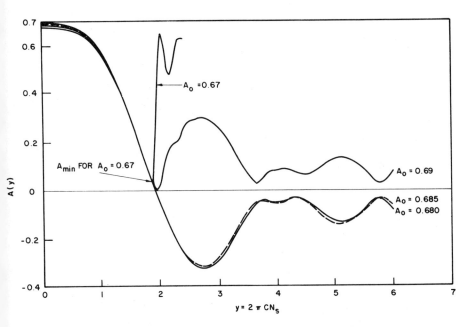

FIG. 1. Rf voltage amplitude versus distance ($C = 0.1$, $QC = 0$, $d = 0$, $b = 1.9$).

A detailed study of the variation of the phase lag across the amplifier $\theta(y)$ versus y is useful in determining the direction of the change necessary in b or A_0 when in the neighborhood of an oscillation condition. The behavior of $\theta(y)$ in the neighborhood of an oscillation condition is illustrated in Fig. 2. This dependence is general for any set of circuit and stream parameters. It is seen that a nearly linear relationship of $\theta(y)$ versus y is obtained at the oscillation condition. This is also true when the values of b and A_0 are far from a start-oscillation condition, and hence one must be near the optimum combination of b and A_0 to use the criterion illustrated in Fig. 2. As the ratio of stream current to minimum starting current is increased, the oscillation condition is more difficult to locate, and it is necessary to examine a plot of the CN_s distance to the minimum value of $A(y)$ versus A_0 as well as the above plot in order to determine the oscillation conditions.

An alternate, but more effective, procedure for determining start-oscillation conditions makes use of the "downhill method", an automatic method for finding the start-oscillation conditions, i.e., combinations of A_0 and b which produce a zero in the rf circuit voltage for $y > 0$.

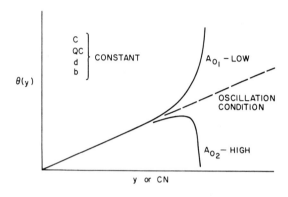

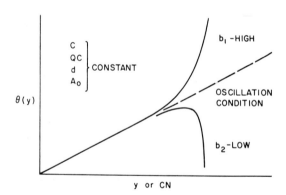

FIG. 2. Indication of proximity to an oscillation condition.

The downhill method has been used to obtain roots of a polynomial $P(\delta)$ by searching the surface over the complex plane generated by

$$S = |\,R[P(\delta)]\,| + |\,I[P(\delta)]\,|. \tag{8}$$

When S becomes zero, the value of δ is then a root of the polynomial. The nature of Eq. (8) is such that if a point δ_1 produces S_1, there is a point δ_2 where $S_2 < S_1$ unless $S_1 = 0$. For a continuous path from δ_1 to

4. SOLUTION PROCEDURE

δ_2 there is a point δ_3 on this path such that $S_2 < S_3 < S_1$. Thus we see that during the search for roots, one always proceeds downhill to the roots. Mechanization of the downhill method requires a systematic scheme for searching the surface S. The method for finding the smallest value of the surface S is summarized below and is illustrated in Fig. 3.

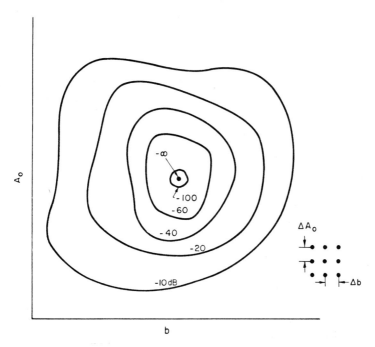

FIG. 3. Downhill method for systematically scanning the A_0-b plane for an oscillation condition.

(1) A "wheel" of points in a b-A_0 plot is selected with a center and a set of four points equally spaced from the center.

(2) The polynomial is evaluated at each of these five points, giving five values of S.

(3) The point for which S is a minimum becomes the center point for the next "wheel." This process is then repeated until the value of S approaches some minimum limit. The value of δ giving this minimum S is considered to be the root.

(4) The initial center point of the wheel is arbitrarily chosen as the origin and the initial spacing of the end points of the wheel is some arbitrary small value.

The above method is easily adapted to the finding of start-oscillation conditions. It is used to search the A_0 versus b plane for those combinations which produce start oscillation. The wheel of points is selected arbitrarily and the systematic procedure forces convergence. There are of course several combinations with larger values of A_0, corresponding to higher oscillator outputs and consequently higher efficiency. At low values of current, oscillations can occur for a small range of A_0 and hence oscillation regions rather than oscillation points can be given.

5 Efficiency Calculations

The CN_s length and value of b required for lowest-order start-oscillation condition may be determined by assuming an arbitrarily small value of the signal level A_0 at $y = 0$ and then varying b for any particular set of C, QC, and d. A plot of the CN_s for this lowest-order oscillation, which corresponds to minimum starting current, and the value of the injection velocity parameter required are shown in Fig. 4. The required length and value of b are seen to increase as the space charge is increased. Identical results are obtained from the linearized equations.

As the value of b is increased from that corresponding to the lowest start-oscillation current, larger values of A_0 are required to produce oscillation. The efficiency increases approximately linearly with increasing b at constant C, QC, and d. The plane (y or CN_s value) at which the circuit rf voltage $A(y)$ goes through zero occurs at a larger value of CN_s in general, which signifies an increase in the ratio of operating current, I_0, to the minimum start-oscillation value, I_s. For moderate values of C the length N_s is nearly independent of the stream current, so that CN_s varies as the one-third power of the current. This relationship is quite accurate for C's as large as 0.1 so that the ratio of stream (operating) current to minimum starting current can be expressed as

$$\frac{I_0}{I_s} = \left(\frac{CN_s}{(CN_s)_{I_s}}\right)^3, \tag{9}$$

where $I_0 =$ stream current, and $I_s =$ minimum start-oscillation current. For constant C the current ratio is equal to the cube of the length ratio.

The efficiency is calculated using the well-known relation,

$$\eta \approx 2CA_0^2, \tag{10}$$

where A_0 is the normalized value of the rf voltage at the gun end of the device. As pointed out above, oscillations occur for a range of A_0 at a

5. EFFICIENCY CALCULATIONS

particular value of I_0/I_s and thus the following efficiency curves are considered to have some breadth.

A composite plot of efficiency versus I_0/I_s from Eq. (9) is shown in Fig. 5 for representative values of C and QC where circuit loss is neglected. The efficiency is seen to increase with an increasing current or

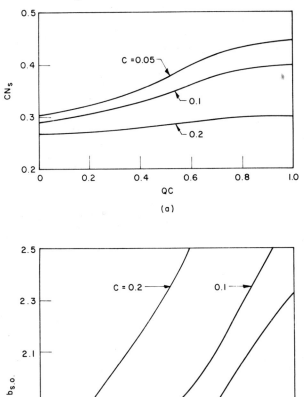

FIG. 4. Length and injection velocity at least start-oscillation current in a nonlinear BWO ($B = 1$, $d = 0$). (a) CN_s versus QC; (b) $b_{s.o.}$ versus QC.

length ratio up to a value of $I_0/I_s \approx 1.8$ and then it levels off for all values of C and QC investigated. Increasing C produces an increased efficiency but increasing the space-charge parameter results in a decreasing efficiency. An increase of C from 0.05 to 0.10 results in an approximate doubling of the efficiency. For the case of $C = 0.05$ and $QC = 0.25$ a start-oscillation condition was found at $I_0/I_s = 4.3$.

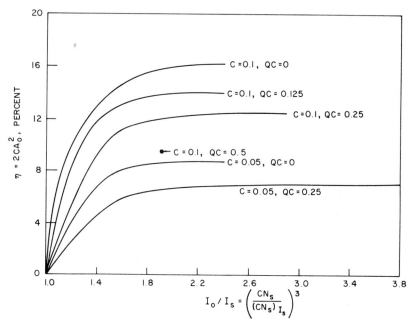

FIG. 5. Efficiency versus I_0/I_s with b adjusted for oscillation ($B = 1$, $d = 0$).

The values of CN_s and b as functions of I_0/I_s corresponding to the data shown in Fig. 5 are shown in Fig. 6. It is seen that as the value of I_0/I_s is increased the value of b approaches a limiting value. These higher-current oscillation conditions indicate increasing output and are extremely difficult to locate, since for large values of b the tube is relatively long and behaves much like a long line in the sense that the oscillation condition is very sensitive to the value of A_0. The value of CN_s at start oscillation is seen to increase smoothly as I_0/I_s is increased and probably becomes asymptotic to some limiting value.

Very high-order oscillation conditions have been observed in the calculations when the circuit voltage goes through zero to negative values and then increases to cross the zero axis again. This occurs at very large values of CN_s; these oscillation conditions would correspond to 80–100 times the minimum start-oscillation current.

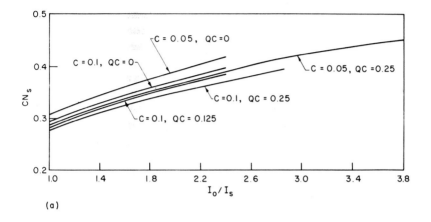

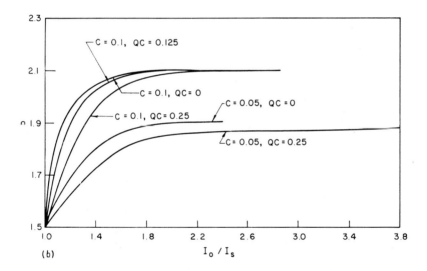

FIG. 6. (a) CN_s and (b) b versus I_0/I_s with b adjusted for oscillation ($B = 1$, $d = 0$).

Solutions were also found that decreased to a nonzero minimum and then increased and went through subsequent maxima and minima, one of which was zero. When $C = 0.2$ and $QC = 0$ it was found that the CN_s length decreased as b was increased (up to $b = 2.3$), indicating that for this large a value of C the length is not independent of the value of current and hence the cube relationship cannot be used. These results are shown in Fig. 7.

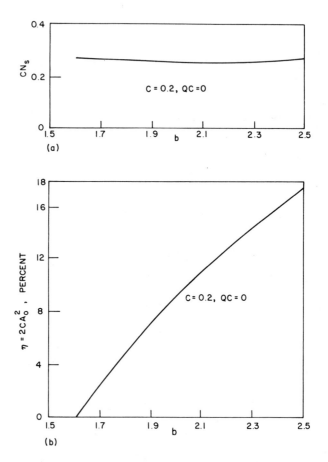

FIG. 7. (a) CN_s and (b) efficiency at start oscillation versus b ($C = 0.2$, $QC = 0$, $d = 0$, $B = 1$).

When C is small, say up to 0.1, it can be determined from a Fourier analysis of the electron velocity variation with distance that the average stream velocity changes slowly with distance and is not significantly reduced at the collector end compared to its initial value. Thus it is seen that the number of stream or circuit wavelengths in a given length is independent of I_0, u_0 being relatively constant. The efficiency for $C = 0.2$ is not appreciably increased over the values obtained when $C = 0.1$. Values of $C = 0.2$ are very difficult to achieve practically and for this reason extensive calculations were not carried out for this high value of the gain parameter.

The value of the ratio of fundamental current to dc current in the stream depends on the level of oscillation, generally increasing as the value of b and hence I_0/I_s is increased. For oscillation levels above the lowest oscillation condition it varies from 1.2 to 1.4, depending upon the values of C and QC. This is approximately midway between the values of 1 and 2 respectively used by Grow and Watkins[3] in computing efficiency from a linear theory. A value of the stream diameter $B = 1$ was used in the calculations, which corresponds to a fairly large beam; hence the efficiency results should be modified to account for variation of the circuit field across the stream.

The problem of solving the backward-wave device equations including the effects of radial variation of the circuit field and the space-charge field is a formidable one. The results of Grow and Watkins[3] in estimating the decrease in efficiency due to radial field variations can probably be applied here; they indicate that the efficiency is reduced by a factor of 0.8 when $B = 1$. It should be recalled here that radial-circuit and space-charge-field effects reduce the TWA efficiency by approximately 20 percent when $B = 1$. Other effects such as circuit loss and velocity spread in the stream will also reduce the efficiency from that calculated here. The efficiency results calculated are in good agreement with the efficiency data of Putz and Luebke[10] as reported by Grow and Watkins.[3] Their experimental values of efficiency indicate a spread of data for η/C versus $\omega_q/\omega C$ between 1.0 and 2.0 with a clustering around 1.5. These experimental data correspond to operating efficiencies around 8–10 per cent and values of QC between 0.2 and 0.6.

6 Relativistic Oscillator Analysis

The relativistic backward-wave oscillator has not yet found a position of practical importance. This is primarily because of the low circuit impedance usually associated with backward-wave circuits and hence the low interaction efficiency. The analysis proceeds along the same lines as that utilized in studying the relativistic forward-wave amplifier.

The final form of the equations (one-dimensional, no initial velocity modulation) is shown below.

Relation Between Variables

$$\frac{\partial \Phi(y, \Phi_0)}{\partial y} + \frac{d\theta(y)}{dy} = \frac{2u(y, \Phi_0)}{1 + 2Cu(y, \Phi_0)}. \tag{11}$$

Circuit Equations Equations (5) and (6) are unchanged.

Force Equation

$$[1 + 2Cu(y, \Phi_0)]\frac{\partial u(y, \Phi_0)}{\partial y} = (1 - k_e^2)^{\frac{3}{2}} \Big\{ -A(y)\Big(1 - C\frac{d\theta(y)}{dy}\Big)$$
$$\cdot \sin \Phi(y, \Phi_0) + C\frac{dA(y)}{dy} \cos \Phi(y, \Phi_0)$$
$$+ \frac{1}{(1 + Cb)}\Big(\frac{\omega_p}{\omega C}\Big)^2 \int_0^{2\pi} \frac{F_{1-z}(\Phi - \Phi') \, d\Phi_0'}{[1 + 2Cu(y, \Phi_0')]}\Big\}. \quad (12)$$

The solution procedure to be utilized is the same as that used in the nonrelativistic case (downhill method) except that one additional parameter $k_e \triangleq u_0/c$ must be selected.

The results of relativistic BWO calculations for efficiency, starting length and electron injection velocity are shown in Figs. 8–10 for a

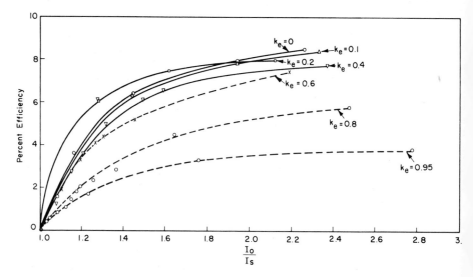

Fig. 8. Efficiency versus current for a relativistic BWO ($C = 0.05$, $\omega_p/\omega \to 0$, $d = 0$).

typical case in which $C = 0.05$ and space-charge forces were neglected. Several efficiency curves are given for increasing values of k_e and it is apparent that the output decreases, the starting length increases and the velocity parameter approaches zero (synchronism) as the electron velocity approaches the velocity of light. A similar behavior occurs for larger values of C and the influence of space-charge forces is to lower the output further. Of course, the overall efficiency can be significantly increased by collector depression techniques. It is apparent that the relativistic BWO has little practical importance in its own right although the phenomena could be of importance in high-energy accelerators.

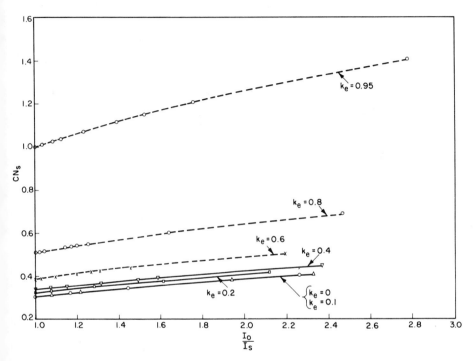

FIG. 9. Starting length versus current for a relativistic BWO ($C = 0.05$, $\omega_p/\omega \to 0$, $d = 0$).

7 Radial and Angular Variations in BWO's

It is, of course, possible to evaluate radial and angular variation effects in O-type backward-wave oscillators in the same way as they were studied in Sections 4–9 Chapter VI. The derivation of the appropriate circuit, force and continuity equations bears the same relationship to the one-dimensional oscillator equations as the multidimensional amplifier analyses bear to their one-dimensional counterpart. Since both the derivation procedure and the results are similar this task is left to the reader.

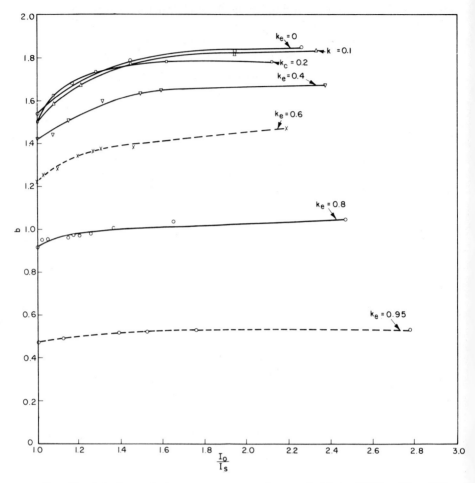

Fig. 10. Injection velocity versus current for a relativistic BWO ($C = 0.05$, $\omega_p/\omega \to 0$, $d = 0$).

References

A. General and Linear Theory

1. Bernier, J., Essai de théorie du tube électronique à propagation d'onde. *Ann. Radioelec.* **2**, 87-101 (1947).
2. Goldberger, A. K., and Palluel, P., The O-type carcinotron. *Proc. IRE* **44**, 333-345 (1956).
3. Grow, R., and Watkins, D. A., Backward-wave oscillator efficiency. *Proc. IRE* **43**, 848-856 (1955).

4. Harman, W. A., "Backward-Wave Interaction in Helix Type Tubes." Stanford Univ. Electron. Lab. Tech. Rept. No. 13 (April 1954).
5. Heffner, H., Analysis of the backward-wave traveling-wave tube. *Proc. IRE* **42**, 930-937 (1954).
6. Johnson, H. R., Backward-wave oscillators. *Proc. IRE* **43**, 684-697 (1955).
7. Kompfner, R., and Williams, N. T., Backward-wave tubes. *Proc. IRE* **41**, 1602-1611 (1953).
8. Muller, M., Traveling-wave amplifiers and backward-wave oscillators. *Proc. IRE* **42**, 1651-1658 (1954).
9. Pierce, J. R., *Traveling-Wave Tubes*. Van Nostrand, Princeton, N.J., 1950.
10. Putz, J. L., and Luebke, W. R., "High-Power S-Band Backward-Wave Oscillator." Stanford Univ. Electron. Lab. Tech. Rept. No. 182-1 (February 1956).
11. Tien, P. K., Bifilar helix for backward-wave oscillators. *Proc. IRE* **42**, 1137-1143 (1954).
12. Walker, L. R., Starting currents in the backward-wave oscillator. *J. Appl. Phys.* **24**, 854-860 (1953).
13. Warnecke, R., Guenard, P., and Doehler, O., Phénomènes fondamentaux dans les tubes à onde progressive. *Onde Elec.* **34**, 323-338 (1954).
14. Watkins, D. A., and Ash, E. A., The helix as a backward-wave circuit structure. *J. Appl. Phys.* **25**, 782-790 (1954).
15. Watkins, D. A., and Rynn, N., Effect of velocity distribution on traveling-wave tube gain. *J. Appl. Phys.* **25**, 1375-1379 (1954).
16. Weglein, R. D., Backward-wave oscillator starting conditions. *IRE Trans. Electron Devices* **4**, No. 2, 177-180 (1957).

B. Nonlinear Theory

17. Nordsieck, A., Theory of the large-signal behavior of traveling-wave amplifiers. *Proc. IRE* **41**, 630-637 (1953).
18. Rapoport, G. N., Preliminary results of the non-linear theory of oscillations in a backward-wave tube with longitudinal field. *Radio Eng. Electron.* (*USSR*) (*Engl. Transl.*) **3**, No. 2, 347-355 (1958).
 Rapoport, G. N., On the mechanism of increasing efficiency of a backward-wave oscillator (carcinotron-O) with increase in space-charge parameter. *Radio Eng. Electron.* (*USSR*) (*Engl. Transl.*) **3**, No. 2, 355-366 (1958).
19. Rowe, J. E., Analysis of nonlinear O-type backward-wave oscillators. *Proc. Symp. Electron. Waveguides, Brooklyn Polytech. Inst. 1958* **8**, 315-339. Wiley (Interscience), New York, 1958.
20. Sedin, J. W., *IRE Conf. Electron Devices, Denver, Color., 1956*.

CHAPTER

VIII | Crossed-Field Drift-Space Interaction

1 Introduction

The various axially symmetric interaction schemes studied in the previous chapters have their counterparts in crossed-field injected-beam interactions, including drift-space, amplifier and oscillator configurations. The drift-space action of a previously modulated stream is investigated in this chapter. Such a configuration arises when either a klystronlike gridded-gap system or an extended severed-circuit region in an amplifier is utilized, possibly in the forms suggested by Fig. 1. It has been shown[1] on a small-signal basis that such drift-space regions can support the exponential growth of rf perturbations in both thin- and thick-stream situations.

The basic analysis of equivalent circuits (not needed in this chapter) and space-charge-field calculations are given in Chapters III and IV and are applied directly in this and the following two chapters to crossed-field problems. In all cases we assume a well-defined injected electron stream which is characterized by its velocity and space-charge distribution at the entrance to the interaction region. The method is directly applicable to emitting-sole configurations. Generally Brillouin flow conditions are assumed.

The only fundamental difference between these discussions and the corresponding ones on O-type devices is that here we do not have a realistic condition corresponding to the confined-flow one-dimensional system assumed in Chapters V, VI, and VII. Since here electrons must move both in the y- and z-directions, the simplest analysis which investigates energy conversion is necessarily two dimensional. This is not a fundamental difficulty but does mean that the simplest system requires somewhat more complicated equations than the simplest O-type system. The ideal stream behavior would be Brillouin flow, in which the stream charge density is uniform in the y-direction and the particle velocity varies uniformly over the stream in this direction.

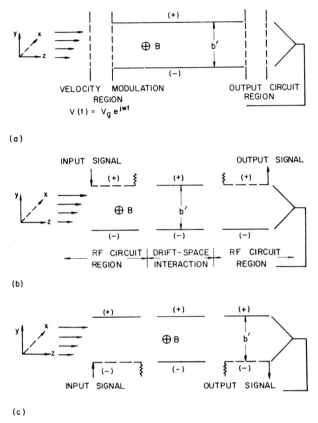

FIG. 1. Crossed-field drift-space amplifiers. (a) Gridded-gap modulation; (b) negative sole-distributed interaction couplers; (c) positive sole-distributed interaction couplers.

2 Two-Dimensional Drift-Space Equations

The drift region is assumed to be defined by smooth, parallel, perfectly conducting sheets separated by a distance b' as illustrated in Fig. 1. The assumed cross-sectional geometry is shown in Fig. 2, indicating that both the beam and the drift plates are of finite length in the magnetic field direction. The large-signal analysis of the drift region was considered by Gandhi[1] and Rowe.[2,3]

The electrons originate from a cathode somewhere outside of the drift-space region and have been premodulated in velocity and/or phase.

Their subsequent motion of interest is governed by the vector Lorentz equation, assuming nonrelativistic conditions:

$$\frac{d(m\mathbf{v})}{dt} = -e[\mathbf{E} + \mathbf{v} \times \mathbf{B}], \tag{1}$$

where $m\mathbf{v}$ = the particle momentum, and $\mathbf{E}$, $\mathbf{B}$ = the electric and magnetic field intensities respectively. We will assume throughout that there is no magnetic field perpendicular to the cathode so that the flow is

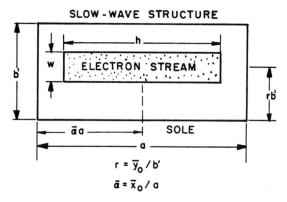

FIG. 2. Cross section of the beam-circuit configuration.

irrotational and thus $\nabla \times (m\mathbf{v}) = e\mathbf{B}$. In the interaction region the static fields to be used in Eq. (1) are written as

$$\mathbf{E}(x, y, z) = [0, -E_0, 0] \tag{2a}$$

and

$$\mathbf{B}(x, y, z) = [B_0, 0, 0]. \tag{2b}$$

Many possible solutions of Eq. (1) subject to Eq. (2) and irrotational flow exist, since the velocity vector has not yet been specified. One solution of great interest in all crossed-field work is that of laminar Brillouin flow, in which the velocity vector is written as

$$\mathbf{v} = [0, 0, v_0]. \tag{3}$$

Solution of this dc system yields the following interesting relationships:

$$v_0 = \omega\, y, \tag{4a}$$

$$E_0 = -\frac{\omega_c^2}{|\eta|} y, \tag{4b}$$

2. TWO-DIMENSIONAL DRIFT-SPACE EQUATIONS

and
$$\omega_p{}^2 = \omega_c{}^2, \tag{4c}$$

where $\omega_c = |\eta| B_0 =$ the cyclotron frequency.

Equation (4c) specifies a constant charge density in the stream and Eq. (4b) indicates a parabolic potential variation within the beam which is matched to a linear variation outside the beam. The solution form of Eqs. (4) is of most interest and will be examined in detail; however it should be noted that the nonlinear theory will handle generalized flows.

In an Eulerian framework the only additional relation necessary is the continuity equation, which may be written as follows along with Eq. (1) in component form:

$$\frac{\partial}{\partial z}[\rho(y,z,t)v_z] + \frac{\partial}{\partial y}[\rho(y,z,t)v_y] + \frac{\partial \rho(y,z,t)}{\partial t} = 0, \tag{5}$$

$$\frac{dv_z}{dt} = -|\eta| E_{sc-z} + \omega_c y, \tag{6}$$

and

$$\frac{dv_y}{dt} = -|\eta| E_0 - |\eta| E_{sc-y} - \omega_c v_z, \tag{7}$$

where variations in the magnetic field direction have been neglected. As pointed out in Chapter II, we must convert the above conservation equation to a Lagrangian one because of the multivalued nature of ρ and v.

The basic Lagrangian procedure is to divide the entering stream charge into a number of representative charge groups and then to integrate along the charge group trajectories at each displacement plane, summing the various forces which exist. In the case of very thick streams the total charge is first divided into a number of layers and subsequently each layer into a number of charge groups. The charge group trajectories, being solutions of Eqs. (6) and (7), may be written as

$$y = f(y_0, z_0, t) \tag{8}$$

and

$$z = g(y_0, z_0, t), \tag{9}$$

where y_0 and z_0 simply indicate the initial coordinates of the particular charge group.

The Lagrangian form of the continuity equation simply expresses the conservation of charge at consecutive displacement planes in the interaction region. The stream charge density $\rho_0(y_0, z_0, 0)$ entering over the area element $dy_0\, dz_0$ must appear as $\rho(y, z, t)$ in $dy\, dz$ at a later time. Thus

$$\rho(y, z, t)|\, dy\, dz\,| = \rho_0(y_0, z_0, 0) \sum |\, dy_0\, dz_0\,|, \tag{10}$$

where the summation carries over all the entering charge which appears within $dy\, dz$ as a result of overtaking and trajectory crossing. The absolute value signs of Eq. (10) indicate that all charge elements make positive contributions at all times. Equation (10) is rewritten as

$$\rho(y, z, t) = \rho_0 \sum \begin{vmatrix} \dfrac{\partial y_0}{\partial y} & \dfrac{\partial y_0}{\partial z} \\ \dfrac{\partial z_0}{\partial y} & \dfrac{\partial z_0}{\partial z} \end{vmatrix} = \rho_0 \sum \left| \dfrac{\partial y_0}{\partial y} \dfrac{\partial z_0}{\partial z} - \dfrac{\partial y_0}{\partial z} \dfrac{\partial z_0}{\partial y} \right|. \tag{11}$$

As was done previously for the klystron, it is convenient to define new independent and dependent variables and to eliminate the initial position z_0 in favor of the initial phase position Φ_0. These variables are defined as

$$q \triangleq \frac{\alpha \omega}{2\bar{u}_0} z = \pi \alpha N_s \equiv X, \quad \alpha = \text{depth of modulation parameter}, \tag{12}$$

$$\Phi(z, t) \triangleq \omega \left(\frac{z}{u_{z0}} - t \right), \tag{13}$$

$$u_z \triangleq \bar{u}_0(1 + \alpha u), \tag{14}$$

$$\Phi_0 \triangleq \frac{\omega z_0}{u_{z0}(y_0)} \tag{15}$$

and

$$v_{y\omega} \triangleq \frac{1}{\omega w} \frac{dy}{dt}, \quad \text{the } rf \text{ } y\text{-directed velocity.} \tag{16}$$

The reader is referred to Chapter IV for definitions of other variables.

The normalized z-position variable indicates the displacement in terms of the stream wavelength $\lambda_s = 2\pi \bar{u}_0/\omega$ and in terms of the initial average stream velocity. This is recognized as corresponding to the bunching parameter X of the klystron. The phase and velocity variables are related to the initial charge position relative to the modulating wave and to the initial average charge velocity respectively. Since the potential varies with y, the initial z-directed velocity u_{z0} is dependent upon y_0, the initial y-position. For a Brillouin beam whose average position is $y = \bar{y}_0$ at the entrance

$$u_{z0} = \omega_c \left(y_0 + \bar{y}_0 - \frac{w}{2} \right). \tag{17}$$

The parameter α indicates the depth of prevelocity modulation. These large-signal variables are illustrated in Fig. 3. The variable Φ serves as an identification coordinate for the charges relative to their initial phase Φ_0.

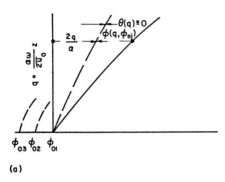

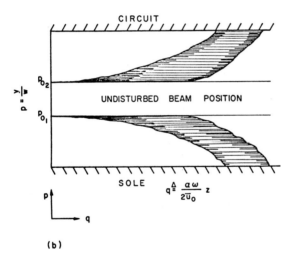

Fig. 3. Flight-line diagrams for crossed-field drift regions. (a) Drift region flight-line diagram at constant p; (b) p-q trajectories for various p_0's.

It is convenient now to eliminate z_0 in favor of Φ_0. The following derivative relations evolve:

$$\left.\frac{\partial \Phi}{\partial z}\right|_{y,t} = \frac{\delta \Phi}{\delta z} + \frac{\delta \Phi}{\delta \Phi_0}\frac{\partial \Phi_0}{\partial z} + \frac{\delta \Phi}{\delta y_0}\frac{\partial y_0}{\partial z}$$

$$= \frac{\omega}{\bar{u}_0} \qquad (18)$$

and

$$\left.\frac{\partial \Phi}{\partial y}\right|_{z,t} = \frac{\delta \Phi}{\delta \Phi_0}\frac{\partial \Phi_0}{\partial y} + \frac{\delta \Phi}{\delta y_0}\frac{\partial y_0}{\partial y} = 0, \tag{19}$$

where $\delta/\delta z$, $\delta/\delta y_0$ and $\delta/\delta \Phi_0$ conveniently indicate differentiation with respect to the variables (y_0, Φ_0, z). Using the definition of Φ_0, Eq. (15), we find after some manipulation that

$$\frac{\delta \Phi}{\delta \Phi_0}\frac{\partial \Phi_0}{\partial z} + \frac{\delta \Phi}{\delta y_0}\frac{\partial y_0}{\partial z} = \frac{\omega}{\bar{u}_0(1+\alpha u)}. \tag{20}$$

Hence we may use Eqs. (18)–(20) to rewrite the conservation equation as

$$\rho = \rho_0 \sum \left| \frac{u_{z0}(y_0)}{\bar{u}_0(1+\alpha u)} \frac{\delta \Phi_0}{\delta \Phi} \frac{\partial y_0}{\partial y} \right|. \tag{21}$$

It is convenient here to define a linear charge density

$$\rho(z, t) = h \int_0^{b'} \rho(y, z, t)\, dy, \tag{22}$$

where h and b' are indicated in Figs. 1 and 2. Typically the stream width in the magnetic field direction h is 0.5–0.7a. Combining with Eq. (21) yields

$$\rho(\Phi, z) = -|\rho_0| h \int \sum \frac{u_{z0}(y_0)}{\bar{u}_0(1+\alpha u)} \frac{\delta \Phi_0}{\delta \Phi} \frac{\partial y_0}{\partial y}\, dy. \tag{23}$$

In order to facilitate later calculation of the fundamental and harmonic current amplitudes in the stream, the charge density $\rho(\Phi, z)$ is expanded in a Fourier series in Φ.

$$\rho(\Phi, z) = \sum_{n=1}^{\infty} \frac{\sin n\Phi}{\pi} \int_0^{2\pi} \rho \sin n\Phi\, d\Phi_0' + \sum_{n=1}^{\infty} \frac{\cos n\Phi}{\pi} \int_0^{2\pi} \rho \cos n\Phi\, d\Phi_0'$$

$$= \sum_{n=1}^{\infty} \rho_{ns} \sin n\Phi + \sum_{n=1}^{\infty} \rho_{nc} \cos n\Phi, \tag{24}$$

where

$$\rho_{ns} = -|\rho_0| h \iint \sum \left[\frac{1+\alpha u_i}{1+\alpha u} \sin n\Phi\right] \frac{\delta y_0}{\delta \Phi} \frac{\partial y_0}{\partial y}\, dy\, d\Phi$$

$$= -|\rho_0| hw \int_0^{2\pi} \int_{\frac{1}{s}-\frac{1}{2}}^{\frac{1}{s}+\frac{1}{2}} \frac{1+\alpha u_i(p_0, \Phi_0', 0)}{1+\alpha u(p_0, \Phi_0', q)} \sin n\Phi\, d\Phi_0'\, dp_0, \tag{25}$$

2. TWO-DIMENSIONAL DRIFT-SPACE EQUATIONS

and

$$\rho_{nc} = -|\rho_0|hw \int_0^{2\pi} \int_{\frac{1}{s}-\frac{1}{2}}^{\frac{1}{s}+\frac{1}{2}} \frac{1 + \alpha u_i(p_0, \Phi_0', 0)}{1 + \alpha u(p_0, \Phi_0', q)} \cos n\Phi \, d\Phi_0' \, dp_0, \qquad (26)$$

where

$\Phi_0'=$ the variable of integration extending over one cycle of the premodulation waveform,

$u_i(p_0, \Phi_0, 0) =$ the initial normalized velocity function,

$s \triangleq w/\bar{y}_0$, stream thickness over average stream position,

$p \triangleq y/w$, normalized y-position variable, and

$p_0 \triangleq y_0/w$, normalized initial y-position variable.

The double integrations of Eqs. (25) and (26) extend over the y-extent of the stream and over all the entering charge groups. The above equations in Lagrangian form are all of those necessary to obtain the final working equations for the crossed-field drift region.

Velocity-Phase and Position Relations. If the dependent phase variable $\Phi(p_0, \Phi_0, q)$ is differentiated with respect to q and the velocity variable introduced, the following results:

$$\frac{\delta\Phi(p_0, \Phi_0, q)}{\delta q} = \frac{2}{\alpha} \left\{ \frac{1}{[1 + \alpha u_i(p_0, \Phi_0, 0)]} - \frac{1}{[1 + \alpha u(p_0, \Phi_0, q)]} \right\}. \qquad (27)$$

The y-position variable is related to the initial position as follows:

$$y = y_0 + \int_0^t \frac{dy}{dt} \, dy$$

$$= y_0 + \int_0^z \frac{dy}{dt} \frac{1}{\bar{u}_0(1 + \alpha u)} \, dz \qquad (28)$$

and after introduction of p, p_0 and $v_{y\omega} = [dp/d(\omega t)] = 1/\omega w(dy/dt)$, Eq. (28) becomes

$$p(p_0, \Phi_0, q) = p_0 + 2 \int_0^q \frac{v_{y\omega}}{\alpha[1 + \alpha u(p_0, \Phi_0, q)]} \, dq. \qquad (29)$$

Lorentz Force Equations. The z- and y-component force equations are obtained in Lagrangian form after direct substitution of the newly defined variables into Eqs. (6) and (7). The results are

$$[1 + \alpha u(p_0, \Phi_0, q)] \frac{\delta u(p_0, \Phi_0, q)}{\delta q} = \frac{2sv_{y\omega}}{\alpha^2 l} + \left(\frac{\omega_p}{\omega}\right)^2 \frac{4rs}{\pi\alpha^2} F_{2-z} \qquad (30)$$

and

$$[1 + \alpha u(p_0, \Phi_0, q)]\left[\frac{\delta v_{y\omega}}{\delta q} + \frac{2l}{s\alpha}\left(\frac{\omega_c}{\omega}\right)^2\right] = \frac{2rl^2}{s\alpha}\left(\frac{\omega_c}{\omega}\right)^2\left(\frac{V_a}{2V_0}\right)$$
$$- \frac{4rl}{\pi\alpha}\left(\frac{\omega_c}{\omega}\right)\left(\frac{\omega_p}{\omega}\right)^2 F_{z-y}, \qquad (31)$$

where

$r \triangleq \bar{y}_0/b'$,

$l \triangleq \bar{u}_0/\omega_c \bar{y}_0$,

$V_a \triangleq$ dc anode potential,

$V_0 \triangleq (\bar{u}_0^2/2\eta)$, voltage equivalent of the average velocity at the entrance,

$$E_{sc-y} = \frac{2|\rho_0|}{\pi\epsilon_0 b'}\frac{w}{\beta} F_{2-y},$$

$$E_{sc-z} = -\frac{2|\rho_0|}{\pi\epsilon_0 b'}\frac{w}{\beta} F_{2-z},$$

and F_{2-y} and F_{2-z} are defined in Eqs. (29) and (30) of Chapter IV.

Since the space-charge fields are expressed in terms of an axial displacement variable

$$\xi \triangleq \frac{\pi}{b'}(z - z'),$$

we shall here express ξ in terms of the stream initial position coordinates. The phase positions for two electrons with initial coordinates $(p_0, \Phi_0, 0)$ and $(p_0', \Phi_0', 0)$ are respectively

$$\Phi(p_0, \Phi_0, z) = \omega\left[\frac{z}{u_{z0}(p_0)} - t\right] \qquad (32a)$$

and

$$\Phi'(p_0', \Phi_0', z) = \omega\left[\frac{z}{u'_{z0}(p_0')} - t'\right] \qquad (32b)$$

and the first-order time separation $t - t'$ is

$$t - t' = \frac{z}{\bar{u}_0}\left[\frac{1}{1 + \alpha u_i} - \frac{1}{1 + \alpha u}\right] - \frac{(\Phi - \Phi')}{\omega}. \qquad (33)$$

If it is assumed that the time distribution at a fixed z-position can be replaced by the spatial distribution at a fixed time, the axial displacement variable is written as

$$\xi = -\frac{2\pi rlq}{\alpha}\left(\frac{\omega_c}{\omega}\right)\left[\frac{1}{1 + \alpha u_i} - \frac{1}{1 + \alpha u}\right]$$
$$+ \pi rl\left(\frac{\omega_c}{\omega}\right)(1 + \alpha u)(\Phi - \Phi'). \qquad (34)$$

2. TWO-DIMENSIONAL DRIFT-SPACE EQUATIONS

The system of equations (27) and (29)–(31), plus the appropriate boundary conditions, completely describes the nonlinear crossed-field drift-space region. This system is solved subject to the imposition of some initial velocity and/or density modulation.

Initial Conditions. At the input plane to the drift region the stream is divided into a number of layers in the y- or p-direction and each layer is further divided into a number of representative charge groups as in previous chapters. We find that each stream layer can effectively represent 4–5% of b' and that at least $m = 32$ charge groups per layer are required. The initial p positions and Φ_0 positions (no density modulation) are given by

$$p_{j,k,0} = \left(\frac{1}{s} - \frac{1}{2}\right) + \frac{k}{(n-1)}, \qquad k = 0, 1, 2, ..., n \tag{35}$$

and

$$\Phi_{j,k,0} = \frac{2\pi j}{m}, \qquad j = 0, 1, 2, ..., m, \tag{36}$$

where

$n \triangleq$ the number of stream layers, and

$m \triangleq$ the number of charge groups per layer.

The initial velocity deviation variables for the z-direction are written as

$$u_{j,k,0} = \frac{\left[p_{j,k,0} - \frac{1}{s}\right]}{\left(\frac{\alpha l}{s}\right)} L_{k,0} = \frac{\left(\frac{k}{n} - \frac{1}{2}\right)}{\left(\frac{\alpha l}{s}\right)} L_{k,0}, \tag{37}$$

where $L_{k,0} = 1$ for all k for a Brillouin beam. Other space-charge flows require $L_{k,0} \neq 1$. For an entering slipping stream with $v_y = 0$ and $v_z = \omega_c[y + \bar{y}_0 - (w/2)]$

$$(v_{y\omega})_{j,k,0} = M_{k,0}, \tag{38}$$

where $M_{k,0} = 0$ for all k for a Brillouin beam. It is also of interest to note that for a Brillouin stream

$$\bar{u}_0 = \omega_c\left(2\bar{y}_0 - \frac{w}{2}\right) = \omega_c \bar{y}_0\left(2 - \frac{s}{2}\right) = \omega_c \bar{y}_0 l \tag{39}$$

and since the input charge density in the stream is uniform

$$I_0 = h\rho_0 \int_{2\bar{y}_0-w}^{2\bar{y}_0+w} \omega_c y \, dy = h\rho_0 w \bar{u}_0. \tag{40}$$

In the case of low-density streams it may be permissible to neglect the space-charge forces and hence some simplification may be introduced in the force equations. If Brillouin flow is again assumed at the input, it is necessary to introduce an additional term $2u_i l/s(\omega_c/\omega)^2$ into the right-hand side of the y-force equation to account for the dc space-charge electric field. The force equations under this assumption are

$$[1 + \alpha u(p_0, \Phi_0, q)] \frac{\delta u(p_0, \Phi_0, q)}{\delta q} = \frac{2sv_{y\omega}}{\alpha^2 l} \qquad (41)$$

and

$$[1 + \alpha u(p_0, \Phi_0, q)] \left[\frac{\delta v_{y\omega}}{\delta q} + \frac{2l}{s\alpha}\left(\frac{\omega_c}{\omega}\right)^2\right] = \frac{2l}{s\alpha}\left(\frac{\omega_c}{\omega}\right)^2 [1 + \alpha u_i(p_0, \Phi_0, 0)]$$

or

$$\frac{\delta v_{y\omega}}{\delta q} = \frac{2l}{s\alpha}\left(\frac{\omega_c}{\omega}\right)^2 \left[\frac{1 + \alpha u_i(p_0, \Phi_0, 0)}{1 + \alpha u(p_0, \Phi_0, q)} - 1\right]. \qquad (42)$$

Solutions of these and the previous equations are discussed in the following sections.

Operating Parameters to be Specified

$\alpha \triangleq$ the depth of modulation parameter,

$s \triangleq w/\bar{y}_0$, stream thickness parameter,

$r \triangleq \bar{y}_0/b'$, stream position parameter,

$l \triangleq$ the ratio of stream average velocity to $\omega_c \bar{y}_0$,

$\omega_c/\omega \triangleq$ the normalized cyclotron frequency,

$\omega_p/\omega \triangleq$ the normalized plasma frequency,

$V_a/V_0 \triangleq$ the ratio of dc anode voltage to dc voltage equivalent of the mean stream velocity.

3 Gap Modulation of a Crossed-Field Stream

For convenience of analysis we will analyze the crossed-field drift region in much the same manner as the O-type klystron was treated. It is assumed that a laminar-flow Brillouin beam exists and is injected into a short-transit-angle velocity modulation gap and thence into the desired crossed-field drift space. Such a configuration is illustrated in Fig. 4

along with the ideal velocity modulation gap. The field in the gap may be purely longitudinal or contain a transverse component, i.e., E_y. Whether or not the transverse field is important depends upon the gap transit angle, the magnetic field strength, i.e., the value of ω_c/ω, and the space-charge flow conditions. Both cases are analyzed below.

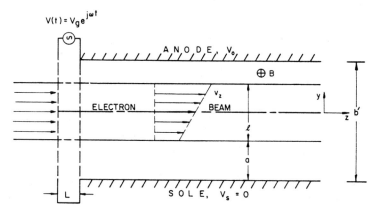

FIG. 4. Crossed-field drift space and velocity modulation gap.

(i) *Transverse Gap Field is Zero*; i.e., $E_y(gap) \equiv 0$. For such an ideal gap the field within the gap is assumed to be given by

$$E_y = 0 \tag{43}$$

and

$$E_z = -\frac{V_g \sin \omega t}{L}. \tag{44}$$

This also assumes that there is no y-component electric field associated with the beam. The equations of motion in this region are thus given by the component vector Lorentz equations, where the forces arise due to both electric and magnetic fields:

$$\frac{d^2y}{dt^2} = -\omega_c \frac{dz}{dt} \tag{45}$$

and

$$\frac{d^2z}{dt^2} = \frac{\eta V_g \sin(\omega t + \theta)}{L} + \omega_c \frac{dy}{dt}, \tag{46}$$

where η = the electron charge-to-mass ratio, and θ = the phase at which the electron enters the gap. At the entrance to the velocity modulation gap, i.e., $z = 0$, $t = 0$, the boundary conditions on electron trajectories are specified by the following:

$$y = y_0, \tag{47a}$$

$$\dot{y}_0 = 0 \tag{47b}$$

and

$$\dot{z}_0 = \omega_c(y_0 + 2a). \tag{47c}$$

Thus integration of Eq. (45) yields

$$\frac{dy}{dt} = -\omega_c z$$

and thus the y-directed velocity component at the output of the gap region is

$$\dot{y}_L = -\omega_c = -\left(\frac{\omega_c}{\omega}\right)\omega L. \tag{48}$$

Integrating Eq. (46) and applying the aforementioned boundary conditions yields the following expression for the z-component electron velocity at the output plane of the gap:

$$\dot{z}_L = \dot{z}_0 + \frac{\eta V_g}{\omega L}[\cos\theta - \cos(\omega t_L + \theta)] + \omega_c(y - y_0). \tag{49}$$

If it is assumed that ωt_L is small, the trigonometric functions are expanded and $\omega_c(y - y_0)$ is neglected, Eq. 49 becomes

$$\dot{z}_L \simeq \dot{z}_0 + \frac{\eta V_g t_L}{L}\sin\theta, \tag{50}$$

where t_L = the transit time across the gap. The transit time t_L may be evaluated by integrating Eq. (50) and applying the boundary conditions. The result is

$$\dot{z}_L = \dot{z}_0\sqrt{1 + [(2\eta V_g)/(\dot{z}_0{}^2)]\sin\theta} = \dot{z}_0\sqrt{1 + (V_g/V_0)\sin\theta}, \tag{51}$$

where $\dot{z}_0$ is given in Eqs. (47). As expected, one notices that Eq. (51) is identical to that obtained for the usual O-type klystron gap.

(ii) $E_y \neq 0$ *in the Gap.* As a typical value of transverse field for the velocity modulation gap assume that E_y is given by the Brillouin beam value:

$$E_y = -\frac{\omega_c{}^2}{\eta}(y + 2a) \tag{52}$$

everywhere in the gap region. The velocity components and transverse displacement at the output plane of the gap, assuming the same input boundary conditions, are now given by

$$y_L \approx y_0, \qquad (53)$$

$$\dot{y}_L = -\omega_c L \left[1 + \frac{2V_0}{V_g \sin \theta} \left(1 - \sqrt{1 + (V_g/V_0) \sin \theta} \right) \right] \qquad (54)$$

and

$$\dot{z}_L = \dot{z}_0 \sqrt{1 + (V_g/V_0) \sin \theta}. \qquad (55)$$

Thus while y_L and $\dot{z}_L$ have been unchanged the transverse velocity $\dot{y}_L$ has been reduced somewhat. Then the effect of the gap is easily evaluated and the output conditions serve as input conditions to the equations of Section 2. The effect of such premodulation on the motion in the drift region is illustrated in the following section.

4 Results for a Two-Dimensional Cf Drift Region

The utility of a crossed-field drift region is evaluated through consideration of the electron trajectories, electron velocities and the harmonic current content of the beam versus distance. The trajectory information is conveniently gained from plots of (prs) versus q and the velocity information from plots of the axial velocity $(1 + \alpha u)$ versus Φ and $v_{y\omega}$ versus Φ.

The harmonic current amplitudes in the drift region are calculated starting from the conservation of charge relationship,

$$\rho(y, z, t) = \rho_0 \sum \begin{vmatrix} \dfrac{\partial y_0}{\partial y} & \dfrac{\partial y_0}{\partial z} \\ \dfrac{\partial z_0}{\partial y} & \dfrac{\partial z_0}{\partial z} \end{vmatrix}. \qquad (56)$$

The procedure of Section 2 is followed: Eq. (56) is transformed by the introduction of Lagrangian variables and then $\rho(\Phi, z)$ is expanded in a Fourier series in the phase variable Φ. The expansion yields Eqs. (25) and (26) for the component charge densities. From the charge relationship the incremental current in any layer is written as

$$dI(q, \Phi, p) = hw \left\{ \rho(0, \Phi_0, p_0) u_0 [1 + 2Du(0, \Phi_0, p_0)] \left| \frac{\partial \Phi_0}{\partial \Phi} \right| \right\} dp_0. \qquad (57)$$

The charge density in Eq. (57) is conveniently written as

$$\rho(0, \Phi_0, p_0) = \rho_0 f(\Phi_0) g(y_0), \tag{58}$$

where $f(\Phi_0)$ and $g(y_0)$ represent the initial axial and transverse charge distributions. After integration of Eq. (57) the current is given by

$$I(q, \Phi) = wh\rho_0 \bar{u}_0 \int_{(\frac{y_0}{w} - \frac{1}{2})}^{(\frac{y_0}{w} + \frac{1}{2})} [1 + 2Du]_i \left| \frac{\partial \Phi_0}{\partial \Phi} \right| dp_0. \tag{59}$$

After expansion of Eq. (59) into a Fourier series the ratio of harmonic current amplitude to the dc current is written as

$$\frac{i_n(q)}{I_0} = \frac{\left\{ \left[\frac{1}{\pi} \int_0^{2\pi} \int_{(\frac{1}{s} - \frac{1}{2})}^{(\frac{1}{s} + \frac{1}{2})} (1 + 2Du_i) \cos n\Phi \, dp_0 \, d\Phi_0 \right]^2 + \left[\frac{1}{\pi} \int_0^{2\pi} \int_{(\frac{1}{s} - \frac{1}{2})}^{(\frac{1}{s} + \frac{1}{2})} (1 + 2Du_i) \sin n\Phi \, dp_0 \, d\Phi_0 \right]^2 \right\}^{1/2}}{\frac{1}{2\pi} \int_0^{2\pi} \int_{(\frac{1}{s} - \frac{1}{2})}^{(\frac{1}{s} + \frac{1}{2})} (1 + 2Du_i) \, dp_0 \, d\Phi_0}, \tag{60}$$

where n denotes the particular harmonic.

An examination of Eqs. (41) and (42) reveals that if $v_{y\omega} = 0$ and $\omega_p/\omega = 0$ then the equations reduce exactly to those for the O-type klystron (drift region) and thus the results do not depend upon ω_c/ω but only on the initial velocity modulation parameter α. The harmonic current amplitude versus distance is shown in Fig. 5 for two different depths of modulation. Whenever $v_{y\omega} \neq 0$ or $\omega_p/\omega \neq 0$ then the results differ markedly from those for $v_{y\omega} = \omega_p/\omega = 0$. The first five harmonic currents in a crossed-field drift region are shown versus distance in Fig. 6 for $v_{y\omega} = 1.0$ when space-charge forces have been neglected, i.e., $\omega_p/\omega = 0$, and a sinusoidal distribution of the modulating potential $v_{y\omega}$ was considered. In order to simplify the trajectory plots only the maxima and minima of each layer are shown. These results clearly indicate both a scalloping and modulation of the beam in the drift region.

The inclusion of space-charge forces in a Brillouin beam makes little difference to the harmonic current amplitudes although the beam profiles and component electron velocities are quite different due to the added forces. For large values of $\omega_c/\omega = \omega_p/\omega$ the scalloping of the beam becomes excessive and collection on either or both the sole and the anode

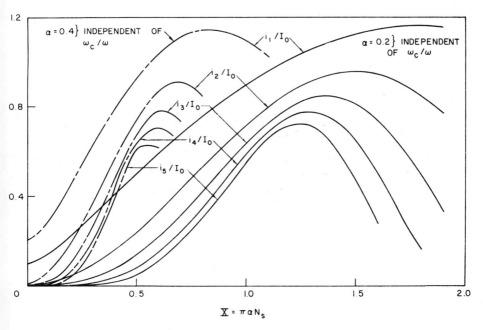

Fig. 5. Harmonic current versus distance and depth of modulation ($r = 0.5$, $\omega_p/\omega = 0$).

occurs. The variation of the fundamental component of current in the beam with $v_{y\omega}$ when $\omega_p/\omega = 0$ is summarized in Fig. 7. It should be recalled that when both $v_{y\omega}$ and ω_p/ω are zero the system degenerates to the O-type drift region.

The magnitude of each harmonic current in the beam was calculated using Eq. (60), which contains integrals over both the sin Φ and cos Φ. It is extremely interesting to calculate the phase angle of the current from the squared terms under the radical. The variation of this phase angle versus distance is shown in Fig. 8 along with the harmonic current amplitude for the first five harmonics. The change of phase angle by approximately π or 2π radians occurs at current minima. Notice that the current peaks do not continually decrease with distance but the various harmonic currents seem to pump one another parametrically. This results from the nonlinear character of the electron beam.

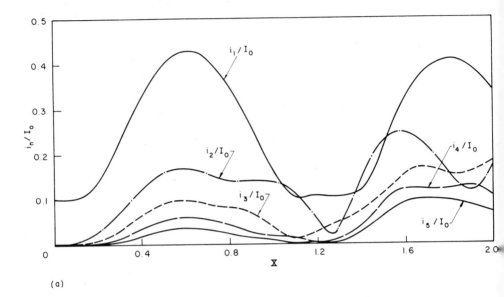

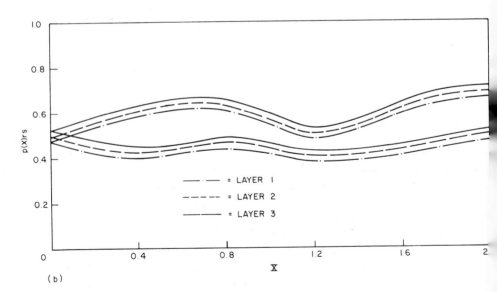

FIG. 6. Currents and trajectories for a crossed-field region ($r = 0.5$, $s = 0.1$, $\alpha = 0.2$, $\omega_c/\omega = 0.5$, $\omega_p/\omega = 0$, $v_{y\omega} = 1$). (a) Harmonic currents; (b) trajectories.

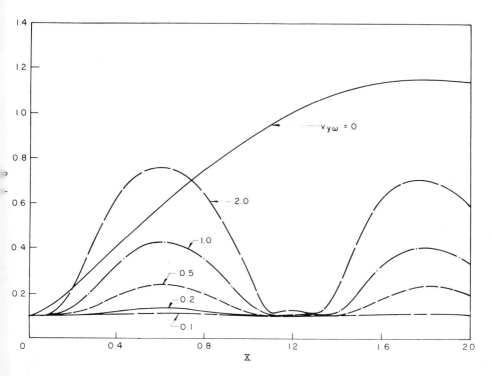

Fig. 7. Fundamental current versus distance and $v_{y\omega}$ ($r = 0.5$, $s = 0.1$, $\alpha = 0.2$, $\omega_c/\omega = 0.5$, $\omega_p/\omega = 0$).

5 Three-Dimensional Drift-Space Equations

In situations where the extent of the stream in the magnetic field direction is significant it may be necessary to consider forces and motion in that direction. There is no magnetic force in the x-direction and thus the only force in that direction is that due to the space-charge field. The component Lorentz force equations are

$$\frac{dv_x}{dt} = -|\eta|E_{sc-x}, \tag{61}$$

$$\frac{dv_y}{dt} = -|\eta|E_0 - |\eta|E_{sc-y} - \omega_c v_z \tag{62}$$

and

$$\frac{dv_z}{dt} = -|\eta|E_{sc-z} + \omega_c v_y. \tag{63}$$

The space-charge-field expressions to be used are given by Eqs. (15)–(17) of Chapter IV. They are shown graphically in Figs. 5–7 of Chapter IV. The procedure for normalization and introduction of Lagrangian variables is similar to that of Section 2. Under most conditions it is appropriate to utilize the two-dimensional drift region equations.

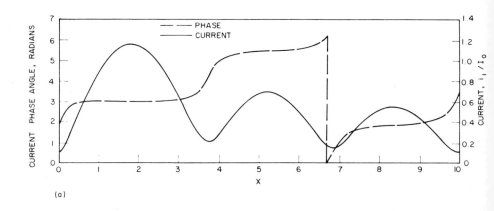

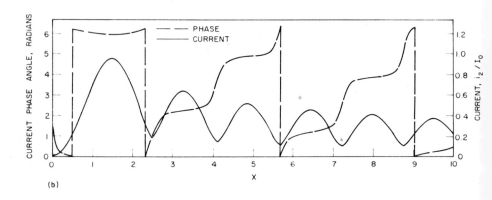

FIG. 8. Harmonic currents and phase angles versus distance ($r = 0.5$, $s = 0.1$, $\alpha = 0.2$, $\omega_c/\omega = 0.5$, $\omega_p/\omega = 0$, $v_{y\omega} = 0$). (a) i_1/I_0; (b) i_2/I_0; (c) i_3/I_0; (d i_4/I_0; (e) i_5/I_0.

5. THREE-DIMENSIONAL DRIFT-SPACE EQUATIONS 301

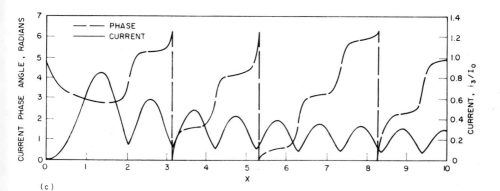
(c)

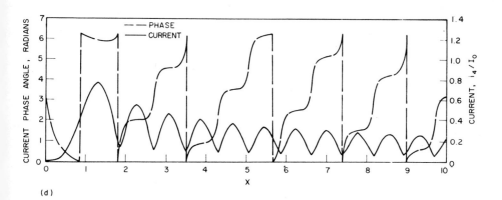
(d)

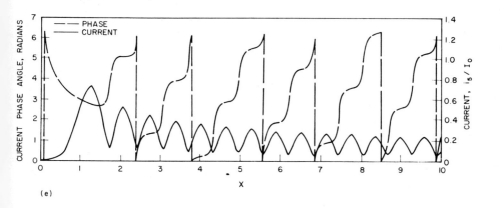
(e)

6 Adiabatic Motion in a Drift Region

An interesting special case arises when the acceleration terms in the equations of motion are neglected. This eliminates the cycloidal component of motion which arises out of the acceleration caused by the electric field. The motion equations so obtained are called the adiabatic equations of motion. Under what conditions may the adiabatic equations be used, i.e., neglecting the cycloidal motion relative to the average motion ? The rate of growth of an rf signal (impressed modulation) in a cyclotron wavelength

$$\lambda_c = \frac{2\pi \bar{u}_0}{\omega_c}$$

is proportional to

$$\text{Growth rate} \sim \exp \frac{\pi \alpha}{(\omega_c/\omega)} \text{Re } \delta,$$

where δ is the complex propagation constant. Thus for small growth rates we must require that

$$\frac{\pi \alpha}{(\omega_c/\omega)} \text{Re } \delta \ll 1,$$

which is well satisfied for conditions under which

$$\frac{2\omega_c}{\omega \alpha} \gg 1. \tag{64}$$

Thus we see that for either very high magnetic field conditions or low amplitude modulation conditions it is permissible to neglect the cycloidal component of motion relative to the average motion. The simplified equations of motion are

$$v_{y\omega} = \frac{2rl}{\pi} \left(\frac{\omega_p}{\omega}\right)^2 F_{2-z} \tag{65}$$

and

$$u(p_0, \Phi_0, q) = \frac{1}{\alpha}\left(\frac{rlV_a}{2V_0} - 1\right) - \left(\frac{\omega_p}{\omega_c}\right)^2 \left(\frac{\omega_c}{\omega}\right) rsF_{2-y}. \tag{66}$$

The results obtained using the adiabatic equations are essentially the same as those obtained using the more general equations except that the cycloidal component of motion is not reflected in the trajectory plots.

References

1. Gandhi, O. P., "Nonlinear Electron-Wave Interaction in Crossed Electric and Magnetic Fields." Univ. of Michigan Electron Phys. Lab. Tech. Rept. No. 39 (October 1960).
2. Gandhi, O. P., and Rowe, J. E., Nonlinear analysis of crossed-field amplifiers. *NTF-Nachrichtentech. Fachber.* **22**, 1-10 (1961).
3. Gandhi, O. P., and Rowe, J. E., Nonlinear theory of injected beam crossed-field devices. In *Crossed Field Microwave Devices*, Vol. 1, Chapter 5, Section 5.2.2, pp. 439-495. Academic Press, New York, 1961.

CHAPTER IX | Crossed-Field Forward-Wave Amplifiers

1 Introduction

The counterpart of the O-type traveling-wave amplifier in crossed E and B fields is the injected-beam crossed-field forward-wave amplifier, which may conveniently be designated as the M-FWA. Such a device naturally evolves from the drift-space device described in the previous chapter when either the bottom or top electrode plate is replaced by a propagating structure such as a vane line or an interdigital line. This arrangement is illustrated in Fig. 1. In such an amplifier the electron stream potential energy is converted to rf energy and therefore a higher efficiency should be obtained than in kinetic energy conversion devices, which are limited in maximum efficiency by a loss of synchronism with the rf wave. This possibility of higher efficiency is the chief reason for interest in crossed-field amplifiers.

As shown in Chapter XIII on phase focusing, however, the efficiency of O-type amplifiers can be made comparable to their M-type counterparts and hence the question of superiority of one device over the other is clouded. In very high power situations the fact that a portion of the dc stream in an M-FWA must usually be dissipated on the rf circuit does represent a disadvantage in comparison to the O-FWA, in which the confining B field can be arranged to realize near 100 per cent beam transmission. Again in Chapter XIII a means for overcoming this difficulty is evaluated.

The superiority of one class of devices over the other can and has been argued many ways without conclusive results. Both are here to stay! In this book it is the aim only to develop the analyses of all such devices so that others may have a basis for deciding which type of device is better for a particular application.

The various configurations outlined in Fig. 1 are treated in this chapter using the basic circuit and space-charge-field equations of Chapters III and IV. The so-called emitting-sole device may be treated in exactly the same manner.

1. INTRODUCTION

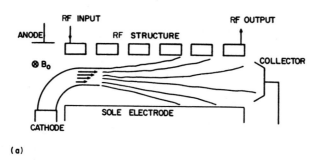

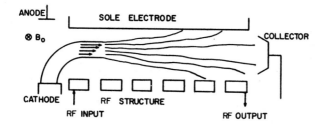

Fig. 1. Schematic diagram of crossed-field forward-wave amplifiers. (a) Negative-sole M-FWA; (b) positive-sole M-FWA; (c) negative-sole M-FWA with a positive-sole prebunching region.

2 Two-Dimensional M-FWA with a Negative Sole

As mentioned in the preceding section and in Chapter I, the energy exchange mechanism involves a displacement of the stream normal (y-direction) to the direction of primary stream flow (z-direction) with little change in the particle kinetic energy. The electrons are essentially phase focused relative to the wave and simply give up energy to the rf wave as they move towards the positive electrode, which is the rf structure in this case. These favorably phased electrons are coupled more tightly to the circuit fields as they move toward the structure, and those which are in unfavorable phase positions take energy from the wave and move towards the sole electrode. Fortunately as they do, their coupling to the circuit wave decreases and they thus extract less energy from the wave. This sorting mechanism is partially responsible for the high efficiency of operation characteristic of these devices.

An interesting approximate treatment of the crossed-field nonlinear problem, M-FWA, was made by Feinstein and Kino[1] in which they developed a quasi-large-signal theory using equivalent circuit methods. They assumed stream-wave synchronism and that the circuit field decayed exponentially in moving from the circuit to the sole, which was thus moved infinitely far away. In order to make the ballistics equations tractable they neglected space-charge fields and assumed the adiabatic equations of motion.

Sedin[2] studied the M-FWA and the M-BWO also using equivalent circuit methods but including space-charge fields, nonlaminar motion, and finite stream thickness. He calculated the induced current at the rf structure from the rate of change of the y-directed space-charge field. The space-charge field was calculated for infinite line charges between infinite, parallel, and perfectly conducting plates as discussed in Chapter IV. As a result of some difficulties with his space-charge model he made calculations only for relatively thin streams. For purposes of calculation he also assumed the adiabatic equations of motion, which implies $D \ll 1$.

A basically similar analysis, yet different in details, has been made by Kooyers and Hull.[3] Yet another study, different in several respects but also similar in many, is due to Gandhi[4] and Rowe.[5-7] The similarities and differences will be more apparent in the following sections of this and the following chapter.

The mathematical theory of the M-FWA is developed by generalizing the considerations of Chapter VIII on the drift region. Rf circuit equations will appear along with circuit field terms in the Lorentz

2. TWO-DIMENSIONAL M-FWA WITH A NEGATIVE SOLE

force equation. Following the pattern of Chapter VIII, the component force equations are written as

$$\frac{dv_z}{dt} = |\eta|\frac{\partial V_c}{\partial z} + |\eta|\frac{\partial V_{sc-z}}{\partial z} + \omega_c v_y \quad (1)$$

and

$$\frac{dv_y}{dt} = |\eta|\frac{\partial V_c}{\partial y} + |\eta|\frac{\partial V_{sc-y}}{\partial y} - \omega_c v_z, \quad (2)$$

where V_c and V_{sc} indicate the circuit and space-charge potentials respectively.

The normalized independent variables are defined as follows for the M-FWA, as in Chapter VIII:

$$q \triangleq \frac{D\omega}{\bar{u}_0} z = 2\pi D N_s, \quad (3)$$

$$\Phi(z, t) \triangleq \omega\left(\frac{z}{u_{z0}} - t\right) + \theta(z), \quad (4)$$

$$u_z \triangleq \bar{u}_0(1 + 2Du), \quad (5)$$

$$\Phi_0 \triangleq \frac{\omega z_0}{u_{z0}(y_0)}, \quad (6)$$

$$v_{y\omega} \triangleq \frac{1}{\omega w}\frac{dy}{dt}, \quad (7)$$

and

$$\bar{u}_0 \triangleq v_0(1 + Db). \quad (8)$$

The dependent variable $\theta(z)$ represents an rf phase difference between the actual wave and a hypothetical wave traveling at u_{z0} and is caused by the loading of the stream on the circuit. The electron phase variable $\Phi(z, t)$ now refers to a phase position relative to the propagating wave on the structure. Note that the depth of modulation parameter α has been replaced by $2D$, the beam circuit coupling or gain parameter. The velocity parameter b measures the difference between the initial average stream velocity and the cold-circuit phase velocity. The large-signal variables are illustrated on the flight-line diagrams of Fig. 2.

We must now consider the rf potential function in the interaction region. Assume that the rf structure impedance is all at the fundamental frequency and that only ρ_1 in the stream produces an effective voltage on the structure. We may write the potential as the product of slowly varying functions in the following form:

$$V(y, z, t) \triangleq \text{Re}\left[\frac{Z_0 I_0}{D} A(z)\psi(y)e^{-j\Phi}\right], \quad (9)$$

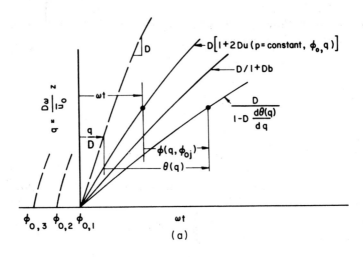

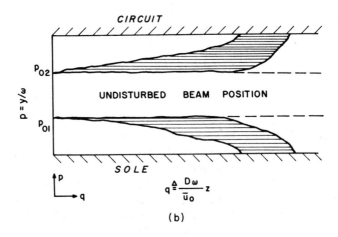

Fig. 2. Flight-line diagram for nonlinear crossed-field devices. (a) q versus $\Phi_{0,j}$ diagram at constant p positions; (b) p-q trajectories for various p_0's.

where $A(z)$ is the amplitude function and $\psi(y)$ is the coupling function which measures the falloff of potential moving away from the structure toward the sole and also indicates the coupling of current in the stream to the circuit. For parallel-plane geometry it is easily shown that

$$\psi(y) = \frac{\sinh \beta y}{\sinh \beta b'}. \tag{10}$$

The above definitions are now used to develop the basic working equations for the negative-sole M-FWA.

Velocity-Phase and Position Equations. Differentiating the phase variable equation (4) with respect to q and introducing the appropriate normalizations yields

$$\frac{\delta \Phi(p_0, \Phi_0, q)}{\delta q} - \frac{d\theta(q)}{dq}$$
$$= \frac{1}{D} \left\{ \frac{1}{[1 + 2Du_i(p_0, \Phi_0, 0)]} - \frac{1}{[1 + 2Du(p_0, \Phi_0, q)]} \right\}, \tag{11}$$

where the total derivative of $\theta(q)$ appears since it does not depend on the stream position; i.e., it is associated purely with the rf structure. The normalized y-position dependent variable is developed, following Chapter VIII directly, with the result

$$p(p_0, \Phi_0, q) = p_0 + \int_0^q \frac{v_{y\omega}}{D[1 + 2Du(p_0, \Phi_0, q)]} dq. \tag{12}$$

Lorentz Force Equations. The M-FWA force equations are generalized from the corresponding drift-region equations by the inclusion of circuit field terms. The potential function in Eq. (9) is written in terms of normalized variables as

$$V(p_0, \Phi_0, q) = \mathrm{Re}\left[\frac{Z_0 I_0}{D} A(q)\psi(p)e^{-j\Phi}\right]. \tag{13}$$

Forming the appropriate derivatives of Eq. (13) and substituting into Eqs. (1) and (2) along with the space-charge-field expressions yields the following force equations after some minor algebra:

$$[1 + 2Du(p_0, \Phi_0, q)] \frac{\delta u(p_0, \Phi_0, q)}{\delta q} = \frac{\omega \psi(p)}{2\omega_c}\left[\frac{dA(q)}{dq}\cos\Phi\right.$$
$$\left. - \frac{A(q)\sin\Phi}{D}\left(1 + D\frac{d\theta(q)}{dq}\right)\right] + \frac{sv_{y\omega}}{2D^2 l} + \left(\frac{\omega_p}{\omega}\right)^2 \frac{rs}{\pi D^2} F_{2-z} \tag{14}$$

and

$$[1 + 2Du(p_0, \Phi_0, q)] \left[\frac{\delta v_{y\omega}}{\delta q} + \frac{l}{sD}\left(\frac{\omega_c}{\omega}\right)^2\right]$$

$$= \frac{l}{s}\left[1 + D\frac{d\theta(q)}{dq}\right] \frac{\cosh\left[\frac{\omega}{\omega_c}\frac{ps}{l}\left(1 + D\frac{d\theta(q)}{dq}\right)\right]}{\sinh\left[\frac{\omega}{\omega_c rl}\left(1 + D\frac{d\theta(q)}{dq}\right)\right]} A(q) \cos\Phi$$

$$+ \frac{rl^2}{sD}\left(\frac{\omega_c}{\omega}\right)^2\left(\frac{V_a}{2V_0}\right) - 2\frac{rl}{\pi D}\left(\frac{\omega_c}{\omega}\right)\left(\frac{\omega_p}{\omega}\right)^2 F_{2-y}, \qquad (15)$$

where the parameters r, l, s, and V_a and the integrals F_{2-y}, F_{2-z} are defined in Chapter VIII. In writing the above equations it is assumed that the wave propagation constant $\beta(q)$ varies only slightly over the length and thus

$$\beta(q) = \frac{\partial \Phi}{\partial z} = \frac{\omega}{u_{z0}} + D\frac{\omega}{\bar{u}_0}\frac{d\theta(q)}{dq} = \bar{\beta}\left(\frac{\bar{u}_0}{u_{z0}} + D\frac{d\theta(q)}{dq}\right)$$

$$\approx \bar{\beta}\left(1 + D\frac{d\theta(q)}{dq}\right). \qquad (16)$$

Rf Circuit Equations. Preliminary to the development of the rf circuit equations we must again consider the space-charge density $\rho(\Phi, z)$, since the coupling function $\psi(p)$ measures the effectiveness of charge elements in the stream in inducing rf currents in the circuit. Equation (18) of Chapter VIII is now

$$\left.\frac{\partial \Phi}{\partial z}\right|_{y,t} = \frac{\omega}{\bar{u}_0} + \frac{d\theta(z)}{dz} \qquad (17)$$

and the induced charge at any plane along the slow-wave structure is

$$\rho(z, t) = h \int \psi(y)\rho(y, z, t)\, dy$$

$$= -|\rho_0| h \int \sum \psi(y) \frac{u_{z0}(y_0)}{\bar{u}_0(1 + 2Du)} \frac{\delta\Phi_0}{\delta\Phi} \frac{\partial y_0}{\partial y}\, dy. \qquad (18)$$

After changing variables, expanding $\rho(\Phi, z)$ into a Fourier series in Φ, and taking the fundamental sine and cosine components, we write the linear charge density coupled to the circuit as

$$\rho_{1s} = -|\rho_0| hw \int_0^{2\pi} \int_{\frac{1}{s}-\frac{1}{2}}^{\frac{1}{s}+\frac{1}{2}} \psi(p) \frac{1 + 2Du_i(p_0, \Phi_0, 0)}{1 + 2Du(p_0, \Phi_0, q)} \sin\Phi\, d\Phi_0\, dp_0 \qquad (19)$$

2. TWO-DIMENSIONAL M-FWA WITH A NEGATIVE SOLE

and

$$\rho_{1c} = -|\rho_0| hw \int_0^{2\pi} \int_{\frac{1}{s}-\frac{1}{2}}^{\frac{1}{s}+\frac{1}{2}} \psi(p) \frac{1 + 2Du_i(p_0, \Phi_0, 0)}{1 + 2Du(p_0, \Phi_0, q)} \cos\Phi \, d\Phi_0 \, dp_0, \quad (20)$$

where

$$\psi(p) = \frac{\sinh(\beta wp)}{\sinh(\beta b')}. \quad (21)$$

The equivalent circuit appropriate to this analysis is the one-dimensional uniform transmission line of Fig. 19 in Chapter III and the appropriate differential equation for the voltage along the line is

$$\frac{\partial^2 V(z,t)}{\partial t^2} - v_0^2 \frac{\partial^2 V(z,t)}{\partial z^2} + 2\omega Dd \frac{\partial V(z,t)}{\partial t}$$
$$= v_0 Z_0 \left[\frac{\partial^2 \rho_1(z,t)}{\partial t^2} + 2\omega Dd \frac{\partial \rho_1(z,t)}{\partial t} \right], \quad (22)$$

where the various parameters are defined in Chapter III.

The gain parameter D, analogous to C in the O-FWA, has been defined as

$$D^2 \triangleq \frac{\omega_c}{\omega} \frac{|I_0| Z_0}{2V_0}, \quad (23)$$

where $Z_0 \triangleq$ the characteristic impedance of the rf circuit, and $I_0 \triangleq$ the dc stream current. It should be pointed out that D^2 as defined above is somewhat different from D^2 as defined by Sedin, who followed Gould's definition (Reference 45, Chapter I). The relationship between the two is

$$D = \left(\frac{\omega_c}{\omega}\right) D_{\text{Sedin}}. \quad (24)$$

Since this is a normalization it makes no essential difference.

Now substitute the appropriate derivatives of Eqs. (19) and (20) into the right-hand side of Eq. (22) and the derivatives of Eq. (13) into the left-hand side, equate coefficients of $\sin\Phi$ and $\cos\Phi$, simplify and obtain the following:

$$\frac{d^2 A(q)}{dq^2} - A(q)\left[\left(\frac{1}{D} + \frac{d\theta(q)}{dq}\right)^2 - \left(\frac{1 + Db}{D}\right)^2\right]$$
$$= -\frac{(1 + Db)}{\pi D}\left[\int_0^{2\pi}\int_{\frac{1}{s}-\frac{1}{2}}^{\frac{1}{s}+\frac{1}{2}} \psi(p) \frac{1 + 2Du_i(p_0, \Phi_0, 0)}{1 + 2Du(p_0, \Phi_0, q)} \cos\Phi \, d\Phi_0 \, dp_0\right.$$
$$\left. + 2dD \int_0^{2\pi}\int_{\frac{1}{s}-\frac{1}{2}}^{\frac{1}{s}+\frac{1}{2}} \psi(p) \frac{1 + 2Du_i(p_0, \Phi_0, 0)}{1 + 2Du(p_0, \Phi_0, q)} \sin\Phi \, d\Phi_0 \, dp_0\right] \quad (25)$$

and

$$2\frac{dA(q)}{dq}\left(\frac{1}{D} + \frac{d\theta(q)}{dq}\right) + A(q)\left[\frac{d^2\theta(q)}{dq^2} + \frac{2d}{D}(1 + Db)^2\right]$$

$$= \frac{(1 + Db)}{\pi D}\left[\int_0^{2\pi}\int_{\frac{1}{s}-\frac{1}{2}}^{\frac{1}{s}+\frac{1}{2}}\psi(p)\frac{1 + 2Du_i(p_0, \Phi_0, 0)}{1 + 2Du(p_0, \Phi_0, q)}\cos\Phi\, d\Phi_0\, dp_0\right.$$

$$\left. - 2dD\int_0^{2\pi}\int_{\frac{1}{s}-\frac{1}{2}}^{\frac{1}{s}+\frac{1}{2}}\psi(p)\frac{1 + 2Du_i(p_0, \Phi_0, 0)}{1 + 2Du(p_0, \Phi_0, q)}\cos\Phi\, d\Phi_0\, dp_0\right], \qquad (26)$$

where the double integration occurs over the full entering charge and its coupling to the circuit measured by $\psi(p)$.

The apparent similarity of these nonlinear crossed-field equations to the corresponding two-dimensional O-FWA is to be expected, since the same Lagrangian method and circuit treatment was used in each. The six working equations to be solved simultaneously are Eqs. (11), (12), (14), (15), (25), and (26).

Initial Conditions. The problem is treated as an initial-value problem rather than as a boundary-value problem for the same reasons as given in Chapter V; thus we are obliged to specify dependent variables and their derivatives at $q = 0$. The stream initial conditions used in Chapter VIII are again appropriate, although here it is not necessary to assume a premodulation or bunching of the stream. These conditions are repeated here for convenience.

$$p_{j,k,0} = \left(\frac{1}{s} - \frac{1}{2}\right) + \frac{k}{(n-1)} \qquad k = 0, 1, 2 \ldots n \qquad (27)$$

and

$$\Phi_{j,k,0} = \frac{2\pi j}{m} \qquad j = 0, 1, 2 \ldots m, \qquad (28)$$

where $n \triangleq$ the number of stream layers, $m \triangleq$ the number of charge groups per layer.

$$u_{j,k,0} = \frac{(k/n) - \frac{1}{2}}{(2Dl)/s} L_{k,0} \qquad (29)$$

and

$$(v_{y\omega})_{j,k,0} = M_{k,0}, \qquad (30)$$

where $L_{k,0} = 1$ and $M_{k,0} = 0$ for all k in a Brillouin stream.

2. TWO-DIMENSIONAL M-FWA WITH A NEGATIVE SOLE

The variables associated with the rf wave must also be specified at $q = 0$:

$A(0) \triangleq A_0$, the arbitrary normalized rf voltage amplitude applied to the circuit. (31)

$$\theta(0) = 0. \tag{32}$$

The rf structure voltage is written as

$$V(y, z, t) = \text{Re}\left[\frac{Z_0 |I_0|}{D} \psi(y) A(z) e^{-j\Phi}\right] = \text{Re}[V_{in}\psi(y) e^{j(\omega t - \beta_0 z)}],$$

where $\beta_0 = \omega/v_0$, the undisturbed-circuit phase constant. The partial derivative of $\bar{V}(y, z, t)$ with respect to z is $[\psi(b') = 1]$

$$\frac{\partial \bar{V}(y, z, t)}{\partial z} = -j\beta_0 V_{in} e^{j(\omega t - \beta_0 z)} = \frac{Z_0 |I_0|}{D} \left[\frac{dA(z)}{dz} - jA(z)\frac{\partial \Phi}{\partial z}\right] e^{-j\Phi}.$$

Equating real and imaginary parts above and evaluating at $z = 0$ for a lossless circuit gives

$$\left.\frac{dA(q)}{dq}\right|_{q=0} = 0, \tag{33}$$

and

$$\left.\frac{\partial \Phi}{\partial z}\right|_{z=0} = \frac{\omega}{\bar{u}_0(1 + 2Du_i)} + \frac{\partial \theta(z)}{\partial z} = \beta_0 \tag{34}$$

which of course must be the case for no circuit loss and for an initially unbunched stream.

Equation (34) thus leads to

$$\left.\frac{d\theta(q)}{dq}\right|_{q=0} = b + \frac{2u_i}{1 + 2Du_i} \tag{35a}$$

$$= b \quad \text{for} \quad u_i \equiv 0. \tag{35b}$$

The system is now complete and we now proceed to studying solutions and investigating various simplifications.

The following operating parameters must be specified:

$D \triangleq$ the gain parameter,
$b \triangleq (u_0 - v_0)/v_0 D$, the velocity parameter,
$d \triangleq$ circuit-loss parameter,
$s \triangleq (w/\bar{y}_0)$, stream-thickness parameter,

$r \triangleq \bar{y}_0/b'$, stream-position parameter,

$l \triangleq$ the ratio of beam average velocity to $\omega_c \bar{y}_0$,

$\omega_c/\omega \triangleq$ the normalized cyclotron frequency,

$\omega_p/\omega \triangleq$ the normalized plasma frequency,

$(V_a/V_0) \triangleq$ the ratio of dc anode voltage to dc voltage equivalent of the mean stream velocity.

The above force equations must be modified as in Eqs. (41) and (42) of Chapter VIII when space-charge forces are neglected.

3 Results for a Two-Dimensional M-FWA with a Negative Sole

a. Introduction

The general crossed-field nonlinear interaction equations of the previous section are solved by digital computer methods in view of the nonlinearities. The conditions under which closed-form solutions may be obtained are examined in Chapter XIII. Complete evaluation of crossed-field amplifier performance requires an evaluation of the dependence of gain and efficiency on such device parameters as D, ω_c/ω, A_0, d, ω_p/ω, r and s. The results are summarized in the following sections.

Before discussing the general results it is worthwhile to consider the manner in which gain is calculated and also to derive expressions for the electronic efficiency and the percentage of available potential energy converted. The normalized rf voltage along the structure was defined in Eq. (13); the power output may be calculated from

$$P_{\text{out}} = \frac{\bar{V}^2(z, t)}{Z_0}\bigg|_{\text{avg.}} \tag{36}$$

and the interaction efficiency from

$$\eta_e \triangleq \text{eff.} = \frac{P_{\text{out}}}{I_0 V_a}, \tag{37}$$

where $I_0 V_a$ represents the dc input power. Combining the above equations gives the electronic efficiency as

$$\eta_e = \frac{I_0 V_0}{I_0 V_a} \frac{A_{\max}^2}{(\omega_c/\omega)} = r\left(1 - \frac{s}{4}\right) \frac{A_{\max}^2}{(\omega_c/\omega)}, \tag{38}$$

where all the normalized variables and parameters have been defined previously. Note that the η_e used in Eq. (38) depends upon the beam entrance position r. A possibly more useful efficiency is obtained by calculating the percentage of available potential energy which has been converted to rf. Such an efficiency is

$$\eta_a \triangleq \frac{\eta_e}{r} = \left(1 - \frac{s}{4}\right) \frac{A_{\max}^2}{(\omega_c/\omega)}. \tag{39}$$

As is illustrated in Fig. 3, the phenomenon of saturation in M-type

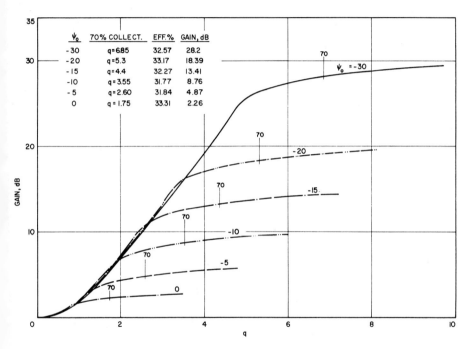

Fig. 3. Gain versus length and drive level ($D = 0.05$, $b = 0$, $r = 0.5$, $s = 0.1$, $\omega_c/\omega = 0.5$, $\omega_p/\omega = 0$).

devices is quite different from that in O-type devices, being spread out over a relatively long distance. As shown this is independent of the initial rf signal amplitude. Thus it is seen that in order to calculate overall efficiencies some criterion for saturation is needed. Rather arbitrarily this has been selected as the displacement plane where 70 per cent of the beam has been collected on the structure. These displacement planes have been noted on Fig. 3 as a function of the initial input-signal level ψ_0.

The initial signal level applied to the rf structure is denoted by A_0 and is specified by the signal-level parameter ψ_0 in decibels relative to $(\omega I_0 V_0/2\omega_c)$. The data of Fig. 3 indicate that the device saturates at a level corresponding to $\omega I_0 V_0/2\omega_c$ independent of ψ_0.

In considering the physical significance of the results presented in the following sections it is helpful to keep in mind the importance of the coupling function $\psi(p)$ and how it depends upon such parameters as ω_c/ω and r. The coupling function is shown in Fig. 4 as a function of the

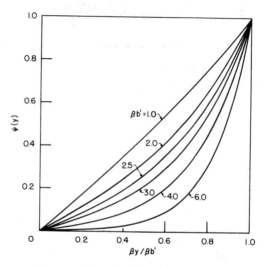

FIG. 4. Beam-circuit coupling function.

radian interaction width $\beta b'$. The beam or individual electron position between the sole and the circuit is specified by $\beta y/\beta b'$, and we see that $\psi(p)$ decreases as $\beta b'$ increases. It is revealing to express $\beta b'$ as follows:

$$\beta b' = \frac{\omega}{\omega_c} \frac{1}{rl}, \tag{40}$$

which indicates that for a fixed beam position $\beta b'$ decreases with increasing ω_c/ω and vice versa. This fact has an important influence on the variation of efficiency with ω_c/ω.

In all cases the beam and circuit wave are made synchronous, i.e., $b = 0$, since this leads to the greatest output.

b. *Dependence of Gain and Efficiency on Operating Parameters*

In most of the calculations to be presented a Brillouin beam condition will be assumed and the stream is divided into a minimum of three layers. Under most conditions the effect of space-charge forces in M-type Brillouin beams is not significant, as shown in Fig. 5.

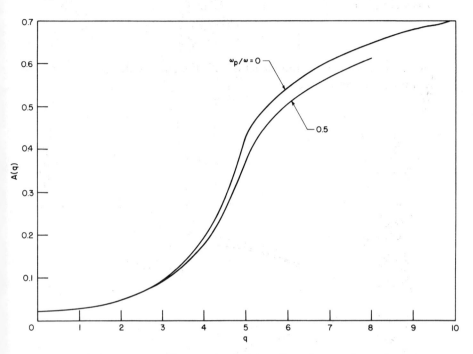

FIG. 5. Rf voltage versus distance showing the effect of space-charge forces in an M-type Brillouin flow ($D = 0.1$, $r = 0.5$, $b = 0$, $\psi_0 = -30$, $\omega_c/\omega = 0.5$, $s = 0.1$).

The gain and efficiency at both the 50 and 70 per cent collection planes for a beam injected midway between the sole and circuit are shown in Figs. 6 and 7. These calculations were made assuming that $\omega_p/\omega \to 0$, i.e., that beam space-charge forces could be neglected. Under conditions in which $D > 0.05$ the gain and efficiency are not significantly changed by the inclusion of space-charge forces, as shown in Figs. 8 and 9. There is, however, an appreciable lowering of both the gain and efficiency due to space-charge forces when $D = 0.05$. The significant influence of space-charge forces at low interaction parameter values is undoubtedly due to the fact that the circuit field forces are too weak to effectively modulate the beam or contain the bunches in the presence of strong debunching forces. Note that available efficiency, η_a, has been plotted; in the cases shown the electronic efficiency is one-half of η_a.

The variation of efficiency with magnetic field as measured by ω_c/ω is quite interesting in that low values of ω_c/ω are indicated to be optimum, although there is a wide range of ω_c/ω for which the gain and efficiency are relatively constant and near the maximum. The interaction parameter D is defined as

$$D^2 \triangleq \frac{\omega_c}{\omega} \frac{I_0 Z_0}{2V_0}$$

and thus in explaining the variation of efficiency with ω_c/ω both D and $\psi(p)$ must be considered.

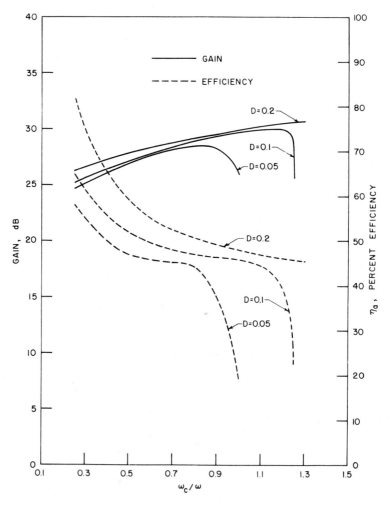

FIG. 6. Gain and efficiency at the 50 per cent collection plane ($r = 0.5$, $s = 0.1$, $\psi_0 = -30$, $b = 0$, $\omega_p/\omega = 0$).

3. TWO-DIMENSIONAL M-FWA WITH A NEGATIVE SOLE

Take as a reference position in the preceding figures a value of $\omega_c/\omega \approx 0.75$. Now as ω_c/ω is reduced from this value while D is maintained constant by increasing the circuit impedance Z_0, it is expected that the rf output and hence the efficiency would increase. This change, however, is modified by the fact that $\beta b'$ is increasing and hence the coupling function $\psi(p)$ is decreasing for a fixed beam position. As ω_c/ω is increased from this mid-range value, the circuit impedance must be decreased in order to maintain D constant and hence a lowering of the

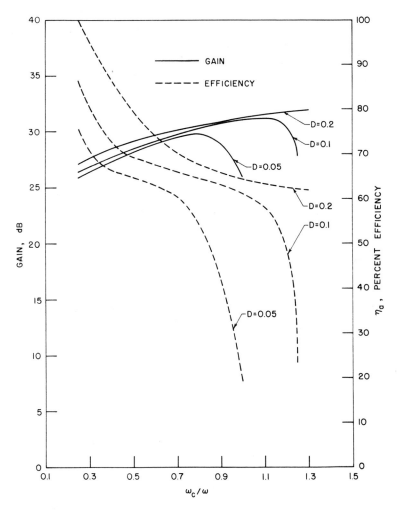

FIG. 7. Gain and efficiency at the 70 per cent collection plane ($r = 0.5$, $s = 0.1$, $\psi_0 = -30$, $b = 0$, $\omega_p/\omega = 0$).

rf conversion efficiency is expected. Again this action is tempered by the fact that $\beta b'$ is decreasing and thus $\psi(p)$ is increasing, which should lead to an increased output and efficiency.

The beam injection position, of course, should have a significant effect on efficiency since the coupling function varies rapidly over the interaction region. The efficiency at the 70 per cent plane is shown in Fig. 10 for $r = 0.75$. The lower gain and efficiency are due to the fact that the beam electrons, starting so close to the structure, are rapidly

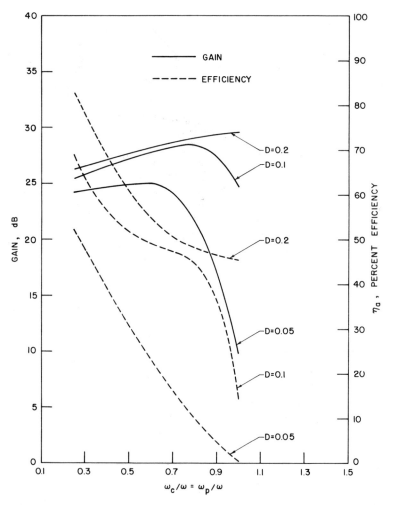

FIG. 8. Effect of space-charge forces on gain and efficiency at the 50 per cent collection plane ($r = 0.5$, $s = 0.1$, $\psi_0 = -30$, $b = 0$, $\omega_p/\omega = \omega_c/\omega$).

collected on the structure before the rf level has built up to the full saturation capability of the interaction process. On the other hand, if the beam has too low an initial position the modulation by the circuit wave is ineffective due to the low value of $\psi(p)$; thus there must be an optimum injection position which results in a balance of these two effects and should lead to a maximum output. The effect of initial beam position at constant ω_c/ω is illustrated in Fig. 11. A value of $r \approx 0.5$ is seen to be near optimum. The variation of the normalized length

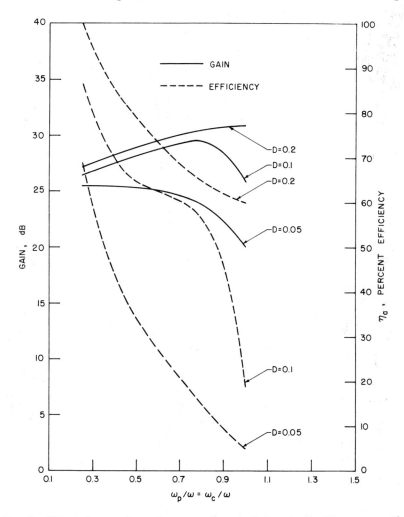

FIG. 9. Effect of space-charge forces on gain and efficiency at the 70 per cent collection plane ($r = 0.5$, $s = 0.1$, $\psi_0 = -30$, $b = 0$, $\omega_p/\omega = \omega_c/\omega$).

DN_s to the 70 per cent collection plane with various parameters is illustrated in Fig. 12.

The stream thickness as measured by $s = w/\bar{y}_0$ does not significantly affect the output or saturation characteristic as shown in Fig. 13. It should be noted that these results are for Brillouin flows. Since D was kept constant while s was varied, the total beam current is constant and the beam current density varies. Similar results are obtained assuming a constant current density in the beam; i.e., D varies.

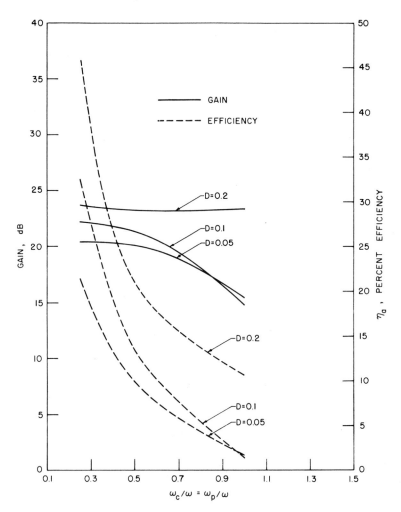

FIG. 10. Gain and efficiency at the 70 per cent collection plane for $r = 0.75$, and $\omega_p/\omega = \omega_c/\omega$ ($b = 0$, $s = 0.1$, $\psi_0 = -30$).

3. TWO-DIMENSIONAL M-FWA WITH A NEGATIVE SOLE

The effect of the initial rf signal level A_0 on the gain of an M-FWA was shown in Fig. 3, where it was evident that the saturated level is approximately equal to $(\omega I_0 V_0/2\omega_c)$. The saturation level at both the 70 and 90 per cent collection planes is plotted versus the input drive level ψ_0 for two values of the gain parameter D in Fig. 14. The efficiency and distance to saturation are shown in Fig. 15.

In high-gain amplifiers a distributed loss along the interaction structure is required in order to insure stability in the presence of reflections.

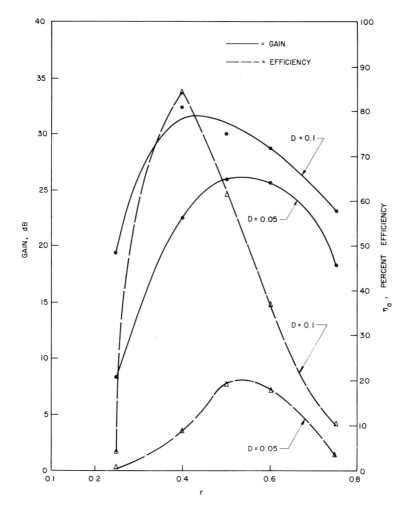

FIG. 11. Gain and efficiency at the 70 per cent collection plane for constant ω_c/ω ($b = 0$, $s = 0.1$, $\omega_p/\omega = 0$, $\omega_c/\omega = 1$, $\psi_0 = -30$).

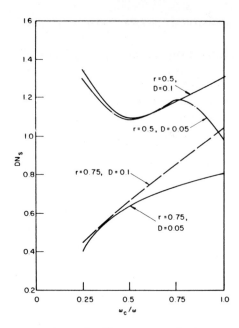

FIG. 12. DN_s versus ω_c/ω at the 70 per cent collection plane ($s = 0.1$, $b = 0$, $\psi_0 = -30$).

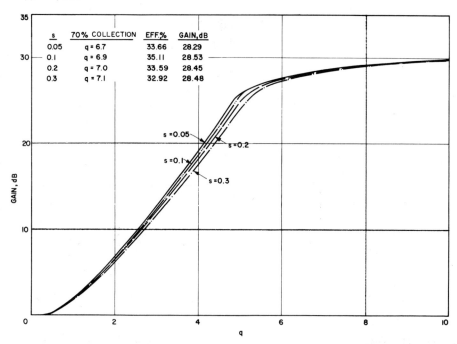

FIG. 13. Gain versus length and beam thickness ($D = 0.1$, $b = 0$, $r = 0.5$, $\omega_c/\omega = 0.5$, $\omega_p/\omega = 0$, $\psi_0 = -30$).

3. TWO-DIMENSIONAL M-FWA WITH A NEGATIVE SOLE

The effect of circuit loss as measured by the loss parameter, d, was included in the circuit equations of Section 2. The reduction in saturated output as a function of d at two different values of ω_c/ω is shown in Fig. 16 for attenuators beginning at $DN_s = 0.3$ and extending to saturation. This type of variation is quite similar to the characteristics of O-type amplifiers with distributed loss. Under high-loss conditions, i.e., $d > 1$, it is possible to encounter saturation effects under the attenuator which further inhibit the energy conversion process.

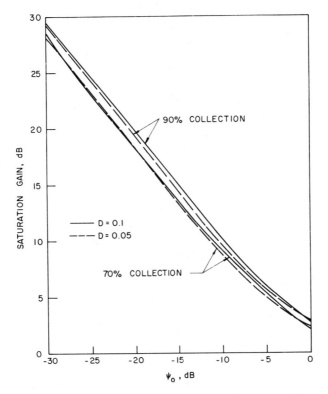

FIG. 14. Saturation gain versus ψ_0 at two different collection planes ($b = 0$, $r = 0.5$, $s = 0.1$, $\omega_c/\omega = 0.5$, $\omega_p/\omega = 0$).

The effect of the amount of free gain beyond the output end of the attenuator on the saturation power level is shown in Fig. 17 for two values of the loss parameter. These results indicate that as long as there is 10–15 dB of saturated gain beyond the attenuator, or 15–20 dB of small-signal gain, there is no appreciable effect on the saturated output level. This result is approximately the same for other operating parameters and agrees well with experimental results.

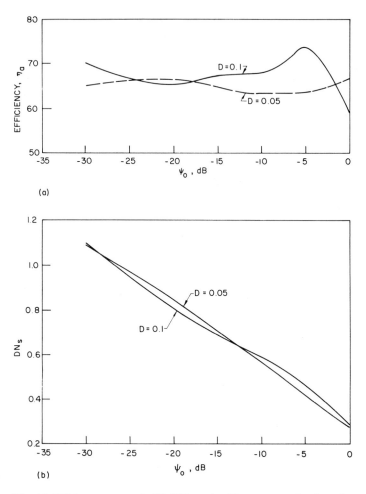

FIG. 15. (a) Efficiency, η_a, and (b) DN_s at the 70 per cent collection plane versus ψ_0 ($b = 0$, $r = 0.5$, $s = 0.1$, $\omega_c/\omega = 0.5$, $\omega_p/\omega = 0$).

c. Beam Collection on the Rf Structure

Significant electron collection on the rf structure begins at a plane approximately 50–60 per cent of the total length along the rf structure and rises rapidly thereafter as shown in Fig. 18. The corresponding interception curves when space-charge forces are included are shown in Fig. 19. The effect of the normalized cyclotron frequency is marked

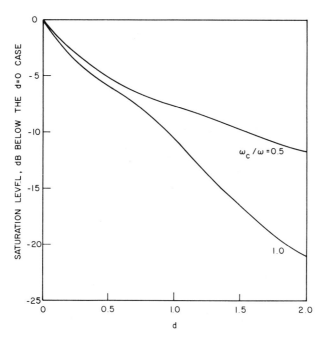

Fig. 16. Saturation level versus loss factor and magnetic field ($D = 0.1$, $r = 0.5$, $b = 0$, $s = 0.1$, $\psi_0 = -30$, $\omega_p/\omega = 0$).

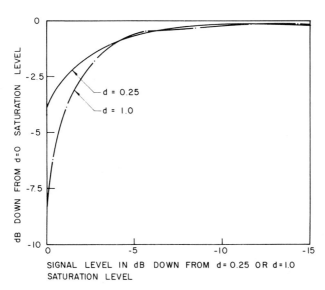

Fig. 17. Effect of attenuator length on saturated output ($D = 0.1$, $r = 0.5$, $s = 0.1$, $b = 0$, $\psi_0 = -30$, $\omega_c/\omega = \omega_p/\omega = 0.5$).

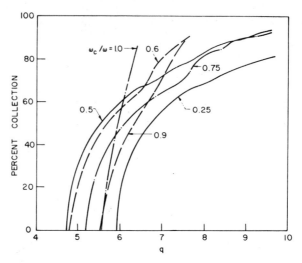

Fig. 18. Electron collection versus distance ($D = 0.05$, $r = 0.5$, $b = 0$, $s = 0.1$, $\psi_0 = -30$, $\omega_p/\omega = 0$).

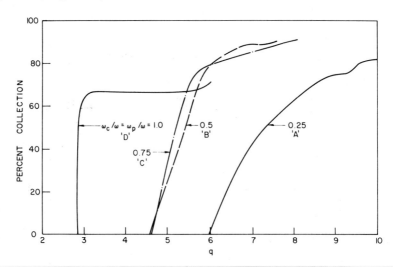

	50% Collection	Gain	Eff.	70% Collection	Gain	Eff.
A.	$q = 7.3$	24.24	26.10	$q = 8.5$	25.43	34.38
B.	$q = 5.4$	24.90	15.30	$q = 5.7$	25.30	17.00
C.	$q = 5.2$	23.00	6.50	$q = 5.5$	24.30	8.80
D.	$q = 2.9$	9.7	0.23	$q = 5.95$	20.00	2.5

Fig. 19. Electron collection versus distance ($D = 0.05$, $r = 0.5$, $b = 0$, $s = 0.1$, $\psi_0 = -30$, $\omega_p/\omega = \omega_c/\omega$).

both at low and high values of D. As expected strong input signals produce a large beam modulation, which results in early and large electron collection.

d. Electron Trajectories

For an infinitely thin beam in crossed fields it is possible to adjust the z-directed velocity to be equal to E_y/B_x, so that laminar flow is achieved. However, in finite-thickness M-type Brillouin beams there is a linear velocity variation across the beam and hence some electrons will be traveling faster and others slower than the velocity indicated by E_y/B_x. If this configuration is considered as a two-layer beam, then the upper electrons (fast) and the lower electrons (slow) experience a circular motion of radius $2Du\bar{u}_0/\omega_c$ due to their drift through the magnetic field. This circular motion, coupled with the z-directed drift velocity, gives rise to cycloidal trajectories. Since the velocity deviation from the mean is usually small, the departure from laminar trajectories is slight. Electron motion in the presence of a strong magnetic field and a relatively weak electric field results in a drift motion perpendicular to the magnetic field and thus along an equipotential line of the magnetic field.

The sinusoidally varying rf field velocity modulates the electron stream, accelerating some electrons and decelerating others. Those that stay in an accelerating rf phase eventually move towards the sole, whereas those that are decelerated gradually move towards the circuit. In evaluating the effect of the rf field, it is necessary to consider the increasing value of the coupling function $\psi(p)$ as the circuit is approached and its decreasing value as the sole electrode is approached. These characteristics of the electron motion are indicated in Fig. 20 for a two-layer beam. Electrons numbered 27_2 to 32_2 and 1_2 to 9_2 of the upper layer and electrons 9_1 to 22_1 of the lower layer are considered as favorably phased electrons. Electrons moving toward the sole extract energy from the circuit wave and these are called unfavorably phased electrons. The net energy transfer which occurs is a result of the better coupling of the favorably phased electrons with the rf structure as a result of their relative closeness to the anode.

If the cyclotron component of motion is neglected, i.e., the acceleration terms are dropped, and the so-called adiabatic equations of motion are used, then the ripple in the trajectories of Fig. 20 is eliminated. The adiabatic electron trajectories for a Brillouin beam with space-charge forces are shown in Fig. 21. The cycloidal component of motion is readily apparent, as is the sorting of unfavorably and favorably phased electrons. It is enlightening to plot the trajectory information prs vs. the

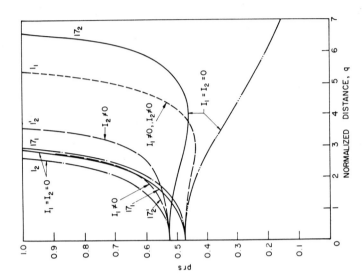

FIG. 21. Adiabatic electron trajectories with and without space-charge forces for a Brillouin beam ($D = 0.05$, $r = 0.5$, $s = 0.1$, $b = 0$, $\omega_c/\omega = \omega_p/\omega = 0.25$).

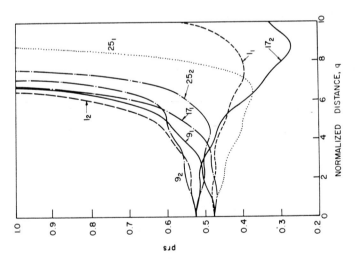

FIG. 20. Nonadiabatic electron trajectories ($D = 0.1$, $r = 0.5$, $s = 0.1$, $b = 0$, $\omega_c/\omega = \omega_p/\omega = 0.25$, $F_{2-z} = F_{2-y} = 0$).

electron phase relative to the rf wave. This information is given in the following section along with the corresponding velocity data.

e. Rf Bunching and Electron Velocities

In all of the data of this chapter a Brillouin beam flow has been assumed and no initial velocity modulation of the beam has been applied. Recall that the z-directed velocity variable is $[1 + 2Du(q)]$ and that the y-directed velocity variable is $v_{y\omega}$. The results presented in the following figures are typical of M-FWA's and are plotted at displacement planes corresponding to beam collections of 20, 50 and 70 per cent respectively. Figures 22–24 give prs, $v_{y\omega}$ and $(1 + 2Du)$ versus phase at these locations.

The formation of the bunch in the maximum decelerating phase $(\pi/2)$ of the rf wave is clearly evident in Figs. 22a, 23a and 24a. Notice also the

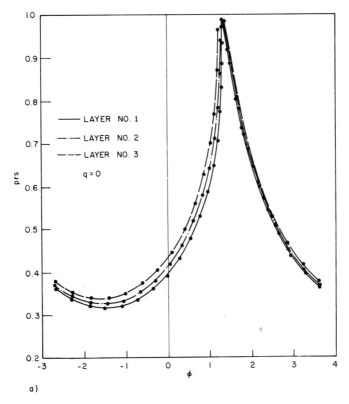

FIG. 22. Electron trajectories and velocities versus phase at the (a) 18, (b) and (c) 20 per cent collection plane ($D = 0.05$, $r = 0.5$, $s = 0.1$, $b = 0$, $\psi_0 = -30$, $\omega_c/\omega = \omega_p/\omega = 0.25$). (a). Trajectory versus phase.

extreme symmetry of the bunch. A *prs* value of unity corresponds to collection on the circuit electrode and *prs* = 0 to collection on the sole. As expected, the electrons in the maximum decelerating phase region experience the greatest y-directed deflection and hence $v_{y\omega}$ is greatest for these electrons. From the plots of z-directed velocity we again note a maximum change for the $\pi/2$ electrons and also that these charge groups have their z-directed kinetic energy reduced by 5–10 per cent, depending directly upon the value of D.

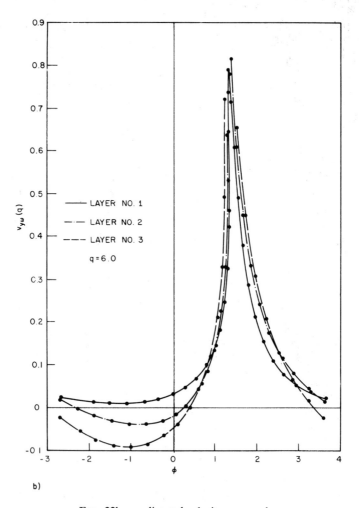

FIG. 22b. y-directed velocity versus phase.

f. Rf Current Density in M-FWA's

In O-type forward-wave amplifiers, which operate on a kinetic energy conversion process, the fundamental to dc current ratio varies from 1.2 to 1.6, depending upon the operating parameters. In M-type amplifiers

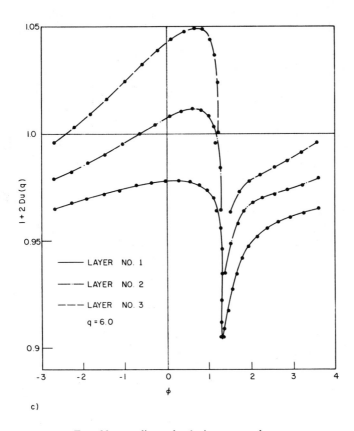

FIG. 22c. z-directed velocity versus phase.

we find that the maximum values of i_1/I_0 are somewhat lower than in O-type amplifiers. Typical results for $D = 0.1$ and two values of the initial beam location parameter r are shown in Fig. 25. The equations for the calculation of harmonic beam currents are defined in Chapter VIII. The greatest values for i_1/I_0 occur for low values of ω_c/ω, which yielded the highest conversion efficiencies as shown earlier.

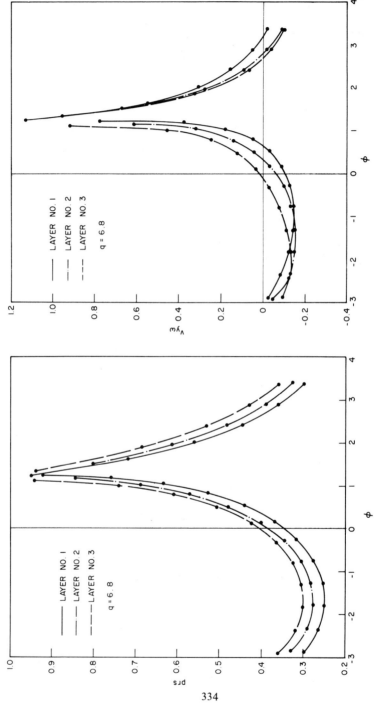

Fig. 23. Electron trajectories and velocities versus phase at the 50 per cent collection plane ($D = 0.05$, $r = 0.5$, $s = 0.1$, $b = 0$, $\psi_0 = -30$, $\omega_c/\omega = \omega_p/\omega = 0.25$). (a) Trajectory versus phase.

Fig. 23b. y-directed velocity versus phase.

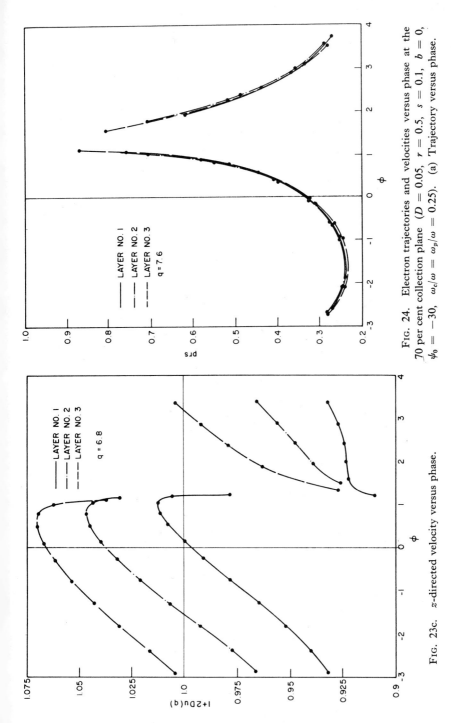

Fig. 23c. z-directed velocity versus phase.

Fig. 24. Electron trajectories and velocities versus phase at the 70 per cent collection plane ($D = 0.05$, $r = 0.5$, $s = 0.1$, $b = 0$, $\psi_0 = -30$, $\omega_c/\omega = \omega_p/\omega = 0.25$). (a) Trajectory versus phase.

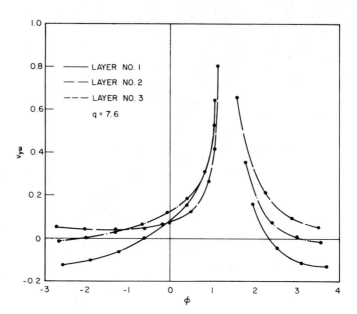

Fig. 24b. *y*-directed velocity versus phase.

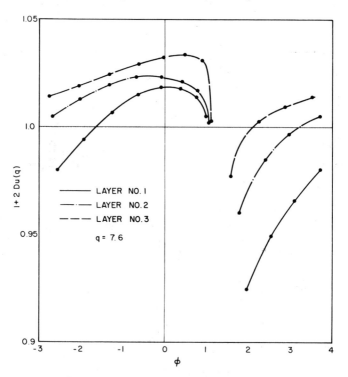

Fig. 24c. *z*-directed velocity versus phase.

g. *Rf Wave Phase Shift versus Distance*

The dependent variable $\theta(q)$ measures the amount of phase lead of the actual rf wave as compared to the hypothetical wave traveling at a velocity corresponding to u_{z0}. It is an interesting and desirable property of M-type devices that the phase lead is very much smaller than the

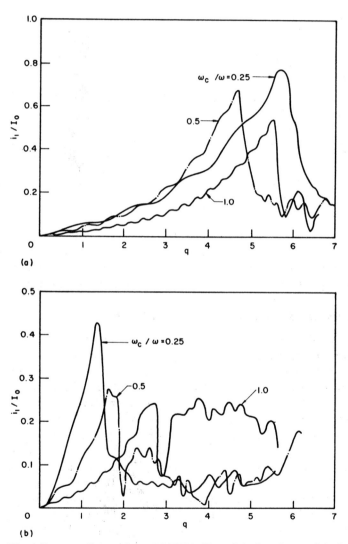

FIG. 25. i_1/I_0 versus distance for an M-FWA ($D = 0.1$, $b = 0$, $s = 0.1$, $\psi_0 = -30$, $\omega_p = \omega_c$). (a) $r = 0.5$; (b) $r = 0.75$.

corresponding phase lag in their O-type counterparts. The variation of $\theta(q)$ with distance and its dependence on the input signal level to the amplifier as measured by ψ_0 is shown in Fig. 26 for a typical set of

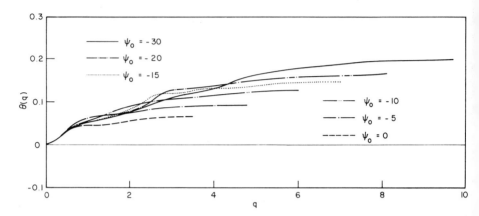

Fig. 26. Rf phase lead versus distance ($D = 0.05$, $r = 0.5$, $b = 0$, $s = 0.1$, $\omega_c/\omega = \omega_p/\omega = 0.5$).

parameters. This minimal phase shift sensitivity results in very desirable amplitude and frequency modulation characteristics. The finite value of $\theta(q)$ arises because of the beam loading on the rf circuit and therefore increases with increasing coupling to the circuit, i.e., increasing D.

h. *Nonlinear Beating-Wave Amplification*

It is easily shown, using a small-amplitude theory, that when $b > b_{x_1=0}$, i.e., there is no growing-wave amplification, amplification does occur due to a beating-wave process. This beating occurs between matched rf waves traveling at different phase velocities on an rf transmission line. A similar process takes place in O-type amplifiers and it has been determined that the amount of beating-wave gain is greater in O-type devices than in M-type devices. Several large-signal solutions are shown in Fig. 27, where the adiabatic assumption has been made since $\omega_c/\omega D \gg 1$. Space-charge forces do not significantly affect the process and have been neglected in these calculations. It is quite apparent from these results that the large-signal performance is considerably enhanced over the predicted small-signal performance. The increased interaction in the linear region of the gain curve is primarily due to the better beam-circuit coupling which results from electrons moving closer to the rf circuit as a consequence of the large initial perturbation in their motion (large A_0). The eventual saturation occurring is primarily due to the collection of an appreciable fraction of the stream on the rf structure.

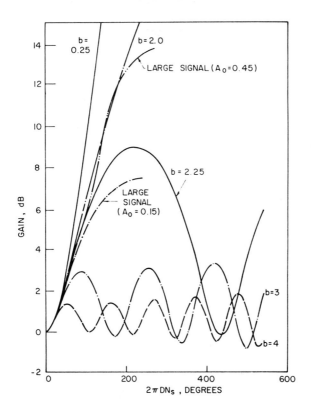

FIG. 27. Beating-wave gain versus distance ($D = 0.075$, $\beta b' = 1.68$, $r = 0.2$).

i. Nonlaminar Stream Solutions

All of the previous information contained in this chapter has dealt with laminar Brillouin streams and their interaction with rf waves. Recall that the general nonlinear working equations are valid for any stream model assumed at the input to the interaction region. Other stream flows are specified through space-charge density and velocity distributions at the input plane.

The effects of beam nonlaminarities on amplifier performance are investigated using an equivelocity stream with the velocity vector directed at an angle γ with respect to the z-direction, with equal probability in the range $-\gamma_m < \gamma < \gamma_m$. As in the Brillouin beam, the space-charge density has been assumed constant across the beam. The results for two values of the angle γ_m and a comparison with the corresponding Brillouin beam result are shown in Fig. 28. Large values of

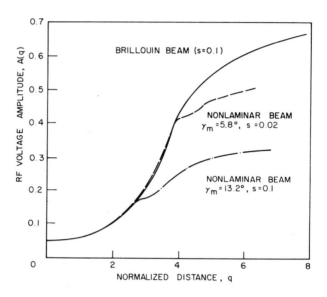

FIG. 28. Rf voltage amplitude versus distance for nonlaminar streams ($D = 0.1$, $r = 0.5$, $\omega_c/\omega = \omega_p/\omega = 0.5$, $F_{2-y} = F_{2-z} = 0$).

γ_m result in significant gain reductions due to the collection of a significant number of electrons on the sole electrode. Note that the small-signal gain value is relatively unaffected by the degree of nonlaminarity. Certainly other nonlaminar models can be generated; although they may be quite different in detail, it is expected that the gross effects will be much the same.

j. Effects of a Circuit Sever

The effect of a sever in the rf circuit on the gain and output of an M-FWA is easily evaluated by simply setting the rf voltage equal to zero at various displacement planes from the output. A convenient indicator of the sever location is the rf level at the sever relative to the saturated power output in the absence of a sever. The gain characteristic for a severed-circuit amplifier is shown in Fig. 29 for several different sever positions. The saturated output is taken as that at the plane of 70 per cent collection of the beam on the circuit. The rf output in decibels relative to the unsevered output is shown in Fig. 30 as a function of the sever location. Thus we see

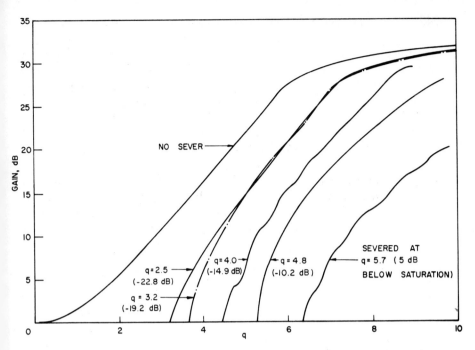

Fig. 29. Gain for a severed-circuit M-FWA ($D = 0.1$, $r = 0.5$, $s = 0.1$, $\omega_c/\omega = \omega_p/\omega = 1$, $\psi_0 = -30$).

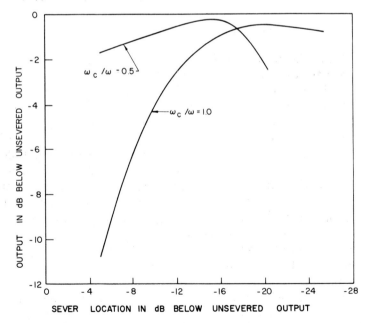

Fig. 30. Output for a severed-circuit M-FWA versus sever position ($D = 0.1$, $r = 0.5$, $s = 0.1$, $\omega_p = \omega_c$, $\psi_0 = -30$).

that the sever must be located at least 16 dB below saturation and since there is approximately 9dB compression in an M-FWA then there must be 25 dB small-signal gain beyond the sever in order to develop the full output. This figure is essentially the same as that for an O-FWA.

4 Two-Dimensional M-FWA with a Positive Sole

To create a forward-wave positive-sole interaction region it is only necessary to interchange the positions of the sole and structure electrodes, leaving the voltage and field arrangements without change as illustrated in Fig. 31. Since the electrons will still continue to move towards the

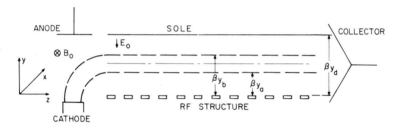

FIG. 31. Positive-sole crossed-field stream-modulation device.

sole electrode, the principal collection will be on the heavy-sole electrode rather than on the circuit. Unfortunately though, the circuit rf fields decay rapidly as one moves toward the sole and hence the coupling between stream and circuit is weak. Those electrons which are sorted out and move towards the circuit continue to extract energy from the wave due to their unfavorable phase positions. This suggests that the positive-sole FWA could serve as an effective means of modulating a high-density stream, as illustrated in Fig. 1c. Gandhi[4] has studied the small-signal performance of positive-sole interaction regions and found that such a region supports *three purely propagating waves*, which beat to yield a strong velocity modulation of a stream passing in close proximity to the circuit. An additional advantage is that this interaction occurs over a very short distance of approximately $DN_s \approx 0.15$.

In analyzing the positive-sole interaction region the stream and circuit equations may be considered separately. Since the dc field configurations have not been changed, only the circuit field coupling function must be changed in the force equations. The circuit equations are also modified by the new coupling function. The circuit field now

4. TWO-DIMENSIONAL M-FWA WITH A POSITIVE SOLE

decays in the positive y-direction and the coupling function is written as:

$$\bar{\psi}(y) = \frac{\sinh \beta(b' - y)}{\sinh \beta b'},$$

or

$$\bar{\psi}(p) = \frac{\sinh \beta w \left(\frac{1}{rs} - p\right)}{\sinh \frac{\beta w}{rs}}. \tag{41}$$

The appropriate large-signal equations for the positive-sole interaction region are given below.

Circuit Equations

$$\begin{cases} \text{Eqs. (25) and (26) with } \psi(p) \\ \text{replaced by } \bar{\psi}(p) \end{cases}. \tag{42} \\ \tag{43}$$

Force Equations

$$[1 + 2Du(p_0, \Phi_0, q)] \frac{\delta u(p_0, \Phi_0, q)}{\delta q}$$

$$= \frac{\omega}{2\omega_c} \frac{\sinh\left[\frac{\omega}{\omega_c} \frac{s}{l} \left(\frac{1}{rs} - p\right) \left(1 + D \frac{d\theta(q)}{dq}\right)\right]}{\sinh\left[\frac{\omega}{\omega_c rl} \left(1 + D \frac{d\theta(q)}{dq}\right)\right]}$$

$$\cdot \left\{ \frac{dA(q)}{dq} \cos \Phi - \frac{A(q)}{D} \sin \Phi \left(1 + D \frac{d\theta(q)}{dq}\right) \right\}$$

$$+ \frac{1}{2D^2} \frac{s}{l} v_{y\omega} + \left(\frac{\omega_p}{\omega}\right)^2 \frac{rs}{\pi D^2} F_{2-z} \tag{44}$$

and

$$[1 + 2Du(p_0, \Phi_0, q)] \left[\frac{\delta v_{y\omega}}{\delta q} + \frac{l}{sD}\left(\frac{\omega_c}{\omega}\right)^2\right]$$

$$= -\frac{l}{s}\left(1 + D\frac{d\theta(q)}{dq}\right) \frac{\cosh\left[\frac{\omega}{\omega_c} \frac{s}{l}\left(\frac{1}{rs} - p\right)\left(1 + D\frac{d\theta(q)}{dq}\right)\right]}{\sinh\left[\frac{\omega}{\omega_c rl}\left(1 + D\frac{d\theta(q)}{dq}\right)\right]} A(q) \cos \Phi$$

$$+ \frac{rl^2}{sD}\left(\frac{\omega_c}{\omega}\right)^2 \left(\frac{V_a}{2V_0}\right) - 2rl\left(\frac{\omega_c}{\omega}\right)\left(\frac{\omega_p}{\omega}\right)^2 F_{2-y}. \tag{45}$$

Velocity-Phase and Position Equations. The equations for the positive-sole interaction region, Eqs. (11), (12), and (42)–(45), may be solved using the same methods outlined previously. Since it is known that the voltage variation along the structure in such a device is similar to that of an M-BWO, initially a large amplitude signal should be considered, i.e., $A_0 \sim 1$. Other boundary conditions are the same as in the negative-sole device.

A typical calculation of the rf voltage variation along the structure in a positive-sole device is shown in Fig. 32 for a representative set of

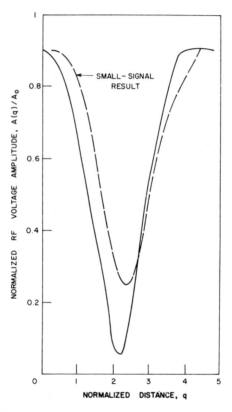

FIG. 32. Large-signal positive-sole rf voltage characteristics ($D = 0.2$, $b = 0$, $r = 0.5$, $s = 0.1$, $\omega_c/\omega = 0.5$).

device operating parameters. The significant energy transfer from circuit wave to electron stream illustrated in Fig. 32 leads to a large modulation of the stream. To illustrate the magnitude of nonlinear effects and also the effects of cyclotron waves included in the nonlinear theory, the small-signal rf voltage characteristic is also plotted in Fig. 32.

In designing an amplifier of the type illustrated in Fig. 1c, the output modulated stream is taken from the positive-sole interaction region and used as the input stream to either a drift region or the normal negative-sole interaction region.

A similar analysis may be carried out for a backward-wave configuration, the only changes in the equations being a change in the signs preceding the right-hand sides of the circuit equations, Eqs. (42) and (43). Such a device, by analogy with the forward-wave behavior, yields backward-wave amplification for proper voltage (b) values.

5 Adiabatic Equations for a Two-Dimensional M-FWA with a Negative Sole

The nonlinear negative-sole crossed-field equations derived in Section 2 include the effects of cycloidal motion of the electrons and some effects of cyclotron waves. It was shown in Chapter VIII that in the case of weak interaction between the stream and wave the acceleration terms in the equations of motion could be neglected. A similar approximation can be introduced in the case of the M-FWA, providing that the growth of the rf wave in a cyclotron wavelength $\lambda_c = 2\pi \bar{u}_0/\omega_c$ is not too large. The condition to be satisfied when using the adiabatic equations of motion is that

$$\frac{\omega_c}{\omega D} \gg 1.$$

Thus $D \ll 1$, the adiabatic approximation, is reasonable. If this assumption is made, the dependent variable and circuit equations are unchanged and the Lorentz force equations are modified as follows:

$$v_{y\omega} = \frac{lD^2}{s}\left(\frac{\omega}{\omega_c}\right)\psi(p)\left[\frac{A(q)\sin\Phi}{D}\left(1 + D\frac{d\theta(q)}{dq}\right) - \frac{dA(q)}{dq}\cos\Phi\right]$$

$$- \frac{2rl}{\pi}\left(\frac{\omega_p}{\omega}\right)^2 F_{2-z} \qquad (46)$$

and

$$u(p_0, \Phi_0, q) = \frac{1}{2}\left(\frac{\omega}{\omega_c}\right)^2\left[1 + D\frac{d\theta(q)}{dq}\right]\frac{\cosh\left[\frac{\omega}{\omega_c}\frac{ps}{l}\left(1 + D\frac{d\theta(q)}{dq}\right)\right]}{\sinh\left[\frac{\omega}{\omega_c rl}\left(1 + D\frac{d\theta(q)}{dq}\right)\right]} A(q)\cos\Phi$$

$$+ \frac{1}{2D}\left(\frac{rlV_a}{2V_0} - 1\right) - \left(\frac{\omega_p}{\omega_c}\right)^2\left(\frac{\omega_c}{\omega}\right)rsF_{2-y}. \qquad (47)$$

The difference between these equations and the corresponding equations of Chapter VIII, Eqs. (65) and (66), is the inclusion of the circuit field terms. The equivalence of the adiabatic results with the nonadiabatic results is in terms of the rf voltage versus distance characteristic. The electron trajectory plots will, however, not be the same, as will be seen in the following results.

The effect of the adiabatic approximation is easily illustrated by making computations both for $\omega_c/\omega D \sim 1$ and for $\omega_c/\omega D \gg 1$. The normalized rf voltage versus distance curve for $\omega_c/\omega D = 1.65$ is shown in Fig. 33, where comparison is made between the adiabatic and non-

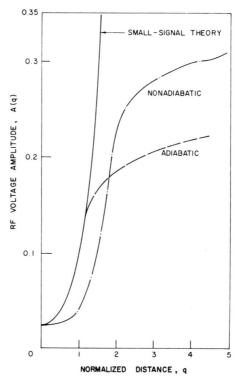

FIG. 33. Rf voltage amplitude versus distance for adiabatic and nonadiabatic solutions with a Brillouin beam ($D = 0.1$, $r = 0.75$, $\omega_p/\omega = \omega_c/\omega = 0.165$, $s = 1/15$, $b = 0$, $F_{2-z} = F_{2-y} = 0$).

adiabatic results and the small-signal calculation. Note that the rf signal growth rate is approximately the same for the adiabatic and nonadiabatic cases, although the actual circuit amplitude in the more exact nonadiabatic case is less in the linear region. This difference is due to the appreciable excitation of cyclotron waves in the nonadiabatic case, which are not accounted for in the other cases. Notice also that the saturation amplitude is appreciably more in the nonadiabatic case, which is in agreement with calculations based on energy changes.

For $\omega_c/\omega D = 5$ the two sets of equations yield essentially the same results, as seen in Fig. 34. The rf wave phase shift is compared in Fig. 35,

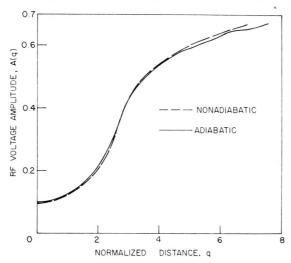

FIG. 34. Rf voltage amplitude versus distance for adiabatic and nonadiabatic equations ($D = 0.1$, $\omega_p/\omega = \omega_c/\omega = 0.5$, $b = 0$, $r = 0.5$, $s = 0.1$, $F_{2-z} = F_{2-y} = 0$).

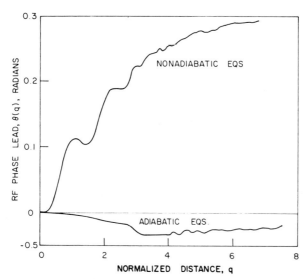

FIG. 35. Phase lead of the rf wave versus distance for adiabatic and nonadiabatic equations ($D = 0.1$, $\omega_p/\omega = \omega_c/\omega = 0.5$, $b = 0$, $r = 0.5$, $s = 0.1$, $F_{2-z} = F_{2-y} = 0$).

where an appreciable difference is noted. Under the adiabatic assumption the beam loading on the circuit is evidently small, since $\theta(q) \approx 0$. Thus we conclude that for $D \ll 1$ the adiabatic approximation may be utilized in calculating gain and efficiency, although in order to determine detailed electron motion and stream loading the nonadiabatic equations must be utilized.

6 Three-Dimensional M-FWA with a Negative Sole

In a general three-dimensional crossed-field interaction region, the effects of space-charge fields and rf circuit voltage variations in the magnetic direction are considered. The general circuit equation to be used is Eq. (96) of Chapter III, and the three components of the vector Lorentz force equation are

$$\frac{dv_x}{dt} = -|\eta|[E_x + E_{sc-x}], \tag{48}$$

$$\frac{dv_y}{dt} = -|\eta|[E_0 + E_y + E_{sc-y}] - \omega_c v_z \tag{49}$$

and

$$\frac{dv_z}{dt} = -|\eta|[E_z + E_{sc-z}] + \omega_c v_y. \tag{50}$$

The development of the general nonlinear interaction equations and the introduction of Lagrangian variables proceeds along the same lines as the three-dimensional O-type nonlinear theory. Such a generalized theory may be required in studying the behavior of M-type amplifiers with biperiodic circuits and wide electron beams.

7 Effect of Cyclotron Waves

We have seen in some of the nonlinear calculations a discrepancy between various results which has been attributed to the appreciable excitation of cyclotron waves. It is, therefore, reasonable to investigate under what conditions cyclotron waves would have an appreciable effect. The phase velocities of the cyclotron waves associated with $\beta_{\pm} \approx \beta_0 \pm \beta_c$ are expressed as

$$v_{\mp} = \frac{\omega}{\omega \pm \omega_c} u_0, \tag{51}$$

where the plus and minus signs refer respectively to the fast and slow cyclotron waves. Thus we see that if $\omega \approx \omega_c$ the cyclotron wave velocities are very different from the stream velocity and hence a negligible interaction takes place. However, if $\omega \ll \omega_c$ in a thick-beam device, some electrons in the stream would be nearly synchronous with the cyclotron wave, resulting in an appreciable energy exchange which would affect the rf wave amplitude along the circuit. Also if $\omega \gg \omega_c$, most of the stream would be in near synchronism with cyclotron waves, meaning that the electrons would experience forces which vary as ω_c and again the rf wave would be modified.

In developing the equations appropriate for cyclotron wave interaction we note that the circuit propagation constant varies with distance. It is thus necessary to allow for a variable phase constant, $\beta(z)$, and a variable impedance $Z_0(\beta)$ as functions of distance along the circuit. Recall that the phase constant (normalized) of the nonlinear theory was written as

$$\beta(q) = \bar{\beta}\left(\frac{\bar{u}_0}{u_{z0}} + D\frac{d\theta(q)}{dq}\right)$$

$$\approx \bar{\beta}\left(1 + D\frac{d\theta(q)}{dq}\right).$$

A problem arises with the uniform equivalent-circuit transmission line, since heretofore the impedance parameter Z_0 has been considered invariant with distance. In studying cyclotron-wave interaction one must therefore use a $Z_0(\beta)$ variation which is characteristic of the periodic structures used in M-FWA's. Such a computation for the first forward space harmonic on an interdigital line is shown in Fig. 36.

If these functional dependencies are introduced into Eq. (22) of this chapter, the following results:

$$\frac{\partial^2 V}{\partial t^2} - \frac{v_0^2}{(1 + D\theta')^2}\frac{\partial^2 V}{\partial z^2} + 2\omega Dd\frac{\partial V}{\partial t}$$
$$= \frac{v_0 Z_0(1 + D\theta')}{1 + D\theta'}\left[\frac{\partial^2 \rho_1}{\partial t^2} + 2\omega Dd\frac{\partial \rho_1}{\partial t}\right], \quad (52)$$

where $Z_0(1 + D\theta')$ indicates the variation of circuit impedance with phase constant.

The force and dependent variable equations are unchanged, and thus only the final two circuit equations need modification. The modified forms of Eqs. (25) and (26) are easily developed from Eq. (52) and the force equations of Section 2.

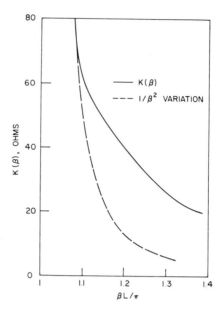

FIG. 36. Variation of a typical slow-wave structure impedance with propagation constant (first forward space harmonic).

8 Comparison with Sedin's Calculations

Sedin[2] has analyzed both injected-beam crossed-field amplifiers and oscillators, although most of his calculations were made for oscillators. The basic methods of analysis are essentially the same as those developed in the preceding sections of this chapter, but some minor differences appear. These differences are summarized as follows:

(1) The origin is taken at the center of the interaction region, which introduces no significant differences. The circuit and sole are at $y = \pm d/2$ respectively.

(2) The space-charge fields are calculated for infinite pencils of charge between the circuit and sole. The charge pencils are free to move in the z- and y-directions.

(3) The beam-circuit coupling function is written as

$$\psi(y) = \frac{\sinh \beta_e[y + (d/2)]}{\sinh \beta_e d}, \qquad (53)$$

the difference being that Sedin uses the stream phase constant β_e whereas

8. COMPARISON WITH SEDIN'S CALCULATIONS

Gandhi and Rowe use $\beta(q) = \bar{\beta}[1 + D(d\theta(q)/dq)]$, the actual wave phase constant.

(4) The driving term, $i_\omega(z, t)$, in the equivalent-circuit transmission-line equation,

$$\frac{\partial^2 V(z, t)}{\partial t^2} - v_0^2 \frac{\partial^2 V(z, t)}{\partial z^2} = v_0 Z_0 \frac{\partial i_\omega(z, t)}{\partial t}$$

$$= \frac{v_0 Z_0}{2d} w \frac{\partial^2}{\partial t^2} \int_{-\infty}^{\infty} \int_{-(d/2)}^{d/2} \rho(y', z', t) F \, dy' \, dz', \quad (54)$$

is taken as the fundamental component of displacement current per unit length flowing into the line.

(5) The gain parameter is defined as

$$D_s^2 \triangleq \frac{\omega}{\omega_c} \frac{I_0 Z_0}{2V_0},$$

whereas Gandhi and Rowe used

$$D^2 \triangleq \frac{\omega_c}{\omega} \frac{I_0 Z_0}{2V_0}.$$

The conversion between the two is

$$D = \left(\frac{\omega_c}{\omega}\right) D_s.$$

Other normalizations introduce no essential differences. The final forms of the equations are somewhat different only in the circuit driving terms.

The general equations of Sedin are as follows:

Circuit Equations

$$\frac{d^2 A(s)}{ds^2} - A(s)\left[\left(\frac{1}{D} - \frac{d\theta(s)}{ds}\right)^2 - \left(\frac{1 + Db}{D}\right)^2\right]$$

$$= \pm \left(\frac{1 + Db}{D}\right) \frac{1}{4\pi^2 \Phi^2 \sinh \beta_e d(\beta_e \tau)}$$

$$\cdot \int_0^{2\pi} \int_{\Phi_0' - \pi}^{\Phi_0' + \pi} \int_{-1}^{1} \frac{i_1}{i_2} F \, dr_0' \, d\Phi_0' \cos \Phi' \, d\Phi', \quad (55)$$

$$A(s)\frac{d^2\theta(s)}{ds^2} - 2\frac{dA(s)}{ds}\left(\frac{1}{D} - \frac{d\theta(s)}{ds}\right) = \mp \left(\frac{1 + Db}{D}\right) \frac{1}{4\pi^2 \Phi^2 \sinh \beta_e d(\beta_e \tau)}$$

$$\cdot \int_0^{2\pi} \int_{\Phi_0' - \pi}^{\Phi_0' + \pi} \int_{-1}^{1} \frac{i_1}{i_2} F \, dr_0' \, d\Phi_0' \sin \Phi' \, d\Phi', \quad (56)$$

where

$$F \triangleq \frac{\cos(\pi y'/d)}{-\sin(\pi y'/d) + \cosh(\pi/d)(z - z')},$$

$$i_1 = |u_1(y_0)\rho_0(y_0)|,$$

and

$$i_2 = \rho_1 u_0.$$

Force Equations

$$[1 + 2Dq(r_0, s, \Phi_0)] \frac{\partial q(r_0, s, \Phi_0)}{\partial s} = \frac{\omega_c}{\omega D} \left[\left\{ \frac{dA(s)}{ds} \cos \Phi \right. \right.$$

$$\left. - A(s) \sin \Phi \left(\frac{1}{D} - \frac{d\theta(s)}{ds} \right) \right\} \frac{D}{\alpha} \sinh \frac{\beta_e d}{2} (1 + r)$$

$$\left. + p + \frac{2g}{\pi} \int_{\Phi_0' - \pi}^{\Phi_0' + \pi} \int_{-1}^{1} \frac{i_1}{i_2} Z \, dr_0' \, d\Phi_0' \right] \tag{57}$$

and

$$[1 + 2Dq(r_0, s, \Phi_0)] \frac{\partial p(r_0, s, \Phi_0)}{\partial s} = \frac{\omega_c}{\omega D} \left[\frac{A(s)}{\alpha} \cosh \frac{\beta_e d}{2} (1 + r) \cos \Phi \right.$$

$$\left. - q + \frac{2g}{\pi} \int_{\Phi_0' - \pi}^{\Phi_0' + \pi} \int_{-1}^{1} \frac{i_1}{i_2} Y \, dr_0' \, d\Phi_0' \right], \tag{58}$$

where Y, Z are the y- and z-directed space-charge-field weighting functions calculated from line charges; see Chapter IV.

Velocity-Phase Equations

$$\frac{\partial r(r_0, s, \Phi_0)}{\partial s} = \frac{2\pi}{\beta_e d} \frac{p(r_0, s, \Phi_0)}{[1 + 2Dq(r_0, s, \Phi_0)]} \tag{59}$$

and

$$\frac{\partial \Phi(r_0, s, \Phi_0)}{\partial s} + \frac{d\theta(s)}{ds} = \frac{2q(r_0, s, \Phi_0)}{[1 + 2Dq(r_0, s, \Phi_0)]}. \tag{60}$$

The normalizations and parameters introduced in Eqs. (55)–(60) are given below:

$$D^2 \triangleq \frac{\omega}{\omega_c} \frac{I_0 Z_0 \Phi^2 \alpha}{2 V_0},$$

$$\Phi^2 \triangleq \frac{Z(y_0)}{Z_0} \triangleq \frac{\sinh^2 \beta_e \left(y_0 + \frac{d}{2} \right)}{\sinh^2 \beta_e d},$$

$$\alpha \triangleq \left| \frac{E_y(y_0)}{E_z(y_0)} \right| \triangleq \coth \beta_e d \left(y_0 + \frac{d}{2} \right),$$

8. COMPARISON WITH SEDIN'S CALCULATIONS

$$\Phi(z, t) \triangleq \omega \left(\frac{z}{u_0} - t\right) - \theta(z),$$

$$s \triangleq \frac{D\omega}{u_0} z = \beta_e D z,$$

$$r \triangleq \frac{\pi y}{d},$$

$$\Phi_0 \triangleq \frac{\omega z_0}{u_1(y)},$$

$$\frac{dy_j}{dt} \triangleq 2u_0 D p_j(r_0, s, \Phi_0),$$

$$\frac{dz_j}{dt} \triangleq u_0[1 + 2D q_j(r_0, s, \Phi_0)],$$

and

$$g \triangleq \frac{\omega_p^2}{8\omega\omega_c D}.$$

The above equations apply to both the crossed-field amplifier and the oscillator; in the case of double signs the upper is used for the amplifier and the lower for the oscillator. Since the function F is known to be a rapidly decreasing function of $z - z'$, the integrals of Eqs. (55)–(58) are carried over only $\Phi_0' - \pi$ to $\Phi_0' + \pi$ rather than $-\infty$ to ∞. This approximation is considered valid since only the charge within $\pm \pi$ of any given charge is effective in producing a field at z.

The above equations, like those previously derived, are readily simplified if the stream-circuit coupling or interaction is weak, i.e., $D \ll 1$. The simplified nonlinear equations are:

Circuit Equations

$$\frac{dA(s)}{ds} = \pm \frac{1}{8\pi^2 \Phi^2 \sinh \beta_e d(\beta_e \tau)} \int_0^{2\pi} \int_{\Phi_0'-\pi}^{\Phi_0'+\pi} \int_{-1}^{1} \frac{i_1}{i_2} F \, dr_0' \, d\Phi_0' \sin \Phi \, d\Phi, \tag{61}$$

$$\frac{d\theta(s)}{ds} + b = \mp \frac{1}{8\pi^2 A(s) \Phi^2 \sinh \beta_e d(\beta_e \tau)} \int_0^{2\pi} \int_{\Phi_0'-\pi}^{\Phi_0'+\pi} \frac{i_1}{i_2} F \, dr_0' \, d\Phi_0' \cos \Phi \, d\Phi. \tag{62}$$

Force Equations

$$q(r_0, s, \Phi_0) = \frac{A(s)}{\alpha} \cosh \frac{\beta_e d}{2} \left[1 + \frac{2}{\pi} r(r_0, s, \Phi_0)\right] \cos \Phi + \frac{S}{2\beta_e d} \int_{\Phi-\pi}^{\Phi+\pi} Y \, d\Phi_0' \tag{63}$$

and

$$p(r_0, s, \Phi_0) = \frac{A(s)}{\alpha} \sinh \frac{\beta_e d}{2} \left[1 + \frac{2}{\pi} r(r_0, s, \Phi_0)\right] \sin \Phi - \frac{S}{2\beta_e d} \int_{\Phi-\pi}^{\Phi+\pi} Z \, d\Phi_0', \quad (64)$$

where $S \triangleq (\sigma_0/2\epsilon_0 B_0 u_0 D)$, the small-signal space-charge parameter.

Velocity-Phase Equations

$$\frac{\partial r(r_0, s, \Phi_0)}{\partial s} = \frac{2\pi}{\beta_e d} p(r_0, s, \Phi_0) \quad (65)$$

and

$$\frac{\partial \Phi(r_0, s, \Phi_0)}{\partial s} + \frac{d\theta(s)}{ds} = 2q(r_0, s, \Phi_0). \quad (66)$$

Both the small-D and finite-D equations of Sedin may be solved in the same numerical manner as used by Gandhi and Rowe. Calculations for similar parameter values using the equations of Gandhi and Rowe and those of Sedin, assuming $D \ll 1$, yield substantially the same results as shown in Fig. 37. The adiabatic equations were used along with a nonslipping laminar stream model.

The principal results of these small-D equations are that the gain is somewhat reduced due to finite space charge, i.e., $S \neq 0$, and the saturation efficiency is relatively unaffected by $S \neq 0$. Also a high percentage of the stream charge is collected on the rf structure just beyond the

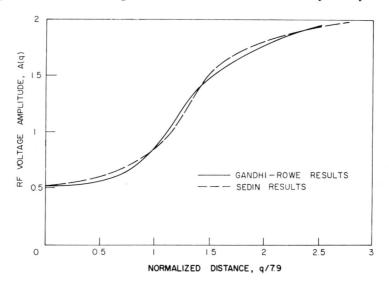

Fig. 37. Comparison with large-signal amplifier results of Sedin ($D = 0.079$, $r = 0.2$, $s = 0.09$, $\omega_p/\omega = \omega_c/\omega = 1$, $F_{2-z} = F_{2-y} = 0$).

halfway point and this continues until virtually all charge is eliminated. Of course, the calculations are meaningless beyond the point of approximately 90–95 per cent collection.

Other results of Sedin are shown in Chapter X on M-type backward-wave oscillators.

9 Results of and Comparison of Various Nonlinear Theories for the M-FWA

a. Introduction

Several nonlinear calculations using somewhat different methods and varying degrees of approximation have been made for the M-FWA and the M-BWO. It is the purpose of this section to relate these studies and to show how the various results compare. The comparison should be taken with a grain of salt in view of the acknowledged difficulty and uncertainty in experimentally determining detailed interaction parameters. The theories to be compared are those of Feinstein-Kino,[1] Sedin,[2] Gandhi-Rowe[4–7] and Hull-Kooyers.[3]

Since the Sedin and Gandhi-Rowe methods are quite similar they were considered in detail in Section 8 for the forward-wave amplifier and nearly identical results were obtained using the two methods. For the development of the backward-wave equations the reader is referred to the following chapter.

b. Feinstein-Kino Calculations

This early investigation of nonlinear effects in the crossed-field interaction process was made using a quasi-large-signal theory in which numerous limiting assumptions were made although quite useful results came forth. The model they considered for the M-FWA, M-BWA, and

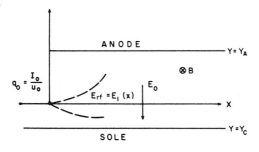

FIG. 38. Model of the magnetron amplifier considered by Feinstein and Kino.[1]

M-BWO was the ideal one shown in Fig. 38 and the following important assumptions were made.

(1) Adiabatic equations of motion;
(2) Neglect of space-charge forces;
(3) Eulerian flow;
(4) Beam-wave synchronism;
(5) Small gain per slow wavelength;
(6) Rf fields decaying exponentially away from the circuit; i.e., sole at $-\infty$;
(7) Single-space-harmonic operation;
(8) Negligible circuit loss;
(9) Nonrelativistic mechanics;
(10) Field of rf signal applied small compared to the dc field, i.e., $E_1 \ll E_0$;
(11) Uniform magnetic field;
(12) Neglect of cyclotron wave interaction.

The normalized force equations were written as

$$\frac{\omega}{\omega_c} \frac{d^2X}{dT^2} = \frac{E_1}{E_0} \cos X \, e^Y - \frac{dY}{dT} \tag{67}$$

and

$$\frac{\omega}{\omega_c} \frac{d^2Y}{dT^2} = \frac{E_1}{E_0} \sin X \, e^Y + \frac{dX}{dT}, \tag{68}$$

where

$$X \triangleq \beta x',$$
$$Y \triangleq \beta y',$$

and

$$T \triangleq \omega t.$$

The x' and y' variables denote motion in a coordinate system moving with the velocity $u_0 = E_0/B$ in the x-direction. Except for the calculation of some electron trajectories they assumed that ω_c/ω is sufficiently large that the left-hand sides of Eqs. (67) and (68) are zero. The circuit field at any position along the structure is calculated by equating the power delivered to the structure to the dc power lost from the beam and introducing the structure impedance as $P = E_1^2/2\beta^2 K$. This procedure led to a general circuit equation of the form

$$\frac{d^2F(x)}{dx^2} \mp \beta^2 D^2 G(F) = 0, \tag{69}$$

9. VARIOUS NONLINEAR THEORIES FOR THE M-FWA

where the upper sign is for forward-wave interaction and the lower is for backward-wave interaction. The function $G(F)$ is shown in Fig. 39 for a particular structure configuration including the effect of electron interception. The effects of interception are considered approximately

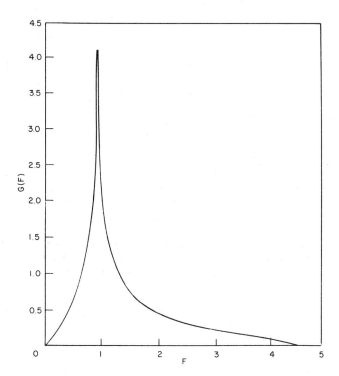

FIG. 39. $G(F)$ for integration limits corrected for finite electrode spacings: $\beta Y_a = 3$, $\beta Y_c = -1/2$. (Feinstein and Kino[1])

in that a fictitious sole electrode is located at $\beta y_c = -(\frac{1}{2})$, i.e., $\beta b' = \beta(|y_a| + |y_c|) = 3$, even though the fields are calculated as though the sole were at $-\infty$. The boundary conditions are taken as

$$\left. \begin{array}{l} F(0) = 0 \\ \dfrac{dF}{dx}\bigg|_{x=0} = \dfrac{\beta E_1(0)}{E_0} \end{array} \right\} \text{M-FWA} \qquad (70)$$

and as

$$\left. \begin{array}{l} F(0) = 0 \\ \dfrac{dF}{dx}\bigg|_{x=L} = \dfrac{\beta E_1(L)}{E_0} \end{array} \right\} \text{M-BWO.} \qquad (71)$$

Of course an oscillation condition is obtained when $E_1(L) = 0$ for $L > 0$. In spite of the limiting assumptions very interesting results were obtained for electron trajectories and rf signal growth. Normalized rf field amplitudes versus distance are shown in Fig. 40 for the M-FWA, M-BWO and M-BWA.

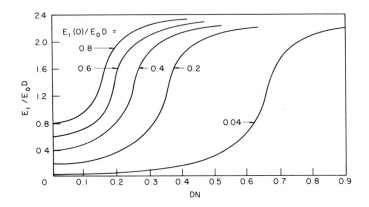

FIG. 40a. The amplitude of the rf field, E_1, of a forward-wave amplifier as a function of position along the circuit measured in wavelengths, N, for the electrode spacings of Fig. 39. The parameter D is proportional to the square root of the ratio of circuit to beam impedance. (Feinstein and Kino[1])

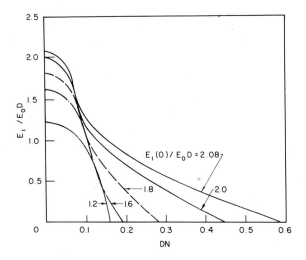

FIG. 40b. The amplitude of the rf field, E_1, of a backward-wave oscillator as a function of position along the circuit for the electrode spacings of Fig. 39. (Feinstein and Kino[1])

9. VARIOUS NONLINEAR THEORIES FOR THE M-FWA

The results of this investigation are conveniently summarized in the following statements.

(1) In a region well below saturation the rf gain increases as the rf drive level is increased. This is a result of beam movement towards the structure with consequent stronger coupling to the circuit.

(2) The above effect is limited by significant collection of electrons on the structure at high signal levels.

(3) For the M-BWO the length DN_s must be considerably greater than the 0.25 small-signal value in order to obtain full saturation output.

(4) Large-signal oscillation can occur for $DN_s < 0.25$ as a result of movement of the beam towards the structure, thereby increasing the coupling impedance with distance traveled. These cases would probably not be self-starting, although if initiated by the application of a large signal they would persist after the initiating signal was removed.

(5) An oscillator too short to yield saturated power output can be driven to saturation by the application of a driving signal in the manner of an injection-locked oscillator.

(6) Backward-wave amplifiers are found to be inherently low-gain devices.

(7) A premodulation circuit in the sole electrode can be used to improve efficiency although no increase in gain is obtained.

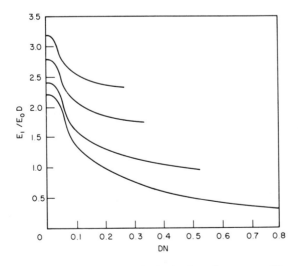

FIG. 40c. The amplitude of the rf field of a backward-wave amplifier as a function of position along the circuit for the electrode spacings of Fig. 39. (Feinstein and Kino[1])

No similar calculations were made using the Lagrangian theory due to the difficulty in satisfying the model and yet obtaining consistency. The general results are in qualitative agreement with Lagrangian calculations and thus it is believed unnecessary to carry out specific calculations to show the similarities.

c. Hull-Kooyers Calculations

Another formally equivalent yet in detail different nonlinear crossed-field theory is used by Hull and Kooyers in studying the M-FWA. Their procedure also uses the equivalent circuit approach and assumes that the stream interacts with only one space harmonic of the total field propagating on the rf structure.

The fundamental assumptions of this analysis for the model of Fig. 41 are enumerated below.

(1) Adiabatic equations of motion;

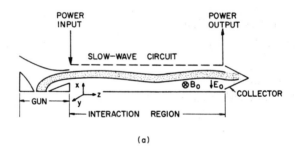

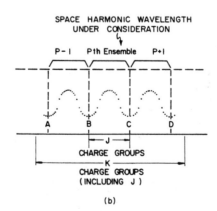

FIG. 41. Hull-Kooyers[3] large-signal crossed-field amplifier model. (a) Interaction configuration; (b) rf beam model.

9. VARIOUS NONLINEAR THEORIES FOR THE M-FWA

(2) Space-charge fields calculated from an array of infinite-length line charges in the manner used in Chapter IV and also used by Sedin;
(3) Lagrangian flow;
(4) Single-space-harmonic interaction;
(5) Nonrelativistic mechanics;
(6) Uniform magnetic field;
(7) Neglect of cyclotron-wave interaction.

The basic approach is to travel in a coordinate system moving in synchronism with the rf wave whose velocity varies with distance along the structure and then to use induced current theory to evaluate the effect of various charge elements in inducing circuit currents. The usual coupling function of $\sinh \beta y/\sinh \beta b'$ is utilized. The rf power flow may be calculated from Poynting's vector

$$P = \tfrac{1}{2} I \cdot I^* Z_c \,, \tag{72}$$

where $I =$ the equivalent rf current flow in the structure, or from the contributions of individual charge groups given by

$$P_j = \sum_j \mathbf{v}_j \cdot \mathbf{E}_{rf_j} Q_j \,, \tag{73}$$

where the summation extends over all charge groups. $\mathbf{E}_{rf_j}$ indicates the rf field acting on Q_j and $\mathbf{v}_j$ denotes the velocity of Q_j in a stationary frame. After much algebra the following working equations evolve.

Circuit Equations. For a current $I(\hat{z}_0)$ flowing in the slow-wave structure the power differential along the structure is

$$P(\hat{z}_0 + \Delta\hat{z}) - P(\hat{z}_0) = \frac{Z_c}{2} \left\{ \left[I(\hat{z}_0) + \sum_{j=1}^{J} (\Delta I_{ip})_j \right]^2 \right.$$
$$\left. + \left[\sum_{j=1}^{J} (\Delta I_{op})_j \right]^2 - I^2(\hat{z}_0) \right\}, \tag{74}$$

where

$\Delta I_{ip_j} = (\Delta I_j) \cos \psi_j$, the component of impressed current in phase with the existing circuit current;

$\Delta I_{op_j} = (\Delta I_j) \sin \psi_j$, the component of impressed current leading the existing current;

$\psi_j = \beta' z_j - \tan^{-1}\left[\dfrac{\hat{v}_{xj}}{\hat{v}_{zj}} \coth(\beta' x_j)\right]$, the phase angle lead of the impressed current relative to the existing structure current; and

$\beta' \Delta \hat{z} = \beta_0 \Delta \hat{z} - \Delta \Phi$.

The phase shift of the current relative to that in the absence of the beam over the interval $\Delta \hat{z}$ is

$$\Delta \Phi = \tan^{-1} \frac{\sum_{j=1}^{J} \Delta I_{op_j}}{I(\hat{z}_0) + \sum_{j=1}^{J} \Delta I_{ip_j}}. \tag{75}$$

The space variables with carats indicate positions in the stationary coordinate system; those without, in the moving frame.

Beam Force Equations. In writing the force equation, cyclotron interaction is neglected and the adiabatic assumption is made. Thus the electron velocity is adjusted so that

$$\mathbf{v}_e = \frac{\mathbf{E}_t \times \mathbf{B}}{B^2}, \tag{76}$$

where $\mathbf{E}_t$ denotes the total electric field acting on the charge group. The circuit electric field components in the moving frame are

$$E_{zj} = E_1 \frac{\sinh(\beta' x_j)}{\sinh(\beta' a)} \cos (\beta' z_j) \tag{77}$$

and

$$E_{xj} = E_1 \frac{\cosh(\beta' x_j)}{\sinh(\beta' a)} \sin (\beta' z_j). \tag{78}$$

The space-charge fields are calculated from the line charge model of Chapter IV. After some manipulation the stationary and moving frame velocity components are

$$\hat{v}_{x_j} = v_{x_j} = \frac{E_{zt}}{B} = \frac{1}{B} \left[\frac{\beta'(2PZ_c)^{\frac{1}{2}} \sinh(\beta' x_j) \cos(\beta' z_j)}{\sinh(\beta' a)} \right.$$
$$\left. + \sum_k E_{s-cz_{j,k}} \right] \tag{79}$$

and

$$\hat{v}_{z_j} = \frac{\omega}{\beta'} + v_{z_j} = \frac{E_{xt}}{B} = \frac{1}{B} \left[E_{dc} - \frac{\beta'(2PZ_c)^{\frac{1}{2}} \cosh(\beta' x_j) \sin(\beta' z_j)}{\sinh(\beta' a)} \right.$$
$$\left. - \sum_k E_{s-cx_{j,k}} \right], \tag{80}$$

where $E_{s-cx_{j,k}}$ and $E_{s-cz_{j,k}}$ are the respective space-charge-field components.

The above system of nonlinear equations appears quite different from the corresponding Lagrangian equations at first glance, although they are based on the same model and assumptions except for the adiabatic motion assumption which was not made in the general theories of Sedin and Gandhi-Rowe. After much manipulation, however, the theories can be shown to be equivalent within the framework of the underlying assumptions.

A possibly more informative and convincing method of establishing equivalence is to compare solutions for the same parameter sets. Several particular calculations of Hull-Kooyers are compared with the corresponding Sedin adiabatic solutions in Fig. 42.

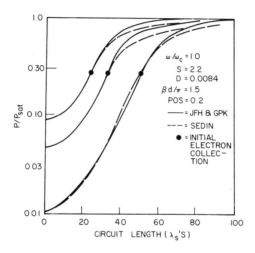

FIG. 42. Rf power versus circuit length. (Space-charge effects on electron motion neglected.) (Hull and Kooyers[3])

REFERENCES

1. Feinstein, J., and Kino, G., The large signal behavior of crossed field traveling-wave devices. *Proc. IRE* **45**, 1364-1373 (1957).
2. Sedin, J. W., "A Large-Signal Analysis of Beam-Type Crossed-Field Traveling-Wave Tubes." Hughes Res. Labs. Tech. Memo No. 520 (July 1958).
  Sedin, J. W., A large-signal analysis of beam-type crossed-field traveling-wave tubes. *IRE Trans. Electron Devices* **9**, No. 1, 41-51 (1962).
3. Hull, J. F., and Kooyers, G. P., Experimental and theoretical characteristics of injected beam type forward-wave crossed-field amplifiers. *NTF-Nachrtech. Fachber.* **22**, 151-158 (1960).
4. Gandhi, O. P., "Nonlinear Electron-Wave Interaction in Crossed Electric and

Magnetic Fields." Univ. of Michigan Electron Phys. Lab. Tech. Rept. No. 39 (October 1960).
5. Gandhi, O. P., and Rowe, J. E., "Nonlinear Analysis of Crossed-Field Amplifiers." Univ. of Michigan Electron Phys. Lab. Tech. Rept. No. 37 (June 1960).
6. Gandhi, O. P., and Rowe, J. E., "Nonlinear Analysis of Crossed-Field Amplifiers." *NTF-Nachrtech. Fachber.* **22**, 135-144 (1961).
7. Gandhi, O. P., and Rowe, J. E., Nonlinear theory of injected-beam crossed-field devices. *Crossed-Field Microwave Devices* (E. Okress, ed.), Vol. 1, pp. 439-495. Academic Press, New York, 1961.

CHAPTER X | Crossed-Field Backward-Wave Oscillators

1 Introduction

The crossed-field backward-wave oscillator, designated the M-BWO, bears the same relationship to the forward-wave amplifier M-FWA as the O-BWO does to the O-FWA. The principal difference arises out of the fact that the stream is designed to interact with a backward space-harmonic component of the rf circuit field. Since this field is described in terms of its forward phase velocity and a backward-directed group velocity, the possibility of a voltage-tunable oscillator exists. Most frequently either vane-type or interdigital-line rf structures are utilized. A schematic diagram of the M-BWO interaction region is shown in Fig. 5b of Chapter I, where a negative sole has been employed.

In view of the fact that the circuits usually used in M-BWO's have a larger backward-wave impedance than the most commonly used helix in O-BWO's, the stream-circuit coupling function is larger. Also, the M-type device is a potential energy conversion device. Hence higher efficiency of operation is expected as compared to the O-BWO, which is a kinetic energy conversion device. As mentioned in Chapter VII, it is possible to use higher-impedance circuits in O-BWO's to obtain somewhat higher efficiencies. Without using some phase-focusing technique, however, the O-type oscillator will still be less efficient than the M-type oscillator. An important limitation on the M-BWO is still the fact that an appreciable percentage of the stream charge is collected on the rf structure. A possible means for reducing or avoiding this collection phenomenon is discussed in Chapter XIII.

In this chapter the nonlinear M-type backward-wave oscillator is analyzed based on the amplifier study of the preceding chapter. Both weak, $D \ll 1$, and strong interaction systems are studied in order to determine required starting conditions and possible saturation efficiencies. Several nonlinear analyses of the M-BWO (Gandhi,[1] Sedin[2,3]) using similar methods have been made and their results are compared here.

2 Two-Dimensional M-BWO with a Negative Sole

The analysis of the M-BWO is greatly facilitated by the use of the backward-wave equivalent circuit procedure outlined in Chapter III and used in Chapter VII, when studying O-type oscillators. The argument that a uniform equivalent transmission-line equation may be obtained by changing the sign of the circuit impedance Z_0 to account for backward-energy flow yields the following differential equation for voltage along the oscillator rf circuit:

$$\frac{\partial^2 V(z,t)}{\partial t^2} - v_0^2 \frac{\partial^2 V(z,t)}{\partial z^2} + 2\omega Dd \frac{\partial V(z,t)}{\partial t}$$
$$= -v_0 Z_0 \left[\frac{\partial^2 \rho(z,t)}{\partial t^2} + 2\omega Dd \frac{\partial \rho(z,t)}{\partial t} \right] \quad (1)$$

where $R/L \triangleq 2\omega Dd$, d being the equivalent loss parameter along the primary direction of electron flow. The minus sign preceding the terms on the right-hand side of Eq. (1) are contrasted with the positive sign in Eq. (22) of Chapter IX, when a forward-wave interaction is studied.

A consideration of the forces acting on the electrons in their interaction with the rf circuit wave reveals that no other changes need be made in the basic operating equations. Thus the only changes in the final working equations for the M-BWO as compared to the M-FWA are in the signs preceding the right-hand sides of the circuit equations. The final oscillator equations are:

Circuit Equations

$$\frac{d^2 A(q)}{dq^2} - A(q) \left[\left(\frac{1}{D} + \frac{d\theta(q)}{dq} \right)^2 - \left(\frac{1+Db}{D} \right)^2 \right]$$
$$= \frac{(1+Db)}{\pi D} \left[\int_0^{2\pi} \int_{\frac{1}{s} - \frac{1}{2}}^{\frac{1}{s} + \frac{1}{2}} \psi(p) \frac{1 + 2Du_i(p_0', \Phi_0', 0)}{1 + 2Du(p_0', \Phi_0', q)} \cos \Phi \, d\Phi_0' \, dp_0' \right.$$
$$\left. + 2\, dD \int_0^{2\pi} \int_{\frac{1}{s} - \frac{1}{2}}^{\frac{1}{s} + \frac{1}{2}} \psi(p) \frac{1 + 2Du_i(p_0', \Phi_0', 0)}{1 + 2Du(p_0', \Phi_0', q)} \sin \Phi \, d\Phi_0' \, dp_0' \right] \quad (2)$$

2. TWO-DIMENSIONAL M-BWO WITH A NEGATIVE SOLE

$$\frac{2\,dA(q)}{dq}\left(\frac{1}{D} + \frac{d\theta(q)}{dq}\right) + A(q)\left[\frac{d^2\theta(q)}{dq^2} + \frac{2d}{D}(1+Db)^2\right]$$

$$= -\frac{(1+Db)}{\pi D}\left[\int_0^{2\pi}\int_{\frac{1}{s}-\frac{1}{2}}^{\frac{1}{s}+\frac{1}{2}} \psi(p)\,\frac{1 + 2Du_i(p_0', \Phi_0', 0)}{1 + 2Du(p_0', \Phi_0', q)}\sin\Phi\,d\Phi_0'\,dp_0'\right.$$

$$\left. -\,2\,dD \int_0^{2\pi}\int_{\frac{1}{s}-\frac{1}{2}}^{\frac{1}{s}+\frac{1}{2}} \psi(p)\,\frac{1 + 2Du_i(p_0', \Phi_0', 0)}{1 + 2Du(p_0', \Phi_0', q)}\cos\Phi\,d\Phi_0'\,dp_0'\right]. \quad (3)$$

Force Equations

$$[1 + 2Du(p_0, \Phi_0, q)]\frac{\delta u(p_0, \Phi_0, q)}{\delta q} = \frac{\omega\psi(p)}{2\omega_c}\left[\frac{dA(q)}{dq}\cos\Phi\right.$$

$$\left. -\frac{A(q)\sin\Phi}{D}\left(1 + D\frac{d\theta(q)}{dq}\right)\right] + \frac{sv_{y\omega}}{2D^2l} + \left(\frac{\omega_p}{\omega}\right)^2\frac{rs}{\pi D^2}F_{2-z}, \quad (4)$$

$$[1 + 2Du(p_0, \Phi_0, q)]\left[\frac{\delta v_{y\omega}}{\delta q} + \frac{l}{sD}\left(\frac{\omega_c}{\omega}\right)^2\right] = \frac{l}{s}\left[1 + D\frac{d\theta(q)}{dq}\right]$$

$$\cdot\frac{\cosh\left[\frac{\omega}{\omega_c}\frac{ps}{l}\left(1 + D\frac{d\theta(q)}{dq}\right)\right]}{\sinh\left[\frac{\omega}{\omega_c rl}\left(1 + D\frac{d\theta(q)}{dq}\right)\right]} A(q)\cos\Phi$$

$$+ \frac{rl^2}{sD}\left(\frac{\omega_c}{\omega}\right)^2\left(\frac{V_a}{2V_0}\right) - 2rl\left(\frac{\omega_c}{\omega}\right)\left(\frac{\omega_p}{\omega}\right)^2 F_{2-y}. \quad (5)$$

Velocity-Phase and Position Equations

$$\frac{\delta\Phi(p_0, \Phi_0, q)}{\delta q} - \frac{d\theta(q)}{dq}$$

$$= \frac{1}{D}\left[\frac{1}{[1 + 2Du_i(p_0, \Phi_0, 0)]} - \frac{1}{[1 + 2Du(p_0, \Phi_0, q)]}\right] \quad (6)$$

and

$$p(p_0, \Phi_0, q) = p_0 + \int_0^q \frac{v_{y\omega}}{D[1 + 2Du(p_0, \Phi_0, q)]}\,dq. \quad (7)$$

As mentioned above, the only differences from the amplifier equations are associated with the right-hand sides of the circuit equations.

Initial Conditions. The oscillator problem is also conveniently solved as an initial-value problem except that an additional requirement for oscillation prevails, i.e., the rf voltage level on the structure must go to zero at some finite displacement plane away from the input. Many of the initial conditions used in obtaining forward-wave amplifier solutions are again applicable to the backward-wave case. These are summarized below.

(*i*) *Stream Initial Conditions*

$$p_{j,k,0} = \left(\frac{1}{s} - \frac{1}{2}\right) + \frac{k}{(n-1)} \quad k = 0, 1, 2, ..., n \tag{8}$$

$$\Phi_{j,k,0} = \frac{2\pi j}{m} \quad j = 0, 1, 2, ..., m \tag{9}$$

$$u_{j,k,0} = \frac{\left(\frac{k}{n} - \frac{1}{2}\right)}{\left(\frac{2Dl}{s}\right)} L_{k,0} \tag{10}$$

$$(v_{y\omega})_{j,k,0} = M_{k,0} \tag{11}$$

where

$n \triangleq$ the number of stream layers,

$m \triangleq$ the number of charge groups per layer,

$L_{k,0} = 1$, for all k in a Brillouin stream, and

$M_{k,0} = 0$, for all k in a Brillouin stream.

(*ii*) *Circuit Wave Initial Values*

$$A(0) \triangleq A_0, \text{ the normalized rf voltage output of the backward-wave device,} \tag{12}$$

$$\frac{dA(0)}{dq} = 0, \text{ for a lossless system,} \tag{13}$$

$$\theta(0) = 0, \tag{14}$$

$$\frac{d\theta(0)}{dq} = b + \frac{2u_i}{1 + 2Du_i}. \tag{15}$$

(*iii*) *Operating Parameters to be Specified*

$D \triangleq$ the gain parameter,

$b \triangleq (u_0 - v_0)/v_0 D$, the injection velocity parameter,

$d \triangleq$ circuit loss parameter,

$s \triangleq w/\bar{y}_0$, stream thickness parameter,

$r \triangleq \bar{y}_0/b$, stream position parameter,

$l \triangleq$ the ratio of stream average velocity to $\omega_c \bar{y}_0$,

$\omega_c/\omega \triangleq$ the normalized cyclotron frequency,

$\omega_p/\omega \triangleq$ the normalized plasma frequency,

$V_a/V_0 \triangleq$ ratio of dc anode voltage to dc voltage equivalent of the mean stream velocity.

Solution Procedure. Perfectly valid solutions to the backward-wave equations can be found by assuming arbitrary combinations of the injection velocity as measured by b and the output rf signal level as measured by A_0. Not all, however, will result in oscillation conditions. If the signal level decreases with distance to some nonzero minimum and then rises, the solution applies to a backward-wave amplifier and only when $A(q) \approx 0$ for $q > 0$ do we have an oscillation condition. The problem is thus to develop a systematic procedure for finding combinations of A_0 and b which yield oscillation while all other parameters are held constant.

The best method known is a modification of the "downhill method" as outlined in Chapter VII in which one systematically scans the A_0-b plane always proceeding towards an oscillation condition, as illustrated in Fig. 1. This method has been used extensively by the author and has been found to minimize the required computer time to find the oscillation conditions.

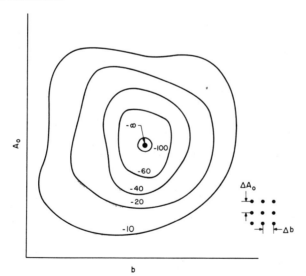

FIG. 1. Downhill method for systematically scanning the A_0-b plane for an oscillation condition.

Another equally valid procedure for zeroing in on proper A_0-b combinations, although considerably more costly of computer time, is to examine the phase variable $\theta(q)$ in the vicinity of the minimum in $A(q)$. The behavior of $\theta(q)$ for a fixed A_0 with b running through the appropriate value for oscillation is illustrated in Fig. 2 for both an adiabatic

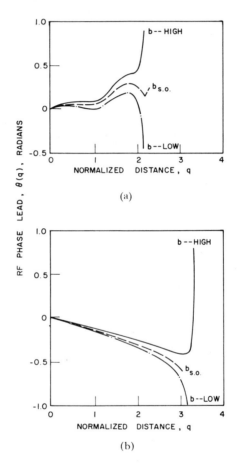

FIG. 2. Rf phase variations near oscillation. (a) Nonadiabatic equations ($D = 0.1$, $r = 0.5$, $s = 0.1$, $A_0 = 0.05$, $\omega_p/\omega = \omega_c/\omega = 0.5$, $F_{2-y} = F_{2-z} = 0$); (b) adiabatic equations ($D = 0.05$, $r = 0.5$, $s = 0.1$, $A_0 = 0.125$, $\omega_p/\omega = \omega_c/\omega = 0.25$, $F_{2-y} = F_{2-z} = 0$).

and a nonadiabatic system. As A_0 is increased from any specific value in combination with $b_{s.o.}$ giving oscillation, the stream loading on the rf circuit increases due to larger perturbations on the stream which cause the electrons to move into regions of higher fields and thus better coupling. This reduction of circuit wave velocity necessitates that b be reduced in order to obtain an oscillation condition again.

3 Results for a Two-Dimensional M-BWO with a Negative Sole

The starting conditions for the M-BWO are calculated using the above downhill procedure for specific values of the operating parameters D, ω_c/ω, r, s, and d. As a direct result of the search method and the evaluation of the space-charge integrals, the computations become quite lengthy when F_{2-y} and F_{2-z} are not zero. In order to make the computation problem manageable the results presented in this section were obtained assuming $F_{2-y} = F_{2-z} = 0$. The nonadiabatic large-signal equations were used even though in most cases $\omega_c/\omega D \geqslant 5$ so that acceleration terms could be neglected.

It is interesting to examine the effect of varying $\beta b'$ and ω_c/ω on the low-order start-oscillation conditions for a fixed initial stream position and thickness. The radian measure of the circuit-sole separation is given by

$$\beta b' = \frac{\omega}{\omega_c} \frac{1}{r} \frac{2}{4-s}. \tag{16}$$

It should be recalled that the stream-circuit coupling function, $\psi(p)$, decreases for increasing $\beta b'$ independent of the initial stream position. Thus we see that a small value of ω_c/ω will lead to a weak coupling between stream and wave and thus it is expected that the oscillation length would be significantly longer than for large ω_c/ω situations. This effect is illustrated in Table I.

TABLE I

DEPENDENCE OF LOW-LEVEL STARTING CONDITIONS ON MAGNETIC FIELD
($D = 0.05$, $r = 0.5$, $s = 0.1$)

ω_c/ω	$\beta b'$	$q_{\min}$	$(DN_s)_{\min}$
0.25	4.10	3.08	0.490
0.50	2.05	2.21	0.352
1.0	1.03	2.44	0.388

There is a slight dip in the $(DN_s)_{\min}$ versus ω_c/ω plot although the trend is definitely towards higher values at large ω_c/ω. The starting conditions of Table I are considerably higher than the small-signal result of $q_{min} = 1.57$ or $(DN_s)_{min} = 0.25$. These differences are attributable to the large value of D and possibly to acceleration effects. Space-charge effects and nonlinearities can be particularly important in backward-wave oscillators since the circuit field is relatively weak in regions of strong rf bunching.

For stream currents greater than the minimum starting current the rf output increases and thus higher rf conversion efficiencies are realized. The efficiency versus the ratio of operating current to minimum starting current for various D and ω_c/ω values is shown in Fig. 3 and the corresponding starting lengths (DN_s versus I_0/I_s) in Fig. 4. As in the

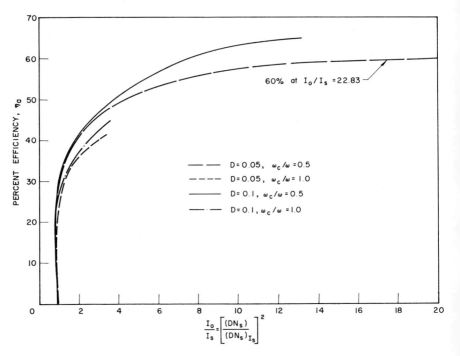

FIG. 3. Backward-wave oscillator efficiency versus current ($r = 0.5$, $F_{2-y} = F_{2-z} = 0$).

O-BWO it is found that N_s is relatively independent of the operating current for moderate D values, so that the current ratio is given by

$$\frac{I_0}{I_s} = \left(\frac{(DN_s)}{(DN_s)_{I_s}}\right)^2$$

3. RESULTS

since $D^2 \propto I_0$. The electronic efficiency of the backward-wave oscillator is calculated from

$$\eta_e = r\left(1 - \frac{s}{4}\right) A_0^2, \quad (17)$$

where A_0 is the rf voltage level at $q = 0$. The efficiency based on the available beam potential energy at $q = 0$ is then

$$\eta_a = \frac{\eta_e}{r} = \left(1 - \frac{s}{4}\right) A_0^2. \quad (18)$$

The efficiency increases rapidly at first but begins to level off when the operating current is a few times the minimum starting current. Efficiency values between 40 and 60% of the available stream energy at entrance appear achievable. The starting lengths (DN_s) are seen to increase continually (although presently leveling off) with increasing current ratio and they are found to be generally independent of D and ω_c/ω over the range investigated.

The start-oscillation value of the injection velocity parameter, b, is found to be zero (synchronism) from the linear theory for zero space-

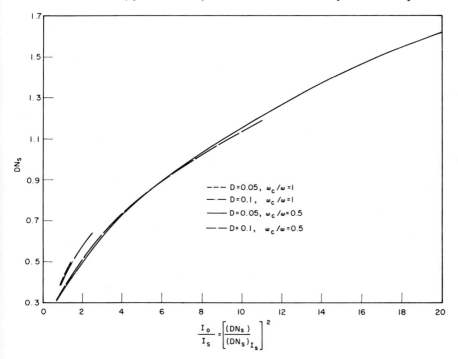

FIG. 4. DN_s versus current ($r = 0.5$, $F_{2-y} = F_{2-z} = 0$).

charge forces independent of D, ω_c/ω, r, etc. The nonlinear theory indicates that the b for oscillation varies between zero and 0.1 for all the data of Fig. 3. The linear theory also shows that the b for start-oscillation increases linearly with the space-charge parameter as indicated later in Fig. 8.

4 M-BWO with a Positive Sole

In Section 4 of Chapter IX, the forward-wave amplifier, M-FWA, was examined wherein the circuit and sole electrode positions were interchanged, giving a so-called positive-sole interaction region. There is of course no mathematical reason why the same procedure cannot be created in a crossed-field device utilizing a backward-wave circuit. An examination of the equations of Section 4, Chapter IX, in the light of the procedure developed in Section 2 for studying backward-wave phenomena indicates that the equations of Section 4, Chapter IX, are applicable here when the signs preceding the right-hand sides of the circuit equations are changed.

Gandhi[4] has examined such a configuration on a small-amplitude basis and found that again a beating-wave phenomenon exists and that the linear rf voltage-distance characteristics are as shown in Fig. 5. It is

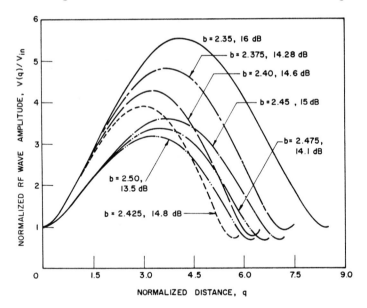

FIG. 5. Positive-sole backward-wave amplifier ($D = 0.1$, $\beta b' = 2.06$, $S = -\sigma_0/(2\epsilon_0\beta_0\bar{u}_0 D) = 0.25$, $\zeta = 2D/(Z_0\omega\epsilon_0 h) = -1.28$).

found that this type of backward-wave interaction is more efficient than the corresponding negative-sole interaction due to the fact that over the length involved the rf current modulation in the stream is quite large. The gains indicated in Fig. 5 are calculated from the ratio of maximum output to the rf voltage value at the minimum.

5 Adiabatic Equations for an M-BWO with a Negative Sole

Again the analysis and equations of Section IX.5 are directly applicable after the circuit equation signs are changed in accordance with oscillator analysis practice. The basic assumption under which the adiabatic equations are developed is that the growth of the rf wave shall be small in a cyclotron wavelength, $\lambda_c = 2\pi \bar{u}_0/\omega_c$. This condition is satisfied when

$$\frac{\omega_c}{\omega D} \gg 1$$

so that we see that for $D \ll 1$, i.e., weak interaction conditions, the adiabatic analysis is appropriate. Generally $D \ll 1$ in all backward-wave oscillators and hence these results are universally applicable. The results for Section 3 are for cases in which $\omega_c/\omega D \geq 5$ and hence either set of equations should yield essentially the same results. The similarities of the adiabatic and nonadiabatic results when $\omega_c/\omega D = 5$ are seen in Fig. 6.

Sedin's[2,3] results were obtained by making the adiabatic approximation and as demonstrated in Sections IX.8 and 9 his final working equations are essentially identical to those of Gandhi and Rowe; thus the same backward-wave solutions are obtained using either theoretical formulation. The nonadiabatic starting conditions were summarized in Section 3 and in this section the detailed results for the adiabatic case as obtained by Sedin[2,3] are presented.

An interesting check on the nonlinear calculations is afforded by determining the lowest-order start-oscillation conditions with and without space-charge forces and comparing these with the corresponding results from the small-signal theory of Gould[5,6] or Dombrowski.[7] The calculations for zero space charge are compared in Table II.

The normalized parameters used by Sedin differ somewhat from those used by Gandhi and Rowe. The conversions appropriate to $\beta b'$ and r are given below.

$$\beta b' = \beta_e d = \frac{\omega}{\omega_c}\left(\frac{1}{r}\right)\left(\frac{2}{4-s}\right)$$

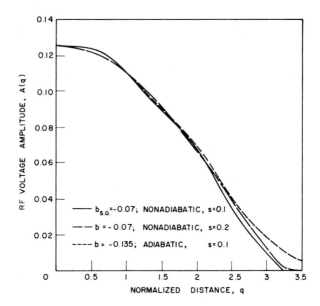

Fig. 6. Comparison of adiabatic and nonadiabatic backward-wave characteristics ($D = 0.05$, $\omega_p\omega = \omega_c/\omega = 0.25$, $r = 0.5$, $F_{2-y} = F_{2-z} = 0$).

TABLE II

Lowest-Order Starting Conditions for $F_{2-y} = F_{2-z} = 0$, $D \ll 1$

$\beta b'$	r	$b_{\text{s.o.}}$	$(DN_s)_{\text{s.o.}}$
2.0	0.5	0	0.250
2.0	0.25	0	0.248
2.5	0.20	0	0.245
3.0	0.17	0	0.239
3.5	0.14	0	0.229

and

$$r = \frac{r_0}{\pi} + 0.5.$$

Since the small-signal theory predicts a normalized starting length of $(DN_s)_{\text{s.o.}} = 0.25$ independent of $\beta b'$ and r, it is surprising to find a dependence on these parameters as noted in the table. Sedin explains that the error in the large-signal calculations at large values of $\beta_e d$ occurs due to the neglect of electrons beyond one-half period in either

5. ADIABATIC EQUATIONS

direction of the phase for which the terminal circuit field is being calculated. This approximation results in a larger beam modulation which in turn leads to a shorter length for oscillation. When $\beta_e d$ is in the range 2 and 3, the effect is negligible.

In calculating the lowest-order starting conditions in the presence of space charge another effect shows up which is purely a computational error associated with the finite-cross-section rods of charge used in calculating the space-charge fields.

Lowest-order starting condition calculations versus the space-charge parameter, $\beta_e d$, and r are shown in Fig. 7 when the adiabatic equations

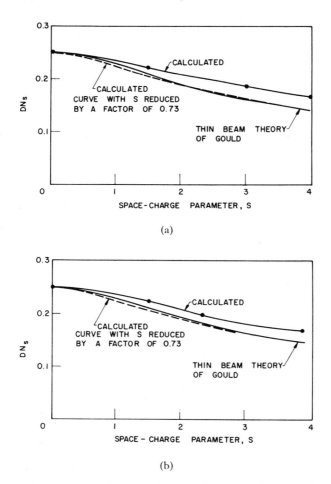

FIG. 7. Backward-wave oscillator starting lengths versus the space-charge parameter. (Sedin[2]) (a) $\beta b' = 2.5$, $r = 0.2$; (b) $\beta b' = 2$, $r = 0.5$.

are used. The space-charge parameter S has been modified by a factor 0.73 to obtain correspondence with the small-signal theory. This need arises due to errors in the space-charge-field calculations. The accompanying velocity parameter at start-oscillation is shown in Fig. 8.

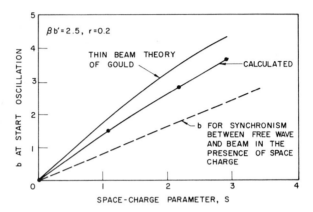

FIG. 8. Relative velocity parameter at start-oscillation versus the space-charge parameter. (Sedin[2])

Figure 7 indicates that the starting conditions are not particularly dependent on either $\beta_e d$ or r when $\omega_c/\omega D \gg 1$. These should be valid for D values up to about 0.05. The nonlinear results also indicate that the required length decreases as the space-charge parameter S increases. The small-signal space-charge parameter S of Gould is expressed in terms of large-signal variables as

$$S = \frac{\omega_p}{\omega} \left(\frac{s}{2lD}\right) \frac{\sinh\left(\frac{\omega}{\omega_c rl}\right)}{\sqrt{\sinh\left(\frac{\omega}{\omega_c l}\right)\cosh\left(\frac{\omega}{\omega_c l}\right)}}, \quad (19)$$

where $l = (4 - s)/2$, for a Brillouin stream in crossed fields.

The discrepancy between the calculated b for start-oscillation and the small-signal values shown in Fig. 8 is attributed to calculation errors in the phase variable θ. Sedin found that the phase velocity of the rf circuit wave was insignificantly affected by the presence of the beam even in the presence of space-charge forces and relatively large rf signal levels. This is to be expected since all of his calculations were made using the adiabatic approximation in which the beam loading is negligible.

In the small-D approximation the power output and efficiency are proportional to A_0^2 and thus the oscillation conditions given above correspond to outputs much less than the saturation capability of the beam. In order to determine maximum efficiencies higher-level oscillation conditions, i.e., larger A_0 values in combination with appropriate b values, must be found. The best technique for determining these is to scan the A_0-b plane using the downhill method. The corresponding required values of the velocity parameter for large A_0 values are shown in Fig. 9. We note as in the nonadiabatic solutions of Section 3 that the

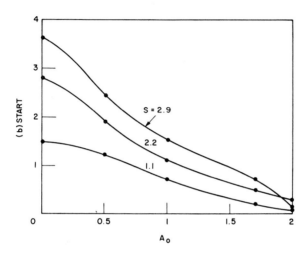

FIG. 9. Velocity parameter for oscillation versus output amplitude with space charge as the parameter. (Sedin[2])

starting length and injection velocity decrease with increasing space charge and A_0. The reduction of DN_s with increasing A_0 is attributed to better coupling due to increased beam modulation, which moves the electrons into regions of higher rf field near the circuit. In Section 3 it is shown that this increases the stream loading on the circuit, which reduces the rf wave velocity, thereby requiring a reduction in the b value for oscillation. In the case of the backward-wave amplifier the gain is seen to be reduced as A_0 is increased and as S increases.

The collection of electrons on the circuit in the adiabatic regime is quite similar to that characteristic of the nonadiabatic results of Section 3. We see in Fig. 10 that a change in the space-charge parameter does not markedly change the collection rate on the structure. Also it is noted that approximately 70% collection is characteristic of the adiabatic regime. The fact that a higher percentage collection is not realized is a result of the rapid fall-off of rf circuit field near the collector end of the oscillator.

Reflecting on the results for the M-FWA with space-charge forces included indicates that they play a much more important role in backward-wave devices than in forward-wave devices.

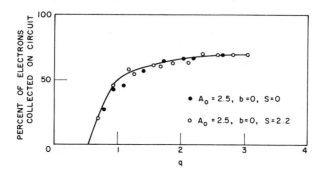

FIG. 10. Percentage electron collection on the rf circuit for large-signal backward-wave interaction. (Sedin[2])

6 Cyclotron Waves in M-BWO's

The study of cyclotron-wave interaction and effects in nonlinear backward-wave oscillators is facilitated by using the equivalent circuit approach outlined for forward-wave cyclotron-wave interaction in Section 7 of Chapter IX. The essential treatment there was to describe the circuit impedance variation with $\beta(z)$ in terms of a function

$$Z_0(1 + D\theta').$$

Such a function was shown in Fig. IX.36 for a forward space-harmonic field component on the slow-wave structure. A similar procedure may be utilized in the case of the backward-wave oscillator, wherein a function similar to that indicated above is obtained for a backward space-harmonic field component. The only other change required is that again the signs in the circuit equations must be changed to account for backward energy flow.

7 Theory versus Experiment

The comparison of nonlinear theoretical results with experimental data is always complicated by the difficulty in knowing accurately various interaction parameters such as the gain parameter, beam thickness and beam position and the effects of various high-order processes such as cyclotron-wave interaction. The most detailed and consistent experimental data on crossed-field interaction were reported by Anderson[8] on a thin-beam backward-wave amplifier. The device studied very closely approximates the model on which calculations were made both by Sedin and by Gandhi-Rowe. The electron beam was obtained from a Kino-Kirstein[9,10] type of gun and the rf circuit was an interdigital line operated on a backward space harmonic.

The normalized rf voltage along the structure is compared with Sedin and Gandhi-Rowe results and shown in Figs. 11–14. Detailed

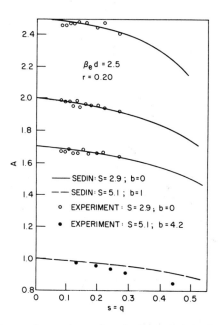

FIG. 11. Comparison of experimental and theoretical backward-wave amplitude characteristics for $S = 2.9$ and 5.1. The parameter b is the usual velocity difference parameter of ordinary traveling-wave tube theory. (Anderson[8])

comparisons for a large space-charge parameter $S = 5.1$ are shown in Fig. 13. The efficiency results on the experimental M-BWA are summarized in Fig. 14. The high conversion efficiencies are in substantial agreement with theoretical calculations.

It should be noted that under all operating conditions the above amplifier satisfied the adiabatic approximation; i.e., $\omega_c/\omega D > 5$. As yet no detailed experimental data have been obtained for the nonadiabatic region of operation. This still remains to be studied.

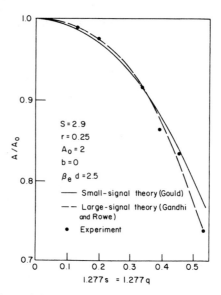

Fig. 12. Comparison of the voltage gain characteristic of experiment, small-signal theory and large-signal theory for $S = 2.9$. A_0 is the value of the amplitude function at the output and of the interdigital circuit. (Anderson[8])

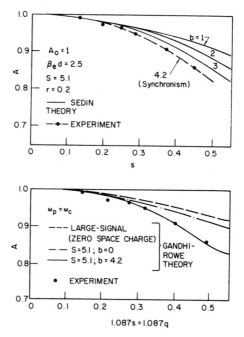

Fig. 13a. Comparison of the voltage gain characteristics of experiment and large-signal theories for $S = 5.1$. The Brillouin flow condition is $\omega_p = \omega_c$. (Anderson[8])

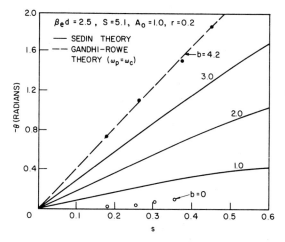

FIG. 13b. Comparison of experimental and theoretical phase shift as a function of the normalized tube length, s, for the case $S = 5.1$. (Anderson[8])

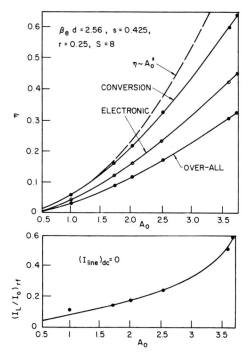

FIG. 14. Efficiency and circuit interception current as a function of the rf output voltage amplitude. The dashed line represents the dependence of efficiency on A_0 according to simple theoretical considerations. (Anderson[8])

References

1. Gandhi, O. P., "Nonlinear Electron-Wave Interaction in Crossed Electric and Magnetic Fields." Univ. of Michigan Electron Phys. Lab. Tech. Rept. No. 39 (October 1960).
2. Sedin, J. W., "A Large-Signal Analysis of Beam-Type Crossed-Field Traveling-Wave Tubes." Hughes Res. Labs. Tech. Memo. No. 520 (July 1958).
3. Sedin, J. W., A large-signal analysis of beam-type crossed-field traveling-wave tubes. *IRE Trans. Electron Devices* **9**, No. 1, 41-51 (1962).
4. Gandhi, O. P., A complementary mode of beam-wave interaction in crossed fields. *J. Electron. Control* **14**, No. 4, 393-401 (1963).
5. Gould, R. W., Space-charge effects in beam-type magnetrons. *J. Appl. Phys.* **28**, 599-604 (1957).
6. Gould, R. W., "Field Analysis of the M-Type Backward-Wave Oscillator." Calif. Inst. Technol. Electron Tube and Microwave Lab. Tech. Rept. No. 3 (September 1955).
7. Dombrowski, G. E., "A Small-Signal Theory of Electron-Wave Interaction in Crossed Electric and Magnetic Fields. Univ. of Michigan Electron Tube Lab. Tech. Rept. No. 22 (October 1957).
8. Anderson, J. R., Experimental investigation of large-signal traveling-wave magnetron theory. *IRE Trans. Electron Devices* **8**, No. 3, 233-240 (1961).
9. Midford, T. A., and Kino, G. S., Some experiments with a new type of crossed-field gun. *IRE Trans. Electron Devices* **8**, No. 4, 324-331 (1961).
10. Kirstein, P. T., On the determination of the electrodes required to produce a given electric field distribution along a prescribed curve. *Proc. IRE* **46**, 1716-1722 (1958).

CHAPTER

XI | Traveling-Wave Energy Converters

1 Introduction

The principal volume of material in this book relates to the efficient conversion of dc electrical energy into rf electrical energy. Devices utilizing the conversion of kinetic energy and/or potential energy have been studied in detail and found to possess desirable characteristics, namely high amplification factors and efficient conversion. Of great interest also is the inverse process, i.e., the efficient conversion of rf energy to dc energy.

It is of course possible to construct many different types of rf to dc converters such as

(1) the thermionic diode,
(2) the inverse magnetron,
(3) the inverse traveling-wave amplifier,
(4) the inverse klystron, and
(5) the inverse injected-beam crossed-field amplifier.

Since we have been primarily concerned with the beam-type dc to rf converters, we will arbitrarily restrict ourselves to a discussion of this type of device in this chapter. In particular, types (3) and (5) listed above are considered here. It should be noted that Thomas[1] has considered in detail the use of the inverted magnetron as an rf to dc energy converter. Schematic diagrams of types (3), (4), and (5), rf to dc converters, are shown in Fig. 1.

The basic principle of operation is common to all three and amounts to the conversion of rf to dc by accelerating electrons and thus increasing their kinetic energy. The electrons are collected at a negative dc potential in order to minimize the energy dissipated as heat in the collector. These devices, although important for some applications, are not without their disadvantages: namely, the need for focusing magnets, anode supplies and auxiliary cathode heater supplies.

The primary limitation in efficiency is due to velocity variations in the

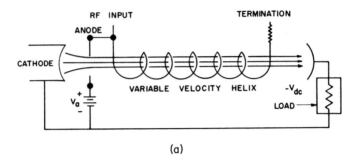
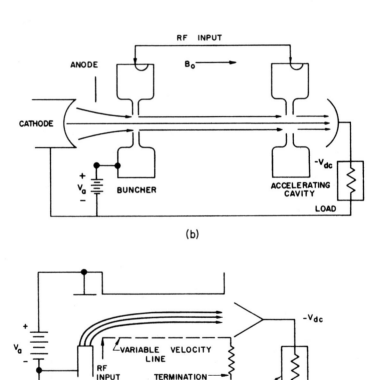

Fig. 1. Schematics of rf to dc energy converters. (a) Inverse traveling-wave converter; (b) inverse klystron converter; (c) inverse crossed-field converter.

electron beam arising from space-charge forces and nonuniform circuit fields acting over the beam. The efficiency can be maximized in such converters by application of phase-focusing techniques (velocity tapering) just as conversion efficiency was enhanced in the dc to rf converters and by collector depression methods. The methods of analyzing variable-phase-velocity circuits developed in Chapter XIII are directly applicable to this problem.

Referring to Fig. 1 we see that the microwave modulated electron stream is directed towards an electrode which has a retarding dc potential so that it will exhibit the proper potential and direction of current flow for the device to behave as a source of power for an external load. As mentioned above it is desirable to collect electrons on this negative electrode with a relatively low velocity in order to avoid the dissipation of large amounts of power at the collection electrode. Not all electrons in the stream will have the same velocity and hence all those not able to overcome the retarding potential will be returned to the interaction region.

2 O-Type Traveling-Wave Energy Converter

In order to convert rf energy to dc in a kinetic energy conversion device it is necessary to run the electron stream at a lower velocity than the rf circuit or wave phase velocity. The action of the wave is then to accelerate the electrons and thereby increase their kinetic energy. The efficiency of the O-type traveling-wave energy converter (O-TWEC), like the inverse O-TWA, will be less than unity due to nonlinear effects such as rf field variation over the bunch and debunching space-charge fields.

The small-signal operation of a traveling-wave amplifier at the Kompfner dip point indicates that this is a mode of operation in which rf energy is converted to beam kinetic energy and thus would seem to be an appropriate mode of operation for an O-TWEC. Coupled-mode analysis indicates that the Kompfner-dip condition (complete energy transfer) results from a complete transfer of energy from the forward-traveling rf circuit wave to the fast beam space-charge wave. Kompfner-dip operation[2] has heretofore been a small-amplitude, low-C operation used to determine the gain parameter C. It is also possible to gain complete energy transfer when C is large, as is shown by the tabulation of small-signal Kompfner-dip conditions for finite C in Appendix A. These are, of course, calculated for infinitesimally small initial rf signal amplitudes.

The large-signal equations of Chapter VI for the traveling-wave amplifier may be used directly to study O-TWEC operation at large-signal levels. To determine large-signal Kompfner-dip conditions the large-signal equations are solved for A_0-b combinations which yield a minimum rf signal level at some $y > 0$, using the "downhill method" discussed in Chapter VII. For small values of A_0 the minimum rf level is zero but as A_0 is increased a smaller fraction of the rf energy is converted to beam kinetic energy. The rf signal level versus distance is shown in Fig. 2, where the b value is that which gives the lowest $A_{\min}$

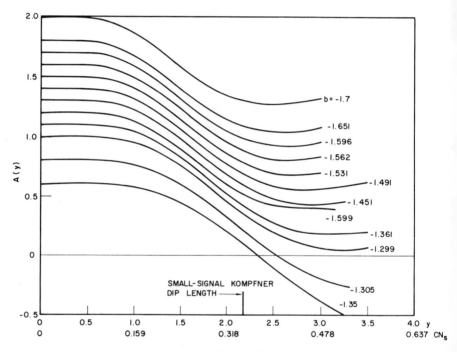

Fig. 2. $A(y)$ versus y for an O-TWEC ($C = 0.1$, $d = 0$, $\omega_p/\omega = 0$).

for a given input signal level A_0. The corresponding decibel plot is shown in Fig. 3. The small-signal Kompfner-dip length is shown on both figures and of course an interesting and important feature of the O-TWEC is its characteristically short length, $CN_s \approx 0.4$. The required stream injection velocity decreases as A_0 increases, as shown in Fig. 4. For an infinitely large A_0 the required b value approaches $(-1/C)$, indicating that the stream has a near zero velocity.

Since the electron beam is injected into the interaction region with a finite velocity there is a beam (electron gun) power of $I_0 V_0$ and thus this

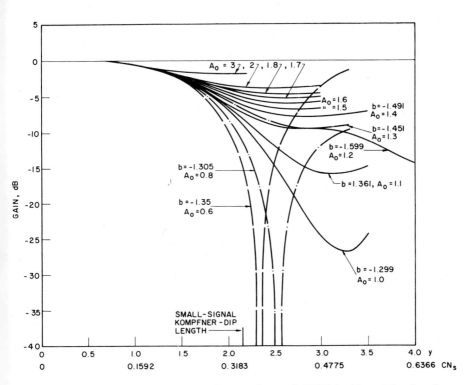

FIG. 3. Relative signal level versus distance for an O-TWEC ($C = 0.1$, $d = 0$, $\omega_p/\omega = 0$).

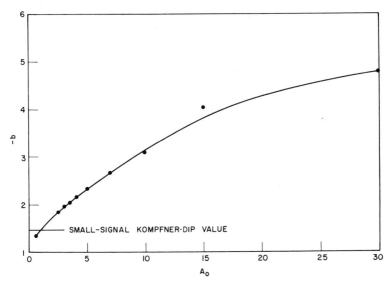

FIG. 4. Injection velocity versus A_0 for an O-TWEC ($C = 0.1$, $d = 0$, $\omega_p/\omega = 0$).

power level provides a convenient reference in calculating conversion efficiencies. The parameter η_a refers the applied rf power level to the electron gun power:

$$\eta_a = \frac{P_{rf}}{I_0 V_0} = 2CA_0^2. \tag{1}$$

The conversion efficiency indicates the fraction of rf energy applied to the input, i.e., converted to beam kinetic energy. This rf conversion efficiency is written as

$$\eta_c = \frac{2C(A_0^2 - A_{\min}^2)}{2CA_0^2} = 1 - \left(\frac{A_{\min}}{A_0}\right)^2 \tag{2}$$

so that if $A_{\min}$ is zero the conversion of rf to dc is complete. The product of η_a and η_c is then an indication of the fraction of rf power converted relative to the electron gun power $I_0 V_0$:

$$\eta_a \eta_c = 2C(A_0^2 - A_{\min}^2). \tag{3}$$

The conversion efficiency η_c is shown in Fig. 5 versus both η_a and

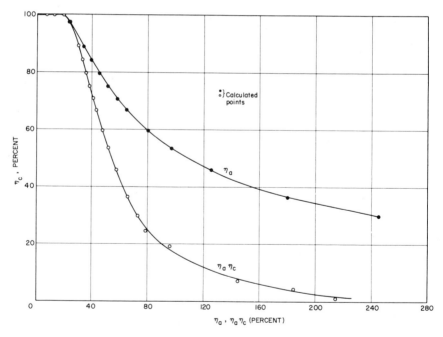

FIG. 5. O-TWEC conversion efficiency versus applied rf level ($C = 0.1$, $d = 0$, $\omega_p/\omega = 0$).

2. O-TYPE TRAVELING-WAVE ENERGY CONVERTER

$\eta_a\eta_c$. At low values of applied rf the conversion is complete; however above $\eta_a \approx 20\%$ the fraction of converted energy drops off rapidly, approaching zero at infinitely large values of A_0^2. The incomplete conversion of rf to dc at high A_0 values is due to the significant number of slow electrons in the beam which deliver energy to the circuit and to the presence of a significant velocity spread in the beam. Under finite space-charge conditions ($QC > 0$), complete conversion ($y_c \approx 100$) will occur for larger values of η_a since the slow and fast space-charge waves will be further separated in velocity than for the case of Fig. 5. Higher conversion efficiencies may be realized by tapering (increasing) the circuit phase velocity to maintain synchronism. The optimum circuit phase velocity profile may be determined using the methods of Chapter XIII.

The acceleration of the stream by the rf wave energy is illustrated in Fig. 6 for different values of applied signal. Since $(1 + 2Cu) = u_l/u_0$,

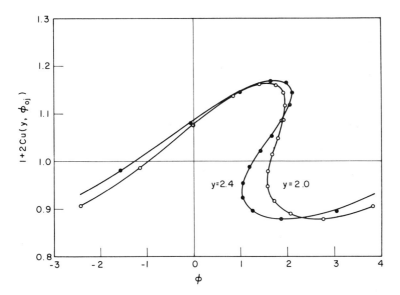

FIG. 6. Velocity versus phase for the O-TWEC. (a) $A_0 = 0.6$, $A_{\min} = 0$ ($-\infty$ dB).

values greater than one indicate an increased velocity and those less than one indicate that electrons are slowed. The degree of acceleration is directly correlated with the $A(y)$ versus y plots of Fig. 2.

Further improvements in the rf conversion efficiency may be realized by varying the circuit phase velocity so as to maintain the proper relative

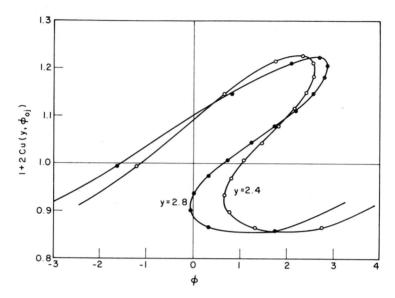

Fig. 6b. $A_0 = 0.8$, $A_{\min} = 0$ ($-\infty$ dB).

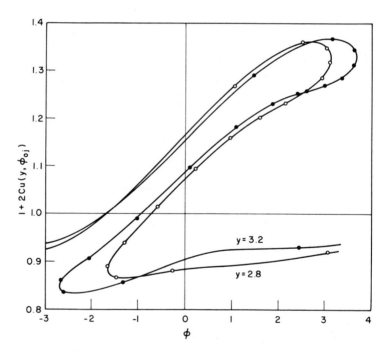

Fig. 6c. $A_0 = 1.3$, $A_{\min} = 0.433$ (-9.54 dB).

velocity between the stream and the rf traveling wave. In the TWEC the circuit phase velocity must be increased with distance, contrary to the situation in the O-TWA for improving efficiency.

After the rf energy has been converted to beam kinetic energy then one must efficiently collect the beam on the collector. Due to the inherent velocity spread, a segmented collector must be utilized. The efficiencies

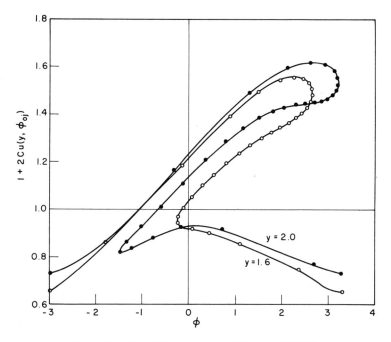

FIG. 6d. $A_0 = 3.0$, $A_{\min} = 2.391$ (-1.97 dB).

of two- and three-stage collectors are shown in Chapter XIV to approach 80 per cent, which will yield a composite O-TWEC efficiency of over 70 per cent.

3 M-Type Traveling-Wave Energy Converter

The crossed-field devices may also be operated as TWEC's by utilizing the positive-sole interaction to effectively transfer energy to the stream. This type of operation has not been explored extensively in M-type devices at either low or high signal levels. The so-called Kompfner-dip condition in the M-type may also be used to transfer energy to the stream but is not so effective as is the positive-sole interaction. The M-type Kompfner-dip conditions are shown in Appendix B for a practical range of operating parameters.

The achievable conversion efficiency for an M-TWEC may be calculated using the M-FWA equations of Chapter IX and the "downhill method" for finding appropriate A_0-b combinations for maximum energy transfer.

REFERENCES

1. Thomas, J., "High Power Microwave Energy Converter." Wright-Patterson Air Force Base Electron Technol. Lab. Tech. Rept. No. ASD-TR-61-476 (Contract AF-33(616)-7378) (1961).
2. Johnson, H. R., Kompfner-dip conditions. *Proc. IRE* **43**, 874 (1955).

CHAPTER XII | Multibeam and Beam-Plasma Interactions

1 Introduction

Not long following the invention and demonstration of traveling-wave interaction, several investigators, Haeff,[1] Nergaard,[2] Pierce,[3] and Hebenstreit,[4] studied the interaction between two drifting electron beams traveling at different velocities. The object was to replace the familiar rf circuit with an electron beam, which could also support the propagation of slow electromagnetic waves, and thereby obtain traveling-wave amplification without the use of an extended periodic circuit. Of course, some form of circuit configuration is necessary to modulate and demodulate the stream. Successful operation was achieved and a dispersion equation of the following form yields the wave propagation constants, assuming that each of the system waves varies as $\exp j[\omega t - \Gamma z]$:

$$\frac{\beta_{p1}^2}{(j\beta_{e1} - \Gamma)^2} + \frac{\beta_{p2}^2}{(j\beta_{e2} - \Gamma)^2} = -1, \qquad (1)$$

where

$\beta_{e1} = \omega/u_{01}$, stream No. 1 phase constant,

$\beta_{e2} = \omega/u_{02}$, stream No. 2 phase constant,

$\beta_{p1} = \omega_{p1}/u_{01}$, stream No. 1 plasma phase constant, and

$\beta_{p2} = \omega_{p2}/u_{02}$, stream No. 2 plasma phase constant.

Solution of the above determinantal equation reveals complex Γ's; thus amplification can be obtained by proper adjustment of beam velocities. It is interesting to note that when one beam or charge field is stationary the above dispersion equation becomes

$$\left(\frac{\omega_{p1}}{\omega}\right)^2 + \frac{\beta_{p2}^2}{(j\beta_{e2} - \Gamma)^2} = -1. \qquad (2)$$

A schematic drawing of this "double-beam amplifier" is shown in Fig. 3 of Chapter I. Since wavelength-dependent circuits are still required at the input and output, the development of useful devices was not actively pursued for some time. The work of Bloom and Peter[5] on the equivalent circuit of an electron beam in conjunction with noise propagation studies is, of course, directly applicable to double-beam interaction studies. One of the later experimental studies on double-beam amplifiers was reported by Bernashevsky, Voronov, Iziumova, and Tchernov[6] and also by Bogdanov, Kislov, and Tchernov.[7]

It is, of course, not necessary that both so-called beams be of the same type of charged particles; i.e., one might consist of electrons and the other of ions or a combination of ions and electrons (plasma). The drifting electron beam interacts with the combined electron-ion (plasma) system in the beam-plasma amplifier and this configuration has been studied by many authors. One such investigation was made by Boyd, Field, and Gould[8] wherein a drifting electron stream was caused to interact with a mercury plasma column. Their device is shown in Fig. 1.

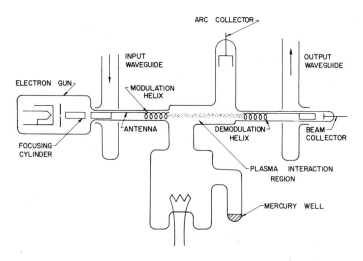

FIG. 1. Schematic drawing of plasma-beam tube used in growing-wave experiment (Boyd, Field, Gould[8]).

The dispersion equation for beam-plasma interactions can be written as

$$\frac{\beta_{pe}^2}{(j\beta_e - \Gamma)^2} + \frac{\beta_{pp}^2}{\left[j\beta_e - \left(\frac{v_T}{u_0}\right)^2 \Gamma\right]} = -1, \qquad (3)$$

1. INTRODUCTION

where

$\beta_{pe} \triangleq$ electron beam plasma phase constant,

$u_0 \triangleq$ electron beam drift velocity,

$v_T \triangleq (3kT/m)^{\frac{1}{2}}$, rms thermal velocity,

$\beta_{pp} \triangleq$ electron plasma frequency for electrons in the plasma, and

$\Gamma \triangleq$ complex propagation constant.

Lampert[9] has shown that for amplification the signal frequency must be less than some critical frequency, $\omega < \omega_{cr}$, as given by

$$\omega_{cr} = \omega_{pp}\left[1 + \left(\frac{v_T}{u_0}\right)\left(\frac{n_e}{n}\right)^{\frac{1}{2}}\right], \tag{4}$$

where n and n_e are electron concentrations in the plasma and beam respectively. The maximum gain occurs at (near the plasma frequency)

$$\omega = \omega_{pp}\left[1 - \left(\frac{v_T}{u_0}\right)\left(\frac{n_e}{n}\right)^{\frac{1}{2}}\right] \tag{5}$$

and is given by

$$G_{\max} = \left(\frac{u_0}{v_T}\right)^{\frac{1}{2}}\left(\frac{n_e}{n}\right)^{\frac{1}{4}}. \tag{6}$$

Field analyses along the lines of the Chu and Jackson field analysis of the traveling-wave amplifier have been made by many authors for wave propagation on a plasma column interacting with an electron beam. An extensive bibliography of articles on this subject is given both at the end of this chapter[10-66] and at the end of Chapter I. Most of the studies are small signal and thereby yield information only on the system dispersion equation and not on the energy exchange between the system elements. Also most authors neglect collision effects due to the difficulty of formulating a useful and manageable form for the collision integral. Lim[68] has treated the beam-plasma interaction problem quite generally using a linear theory and has accounted for plasma binary collisions in determining the dispersion equation.

The basic Lagrangian nonlinear method utilized in previous chapters can also be adapted to the study of both double-beam interactions and beam-plasma interactions. In this chapter a one-dimensional nonlinear analysis of these interaction configurations is made accounting for space-charge forces within each beam or plasma. The interest is primarily in the macroscopic energy exchange phenomena taking place and therefore microscopic collisions within each charge unit are ignored. It is not necessary to make the cold plasma approximation in this analysis.

2 Nonlinear Equations for Combined One-Dimensional Beam-Plasma Circuit

The general nonlinear Lagrangian theory developed here applies to the model illustrated in Fig. 2, where either beam may be made up of

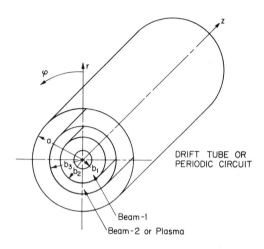

FIG. 2. Beam-plasma-circuit interaction configuration. $B_{10} \triangleq \beta_1 b_0$, $B_{20} \triangleq \beta_2 b_0$, $B_{22} \triangleq \beta_2 b_2$, $B_{23} \triangleq \beta_2 b_3$, $B_{11} \triangleq \beta_1 b_1$, $B_{1a} \triangleq \beta_1 a$, $B_{2a} \triangleq \beta_2 a$.

electrons or be an ideal plasma. The combination might also be surrounded by a wave propagating circuit. A one-dimensional finite-diameter axially symmetric system is assumed. The plasma field is also assumed to be fully ionized and any effects of neutrals on the interaction process are neglected. Nonrelativistic particle and wave velocities are also assumed.

a. Definitions

Since it is desired to allow each of the charge fields of Fig. 2 to have an arbitrary sign we define appropriate charge-to-mass ratios as

$$\eta_1 \triangleq \epsilon_1 \mid \eta_1 \mid \tag{7a}$$

and

$$\eta_2 \triangleq \epsilon_2 \mid \eta_2 \mid, \tag{7b}$$

where ϵ_1, $\epsilon_2 = \pm 1$ according to the negative or positive charge of the particular field. The following fundamental definitions pertain to

2. NONLINEAR EQUATIONS

combined charge-to-mass ratios and the respective "charge field" currents:

$$\eta_0 \triangleq \frac{|\eta_1| + |\eta_2|}{2}, \quad \eta_0 > 0 \tag{8a}$$

and

$$\eta' \triangleq \frac{|\eta_1| - |\eta_2|}{2\eta_0}, \quad -1 < \eta' < 1. \tag{8b}$$

Solving for η_1 and η_2 in these expressions yields

$$\eta_1 = \epsilon_1 \eta_0 (1 + \eta') \tag{9a}$$

and

$$\eta_2 = \epsilon_2 \eta_0 (1 - \eta'). \tag{9b}$$

The dc field currents are given by

$$I_{01} = \epsilon_1 |I_{01}| \tag{10a}$$

and

$$I_{02} = \epsilon_2 |I_{02}|. \tag{10b}$$

Define

$$I_0 \triangleq \frac{|I_{01}| + |I_{02}|}{2}, \quad I_0 > 0 \tag{11a}$$

and

$$I' \triangleq \frac{|I_{01}| - |I_{02}|}{2I_0}, \quad -1 < I' < 1 \tag{11b}$$

such that

$$I_{01} = \epsilon_1 I_0 (1 + I') \tag{12a}$$

and

$$I_{02} = \epsilon_2 I_0 (1 - I'). \tag{12b}$$

The average velocity of the two drifting charge fields is given by

$$u_0 \triangleq \frac{u_{01} + u_{02}}{2}, \quad u_0 > 0 \tag{13}$$

and a relative velocity parameter is then defined as

$$b \triangleq \frac{u_{01} - u_{02}}{2u_0} \quad |b| \neq 1. \tag{14}$$

Thus the individual initial average velocities are given by

$$u_{01} = u_0(1 + b) \tag{15a}$$

and

$$u_{02} = u_0(1 - b). \tag{15b}$$

Since a one-dimensional system is assumed any possible potential depression across the charge field is ignored.

b. Rf Circuit Equation

As in previous one-dimensional analyses, the presence of an rf propagating circuit is accounted for by the following one-dimensional equivalent-circuit transmission-line equation:

$$\frac{\partial^2 V(z, t)}{\partial t^2} - v_0^2 \frac{\partial^2 V(z, t)}{\partial z^2} = \pm v_0 Z_0 \frac{\partial^2 \rho(z, t)}{\partial t^2}, \tag{16}$$

where possible circuit loss has been neglected; the plus sign on the right of Eq. (16) assumes a forward space-harmonic interaction and the minus sign a backward-wave interaction. The rf voltage function is again written as the product of slowly varying functions

$$V(z, t) = V(y, \Phi) \triangleq \mathrm{Re}\left[\frac{Z_0 I_0}{C} A(y) e^{-j\Phi}\right], \tag{17}$$

where

and

$$C^3 \triangleq \frac{\eta_0 I_0 Z_0}{2u_0^2}$$

$$D^2 \triangleq \left(\frac{v_0}{u_0}\right)^2 C^2.$$

It is convenient to introduce another velocity parameter which indicates the difference in the average stream velocity u_0 and the circuit velocity v_0. This parameter is defined by

$$D^2 = \frac{C^2}{(1 + Cp)^2},$$

where

$$\frac{u_0}{v_0} = 1 + Cp.$$

2. NONLINEAR EQUATIONS

The normalized axial-distance and particle phase-position variables are defined in a similar fashion to those in Chapters V, VI, and VII.

$$y \triangleq \frac{C\omega}{u_0} z = 2\pi C N_s \tag{18}$$

and

$$\Phi(z, t) \triangleq \omega \left(\frac{z}{u_0} - t \right) - \theta(z), \tag{19}$$

where $N_s = z/\lambda_s = zf/u_0$. The definitions are completed by introducing the particle velocities as

$$\left. \frac{dz}{dt} \right|_1 = v_1 \triangleq u_{01}[1 + 2Cu_1(y, \Phi_0)] \tag{20a}$$

and

$$\left. \frac{dz}{dt} \right|_2 = v_2 \triangleq u_{02}[1 + 2Cu_2(y, \Phi_0)]. \tag{20b}$$

The values of $A(y)$, Φ, u_1, and u_2 that a particular charged particle sees at a given displacement plane depend upon the initial phase characterizing its entry into one of the charge fields. Define initial phases as $\Phi_{01,j}$ and $\Phi_{02,j}$ for the particles in the first and second charge fields, respectively. After expansion of ρ_1 and ρ_2, the charge densities associated with the individual beams or fields, into a Fourier series in the variable Φ we may express the *fundamental components* ρ_{11} and ρ_{21} of ρ_1 and ρ_2 as

$$\rho_{11} = \operatorname{Re} \frac{I_{01}}{u_{01}\pi} e^{-j\Phi} \left\{ \int_0^{2\pi} \frac{\cos \Phi' \, d\Phi'_{01}}{1 + 2Cu_1'} + j \int_0^{2\pi} \frac{\sin \Phi' \, d\Phi'_{01}}{1 + 2Cu_1'} \right\}$$

and

$$\rho_{21} = \operatorname{Re} \frac{I_{02}}{u_{02}\pi} e^{-j\Phi} \left\{ \int_0^{2\pi} \frac{\cos \Phi'' \, d\Phi''_{02}}{1 + 2Cu_2''} + j \int_0^{2\pi} \frac{\sin \Phi'' \, d\Phi''_{02}}{1 + 2Cu_2''} \right\}, \tag{21}$$

where

$$\Phi' \triangleq \Phi_1(y, \Phi'_{01}), \qquad \Phi'' \triangleq \Phi_2(y, \Phi''_{02}), \qquad u_1' \triangleq u_1(y, \Phi'_{01})$$

and

$$u_2'' \triangleq u_2(y, \Phi''_{02}). \tag{22}$$

Equations (21) and (22) are now substituted into the circuit differential equation and the resulting equation is separated into the following two inhomogeneous circuit equations:

$$A(y) + D^2 \frac{d^2 A(y)}{dy^2} - D^2 \left(\frac{1}{C} - \frac{d\theta(y)}{dy}\right)^2 A(y)$$

$$= \frac{\epsilon_1 D(1+I')}{\pi(1+b)} \int_0^{2\pi} \frac{\cos \Phi' \, d\Phi'_{01}}{1 + 2Cu_1'} + \frac{\epsilon_2 D(1-I')}{\pi(1-b)} \int_0^{2\pi} \frac{\cos \Phi'' \, d\Phi''_{02}}{1 + 2Cu_2''} \quad (23)$$

and

$$D \frac{d^2\theta(y)}{dy^2} A(y) - 2D \left(\frac{1}{C} - \frac{d\theta(y)}{dy}\right) \frac{dA(y)}{dy}$$

$$= \frac{\epsilon_1(1+I')}{\pi(1+b)} \int_0^{2\pi} \frac{\sin \Phi' \, d\Phi'_{01}}{1 + 2Cu_1'} + \frac{\epsilon_2(1-I')}{\pi(1-b)} \int_0^{2\pi} \frac{\sin \Phi'' \, d\Phi''_{02}}{1 + 2Cu_2''}. \quad (24)$$

Velocity-Phase Equations. After taking the appropriate differentials of the dependent phase variables the following relations between Φ and u are evolved:

$$\frac{\partial \Phi_1(y, \Phi_{01})}{\partial y} + \frac{d\theta(y)}{dy} = \frac{1}{C}\left[1 - \frac{1}{(1+b)[1 + 2Cu_1(y, \Phi_{01})]}\right] \quad (25)$$

and

$$\frac{\partial \Phi_2(y, \Phi_{02})}{\partial y} + \frac{d\theta(y)}{dy} = \frac{1}{C}\left[1 - \frac{1}{(1-b)[1 + 2Cu_2(y, \Phi_{02})]}\right]. \quad (26)$$

Each of the above equations is m in number when m individual charge groups are considered.

Lorentz Force Equations. We have assumed that all microscopic collision effects may be neglected and since all wave and particle velocities are small compared to the velocity of light the particle self-magnetic field is negligible and thus for confined one-dimensional flow the Lorentz force equations become

$$\frac{dv_1}{dt} = \eta_1 \left[-\frac{\partial V_c}{\partial z} + E_{scz-1}\right] \quad (27)$$

and

$$\frac{dv_2}{dt} = \eta_2 \left[-\frac{\partial V_c}{\partial z} + E_{scz-2}\right], \quad (28)$$

where E_{scz-1} and E_{scz-2} represent the total space-charge fields acting

2. NONLINEAR EQUATIONS

upon each particle individually. The previously defined normalized variables are used to write Eqs. (27) and (28) as

$$[1 + 2Cu_1(y, \Phi_{01})] \frac{\partial u_1(y, \Phi_{01})}{\partial y} = -\frac{\epsilon_1 C(1 + \eta')}{(1 + b)^2} \left[\frac{dA(y)}{dy} \cos \Phi_1(y, \Phi_{01}) \right.$$

$$\left. -A(y)\left(\frac{1}{C} - \frac{d\theta(y)}{dy}\right) \sin \Phi_1(y, \Phi_{01}) \right] + \frac{\epsilon_1(1 + \eta')}{(1 + b)^2} \left(\frac{\eta_0}{2C^2 \omega u_0}\right) E_{scz-1} \quad (29)$$

and for the second charge field

$$[1 + 2Cu_2(y, \Phi_{02})] \frac{\partial u_2(y, \Phi_{02})}{\partial y} = -\frac{\epsilon_2 C(1 - \eta')}{(1 - b)^2} \left[\frac{dA(y)}{dy} \cos \Phi_2(y, \Phi_{02}) \right.$$

$$\left. -A(y)\left(\frac{1}{C} - \frac{d\theta(y)}{dy}\right) \sin \Phi_2(y, \Phi_{02}) \right] + \frac{\epsilon_2(1 - \eta')}{(1 - b)^2} \left(\frac{\eta_0}{2C^2 \omega u_0}\right) E_{scz-2}. \quad (30)$$

The space-charge-field expressions are calculated from electrostatics for an axially symmetric system considering the charge fields to be interpenetrating disks of charge. In this analysis it is assumed that $\epsilon \approx \epsilon_0$ and that only two streams interact when calculating E_{sc}. It is convenient to define a mean radius for the outer charge field. Thus

$$b_0 \triangleq \frac{b_2 + b_3}{2}.$$

The total coulomb field is the sum of that produced by each charge field acting separately. Poisson's equation is written as

$$\nabla^2 V(r, z) = -\frac{\rho}{\epsilon_0 \alpha},$$

where $\alpha \triangleq$ cross-sectional area of the particular charge field.

The complementary Laplace equation must also be satisfied for $b_3 < r < a$ and the field matching at the various boundaries defines the eigenvalue problem, the eigenvalues being the space-charge-wave propagation constants designated as β_1 and β_2, respectively.

The harmonic method outlined in Chapter IV for deriving a one-dimensional space-charge-field expression is used again. If the charge density is expanded in a Fourier series in phase space, and the tangential E field is taken as zero at the drift-tube wall, the electric fields at $r = 0$ and b_0 are given by

$$E_1(0) = \frac{jA_1\rho_1}{\beta_1} \left\{ 1 - \frac{B_{11}}{I_0(B_{1a})} [K_1(B_{11})I_0(B_{1a}) + K_0(B_{1a})I_1(B_{11})] \right\}, \quad (31)$$

$$E_1(b_0) = \frac{jA_1\rho_1}{\beta_1} \left\{ B_{11}I_1(B_{11}) \left[K_0(B_{10}) - \frac{K_0(B_{1a})}{I_0(B_{1a})} I_0(B_{10}) \right] \right\}, \quad (32)$$

$$E_2(0) = \frac{jA_2\rho_2}{\beta_2} \left\{ \frac{B_{22}}{I_0(B_{2a})} [K_0(B_{2a})I_1(B_{22}) + K_1(B_{22})I_0(B_{2a})] \right.$$
$$\left. - \frac{B_{23}}{I_0(B_{2a})} [K_0(B_{2a})I_1(B_{23}) + K_1(B_{23})I_0(B_{2a})] \right\}, \tag{33}$$

and

$$E_2(b_0) = \frac{jA_2\rho_2}{\beta_2} \left\{ 1 + \frac{I_0(B_{20})K_0(B_{2a})}{I_0(B_{2a})} [B_{22}I_1(B_{22}) - B_{23}I_1(B_{23})] \right.$$
$$\left. - B_{23}I_0(B_{20})K_1(B_{23}) - B_{22}I_1(B_{22})K_0(B_{20}) \right\}, \tag{34}$$

where the following definitions have been utilized:

$$A_1 \triangleq \frac{1}{\epsilon_0 \alpha_1} \qquad A_2 \triangleq \frac{1}{\epsilon_0 \alpha_2} \qquad B_{11} \triangleq \beta_1 b_1,$$

$$B_{10} \triangleq \beta_1 b_0 \qquad B_{20} \triangleq \beta_2 b_0 \qquad B_{1a} \triangleq \beta_1 a,$$

and

$$B_{22} \triangleq \beta_2 b_2 \qquad B_{23} \triangleq \beta_2 b_3 \qquad B_{2a} \triangleq \beta_2 a.$$

Equations (32) and (34) give the coulomb fields at $r = b_0$ due to charge fields "1" and "2" respectively. The arguments of the Bessel functions contain the charge-field phase constants and the various radial boundary conditions. The respective phase constants are

$$\beta_1 \triangleq \frac{\omega}{u_{01}} \qquad \text{and} \qquad \beta_2 \triangleq \frac{\omega}{u_{02}}.$$

In view of the above geometrical dependence of the charge fields the following radian plasma-frequency reduction factors are defined:

$${}_{11}R_n^2 \triangleq 1 - \frac{(nB_{11})}{I_0(nB_{1a})} [K_1(nB_{11})I_0(nB_{1a}) + K_0(nB_{1a})I_1(nB_{11})], \tag{35}$$

$${}_{12}R_n^2 \triangleq (nB_{11})I_1(nB_{11}) \left[K_0(nB_{10}) - \frac{K_0(nB_{1a})}{I_0(nB_{1a})} I_0(nB_{10}) \right], \tag{36}$$

$${}_{21}R_n^2 \triangleq \frac{(nB_{22})}{I_0(nB_{2a})} [K_0(nB_{2a})I_1(nB_{22}) + K_1(nB_{22})I_0(nB_{2a})]$$
$$- \frac{(nB_{23})}{I_0(nB_{2a})} [K_0(nB_{2a})I_1(nB_{23}) + K_1(nB_{23})I_0(nB_{2a})], \tag{37}$$

and

$${}_{22}R_n^2 \triangleq 1 + \frac{I_0(nB_{20})K_0(nB_{2a})}{I_0(nB_{2a})} [I_1(nB_{22})(nB_{22}) - I_1(nB_{23})(nB_{23})]$$
$$- I_0(nB_{20})K_1(nB_{23})(nB_{23}) - I_1(nB_{22})K_0(nB_{20})(nB_{22}). \tag{38}$$

2. NONLINEAR EQUATIONS

In the above Lagrangian description of the system interaction, the charge densities, ρ_1 and ρ_2, are written in terms of the entering charge densities, respectively, after invoking the conservation of charge. The following expressions evolve:

$$\rho_1 = \frac{I_{01}}{u_{01}} \left| \frac{\partial \Phi_{01}}{\partial \Phi_1} \right| \frac{1}{[1 + 2Cu_1(y, \Phi_{01})]} \tag{39}$$

and

$$\rho_2 = \frac{I_{02}}{u_{02}} \left| \frac{\partial \Phi_{02}}{\partial \Phi_2} \right| \frac{1}{[1 + 2Cu_2(y, \Phi_{02})]}. \tag{40}$$

The one-dimensional space-charge-field weighting functions are now derived following the procedure outlined in Chapter IV, Section 12.

$$_{11}F(\Phi_1 - \Phi_1') \triangleq \sum_{n=1}^{\infty} \frac{\sin\left[n(\Phi_1 - \Phi_1')\left(\frac{1}{1+b}\right)\right] {}_{11}R_n^2}{2\pi n},$$

$$_{21}F(\Phi_1 - \Phi_2'') \triangleq \sum_{n=1}^{\infty} \frac{\sin\left[n(\Phi_1 - \Phi_2'')\left(\frac{1}{1-b}\right)\right] {}_{21}R_n^2}{2\pi n},$$

$$_{12}F(\Phi_2 - \Phi_1') \triangleq \sum_{n=1}^{\infty} \frac{\sin\left[n(\Phi_2 - \Phi_1')\left(\frac{1}{1+b}\right)\right] {}_{12}R_n^2}{2\pi n},$$

and

$$_{22}F(\Phi_2 - \Phi_2'') \triangleq \sum_{n=1}^{\infty} \frac{\sin\left[n(\Phi_2 - \Phi_2'')\left(\frac{1}{1-b}\right)\right] {}_{22}R_n^2}{2\pi n}, \tag{41}$$

where the symbols Φ_2'' and Φ_1' mean "taking the phase of the particles in the second (first) beam." Now define the appropriate radian plasma frequencies as

$$\omega_{p1}^2 \triangleq \frac{\eta_1 I_{01}}{\epsilon_0 \alpha_1 u_{01}} \quad \text{and} \quad \omega_{p2}^2 \triangleq \frac{\eta_2 I_{02}}{\epsilon_0 \alpha_2 u_{02}}.$$

The complete forms for the electric fields associated with the charge fields are now written as

$$E_{scz-1} = \frac{2\epsilon_1 u_{01}}{(1 + \eta')(1 + b)} \left(\frac{\omega_{p1}^2}{\eta_0 \omega}\right) \int_0^{2\pi} \frac{{}_{11}F(\Phi_1 - \Phi_1') \, d\Phi_{01}'}{[1 + 2Cu_1(y, \Phi_{01}')]}$$

$$+ \frac{2\epsilon_2 u_{01}}{(1 - \eta'))1 - b)} \left(\frac{\omega_{p2}^2}{\eta_0 \omega}\right) \int_0^{2\pi} \frac{{}_{21}F(\Phi_1 - \Phi_2'') \, d\Phi_{02}''}{[1 + 2Cu_2(y, \Phi_{02}'')]} \tag{42}$$

and

$$E_{scz-2} = \frac{2\epsilon_1 u_{02}}{(1+\eta')(1+b)} \left(\frac{\omega_{p1}^2}{\eta_0 \omega}\right) \int_0^{2\pi} \frac{{}_{12}F(\Phi_2 - \Phi_1') \, d\Phi_{01}'}{[1 + 2Cu_1(y, \Phi_{01}')]}$$
$$+ \frac{2\epsilon_2 u_{02}}{(1-\eta')(1-b)} \left(\frac{\omega_{p2}^2}{\eta_0 \omega}\right) \int_0^{2\pi} \frac{{}_{22}F(\Phi_2 - \Phi_2'') \, d\Phi_{02}''}{[1 + 2Cu_2(y, \Phi_{02}'')]}. \quad (43)$$

Substitution of the above expressions for the fields and the plasma frequencies into Eqs. (29) and (30) yields the final form of the force equations:

$$[1 + 2Cu_1(y, \Phi_{01})] \frac{\partial u_1(y, \Phi_{01})}{\partial y} = -\frac{\epsilon_1 C(1+\eta')}{(1+b)^2} \left[\frac{dA(y)}{dy} \cos \Phi_1(y, \Phi_{01})\right.$$
$$\left. - A(y) \left(\frac{1}{C} - \frac{d\theta(y)}{dy}\right) \sin \Phi_1(y, \Phi_{01})\right] + \frac{1}{(1+b)^2} \left(\frac{\omega_{p1}}{\omega C}\right)^2$$
$$\times \int_0^{2\pi} \frac{{}_{11}F(\Phi_1 - \Phi_1') \, d\Phi_{01}'}{[1 + 2Cu_1(y, \Phi_{01}')]} + \frac{\epsilon_1 \epsilon_2 (1+\eta')}{(1+b)^2(1-\eta')} \left(\frac{\omega_{p2}}{\omega C}\right)^2 \int_0^{2\pi} \frac{{}_{21}F(\Phi_1 - \Phi_2'') \, d\Phi_{02}''}{[1 + 2Cu_2(y, \Phi_{02}'')]}$$

(44)

and

$$[1 + 2Cu_2(y, \Phi_{02})] \frac{\partial u_2(y, \Phi_{02})}{\partial y} = -\frac{\epsilon_2 C(1-\eta')}{(1-b)^2} \left[\frac{dA(y)}{dy} \cos \Phi_2(y, \Phi_{02})\right.$$
$$\left. - A(y) \left(\frac{1}{C} - \frac{d\theta(y)}{dy}\right) \sin \Phi_2(y, \Phi_{02})\right] + \frac{1}{(1-b)^2} \left(\frac{\omega_{p2}}{\omega C}\right)^2$$
$$\times \int_0^{2\pi} \frac{{}_{22}F(\Phi_2 - \Phi_2'') \, d\Phi_{02}''}{[1 + 2Cu_2(y, \Phi_{02}'')]} + \frac{\epsilon_1 \epsilon_2 (1-\eta')}{(1-b)^2(1+\eta')} \left(\frac{\omega_{p1}}{\omega C}\right)^2 \int_0^{2\pi} \frac{{}_{12}F(\Phi_2 - \Phi_1') \, d\Phi_{01}'}{[1 + 2Cu_1(y, \Phi_{01}')]}.$$

(45)

The above system of equations, (23)–(26), (44), and (45), constitutes the nonlinear equations for the beam-plasma or double-beam system including the presence of rf circuit fields.

The solution of the above system of equations is obtained using the same methods as in Chapters V, VI, and VII. It is solved as an initial-value problem where the dependent variables and their derivatives are specified at the input plane. It is also necessary to specify the following system operating parameters:

$C, D \triangleq$ gain and coupling parameters,

$\epsilon_1, \epsilon_2 \triangleq$ charge signs for individual fields,

$I' \triangleq$ normalized current,

$\eta' \triangleq$ normalized charge-to-mass ratio,

$b \triangleq$ stream relative injection velocity parameter,

$p \triangleq$ circuit-stream relative velocity parameter, and

$\left.\begin{array}{l}B_{11}, B_{10}, B_{20}, B_{1a}, \\ B_{22}, B_{23}, B_{2a}\end{array}\right\} \triangleq$ radian beam or plasma diameters.

The radian plasma-frequency reduction factors which determine the effective plasma frequency are shown in Fig. 3. The space-charge-field weighting functions $_{11}F$, $_{12}F$, $_{21}F$, and $_{22}F$ indicate the forces between various charge groups as a function of the difference of their phase positions. These functions are shown in Figs. 4–7 for a particular geometrical configuration.

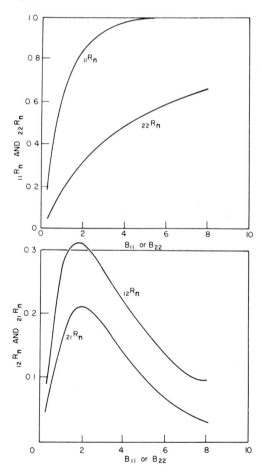

FIG. 3. Plasma-frequency reduction factors for a double-charge-field configuration.

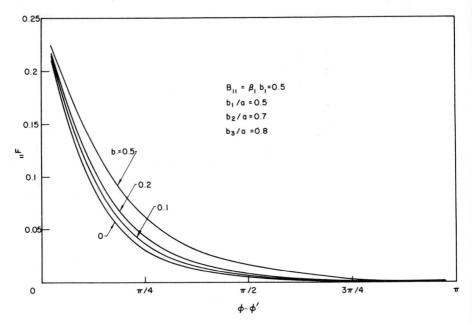

Fig. 4. Space-charge-field weighting function $_{11}F$.

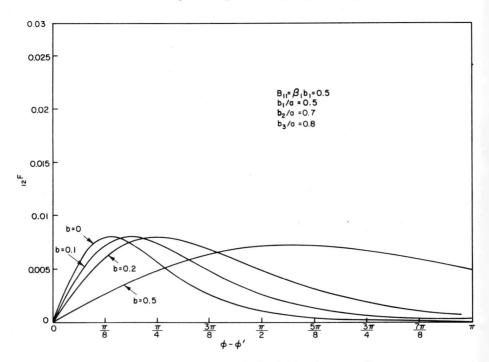

Fig. 5. Space-charge-field weighting function $_{12}F$.

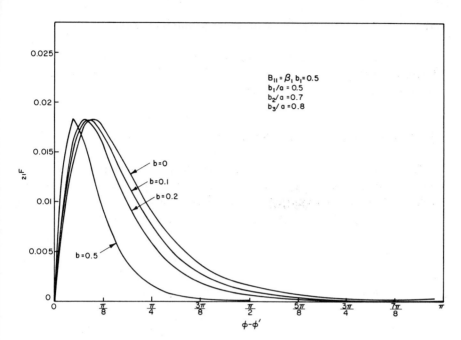

FIG. 6. Space-charge-field weighting function $_{21}F$.

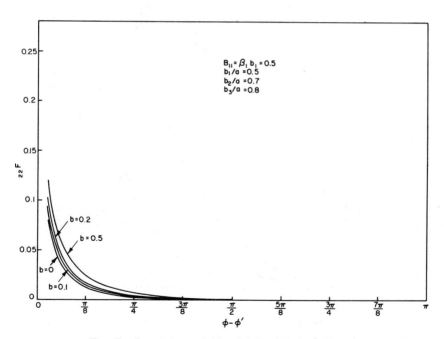

FIG. 7. Space-charge-field weighting function $_{22}F$.

410 XII. MULTIBEAM AND BEAM-PLASMA INTERACTIONS

Solutions of the above nonlinear equations for both double-beam and beam-plasma systems are outlined and discussed in the following sections of this chapter.

3 Double-Beam Circuit Solutions

In evaluating a double-beam interaction in the presence of a propagating rf circuit wave the principal operating parameters to vary are the relative stream velocity, b, the beam-circuit relative velocity, p, and the

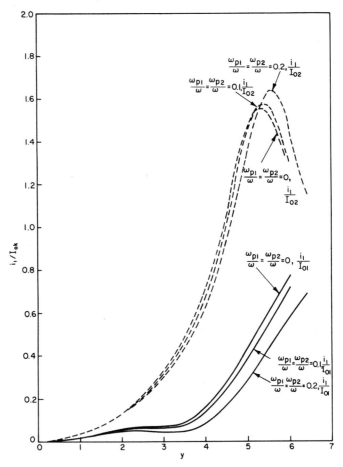

FIG. 8. Double beam and circuit wave. Fundamental component of beam current versus distance and space charge. ($C = 0.1$, $b = 0.1$, $p = b$, $\epsilon_1 = \epsilon_2 = -1$, $\varGamma' = 0$, $\eta' = 0$, $\psi_0 = -30$.)

space-charge parameter ω_p/ω. The fundamental rf current amplitude plots of Figs. 8 and 9 indicate, as expected, that the current reaches a maximum first in one stream and later in the other. The maximum value of approximately 1.6 is somewhat higher than that typical of conventional traveling-wave amplifiers as shown in Chapter VI.

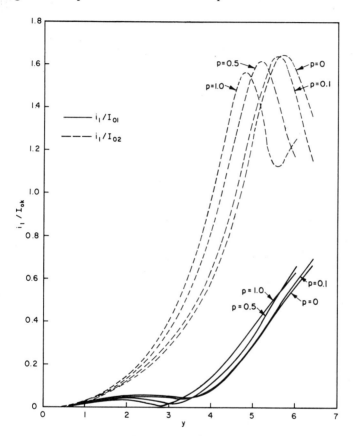

FIG. 9. Double beam and circuit wave. Fundamental current amplitude versus distance and p. ($C = 0.1$, $b = 0.1$, $\epsilon_1 = \epsilon_2 = -1$, $I' = \eta' = 0$, $\psi_0 = -30$, $\omega_{p1}/\omega = \omega_{p2}/\omega = 0.2$.)

The gain and fundamental current amplitude in a double-beam plus circuit-wave device have been examined for a range of the beam-circuit relative velocity parameter and it was found that varying both I' and η' over a reasonable range had little effect on either the gain or the efficiency.

Since the two beams may have both different average currents I_{0k} and different velocities u_{0k}, a more general efficiency expression than that used for a single-beam device must be developed for the double-beam interaction. The total power carried by the two beams is written as follows:

$$P_b = \frac{u_{01}^2}{2\eta_1} I_{01} + \frac{u_{02}^2}{2\eta_2} I_{02} = I_0 V_0 \left[\frac{(1+b)^2(1+I')}{(1+\eta')} + \frac{(1-b)^2(1-I')}{(1-\eta')} \right]. \quad (46)$$

From a derivation procedure quite similar to that used in Chapter VI, the rf conversion efficiency for the double-beam plus circuit-wave device is written as

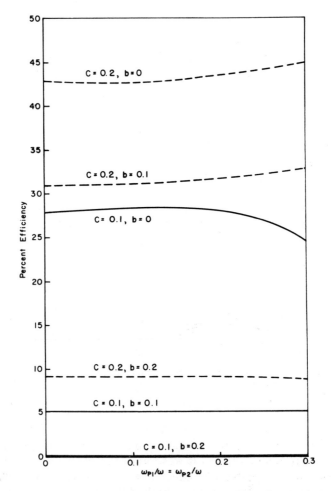

FIG. 10. Efficiency versus ω_p/ω for a double beam and circuit wave ($\epsilon_1 = \epsilon_2 = -1$, $I' = \eta' = 0$, $\psi_0 = -30$, $p = b$).

3. DOUBLE-BEAM CIRCUIT SOLUTIONS

$$\eta_e \triangleq \frac{2C[A^2(y) - A_0^2]\dfrac{(1 - Cd\theta/dy)}{(1 + Cb)}}{\left[\dfrac{(1 + b)^2(1 + I')}{(1 + \eta')} + \dfrac{(1 - b)^2(1 - I')}{(1 - \eta')}\right]} . \quad (47)$$

Clearly Eq. (47) reduces to the proper form as I', η', and b tend to zero.

A summary of the efficiency results for various relative beam velocities and $p = b$ is shown in Fig. 10. The velocity spread between the streams clearly reduces both the gain and the efficiency. The $b = 0$ results are effectively those for a single-beam device. Holding the beam relative velocity b fixed and varying the beam-circuit relative velocity p gives the efficiency results of Fig. 11. Again a large velocity spread is deleterious

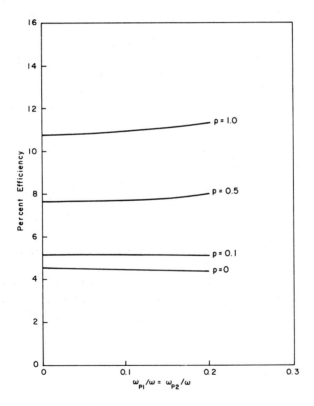

FIG. 11. Double beam and circuit wave. Efficiency versus ω_p/ω and p ($C = 0.1$, $b = 0.1$, $\epsilon_1 = \epsilon_2 = -1$, $I' = \eta' = 0$, $\psi_0 = -30$).

to the gain mechanism. It is easily shown that when there is an rf wave present the performance of a double-beam device is no better than that of a single-beam device. This is independent of the charge signs, i.e., either for $\epsilon_1 = \epsilon_2$ or for $\epsilon_1 \neq \epsilon_2$.

4 Interaction Equations in the Absence of a Circuit

a. Derivation of Equations

The general system analyzed in Section 2 is considerably simplified in the absence of an rf circuit. The circuit equations disappear and the others considerably simplify. The system is excited by the application of a modulation to either the double-beam or the beam-plasma system. This may be accomplished by using grids, cavities, or short-length propagating circuits. The simplification of the general system is facilitated using the following definitions: $\alpha \triangleq 2C$, a depth of modulation parameter, and $X \triangleq \pi \alpha N_s$. The final equations are then obtained from Eqs. (25), (26), (44) and (45). This exercise is left to the reader.

The above system of nonlinear double-beam and beam-plasma equations is quite similar to the klystron equations and is solved in a similar fashion. The following information denotes the essential features of each type of interaction.

b. Double-Beam Interaction

In the case of two electron beams interacting, i.e., $\epsilon_1 = \epsilon_2 = -1$, we would expect to obtain results much like the ones for the klystron. If $b = 0$, i.e., no velocity slip between beams, then the results are identical to those for the klystron as shown previously. If $b \gtrless 0$ then the current characteristics of the two beams are displaced with respect to one another as a result of the phasing of currents in the two beams. The harmonics are of lower amplitude and reach a maximum in a shorter distance from the velocity modulation cavity. The maximum value of i_1/I_{0k} (k denotes the particular beam) is approximately 1.16, which corresponds to the ballistic theory value of

$$\frac{i_1}{I_0} = 2J_1 \big|_{max}$$
$$= 2(0.58) = 1.16.$$

The distance scale is again plotted in terms of the bunching parameter $X = \pi \alpha N_s$, so that when $X \approx 1.84$ the initial velocity modulation has been converted to a current modulation, i.e., in $\lambda_q/4$ as predicted by the ballistic theory.

4. INTERACTION EQUATIONS IN THE ABSENCE OF A CIRCUIT

The effects of space charge, i.e., finite ω_p/ω, on the fundamental current amplitude is depicted in Fig. 12 for $b = 0.1$. The principal effect is to increase the fundamental current amplitude above the 1.16 value in one beam at the expense of the other beam. There is also a slight shifting of the position at which the current is a maximum; again the average is approximately that for the $b = 0$ case. Similar shifts appear when there is a velocity slip between the beams.

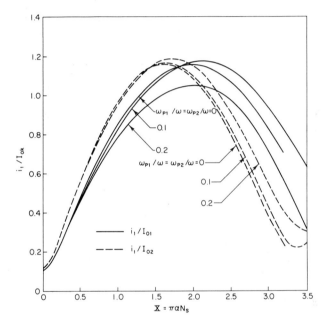

FIG. 12. Double-beam interaction. Effect of space charge and velocity slip on fundamental current amplitude. ($\epsilon_1 = \epsilon_2 = -1$, $\alpha = 0.2$, $I' = \eta' = 0$, $b = 0.1$.)

The characteristics of double-beam interactions are conveniently summarized in Fig. 13, where the maximum fundamental current amplitude is displayed as a function of ω_p/ω and b.

c. Beam-Plasma (Ions) Interaction

The characteristics of the beam-plasma (electron-ion) interaction are quite similar to those of the double-beam (electrons) case, the fundamental current amplitudes being in the vicinity of 1.16. The same equations apply to this problem and the difference in charge signs is accounted for by the specification of ϵ_1 and ϵ_2. In this section $\epsilon_1 = -1$ and $\epsilon_2 = 1$. Representative results as a function of the velocity slip b, the plasma

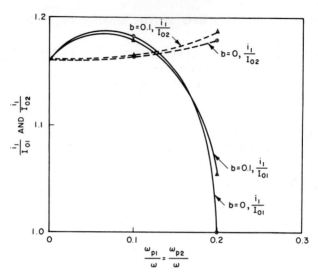

FIG. 13. Double-beam interaction. Fundamental current amplitude versus ω_p/ω. ($\epsilon_1 = \epsilon_2 = -1$, $\alpha = 0.2$, $I' = \eta' = 0$.)

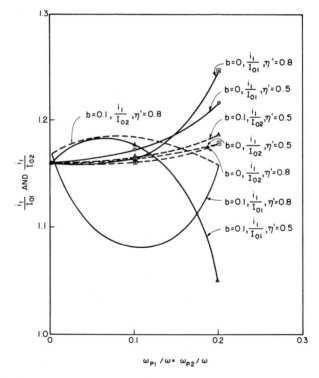

FIG. 14. Beam-plasma interaction. Fundamental current amplitude versus ω_p/ω, b, and η'. ($\epsilon_1 = -1$, $\epsilon_2 = +1$, $\alpha = 0.2$, $I' = 0$.)

frequency ω_p/ω, and the relative charge-to-mass ratio η' are summarized in Fig. 14. These data indicate that the fundamental current amplitude in the electron beam can rise well above the 1.16 value while the corresponding value for the ion beam is approximately 1.05. The exact values are quite dependent upon b, I', and η'.

5 Velocity Distributions

In the previous sections of this chapter the "cold plasma" approximation was made; i.e., the thermal spread in velocities was ignored. Such velocity distributions are important particularly in charge fields with low absolute drift velocities. The Lagrangian nature of the formulation conveniently allows the introduction of velocity distributions in any charge field by simply specifying the dependent velocity variable at the input plane.

The total charge group velocity was specified by

$$u_t(y, \Phi_{0j}) = u_0[1 + 2Cu(y, \Phi_{0j})]$$
$$= u_0[1 + \alpha u(y, \Phi_{0j})] \tag{48}$$

and hence an initial distribution of velocities may be incorporated by simply specifying $u(0, \Phi_{0j})$. It is convenient to define

$$u(y, \Phi_{0j}) \triangleq \frac{S}{2C}\left(\frac{n - \frac{1}{2}}{N} - \frac{1}{2}\right), \tag{49}$$

where $S \triangleq (v_2 - v_1)/u_0 = \Delta v/u_0$, and $N = $ the number of velocity classes with $n = 1, 2, ..., N$. Also define $m_i =$ the number of charge groups per velocity class. Thus

$$u_t(y, \Phi_{0j}) = u_0\left[1 + S\left(\frac{n - \frac{1}{2}}{N} - \frac{1}{2}\right)\right]. \tag{50}$$

Two of the most interesting and physically realizable velocity distributions are illustrated in Fig. 15.

If the velocity spread parameter $S \ll 1$ then the effect on the current amplitude in the beam would be small.

The velocity-phase curves show the relative motion induced in each beam due to the velocity spread (Figs. 16 and 17). Eventually this phenomenon could lead to severe beam breakup. Such a technique is useful in studying beam breakup and multistream amplification in low-

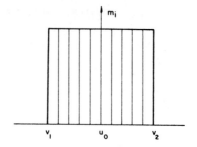

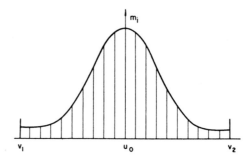

Fig. 15. Velocity distributions. m_i = the number of charge groups per velocity class.

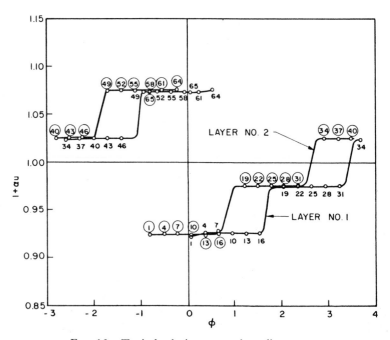

Fig. 16. Typical velocity versus phase diagram.

6. TWO-DIMENSIONAL EFFECTS IN BEAM-PLASMA INTERACTIONS

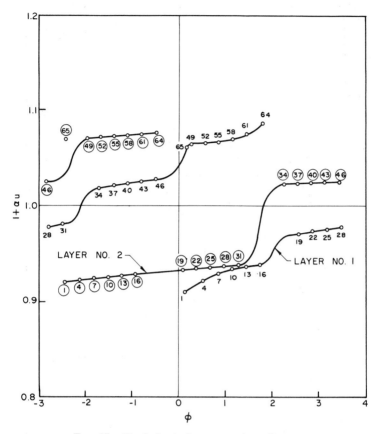

FIG. 17. Typical velocity versus phase diagram.

velocity beams such as exist in the electron gun of a low-noise amplifier. A summary of such velocity distribution calculations is given in Fig. 18.

6 Two-Dimensional Effects in Beam-Plasma Interactions

In both double-beam and beam-plasma interaction systems it is useful to explore the effects of radial rf field and space-charge-field variations on the gain and conversion efficiency. The development of appropriate nonlinear Lagrangian equations can proceed directly along the lines of Chapter VI and Section 2 of this chapter. Since the detailed procedures follow directly along the lines outlined previously they will not be treated here.

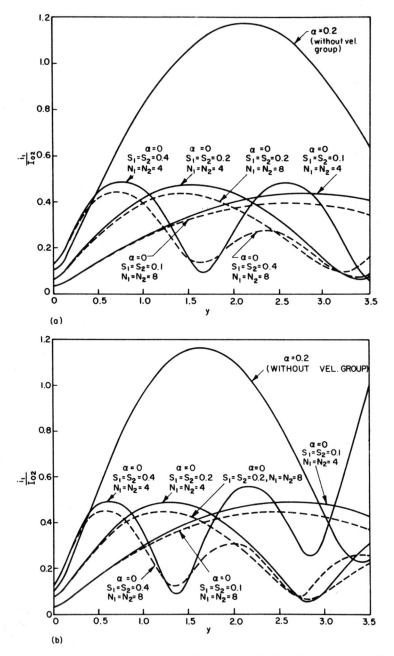

FIG. 18. Double-beam interaction. Velocity distribution characteristics. ($b = 0.1$, $\omega_{p1}/\omega = \omega_{p2}/\omega = 0.1$, $\epsilon_1 = \epsilon_2 = -1$, $I' = \eta' = 0$.) (a) i_1/I_{01}; (b) i_1/I_{02}.

References

1. Haeff, A. V., The electron-wave tube—A novel method of generation and amplification of microwave energy. *Proc. IRE* **37**, No. 1, 4-10 (1949).
2. Nergaard, L. S., Analysis of a simple model of a two-beam growing-wave tube. *RCA Rev.* **9**, No. 4, 585-602 (1948).
3. Pierce, J. R., Double-stream amplifiers. *Proc. IRE* **37**, No. 9, 980-986 (1949).
4. Pierce, J. R., and Hebenstreit, W. B., A new type of high-frequency amplifier. *Bell System Tech. J.* **28**, No. 1, 33-52 (1949).
5. Bloom, S., and Peter, R. W., Transmission-line analog of a modulated electron beam. *RCA Rev.* **15**, No. 1, 95-113 (1954).
6. Bernashevsky, G. A., Voronov, Z. S., Iziumova, T. I., and Tchernov, Z. S., Investigation of Double Stream Electron Wave Systems. *Proc. P.I.B. Symp. Electron. Waveguides* **8**, 249-254 (1958).
7. Bogdanov, E. V., Kislov, V. J., and Tchernov, Z. S., Interaction between an electron stream and plasma. *Proc. P.I.B. Symp. Millimeter Waves* **9**, 57-73 (1959).
8. Boyd, G. D., Field, L. M., and Gould, R. W., Excitation of plasma oscillations and growing plasma waves. *Phys. Rev.* **109**, 1393-1394 (1958).
9. Lampert, M. A., Plasma oscillations at extremely high frequencies. *J. Appl. Phys.* **27**, No. 1, 5-12 (1956).
10. Kino, G. S., and Allen, M. A., "The Effects of Fluctuations on Propagation through a Plasma Medium." Stanford Univ. Microwave Lab. Tech. Rept. No. 802 (April 1961); *Proc. 5th Intern. Conf. Ionization Phenomena in Gases, Munich, 1961*. North Holland, Amsterdam, 1962.
11. Boyd, G. D., "Experiments on the Interaction of a Modulated Electron Beam with a Plasma." Calif. Inst. Technol. Electron Tube and Microwave Lab. Tech. Rept. No. 11 (May 1959).
12. Boyd, G. D., Field, L. M. and Gould, R. W., Interaction between an electron stream and an arc discharge plasma. *Proc. Symp. Electronic Waveguides, Brooklyn Polytech. Inst., 1958* **8**, 367-375. Wiley (Interscience), New York, 1958.
13. Sturrock, P. A., Nonlinear effects in electron plasmas. *Proc. Roy. Soc. (London)* **A242**, 277-299 (1957).
14. Sturrock, P. A., A variation principle on an energy theorem for small-amplitude disturbances of electron beams and electron-ion plasmas. *Ann. Physik* [7] **4**, 306-324 (1958).
15. Sturrock, P. A., Action-transfer of frequency-shift relations in the nonlinear theory of waves and oscillations. *Ann. Physik* [7] **9**, 422-434 (1960).
16. Sturrock, P. A., Nonlinear effects in electron plasmas. *J. Nucl. Energy, Pt. C* **2**, 158-163 (1961).
17. Gould, R. W., and Trivelpiece, A. W., Electro-mechanical modes in plasma waveguides. *Proc. IEE (London)* **B-105**, Suppl. 10, 516-519 (1958).
18. Gould, R. W., and Trivelpiece, A. W., Space charge waves in cylindrical plasma columns. *J. Appl. Phys.* **30**, 1784-1792 (1959).
19. Smullin, L. D., and Chorney, P., Wave propagation in ion-plasma loaded waveguides. *Proc. Symp. Electronic Waveguides, Brooklyn Polytech. Inst., 1958* **8**, 229-247. Wiley (Interscience), New York, 1958.
20. Fainberg, Y. B., and Gorbatenko, M. F., Electromagnetic waves in a plasma situated in a magnetic field. *J. Tech. Phys. USSR* **29**, 487-500 (1959); Translation in *Soviet Phys.—Tech. Phys.* **4**, 487-500 (1959).

21. Schumann, W. O., On the propagation of electric waves along a dielectrically confined plasma layer. *Z. Angew. Phys.* **10**, 26-31 (1958).
22. Schumann, W. O., On wave propagation in a plasma between two perfectly conducting planes in an externally impressed magnetic field. *Z. Angew. Phys.* **8**, 482-485 (1956).
23. Pierce, J. R., and Morrison, J. A., Disturbances in a multivelocity plasma. *IRE Trans. Electron Devices* **6**, 231-236 (1959).
24. Walker, L. R., The dispersion formula for plasma waves. *J. Appl. Phys.* **25**, 131-132 (1954).
25. Vlasov, A. A., M.S. Thesis, *Zh. Eksperim. i Teor. Fiz.* **8**, 291 (1938).
26. Vlasov, A. A., On the kinetic theory of an assembly of particles with collective interaction. *J. Phys. USSR* **9**, 25-40 (1945).
27. Van Kampen, N. G., On the theory of stationary waves in plasmas. *Physica* **21**, 949-963 (1955).
28. Van Kampen, N. G., The dispersion equation for plasma waves. *Physica* **23**, 641-650 (1957).
29. Bohm, D., and Gross, E. P., Theory of plasma oscillations, A: Origin of medium-like behavior. *Phys. Rev.* **75**, 1851-1864 (1949).
30. Bohm, D., and Gross, E. P., Theory of plasma oscillations, B: Excitation and damping of oscillations. *Phys. Rev.* **75**, 1864-1876 (1949).
31. Bohm, D., and Gross, E. P., Effects of plasma boundaries in plasma. *Phys. Rev.*, **79**, 992-1001 (1950).
32. Bernstein, I. B., Greene, J. M., and Kruskal, M. D., Exact nonlinear plasma oscillations. *Phys. Rev.* **108**, 546-550 (1957).
33. Bohm, D., and Pines, D., A collective description of electron interactions, I: Magnetic interactions. *Phys. Rev.* **82**, 625-634 (1951).
34. Bohm, D., and Pines, D., A collective description of electron interactions, II: Collective vs. individual particle aspects of the interactions. *Phys. Rev.* **85**, 338-353 (1952).
35. Bohm, D., and Pines, D., A collective description of electron interactions, III: Coulomb interactions in a degenerate electron gas. *Phys. Rev.* **92**, 609-625 (1953).
36. Twiss, R. Q., On Bailey's theory of amplified circularly polarized waves in an ionized medium. *Phys. Rev.* **84**, 448-457 (1951).
37. Piddington, J. H., Growth electric space-charge waves. *Australian J. Phys.* **9**, 31-43 (1956).
38. Piddington, J. H., Growing electric space-charge waves and Haeff's electron-wave tube. *Phys. Rev.* **101**, 14-16 (1956).
39. Bernstein, I. B., Waves in a plasma in a magnetic field. *Phys. Rev.* **109**, 10-21 (1958).
40. Case, K. M., Plasma oscillations. *Ann. Physik* [7] **7**, 349-364 (1959).
41. Pierce, J. R., Possible fluctuations in electron streams due to ions. *J. Appl. Phys.* **19**, 231-241 (1948).
42. Buneman, O., How to distinguish between attenuating and amplifying waves. In *Plasma Physics* (J. E. Drummond, ed.), 143-164. McGraw-Hill, New York, 1961.
43. Trivelpiece, A. W., "Slow Wave Propagation in Plasma Waveguides." California Inst. Technol. Electron Tube and Microwave Lab. Tech. Rept. No. 7 (May 1958).
44. Allen, M. A., and Kino, G. S., Interaction of an electron beam with a fully ionized plasma. *Phys. Rev. Letters* **6**, 163-165 (1961).
45. Akhiezer, A. I., and Fainberg, Y. B., High-frequency oscillations of an electron plasma. *Zh. Eksperim. i Teor. Fiz.* **21**, 1262-1269 (1951).
46. Feinstein, J., and Sen, H. K., Radio wave generation by multi-stream charge interaction. *Phys. Rev.* **83**, 405-412 (1951).

REFERENCES

47. Twiss, R. Q., Discussion on the generation of plasma oscillations. Presented at *Conf. Ionized Media, University College, London* (1951).
48. Sumi, M., Theory of excited plasma waves. *J. Phys. Soc. (Japan)* **13**, 1476-1485 (1958).
49. Filiminov, G. F., Growing wave propagation in a plasma. *Radiotekhn. i Elektron.* **4**, 75-87 (1959) (in Russian).
50. Demirkhanov, R. A., Gevorkov, A. K., and Popov, A. P., The interaction of a beam of charged particles with a plasma. *Proc. 4th Intern. Conf. Ionization Phenomena in Gases, Uppsala, 1959* Vol. 2, 665-669. North-Holland, Amsterdam, 1960.
51. Gordon, E. I., " 'Plasma Oscillations,' The Interaction of Electron Beams with Gas Discharge Plasmas." Ph.D. Dissertation, Phys. Dept., Mass. Inst. Technol. (January 1957).
52. Wehner, G., Electron plasma oscillations. *J. Appl. Phys.* **22**, 761-765 (1951).
53. Fried, B. D., and Gould, R. W., "On the Detection of Ion Oscillations in a Mercury Discharge." Space Tech. Labs., Los Angeles, Rept. No. STL-TR-60-0000-GR-413 (December 1960).
54. Crawford, F. W., Electrostatic sound wave modes in a plasma. *Phys. Rev. Letters* **6**, 663-665 (1961).
55. Bernstein, I. B., Frieman, E. A., and Kulsrud, R. M., Ion wave instabilities. *Phys. Fluids* **3**, 136-137 (1960).
56. Bernstein, I. B., and Kulsrud, R. M., Ion wave instabilities. *Phys. Fluids* **3**, 937-945 (1960).
57. Buneman, O., Instability, turbulence, and conductivity in current-carrying plasma. *Phys. Rev. Letters* **1**, 8-9 (1958).
58. Pierce, J. R., Possible fluctuations in electron streams due to ions. *J. Appl. Phys.* **19**, 231-236 (1948).
59. Hernqvist, K. G., Plasma ion oscillations in electron beams. *J. Appl. Phys.* **26**, 544-548 (1955).
60. Ettenberg, M., and Targ, R., Observations of plasma and cyclotron oscillations. *Proc. Symp. Electronic Waveguides, Brooklyn Polytech. Inst. 1958* **8**, 379-388. Wiley (Interscience), New York, 1958.
61. Chorney, P., "Electron-Stimulated Ion Oscillations." M.I.T. Res. Lab. Electronics Tech. Rept. No. 277 (May 1958).
62. Revans, R. W., The transmission of waves through an ionized gas. *Phys. Rev.* **44**, 798-802 (1933).
63. Crawford, F. W., and Lawson, J. D., "Some Measurements of Fluctuations in a Plasma." Stanford Univ. Microwave Lab. Tech. Rept. No. ML-753 (October 1960); *J. Nucl. Energy, Pt. C* **3**, No. 3, 179-185 (1961).
64. Allen, M. A., and Kino, G. S., Beam plasma amplifiers. Presented at *IRE WESCON, San Francisco, 1961*.
65. Anderson, J. M., Possible low noise beam-plasma amplifier. *J. Appl. Phys.* **30**, 1624-1625 (1959).
66. Kino, G. S., A proposed millimeter wave generator. *Proc. Symp. Millimeter Waves, Brooklyn Polytech. Inst. 1959* **9**, 233-248. Wiley (Interscience), New York, 1959.
67. *Proc. IRE* **49**, No. 12: Plasmas (1961).
68. Lim, Y. C., "Linear Theory of Beam-Plasma Interaction." Univ. of Michigan Elec. Phys. Lab. Tech. Rept. No. 60 (March 1963).

CHAPTER

XIII | Phase Focusing of Electron Bunches

1 Introduction

In the previous chapters the interaction of an electron stream with a wave-carrying medium, such as an rf structure or plasma, has been analyzed for a number of geometrical configurations and static field arrangements. The results indicate the detailed motion of charge groups through the field configurations and tell the amount of initial dc energy converted to rf energy. Even simplified one-dimensional analyses indicate that conversion efficiencies fall far short of the desired 100%.

In general crossed-field devices which depend on a potential energy conversion process are more efficient than the so-called linear-beam devices in which kinetic energy is converted to rf. The principal reason for this is that in the crossed-field device the electron bunches become phase locked with the rf wave, and since their kinetic energy remains nearly constant they do not lose synchronism with the wave as dc energy is converted to rf. Thus it is wondered whether the energy conversion process in linear-beam devices might be improved by some phase-focusing technique, and also whether the efficiency of crossed-field devices might be improved by converting some of the kinetic energy of the charges. These questions are the basis of the consideration in this chapter, and in general the answer is shown to be "yes" in both cases cited. Phase focusing may be accomplished either by velocity tapering of the circuit or by application of a dc gradient to the flow. Both are examined for situations of interest. Considerable improvement can be accomplished although 100% conversion efficiency is still far off.

In making phase-focusing studies it is, of course, desirable to be able to solve the nonlinear equations in closed form so that the ideal focusing conditions can be found directly rather than by a roundabout trial and error process. If ideal bunches are assumed and space-charge forces are neglected, then it is possible to solve the nonlinear system directly. In this chapter the closed-form solutions are developed along with general computer solutions and the results compared. The approximate analytical treatment gives results much like the more general computer-determined results. The limit in improvement of efficiency is still found to be primarily due to velocity spread in the bunches resulting from non-uniform circuit fields and strong space-charge defocusing forces. Further over-all efficiency improvement by collector potential depression is discussed in Chapter XV.

2 Historical Background and Experimental Work

Almost since the invention of the traveling-wave amplifier, efforts have been made to improve both the electronic conversion efficiency and the over-all efficiency. It has been apparent from the results of the preceding calculations that the loss of a significant amount of electron kinetic energy leads to a loss of synchronism and a limit on the conversion efficiency. In Chapter XV it is shown that the over-all or plate circuit efficiency can be raised by collecting electrons at a low potential and saving energy normally dissipated as heat at the collector electrode. In this chapter we are concerned with improving the fundamental conversion efficiency.

Early suggestions by Pierce,[1] Slater,[2] and many others too numerous to list dealt with slowing the wave as the electrons gave up energy to maintain synchronism or speeding up the electrons by application of a dc gradient in order to improve the efficiency. The suggestions were basically sound, but unfortunately a nonlinear interaction theory was lacking so that little design or analysis could be carried out. Many early experiments were made with little or no success (due to improper tapering), although an experimental study by Cutler[3] did reveal several important facts and some improvement in operation. Cutler's approximate approach to determining an appropriate wave phase velocity variation was to assume in the nonlinear region that $i_1/I_0 = 1$ and calculate the power given up by the beam from the product of circuit field amplitude, beam current and bunch phase position. The power along the line is related to the electric field and hence from information on the rate of change of power along the line the velocity variation could be approximated. This approximate method suggested a slow and gradual taper (which is now known to be too slow) and only a slight improvement in absolute saturated output was obtained. The important result was that increased gain and power output resulted: the output power at a voltage for maximum gain was some 4 dB higher and at a voltage for maximum power the gain was increased by some 15 dB.

The advent of various nonlinear theories and high-speed computers resulted in a closer examination of the problem by many workers, using both linear and nonlinear theories and various experimental approaches. The primary objectives were to increase efficiency and phase linearity in forward-wave amplifiers and to increase efficiency and decrease starting current in backward-wave oscillators. Geppert[4] studied the backward-wave oscillator using a linear theory, neglecting space-charge forces, and assuming a linear velocity taper. Although he found that a decrease in starting current could occur he did not give the required form and in view of the linear theory used no information on efficiency could be obtained. A more comprehensive treatment of the small-signal problem has been given by Haddad-Bevensee[5] and Haddad-Rowe.[6]

Filimenov[7] investigated the effect of velocity tapering using the nonlinear equations of Vainshtein[8] and assuming $C \ll 1$. Hess,[9] utilizing the basic theories of Tien and Rowe, studied the use of a dc gradient in enhancing forward-wave amplifier performance. His studies indicate that the saturated output is increased markedly by application of a dc gradient although the efficiency is increased only modestly. Theoretical and experimental work by Meeker and Rowe[10,11] using the general nonlinear theory of Rowe indicated that velocity tapering of an rf structure could and did give rise to efficiency improvement along with other benefits such as lower-voltage operation, shorter length, and improved stability. Clarke,[12] following a similar approach to that of Meeker and Rowe, has also investigated tapering in large-signal traveling-wave amplifiers. Haddad[13] and Rowe[14] have found that velocity tapering in a backward-wave oscillator can give efficiency improvement and reduced starting current.

In support of the theoretical studies several experimental investigations have been made which generally corroborate the theoretical work. Spangenberg and co-workers[15] applied dc gradients to an external lumped-constant circuit traveling-wave amplifier and found that an accelerating dc gradient did in fact improve the power output and efficiency. They did not determine the optimum potential distribution, although a quadratic one did work well and as is shown theoretically in Section 3.b this is near-optimum.

Ruetz[16] applied a linear velocity taper to the output of a high-power TWA and found that the efficiency increased from 28 to 47% along with some bandwidth improvement and enhanced stability. Had a more nearly optimum tapering been utilized, a further improvement in efficiency would have resulted. Ruetz's results are shown in Fig. 1.

Extensive studies by Meeker, Rowe and Brackett[10,11,17] on a tapered velocity helix TWA which was designed according to theoretical predictions have yielded an improvement in a 30% efficient tube to some

2. HISTORICAL BACKGROUND AND EXPERIMENTAL WORK

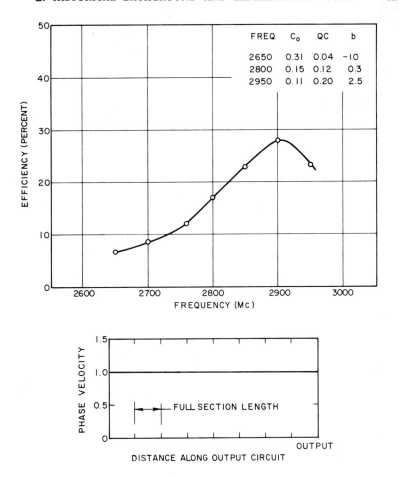

FIG. 1a. Untapered structure of Varian Associates high-power traveling-wave amplifier. (Reported by Ruetz.[16])

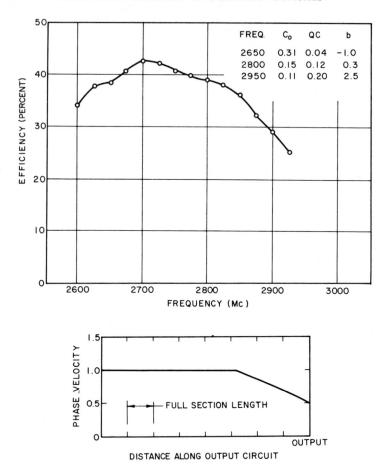

Fig. 1b. Tapered cavity structure of Varian Associates high-power traveling-wave amplifier. (Reported by Ruetz.[16])

50%. These results are summarized in Fig. 2. Some of the backward-wave oscillator results of Haddad[13,14] on efficiency enhancement are shown in Fig. 3. Even though the rf power level and efficiency are low on the absolute scale the improvement as a result of phase focusing is apparent.

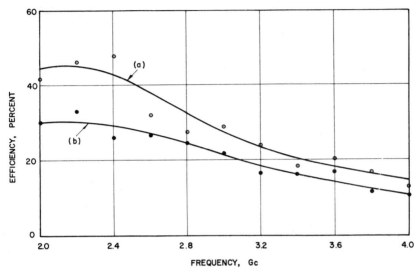

FIG. 2. Maximum efficiency versus frequency before and after tapering. (a) Variable pitch. Voltage and attenuator position optimized for maximum power output. (b) Uniform pitch. Voltage and attenuator position optimized for maximum power output.

All these studies indicate that by creating an environment (phase focusing) for the electron bunches in a linear-beam device such that the favorably phased electrons do not fall out of synchronism, the efficiency of linear-beam devices can be made more competitive with crossed-field devices in which there is built in a phase-locking mechanism. We note also that the kinetic energy of the output beam in a crossed-field device is relatively unchanged from its input value and this suggests that further efficiency improvement might be obtained there by velocity tapering. This is shown to be in fact true in Sections 5 and 6.

3 Efficiency Improvement in Traveling-Wave Amplifiers

a. Introduction

We now address ourselves to the problem of determining the velocity or potential profile which will give efficiency improvement. The most

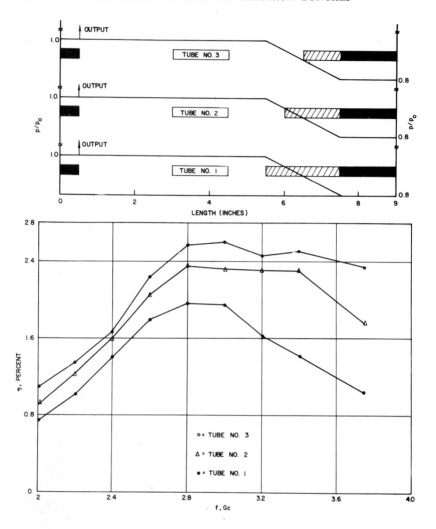

FIG. 3. Pitch variation and efficiency for tapered backward-wave oscillator ($p_0 = 0.133$ in. is the pitch of the uniform section; $I_{\text{collector}} = 9$ mA).

■ Designates the length over which "Dag Dispersion No. 226" is placed inside and outside the vacuum envelope.

▨ Designates the length over which "Dag Dispersion No. 226" is placed outside the vacuum envelope only.

desirable method is, of course, to solve the general nonlinear equations in closed form including finite C and space-charge forces. To date little success has resulted from this approach. It is, however, possible to solve the nonlinear equations in closed form if certain assumptions are made; namely the neglect of space-charge forces and the assumption of ideal bunches. In fact a so-called ideal "hard-kernel" bunch is used in which the charge is assumed concentrated at the centroid and internal space-charge fields and velocity spread are neglected. Solutions thus obtained are shown to be in agreement with more exact computer solutions and, possibly more significant, have led to experimental improvements.[10,11,13,18]

An examination of the velocity phase curves of Chapter VI indicates that the tapering or gradient should not be applied prior to a signal level of 6–10 dB below saturation, which occurs some three-fourths or more of the distance along the structure, since it is in this region that the bunches become well defined and are in the proper phase position relative to the rf wave. Thus our analysis is concerned with maintaining the phase position of this bunch fixed and neglects the effects of the few remaining fast electrons. The following work proceeds using the general large-signal equations of Rowe presented in Chapter VI. If the phase velocity of a structure such as a helix is varied with distance, then the impedance will vary unless something else is done. In the case of a helical line the impedance may be kept constant by decreasing the diameter as the pitch is varied.

b. Nonlinear One-Dimensional Equations for the O-Type Device Including Velocity Tapering and a Dc Gradient

In treating the problem of phase focusing generally one must consider nonideal electron bunches with velocity spreads and inter-bunch space-charge forces. Also the impedance $Z_0(y)$ and coupling function $\zeta(y)$ are to be included. The transmission-line equivalent circuit differential equation for a line with variable parameters was derived in Chapter IV as

$$\frac{\partial^2 V(z,t)}{\partial z^2} - \frac{1}{v_0^2(z)}\frac{\partial^2 V(z,t)}{\partial t^2} - \frac{\partial}{\partial z}\ln\left(\frac{Z_0(z)}{v_0(z)}\right)\frac{\partial V(z,t)}{\partial z}$$
$$- \frac{2\omega C_0 d(z)}{v_0^2(z)}\frac{\partial V(z,t)}{\partial t} = \mp \frac{Z_0(z)}{v_0(z)}\left[\frac{\partial^2 \rho(z,t)}{\partial t^2} + 2\omega C_0 d(z)\frac{\partial \rho(z,t)}{\partial t}\right] \quad (1)$$

where the top sign on the right-hand side refers to the FWA and the lower sign to the BWO.

The large-signal equations are developed in exactly the same manner as in the case of the uniform line except that normally invariant parameters are functions of distance. The velocity variable is defined as

$$\frac{u_0}{v_0(z)} = 1 + C_0 b(z). \tag{2}$$

The flight-line diagrams illustrating the definitions of variables in Chapter VI are again appropriate and the reader is referred to them.

The final working equations, including a dc voltage gradient, evolve as:

Circuit Equations

$$\frac{d^2 A(y)}{dy^2} - A(y)\left[\left(\frac{d\theta(y)}{dy} - \frac{1}{C_0}\right)^2 - \left(\frac{1 + C_0 b(y)}{C_0}\right)^2\right]$$
$$- \frac{dA(y)}{dy}\left[\frac{d}{dy} \ln\left(\frac{Z_0(y)}{Z_0}\right) + \frac{C_0}{1 + C_0 b(y)} \frac{db(y)}{dy}\right]$$
$$= \mp \left(\frac{1 + C_0 b(y)}{\pi C_0}\right)\left(\frac{Z_0(y)}{Z_0}\right) \zeta(y) \left[\int_0^{2\pi} \frac{\cos \Phi(y, \Phi'_{0j})\, d\Phi'_{0j}}{1 + 2C_0 u(y, \Phi'_{0j})}\right.$$
$$\left. + 2C_0 d(y) \int_0^{2\pi} \frac{\sin \Phi(y, \Phi'_{0j})\, d\Phi'_{0j}}{1 + 2C_0 u(y, \Phi'_{0j})}\right], \tag{3}$$

$$A(y)\left[\frac{d^2\theta(y)}{dy^2} - \left(\frac{d\theta(y)}{dy} - \frac{1}{C_0}\right)\left(\frac{d}{dy} \ln \frac{Z_0(y)}{Z_0} + \frac{C_0}{1 + C_0 b(y)} \frac{db(y)}{dy}\right)\right.$$
$$\left. - \frac{2d(y)}{C_0}(1 + C_0 b(y))^2\right] + 2\left(\frac{d\theta(y)}{dy} - \frac{1}{C_0}\right)\frac{dA(y)}{dy}$$
$$= \mp \left(\frac{1 + C_0 b(y)}{\pi C_0}\right)\left(\frac{Z_0(y)}{Z_0}\right) \zeta(y) \left[\int_0^{2\pi} \frac{\sin \Phi(y, \Phi'_{0j})\, d\Phi'_{0j}}{1 + 2C_0 u(y, \Phi'_{0j})}\right.$$
$$\left. - 2C_0 d(y) \int_0^{2\pi} \frac{\cos \Phi(y, \Phi'_{0j})\, d\Phi'_{0j}}{1 + 2C_0 u(y, \Phi'_{0j})}\right]. \tag{4}$$

Force Equation

$$\frac{\partial u(y, \Phi'_{0j})}{\partial y}[1 + 2C_0 u(y, \Phi'_{0j})] = \zeta(y)\left[-A(y)\left(1 - C_0 \frac{d\theta(y)}{dy}\right) \sin \Phi(y, \Phi'_{0j})\right.$$
$$\left. + C_0 \frac{dA(y)}{dy} \cos \Phi(y, \Phi'_{0j})\right] + C_0 \frac{dA_{dc}(y)}{dy}$$
$$+ \frac{1}{1 + C_0 b_0}\left(\frac{\omega_p}{\omega C_0}\right)^2 \int_0^{2\pi} \frac{F(\Phi - \Phi')\, d\Phi'_{0j}}{1 + 2C_0 u(y, \Phi'_{0j})}.$$
$$\tag{5}$$

3. EFFICIENCY IMPROVEMENT IN TRAVELING-WAVE AMPLIFIERS 433

Dependent Variable Relation (No Initial Velocity Modulation)

$$\frac{d\theta(y)}{dy} + \frac{\partial \Phi(y, \Phi_{0j})}{\partial y} = \frac{2u(y, \Phi_{0j})}{1 + 2C_0 u(y, \Phi_{0j})}. \tag{6}$$

The above system of nonlinear equations may be used for both the FWA and BWO, is subject to the assumptions discussed in Chapter VI, and may be solved in exactly the same manner. In addition to the usual parameters the functions $Z_0(y)/Z_0$ and $\zeta(y)$ must be specified. The nonlinear equations are readily simplified to the small-C system as shown in Chapter VI.

When $dA_{dc}(y)/dy$ in Eq. (5) is positive the dc gradient is accelerating; when negative, it is decelerating.

It is of course desirable to solve the above system in closed form for both the velocity taper and the dc gradient. An examination reveals that this is not generally possible and one must therefore resort to making certain approximations or to computer techniques. The solution of the above system for both velocity tapering and dc gradients is outlined in the following sections of this chapter.

c. Closed-Form Solution of FWA Nonlinear Equations for Finite C

A particular form of the circuit equation convenient to use for this analysis is the conservation of energy equation which relates the energy given up by the beam to energy gained by the circuit. This conservation law may be developed from Eqs. (4) and (5) using the hard-kernel-bunch assumptions. It is convenient to define a velocity parameter as

$$X(y, \phi_f) \triangleq 1 + 2C_0 u(y, \phi_f), \tag{7}$$

where ϕ_f denotes the phase position of the so-called hard-kernel bunch with respect to the rf wave. For maximum conversion efficiency it is desired to maintain the hard-kernel bunch in the maximum decelerating phase, $\phi_f = \pi/2$, with respect to the rf wave. For constant phase (synchronism)

$$u_0[1 + 2C_0 u(y, \phi_f)] = \frac{u_0}{[1 + C_0 b(y)]} = \frac{u_0}{\left[1 - C_0 \dfrac{d\theta(y)}{dy}\right]}. \tag{8}$$

Under these conditions and with space-charge forces, dc gradients and terms of order C_0^2 neglected, the bunch force equation is written as

$$\frac{dX(y, \phi_f)}{dy} = -\frac{2C_0 \zeta(y) A(y) \sin \phi_f}{X^2(y, \phi_f)}, \tag{9}$$

where $\zeta(y)$ is the circuit field reduction factor in moving from the circuit to the beam (bunch) position. Under these assumptions, and for zero circuit loss, Eq. (4) becomes

$$\frac{dA(y)}{dy} = \frac{\zeta(y)Z_0(y)}{Z_0} \frac{\sin\phi_f}{X(y,\phi_f)} + \frac{A(y)}{2} \frac{1}{Z_0(y)} \frac{dZ_0(y)}{dy}. \quad (10)$$

Equations (9) and (10) are now combined to yield

$$2C_0 \frac{d}{dy}\left[\frac{Z_0}{Z_0(y)} A^2(y)\right] = -\frac{dX^2(y,\phi_f)}{dy}. \quad (11)$$

Integration of Eq. (11) and applying the boundary conditions, i.e., $A(0) = A_0$, $Z_0(0) = Z_0$, $X(0) = 1$, gives

$$\frac{2C_0 A^2(y)}{Z_0(y)/Z_0} = 2C_0 A_0^2 + 1 - X^2(y,\phi_f), \quad (12)$$

which is the desired power conservation law ($2C_0 A_0^2 = $ the ratio of the rf power at the start of the taper to the dc beam power). Combining Eqs. (9) and (12) yields

$$\sqrt{2C_0}\sin\phi_f\, dy = \frac{-X^2(y,\phi_f)\, dX(y,\phi_f)}{\zeta(y)\sqrt{(Z_0(y)/Z_0)[2C_0 A_0^2 + 1 - X^2(y,\phi_f)]}} \quad (13)$$

and

$$A(y) = \sqrt{[Z_0(y)/2C_0 Z_0][2C_0 A_0^2 + 1 - X^2(y,\phi_f)]}. \quad (14)$$

The above results apply to any phase angle ϕ_f. Appropriate velocity profiles for phase focusing may now be found for specific forms of $\zeta(y)$ and $Z_0(y)$. Several cases are considered below.

(i) $Z_0(y) = Z_0$ and $\zeta(y) = [C(y)/C_0]^{\frac{3}{2}} = 1$. The constant-impedance assumption is strictly valid only for small tapers. Substituting the above functions and integrating Eq. (13) yields

$$\sqrt{2C_0}\sin\phi_f\, y = \tfrac{1}{2}\Big\{X(y,\phi_f)\sqrt{2C_0 A_0^2 + 1 - X^2(y,\phi_f)} - \sqrt{2C_0 A_0^2}$$
$$- (1 + 2C_0 A_0^2)\left[\sin^{-1}\frac{X(y,\phi_f)}{\sqrt{1 + 2C_0 A_0^2}} - \sin^{-1}\frac{1}{\sqrt{1 + 2C_0 A_0^2}}\right]\Big\}. \quad (15)$$

The electron velocity or circuit phase velocity (they are equal) is shown versus $\sqrt{2C_0}\sin\phi_f\, y$ for particular values of the initial amplitude $2C_0 A_0^2$ in Fig. 4. Making the small-C assumption, i.e., $C_0 \ll 1$, and letting $\phi_f = \pi/2$ simplifies Eq. (15) to the following:

$$u(y,\phi_f) = 1 - 2C_0 A_0 y - C_0 y^2. \quad (16)$$

3. EFFICIENCY IMPROVEMENT IN TRAVELING-WAVE AMPLIFIERS

(ii) Helix Impedance Variation—Solid Beam. (See Appendix A.)
Assume that

$$\frac{Z_0(y)}{Z_0} = \left[\frac{C(y)}{C_0}\right]^3 = \left[\frac{v_0(y)}{v_0}\right]^{\frac{3}{2}} = [X(y, \phi_f)]^{\frac{3}{2}},$$

$$\zeta(y) = \left[\frac{C(y)}{C_0}\right]^{\frac{3}{2}} = \left[\frac{v_0(y)}{v_0}\right]^{\frac{3}{4}} = [X(y, \phi_f)]^{\frac{3}{4}}.$$

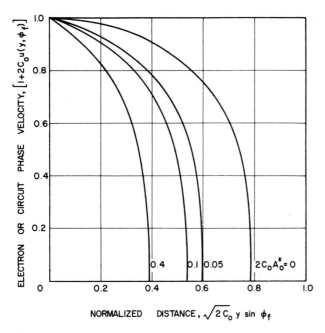

FIG. 4. Velocity profiles based upon hard-kernel-bunch approximation. Interaction impedance is constant.

Applying these functions to Eq. (15) and integrating gives

$$\sqrt{2C_0} \sin \phi_f \, y = \frac{(1 + 2C_0 A_0^2)^{\frac{1}{4}}}{\sqrt{2}} \left\{ 4E\left(\frac{1}{\sqrt{2}}, z'\right) - 2F\left(\frac{1}{\sqrt{2}}, z'\right) \right\} \Big|_{z'=z_2'}^{z'=z_1'}, \tag{17}$$

where E and F are elliptic integrals of the second and first kinds respectively and the limits are

$$z_1' = \sin^{-1}\left\{\sqrt{2} \sin\left[\frac{\cos^{-1}(X(y, \phi_f)/\sqrt{1 + 2C_0 A_0^2})}{2}\right]\right\} \tag{18a}$$

and

$$z_2' = \sin^{-1}\left\{\sqrt{2}\sin\left[\frac{\cos^{-1}(1/\sqrt{1+2C_0A_0^2})}{2}\right]\right\}. \tag{18b}$$

The velocity profiles for this impedance variation are shown in Fig. 5.

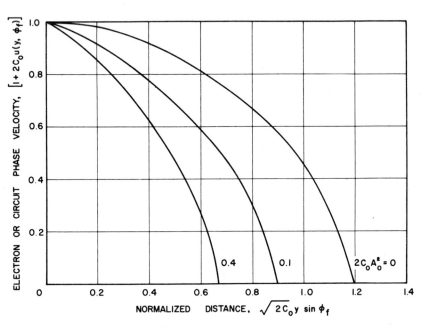

Fig. 5. Velocity profiles based upon hard-kernel-bunch approximation. Interaction impedance varies as $[1 + 2C_0u(y, \phi_f)]^{\frac{3}{2}}$.

(*iii*) *Cutler Impedance Variation—Helix with Solid Beam.*

$$\frac{Z_0(y)}{Z_0} = \left[\frac{C(y)}{C_0}\right]^3 = \left[\frac{v_0(y)}{v_0}\right]^2 = X^2(y, \phi_f),$$

$$\zeta(y) = \left[\frac{C(y)}{C_0}\right]^{\frac{3}{2}} = \frac{v_0(y)}{v_0} = X(y, \phi_f).$$

Proceeding in exactly the same manner as in the previous cases, the velocity profile is obtained as

$$\sqrt{2C_0}\sin\phi_f\, y = \cos^{-1}\left(\frac{X(y,\phi_f)}{\sqrt{1+2C_0A_0^2}}\right) - \cos^{-1}\left(\frac{1}{\sqrt{1+2C_0A_0^2}}\right). \tag{19}$$

3. EFFICIENCY IMPROVEMENT IN TRAVELING-WAVE AMPLIFIERS

The solution forms are shown in Fig. 6 and, as expected, the taper extends over a longer distance.

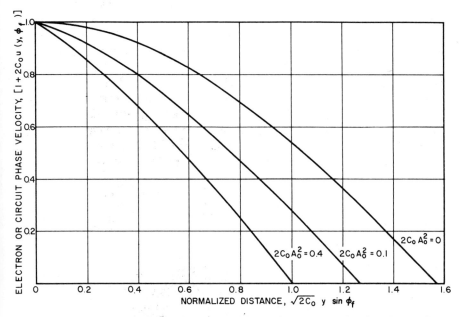

FIG. 6. Velocity profiles based upon hard-kernel-bunch approximation. Interaction impedance varies as $[1 + 2C_0 u(y, \phi_f)]^2$.

d. Computer Solutions for the Velocity-Tapered FWA

The nonlinear equations of Section 3.b may be used in basically two different ways after impedance and coupling function variations have been decided upon: (1) they can be used to evaluate and compare specific forms of velocity taper or dc gradients such as those developed using approximate theories; and (2) they can be solved in the computer using some criterion for determining an optimum or near-optimum taper or gradient. It is impractical to try to make the computer select an absolute optimum taper or gradient, although there are a number of ways in which to find near-optimum variations.

In order for the computer to determine a near-optimum profile it is necessary to establish a selection criterion. These may relate either to the rf amplitude on the structure or to the phase positions of the electrons in the beam. It is possible to require that the computer optimize the rf growth rate or gain along the device by selecting a value of $b(y)$ over a specified interval (usually large) to give maximum gain. This necessarily involves an iterative procedure and is quite time-consuming and costly: so much so as to be impractical in competition with other methods. A more direct method, i.e., not involving an iterative process, of reaching the same optimum profile is to make use of the fact that the bunch should be held close to the phase position for maximum decelerating rf electric field in order to maximize the conversion of kinetic energy incrementally. This is easily accomplished without iteration since the computer evaluates the phase position and velocity of each charge group at each displacement plane in the general course of things. If it holds the bunch in a fixed phase position and the bunch is slowing down then the circuit phase velocity is slowing at the same rate. Holding a bunch in a given phase position may be accomplished in either of two ways: (1) by maintaining the phase of a center of the bunch electron, or (2) by maintaining the average phase of the bunch. Either process works equally well.

Before viewing the effect of variously determined tapers on saturation efficiency it is interesting to compare velocity-phase diagrams for various tapers with one another and with the untapered one. In Fig. 7

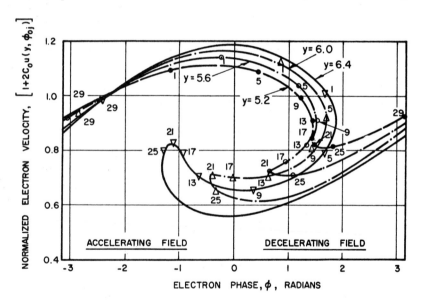

FIG. 7. Phase diagrams for untapered and tapered amplifiers ($C_0 = 0.1$, $QC = 0$, $b_0 = 1$, $d_0 = 0$, $\psi_0 = -30$). (a) Untapered amplifier.

these phase diagrams are shown. Notice the considerable improvement in minimum velocity as a result of tapering when compared with the phase diagram of the untapered device. The bunch is better held together

3. EFFICIENCY IMPROVEMENT IN TRAVELING-WAVE AMPLIFIERS   439

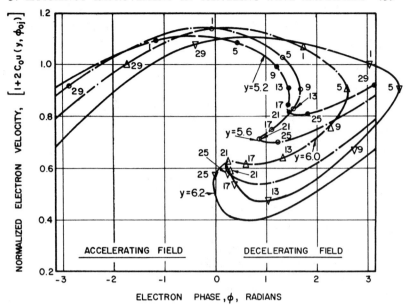

FIG. 7b. Preprogrammed taper, $b_0 = 1$, to $y = 5.2$; then $b(y)$ polynomial followed to maintain phase focusing.

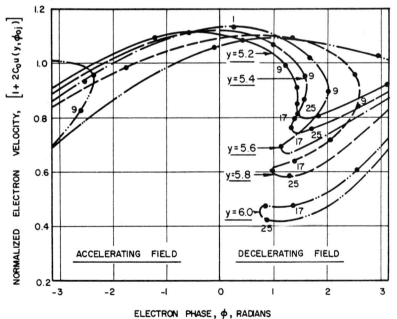

FIG. 7c. Computer-chosen profile from $y = 5.2$ to hold electron No. 17 at $\phi = 1.43$ radians.

and more efficiency results. It will be seen that small differences in the velocity-phase diagrams are demagnified in terms of rf signal level differences.

The comparison of various taper forms is illustrated in Fig. 8, where the rf voltage level is shown versus distance for several types of variations. We see that all tapers give improvement and that the differences in terms of efficiency are slight. The most important fact is that the taper calculated from approximate closed-form solutions does virtually as well as the computer can do in accounting for all effects. This stems from the fact that you cannot phase focus all charge groups, but rather only one or

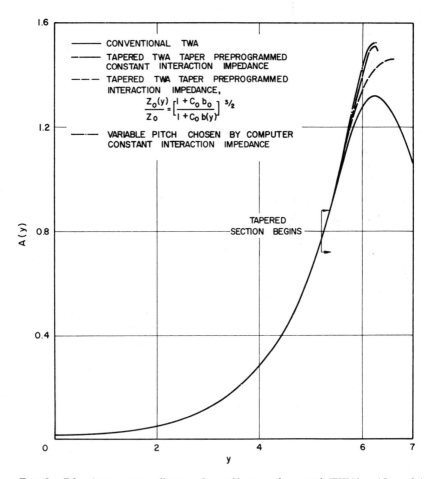

FIG. 8. Rf voltage versus distance for uniform and tapered TWA's ($C_0 = 0.1$, $d_0 = 0$, $QC = 0$, $b_0 = 1$, $\psi_0 = -30$).

some average phase. We see that tapering is effective only so long as the bunch is well defined; intense space-charge forces counter this condition.

The results of numerous calculations on this problem are summarized in Figs. 9 and 10, where maximum gain and maximum power conditions have been studied. We see that in the case of untapered tubes operating at maximum power output in which there is a large velocity spread in the beam, tapering gives only a slight improvement, whereas for tubes operating less efficiently at a voltage of maximum gain the improvement is large. The conclusion is important that through velocity tapering one

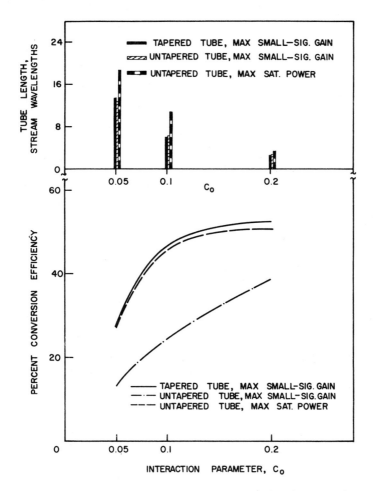

FIG. 9. Efficiency and tube length versus C_0 for tapered and untapered amplifiers ($QC = 0$, $d_0 = 0$, $\psi_0 = -30$).

can operate under conditions of highest gain and efficiency simultaneously with a reduced length and, of course, weight. It is also seen that the degree of improvement decreases as C and QC increase, which is to be expected since increasing either of these parameters increases the velocity spread in the beam, thereby making phase focusing more difficult. It is desirable in both tapered and untapered tubes to keep the space-charge forces to a minimum, i.e., $QC < \frac{1}{4}$, in order to obtain highest operating efficiencies.

At this point it is pertinent to wonder whether or not a continuous, smoothly varying velocity taper is required or whether a one- or two-step

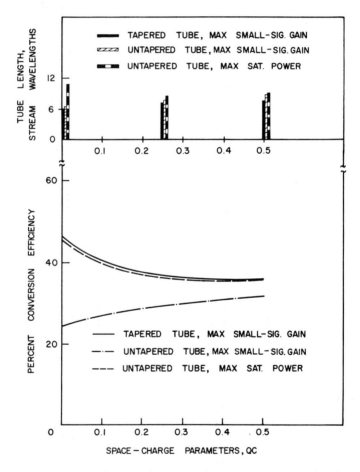

Fig. 10. Saturation efficiency and tube length versus QC for tapered and untapered tubes ($C_0 = 0.1$, $d_0 = 0$, $B = 1$, $a/b' = 2$, $\psi_0 = -30$).

3. EFFICIENCY IMPROVEMENT IN TRAVELING-WAVE AMPLIFIERS 443

change in circuit phase velocity would suffice. Such calculations are rather easily made by simply invoking step changes in $b(y)$ at various displacement planes. Such one- and two-step changes in $b(y)$ are illustrated in Fig. 11. The improved efficiency is found to be 42.5% for the one-step change and 45% for the two-step change. These are lower but

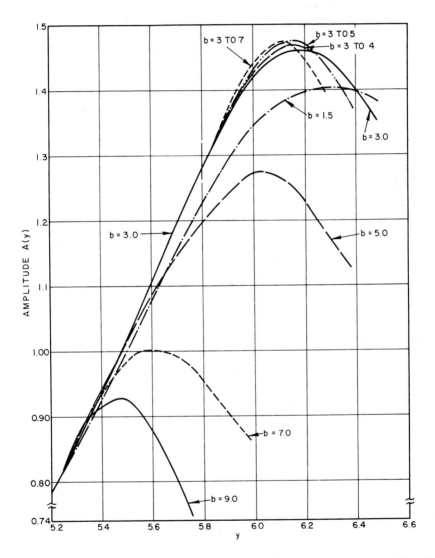

FIG. 11. Rf voltage versus distance for TWA's with one- and two-step shifts of velocity ($C_0 = 0.1$, $d_0 = 0$, $QC = 0$, $\psi_0 = -30$, $b_0 = 1$ up to $y = 5.2$).

compare favorably with the approximately 50% efficiency obtained in the uniform-taper case. The actual velocity profiles for the various taper situations discussed above are illustrated in Fig. 12.

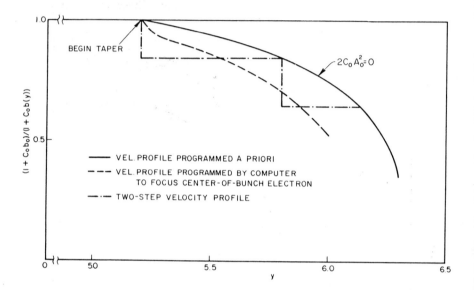

FIG. 12. Velocity profiles for different forms of velocity tapering ($C_0 = 0.1$, $d_0 = 0$, $QC = 0$, $\psi_0 = -30$, $b_0 = 1$ up to $y = 5.2$).

e. Experimental Verification

The experimental confirmation of efficiency improvement by phase focusing was touched upon in the opening to this chapter (see Figs. 1 and 2). Efficiency improvement has been obtained experimentally under both growing-wave and beating-wave conditions. In experimental tubes one does not quite realize the predicted untapered efficiencies, due primarily to radial velocity and radial field effects, and thus also efficiencies in tapered tubes fall short of predicted values. This is to be expected since the one-dimensional theory was used. However, the important fact still remains that it has been possible to increase the saturated efficiency experimentally from 30 to over 50% using a taper based on the hard-kernel-bunch theory.[11]

It is interesting to note that efficiency improvement is realized over an extremely wide frequency range and range of drive powers. This indicates that the exact form of the taper is not too critical as long as it is located in the region of a well-defined and phase-positioned bunch. Another important point is that the power output versus power input curve for a tapered tube indicates dynamic limiting over a wide range of drive powers as a result of phase-focusing. This problem is discussed theoretically in Section 3.f.

f. Computer Solutions for the Dc Gradient FWA

Extensive numerical calculations on the effects of dc gradients applied to amplifiers for phase-focusing purposes have been made and analyzed

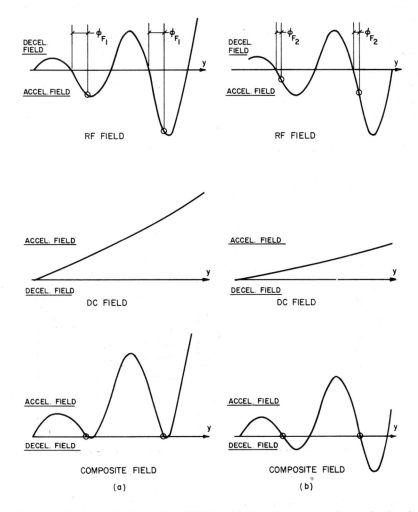

FIG. 13. Focusing in dc gradient TWA's. (a) Focusing near maximum decelerating rf field. (b) Focusing far from maximum decelerating rf field.

by Hess[9] and Meeker.[10,11] Although the theories used were different in details they have been shown to be identical in Chapter VI, and as expected the same results are obtained using either theory.

The determination of near-optimum dc gradient profiles for performance improvement may proceed in a completely analogous fashion to those methods used in evaluating velocity profiles in Section 3.d; i.e., both calculated and computer-determined profiles can be studied. Optimization criteria in the computer are again associated with the velocity-phase characteristics of charge groups. The action of the dc gradient is to counteract the phase slipping of electrons by holding as many electrons as possible in the decelerating field region as they give up their kinetic energy to the rf wave. This action is illustrated in Fig. 13, where both of the fields are shown for different focusing conditions. The optimum focus position of the bunch must be chosen short of that for a maximum decelerating field in order to trap a maximum number of electrons and realize an efficiency improvement. Focusing at the maximum decelerating field position leads to little or no increase in efficiency. Both, however, do lead to higher gain and power output. The

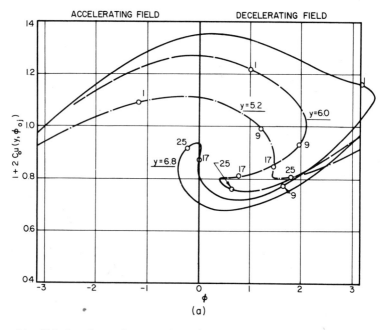

Fig. 14. Velocity-phase diagrams for TWA's ($C_0 = 0.1$, $QC = 0$, $d_0 = 0$, $b = 1$, $\psi_0 = -30$). (a) Constant dc gradient, $\Delta(V_{dc}/V_0) = 0.32$ per y-unit from $y = 5.2$. (b) Dc gradient chosen by computer to hold electron No. 17 at $\phi = 1.45$ radians.

3. EFFICIENCY IMPROVEMENT IN TRAVELING-WAVE AMPLIFIERS

application of a dc gradient tends to reduce the velocity spread in the beam at the output and therefore such tubes are well suited to subsequent depressed collector operation.

Both the derived gradient profiles and computer-determined profiles are studied. Again it is instructive to compare velocity-phase diagrams for various dc gradient types and with the uniform device velocity-phase characteristic (Fig. 7a). These are shown in Fig. 14. The hard-kernel-bunch theory is again quite good, although uniform computer optimization is the best. Step changes are also a good approximation. These various profiles are illustrated in Fig. 15 and the corresponding rf solutions in Figs. 16 and 17.

The conversion efficiency in the case of the hard-kernel-bunch profile increased significantly in view of the fact that nearly all the dc gradient

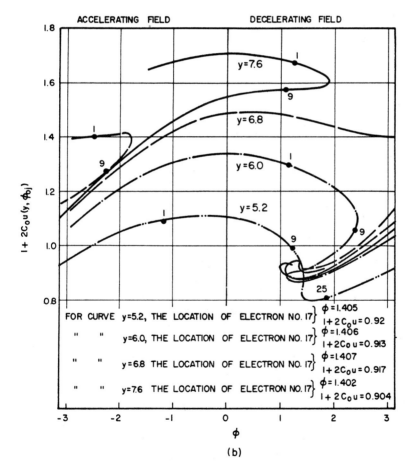

(b)

energy that went into the bunch was transferred to the rf field. In the case of finite bunches with velocity spreads and space-charge debunching forces, the efficiency is relatively unchanged if focusing is maintained in the maximum decelerating region. This, however, is overcome to some extent, as pointed out by Hess,[9] if focusing is short of $\pi/2$, resulting in the trapping of more electrons. He found in a zero-space-charge force case for a uniform tube efficiency of 43% that if $\phi_f = 45°$ and a gradient is applied, the efficiency is increased to 70%, whereas if $\phi_f = 90°$, even though a higher potential is used, the efficiency is 35%.

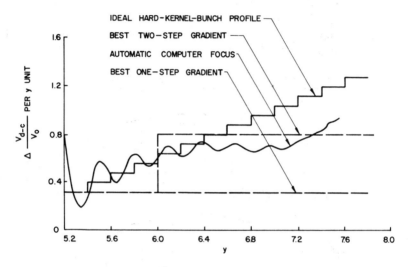

Fig. 15. Comparison of dc gradient profiles ($C_0 = 0.1$, $QC = 0$, $d_0 = 0$, $b = 1$, $\psi_0 = -30$ at $y = 0$).

The dc gradient also extends the saturation gain to large values and increases power output. It should also be pointed out that it is also accompanied by an increase in length, which is certainly a disadvantage when compared to a tapered-circuit velocity device.

g. *Multiple Bunch Focusing in an FWA*

The experimental data on phase-focused traveling-wave amplifiers have indicated that these devices have a relatively constant power output for a wide range of drive powers. This wide-range limiting is illustrated in Fig. 18. Since such a characteristic is not usually associated with amplifiers without phase focusing, it appears that velocity tapering could be used as a means of obtaining wide-range power limiting. Since the saturation power output in the nonlinear theory is relatively independent

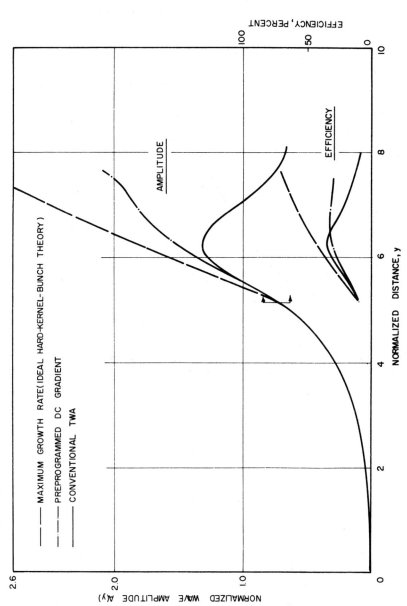

FIG. 16. Rf voltage and efficiency versus distance for a TWA with a preprogrammed dc gradient ($C_0 = 0.1$, $d_0 = 0$, $QC = 0$, $b_0 = 1$, $\psi_0 = -30$).

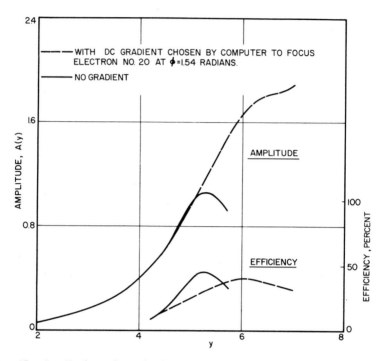

Fig. 17. Amplitude and efficiency versus distance for TWA with and without "optimum" dc gradient ($C_0 = 0.2$, $QC = 0$, $d_0 = 0$, $b_0 = 1$, $\psi_0 = -30$).

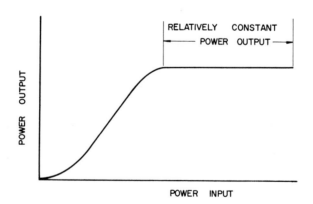

Fig. 18. Desired power output versus power input curve.

3. EFFICIENCY IMPROVEMENT IN TRAVELING-WAVE AMPLIFIERS

of drive level (only the length changes), we may approach the problem of obtaining a characteristic such as shown in Fig. 18 by considering means for keeping the rf power level on the structure invariant with distance.

The theoretical analysis proceeds using the small-C, negligible-space-charge, nonlinear equations obtained by simplifying Eqs. (3)–(6) in the manner outlined in Chapter VI. Consider that the electron beam has been bunched into two ideal bunches and that saturation exists or is impending. The two hard-kernel bunches are assumed to be located in phase positions relative to the field as illustrated in Fig. 19. Since there is

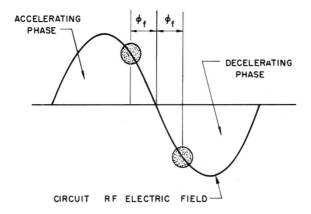

FIG. 19. Hard-kernel-bunch focusing.

no net transfer of power between the beam and the circuit wave under saturated conditions, the phase angles with respect to the wave are the same for each bunch. The condition we wish to maintain is thus

$$\frac{dA(y)}{dy} = 0 \tag{20}$$

from saturation on. From Eqs. (3) and (20) (assume $Z_0(y) \equiv Z_0$)

$$\frac{dA(y)}{dy} = 0 = \sin\phi_{f+} + \sin\phi_{f-}, \tag{21}$$

where $\Phi = \pm\phi_f$. Thus

$$A(y) = \text{constant} \triangleq A_{\text{sat.}}. \tag{22}$$

Note that $\phi_{f+} = -\phi_{f-}$ and that these may be functions of distance in order to maintain $A(y) \equiv A_{\text{sat.}}$. The derivative of Eq. (5) is combined with Eq. (6) and rearranged to give

$$\frac{d^2\phi_f(y)}{dy^2} \pm \frac{d^2\theta(y)}{dy^2} = -2A_{\text{sat.}} \sin \phi_f(y). \tag{23}$$

Since $d^2\theta(y)/dy^2 = 0$,

$$\frac{d^2\phi_f(y)}{dy^2} = -2A_{\text{sat.}} \sin \phi_f(y). \tag{24}$$

Equation (24) is multiplied by $2d\phi_f(y)/dy$ and integrated, with the result

$$\left(\frac{d\phi_f(y)}{dy}\right)^2 = 4A_{\text{sat.}} \cos \phi_f(y) + K_1, \tag{25}$$

where K_1 is determined from known values of $A_{\text{sat.}}$ and $\phi_f(y)$ at saturation.

The integration of Eq. (25) involves an elliptic integral and will be evaluated presently. The required velocity parameter variation is obtained from the simplified version of Eq. (4) as

$$b(y) = -\frac{2 \cos \phi_f(y)}{A_{\text{sat.}}} - \frac{d\theta(y)}{dy}\bigg|_{\text{sat.}}$$

$$= -\frac{2 \cos \phi_f(y)}{A_{\text{sat.}}} - K_2. \tag{26}$$

An explicit form for $\phi_f(y)$ must now be obtained from the solution of Eq. (25); K_1 and K_2 are known constants. Equation (25) is conveniently written in near standard elliptic integral form using $\cos \phi_f(y) = 1 - 2 \sin^2[\phi_f(y)/2]$:

$$y = \frac{2}{K_3^{\frac{1}{2}}} \int_{\phi_{f,\text{sat.}}/2}^{\phi_f/2} \frac{d\phi_f(y)/2}{\sqrt{1 - \frac{K_4}{K_3} \sin^2 \frac{\phi_f(y)}{2}}}, \tag{27}$$

where $K_3 \triangleq K_1 + 4A_{\text{sat.}}$, $K_4 \triangleq 8A_{\text{sat.}}$.

In order to make use of standard elliptic integral forms, Eq. (27) is rewritten as follows: where $y = 0$ at saturation and at this displacement plane $\phi_f \equiv \phi_{f,\text{sat.}}$,

$$y = \frac{2}{K_3^{\frac{1}{2}}} \left[\int_0^{\phi_f/2} \frac{d\phi_f(y)/2}{\sqrt{1 - k^2 \sin^2 \frac{\phi_f(y)}{2}}} - \int_0^{\phi_{f,\text{sat.}}/2} \frac{d\phi_f(y)/2}{\sqrt{1 - k^2 \sin^2 \frac{\phi_f(y)}{2}}} \right], \tag{28}$$

3. EFFICIENCY IMPROVEMENT IN TRAVELING-WAVE AMPLIFIERS

where $k^2 \triangleq K_4/K_3$. Equation (28) is further simplified to

$$y + K_5 = \frac{2}{K_3^{\frac{1}{2}}} \int_0^{\phi_f/2} \frac{d\phi_f(y)/2}{\sqrt{1 - k^2 \sin^2 \frac{\phi_f(y)}{2}}}, \tag{29}$$

where

$$K_5 \triangleq \frac{2}{K_3^{\frac{1}{2}}} \int_0^{\phi_{f,\text{sat.}}/2} \frac{d\phi_f(y)/2}{\sqrt{1 - k^2 \sin^2 \frac{\phi_f(y)}{2}}}. \tag{30}$$

Both $\phi_{f,\text{sat.}}$ and k^2 are known for a particular situation and therefore K_5 is known. Equation (26) is now in standard elliptic integral form and thus

$$\sin \frac{\phi_f(y)}{2} = \text{Sn} \left\{ \frac{K_3^{\frac{1}{2}}}{2} (y + K_5) \right\}, \tag{31}$$

where Sn is the elliptic sine function. Two cases are of interest, $k^2 < 1$ and $k^2 > 1$.

Case A: $k^2 < 1$. Under this condition Sn u is written in series form as

$$\text{Sn } u = u - (1 + k^2) \frac{u^3}{3!} + (1 + 14k^2 + k^4) \frac{u^5}{5!} + \cdots . \tag{32}$$

Expanding the arcs in a power series gives $\phi_f(y)$ as

$$\phi_f(y) = 2 \left\{ \text{Sn} \left[\frac{K_3^{\frac{1}{2}}}{2} (y + K_5) \right] + \frac{\text{Sn}^3[(K_3^{\frac{1}{2}}/2)(y + K_5)]}{2 \cdot 3} + \cdots \right\}. \tag{33}$$

After substituting the series expression for Sn and simplifying, $\phi_f(y)$ and $b(y)$ are written as

$$\phi_f(y) = K_3^{\frac{1}{2}}(y + K_5) - \frac{k^2 K_3^{\frac{3}{2}}}{2^3 \cdot 3!} (y + K_5)^3$$

$$+ \frac{(-9 + 4k^2 + k^4) K_3^{5/2}}{2^5 \cdot 5!} (y + K_5)^5 + \cdots . \tag{34}$$

and

$$b(y) = \left(-\frac{2}{A_{\text{sat.}}} - K_2 \right) + \frac{K_3}{A_{\text{sat.}}} (y + K_5)^2 - \frac{K_3^2(1 + k^2)}{2^2 \cdot 3 A_{\text{sat.}}} (y + K_5)^4$$

$$+ \frac{K_3^3(2 + 13k^2 + 2k^4)}{2^4 \cdot 45} (y + K_5)^6 + \cdots . \tag{35}$$

Case B: $k^2 > 1$. In this case write $\text{Sn}(u, k) = (1/k)\,\text{Sn}(uk, 1/k^2)$; if $k^2 > 1$ then $1/k^2 < 1$ and Eq. (31) is rewritten as

$$\sin \frac{\phi_f(y)}{2} = \frac{1}{k}\,\text{Sn}\left[\frac{K_4^{\frac{1}{2}}}{2}(y + K_5),\,\frac{1}{k^2}\right]. \tag{36}$$

The series expression for $\text{Sn}[(K_4^{\frac{1}{2}}/2)(y + K_5),\,1/k^2]$ is the same as before except for replacing $K_3^{\frac{1}{2}}$ by $K_4^{\frac{1}{2}}$ and k^2 by $1/k^2$. Following a procedure similar to that above one obtains

$$\phi_f(y) = \frac{K_4^{\frac{1}{2}}}{k}(y + K_5) - \frac{K_4^{\frac{3}{2}}}{2^2 \cdot 3!\,k}(y + K_5)^3 + \frac{(-9 + 4k^2 + k^4)K_4^{\frac{5}{2}}}{2^4 \cdot 5! \cdot k^5}$$
$$\cdot (y + K_5)^5 + \cdots. \tag{37}$$

and

$$b(y) = \left(-\frac{2}{A_{\text{sat.}}} - K_2\right) + \frac{K_4}{A_{\text{sat.}}\,k^2}(y + K_5)^2 - \frac{K_4^2(1 + k^2)}{2^2 \cdot 3 \cdot k^4\,A_{\text{sat.}}}(y + K_5)^4$$
$$+ \frac{K_4^3(2 + 13k^2 + 2k^4)}{2^4(45)k^6 A_{\text{sat.}}}(y + K_5)^6 + \cdots. \tag{38}$$

Thus in order to keep the rf voltage level on the circuit constant, either the circuit phase velocity or the electron beam kinetic energy must vary as prescribed by Eq. (38). Inspection of Eq. (35) or (38) reveals that $b(y)$ increases with distance beyond saturation and thus either $v_0(y)$ must decrease or $u_0(y)$ must increase with distance. The rate of change is not appreciably different from that needed for phase focusing and thus it is not surprising that one observes wide-range power limiting in experiments on velocity-tapered amplifiers.

4 Efficiency Improvement in O-Type Backward-Wave Oscillators

a. General Background

The O-type backward-wave oscillator was treated with a nonlinear Lagrangian analysis in Chapter VII assuming that the rf structure and potential along the beam were uniform. As would be expected from a qualitative study of the interaction process, the efficiency of the BWO is considerably lower than that of the forward-wave amplifier. This occurs for two reasons, namely (1) the backward-wave space-harmonic impedance is generally low, and (2) the rf field due to the circuit field is weak in regions where the beam bunching is greatest and is thus unable to maintain the beam bunching.

4. EFFICIENCY IMPROVEMENT IN O-TYPE OSCILLATORS

An examination of the velocity-phase plots for the spent beam of a BWO reveals that it also is a good candidate for phase-focusing techniques in view of the high remaining kinetic energy of the beam. The same procedures of phase focusing by velocity tapering or the application of a dc gradient may be applied to the BWO. In the FWA we found that the gain and efficiency were both increased by the use of phase-focusing techniques. In the BWO, then, the question is, what happens to the efficiency, start-oscillation current and tuning range if velocity tapering or voltage gradient focusing is used?

A limited study of velocity tapering in a BWO was made by Geppert[4] using a linear theory, which clearly yields no information on efficiency. He found that under certain conditions the starting current would in fact decrease. A general study was made by Haddad[13] in which he considered both linear and nonlinear effects. The results indicated that appropriate velocity tapering would yield efficiency improvement along with some decrease in starting current. The tuning range of the oscillator is relatively unaffected. These predictions have generally been substantiated with experimental results as shown in Fig. 3.

In view of the fact that the only differences between the nonlinear FWA and BWO equations occur in the signs of the terms on the right-hand side of the circuit equation, the solution methods developed for the FWA may be applied to the BWO. The following subsections are devoted to this end.

b. Closed-Form Solution of BWO Nonlinear Equations for Finite C

Again a convenient method for solving the nonlinear equations is to write an energy balance equation expressing conservation in a zero-space-charge-field, lossless-circuit environment. The procedure used here is exactly parallel to that used in Section 3.c for the forward-wave amplifier except that the lower set of signs on the right-hand side of the circuit equations, Eqs. (3) and (4), is used. A balance of the rf increase in power with the loss in beam kinetic power is expressed as follows for the BWO:

$$\frac{2C_0 A^2(y)}{Z_0(y)/Z_0} = 2C_0 A_0^2 - X^2(0,\phi_f) + X^2(y,\phi_f), \quad (39)$$

where again $X(0,\phi_f)$ denotes the rf state of the hard-kernel bunch at $y = 0$ (rf output, beam input). The normalized rf signal level A_0 now denotes the magnitude of the rf output, i.e., at $y = 0$.

Applying the same synchronism conditions as in the amplifier and solving the energy equation and the force equation yields the following results for the BWO, corresponding to Eqs. (13) and (14) for the FWA $[X(0, \phi_f) = 1]$:

$$\sqrt{2C_0} \sin \phi_f \, dy = \frac{-X^2(y, \phi_f) \, dX(y, \phi_f)}{\zeta(y) \sqrt{[Z_0(y)/Z_0]} \, [2C_0 A_0^2 - 1 + X^2(y, \phi_f)]} \qquad (40)$$

and

$$A(y) = \sqrt{[Z_0(y)/2C_0 Z_0][2C_0 A_0^2 - 1 + X^2(y, \phi_f)]}. \qquad (41)$$

It is noticed, as expected, that the only differences are sign changes which arise out of the circuit equation since the energy travels in a direction opposite to the modulation. Solutions of Eqs. (40) and (41) are outlined below for representative impedance variations.

(*i*) $Z_0(y) = Z_0$ and $\zeta(y) = 1$. Solution of the integral equation yields

$$\sqrt{2C_0} \sin \phi_f y = \tfrac{1}{2} [\sqrt{2C_0 A_0^2} - X(y, \phi_f) \sqrt{(2C_0 A_0^2 - 1) + X^2(y, \phi_f)}]$$
$$+ \frac{2C_0 A_0^2 - 1}{2} \left[\ln \frac{X(y, \phi_f) + \sqrt{(2C_0 A_0^2 - 1) + X^2(y, \phi_f)}}{1 + \sqrt{2C_0 A_0^2}} \right]. \qquad (42)$$

The above result is shown in Fig. 20 for a range of initial amplitudes $2C_0 A_0^2$ ($2C_0 A_0^2$ = ratio of rf power at start of taper to dc beam power).

(*ii*) *Particular Impedance Variation for the Backward Wave.* (See Appendix A.)

$$\frac{Z_0(y)}{Z_0} = \left[\frac{v_0(y)}{v_0}\right]^2 = X^2(y, \phi_f)$$

and the field varies as $\zeta(y) = 1$. Following the above procedure gives the result for this case as

$$\sqrt{2C_0} \sin \phi_f y = \sqrt{2C_0 A_0^2} - \sqrt{(2C_0 A_0^2 - 1) + X^2(y, \phi_f)}. \qquad (43)$$

A plot of Eq. (43) is shown in Fig. 21, again as a function of $2C_0 A_0^2$. The same results are obtained if the nonlinear theory circuit equations are used in place of the conservation of energy equation without making the small-C assumption.

c. Computer Solutions for the Velocity-Tapered BWO

In order to solve the nonlinear backward-wave oscillator equations of Section 3.b by the downhill method certain parameters and initial conditions must be specified. These are

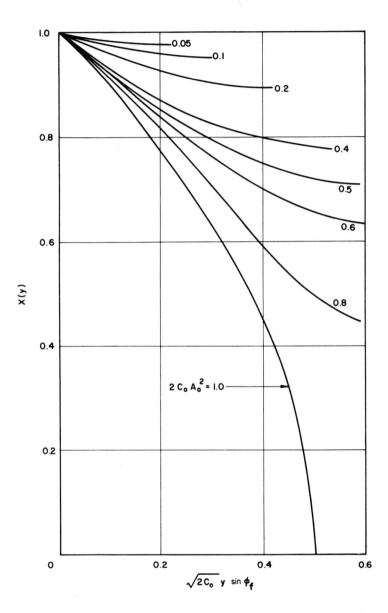

FIG. 20. Velocity profiles based upon a hard-kernel-bunch approximation (constant interaction impedance).

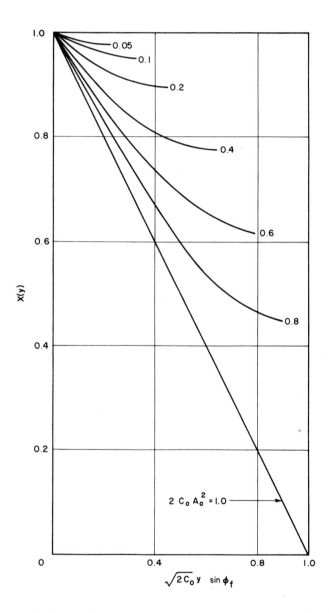

Fig. 21. Velocity profiles based upon hard-kernel-bunch approximation (interaction impedance variation as $X^2(y)$).

4. EFFICIENCY IMPROVEMENT IN O-TYPE OSCILLATORS

(1) operating parameters C, ω_p/ω, d, and $\gamma b'$;
(2) the point along the rf structure where the velocity tapering begins; and
(3) the variation of phase velocity with distance.

After these have been specified, the downhill method can be used to find sets of values of A_0, b which yield an oscillation condition. The efficiency of the device is given by

$$\text{eff.} = 2C_0[A_0^2 - A_{\min}^2], \tag{44}$$

where $A_{\min} \approx 0$ in the oscillator case and $A_{\min} \neq 0$ for the backward-wave amplifier.

A more general procedure would be to require that the computer determine the optimum profile and starting position for the taper. This is unfortunately extremely costly in computer time and thus has not been explored extensively. An examination of the previously outlined hard-kernel-bunch results indicates that an exponential phase velocity variation with distance would be appropriate for a wide range of oscillator operating parameters. The form of the velocity taper studied extensively is

$$\frac{v_0(y)}{v_0(0)} = \exp\left[-\sqrt{2C_0}\, y \sin \phi_f\right]. \tag{45}$$

Such a velocity taper is used to write the velocity parameter versus distance as

$$b(y) = \left\{\left(\frac{1 + C_0 b_0}{C_0}\right) \exp\left[\sqrt{2C_0}\,(y - y_{s,t}) \sin \phi_f\right] - \frac{1}{C_0}\right\}, \tag{46}$$

where
$b_0 \triangleq b(0)$,

$y_{s,t} \triangleq$ the displacement plane at which the taper begins,
and
$y \triangleq$ displacement plane beyond $y_{s,t}$.

Representative results obtained using the above velocity variation are shown in Figs. 22–24 for particular sets of operating parameters. It is believed that these results are generally representative of tapered-velocity BWO characteristics. Both finite and zero space-charge conditions are shown. The efficiency is seen to be significantly greater in the tapered tube particularly at low values of the injection velocity parameter, b_0. It is noted also that the value of CN_s required for oscillation in the tapered case is significantly greater than that in the untapered oscillator. The position at which the taper starts is shown in Fig. 23.

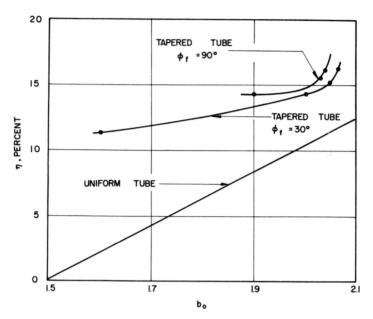

FIG. 22. Comparison of efficiency in uniform and tapered backward-wave oscillators. Interaction impedance is assumed constant. ($C_0 = 0.1$, $QC = 0.25$, $d_0 = 0$, $b = 1$, $a/b' = 2$.)

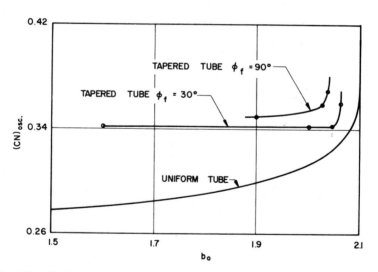

FIG. 23. Uniform and tapered normalized tube lengths versus b_0. These normalized oscillation lengths correspond to the efficiencies shown in Fig. 22. ($C_0 = 0.1$, $QC = 0.25$, $d_0 = 0$, $B = 1$, $a/b' = 2$.)

In comparing conditions for start-oscillation in tapered and untapered devices it is convenient to assume that the oscillators are characterized by the same current, voltage and frequency. Thus, if constant interaction impedance is assumed in addition to the above, then the value of C_0 is the same in the two devices. This is the basis of Figs. 22–24, where the results are plotted versus b_0. The higher efficiency is accompanied by an increased length in the tapered device.

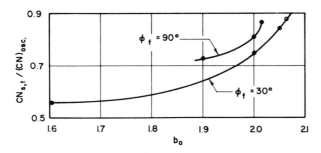

FIG. 24. Ratio of the length at which the taper starts to the total length of the tube. These curves correspond to the tapered tubes of Figs. 22 and 23.

The greater the interaction parameter C_0 the greater can be the strength of taper, although if the taper is too strong the tube will never start oscillating. This can be used to advantage in the forward-wave amplifier to suppress backward-wave oscillations.

Haddad[13] has found from a linear analysis of start-oscillation conditions in tapered-velocity oscillators that for relatively large values of C_0 the starting current is decreased when both the uniform and tapered tubes oscillate at the same frequency and have the same physical length. Also the tapered tube will oscillate at a somewhat lower voltage.

5 Efficiency Improvement in Crossed-Field Amplifiers

a. *General Background*

In Chapters IX and X we saw that the fundamentally different interaction mechanism in the crossed-field device led to significantly higher efficiencies than are usually obtained in O-type devices because of an inherent phase-focusing mechanism in which an electron bunch remains relatively fixed in phase position as its potential energy is converted to rf. The velocity-phase diagrams at the output plane of an M-FWA indicate that little change has taken place in the electron kinetic energy.

In the preceding sections of this chapter phase-focusing techniques were applied with good success to both O-type amplifiers and oscillators to gain efficiency improvement. Efficiencies are thus competitive between the two classes of devices.

The question thus arises as to what, if any, efficiency improvement can be realized in M-type amplifiers and oscillators? More precisely, can some of the electron kinetic energy in an M-type interaction configuration be converted to rf energy? The answer is found to be "yes." Basically the process to be used is the adiabatic deceleration of an electron bunch, which is realized by tapering either the dc electric or magnetic field together with a slowing of the rf circuit wave in order to maintain phase locking as the kinetic energy is converted to potential energy and thence to rf.

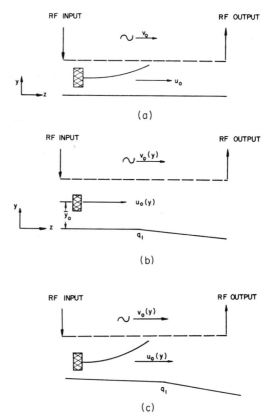

FIG. 25. Crossed-field energy conversion regions. (a) Potential energy conversion region, $v_0 = u_0 =$ constant. (b) Kinetic energy conversion region $\bar{y}_0 =$ constant and $u_0(y) = v_0(y)$. (c) Potential and kinetic energy conversion, $y \neq$ constant, $u_0(y) = v_0(y)$.

In this section it is shown that efficiency and gain improvements can be realized through such a process. The nonlinear interaction equations have been solved both by digital computer methods and in closed form for a hard-kernel-bunch model by Volkholz[19] and Rowe.[20]

It is possible to solve these nonlinear equations for the following three interesting cases:

(1) a potential energy conversion region in which there is beam-wave synchronism and the kinetic energy remains unchanged.

(2) a kinetic energy conversion region in which the electron y position remains invariant and thus so is its potential energy.

(3) a region in which both kinetic and potential energies are converted to rf, accomplished by tapering both the circuit-sole spacing and the circuit phase velocity.

The above three situations are illustrated in Fig. 25 for a crossed-field amplifier.

Whether or not the equivalent circuit transmission line impedance Z_0 and the interaction parameter D vary with distance depends on the type of rf structure used and the degree of velocity tapering. In the case of the interdigital line it is necessary to vary both the backwall spacing and the spacing between the fingers and the sole plate to obtain a coincident variation of phase and group velocity. It is shown in Appendix A that for a 50% phase velocity variation the interdigital line impedance varies less than 15%. As in the corresponding O-type analyses, the circuit equations or the equivalent conservation of energy expressions may be utilized.

b. Closed-Form Solution of M-FWA Nonlinear Equations in a Potential Energy Conversion Region

As in the case of O-type phase-focusing studies, a hard-kernel bunch is assumed throughout the interaction region and space-charge forces are neglected. The justification for such a model is again given by the results, which compare very favorably with those obtained from computer solutions where it is not necessary to make such approximations. The starting point for the analysis, developed by Volkholz,[19] is the nonlinear M-FWA equations of Chapter IX with the assumption $D \ll 1$. These are

$$\frac{dA(q)}{dq} = \frac{1}{2\pi} \int_0^{2\pi} \int_{\frac{1}{s}-\frac{1}{2}}^{\frac{1}{s}+\frac{1}{2}} \psi(p) \sin \Phi(p_0', \Phi_0', q) \, d\Phi_0' \, dp_0', \qquad (47)$$

$$\frac{d\theta(q)}{dq} - b = \frac{1}{2\pi A(q)} \int_0^{2\pi} \int_{-\frac{1}{s}-\frac{1}{2}}^{\frac{1}{s}+\frac{1}{2}} \psi(p) \cos \Phi(p_0', \Phi_0', q) \, d\Phi_0' \, dp_0', \qquad (48)$$

$$\frac{\partial \Phi(p_0, \Phi_0, q)}{\partial q} - \frac{d\theta(q)}{dq} = 2[u(p_0, \Phi_0, q) - u_i(p_0, \Phi_0, 0)], \qquad (49)$$

$$p(p_0, \Phi_0, q) = p_0 + \frac{1}{D} \int_{q_0}^{q} v_{y\omega}(p_0, \Phi_0, q) \, dq, \qquad (50)$$

$$\frac{\partial u(p_0, \Phi_0, q)}{\partial q} = -\frac{\omega}{2\omega_c D} \psi(p) A(q) \sin \Phi(p_0, \Phi_0, q)$$
$$+ \frac{s}{2lD} v_{y\omega}(p_0, \Phi_0, q) \qquad (51)$$

and

$$\frac{\partial v_{y\omega}(p_0, \Phi_0, q)}{\partial q} = \frac{l}{s} \varphi(p) A(q) \cos \Phi(p_0, \Phi_0, q)$$
$$- \frac{2l}{s} \left(\frac{\omega_c}{\omega}\right)^2 [u(p_0, \Phi_0, q) - u_i(p_0, \Phi_0, 0)], \qquad (52)$$

where

$$\psi(p) = \frac{\sinh\left[\frac{\omega ps}{\omega_c l}\left(1 + D\frac{d\theta(q)}{dq}\right)\right]}{\sinh\left[\frac{\omega}{\omega_c rl}\left(1 + D\frac{d\theta(q)}{dq}\right)\right]}$$

and

$$\varphi(p) = \frac{\cosh\left[\frac{\omega ps}{\omega_c l}\left(1 + D\frac{d\theta(q)}{dq}\right)\right]}{\sinh\left[\frac{\omega}{\omega_c rl}\left(1 + D\frac{d\theta(q)}{dq}\right)\right]}.$$

In these equations the dependence on the independent variables is noted for clarity and convenience. Hereafter these will be dropped in order to simplify the writing of equations.

In this purely potential energy conversion region we assume beam wave synchronism throughout and that a hard-kernel bunch is focused at the maximum decelerating field position of $\pi/2$ radians. These conditions are summarized below.

$$\frac{\partial \Phi}{\partial q} = 0, \quad b = 0, \quad \text{and} \quad u = 0, \qquad (53)$$

since there is no kinetic energy conversion.

5. EFFICIENCY IMPROVEMENT IN CROSSED-FIELD AMPLIFIERS

If we further assume that the acceleration terms in the nonlinear equations are small then the system reduces to the following set:

$$\frac{dA(q)}{dq} = \psi(p), \tag{54}$$

$$v_{y\omega} = \frac{\omega D}{\omega_c}\left(\frac{l}{s}\right)\psi(p)A(q), \tag{55}$$

and

$$p = p_0 + \int_{q_0}^{q} \frac{v_{y\omega}}{D} \, dq. \tag{56}$$

In Eq. (56), q_0 denotes the displacement plane at which the bunch is injected and the y-position of the bunch is denoted by p_0. Solving the above system yields

$$\frac{dp}{dq} = \frac{\omega l}{\omega_c s}\psi(p)\int_{q_0}^{q}\psi(p)\,dq. \tag{57}$$

This differential equation may be conveniently solved after appropriate definitions and transformations are made. The result is

$$q - q_0 = \frac{1}{\sqrt{2C}}\int \frac{d\pi(q)}{\sinh \pi(q)\sqrt{\frac{\alpha\pi(q)}{C} - 1}}, \tag{58}$$

where

$$\pi(q) \triangleq \frac{\omega s}{\omega_c l} p(q),$$

$$\alpha \triangleq \frac{\left(\frac{\omega}{\omega_c}\right)^2}{\left[\sinh\left(\frac{\omega}{\omega_c rl}\right)\right]^2}$$

and

$$C \triangleq \alpha \frac{\omega}{\omega_c}\frac{s}{l}p_0,$$

an integration constant determined by assuming $v_{y\omega} = 0$ at the initial bunch position q_0, p_0.

The following velocity and position equations are thus obtained for the bunch:

$$v_{y\omega} = \frac{Dl}{s}\left[2\frac{\omega}{\omega_c}\frac{s}{l}(p - p_0)\right]^{\frac{1}{2}}\psi(p) \tag{59}$$

and

$$q(p) - q_0 = \frac{s}{l} \int_{p'=p_0}^{p'=p} \frac{dp'}{\psi(p') \left[2 \frac{\omega}{\omega_c} \frac{s}{l}(p' - p_0)\right]^{\frac{1}{2}}}. \tag{60}$$

The rf wave amplitude function is then found to be

$$A(p) = \left[2 \frac{\omega_c}{\omega} \frac{s}{l}(p - p_0)\right]^{\frac{1}{2}}. \tag{61}$$

Fortunately the integral in the hard-kernel-bunch trajectory equation converges for $p > p_0$. Hence it is easily computed, and is shown in Fig. 26.

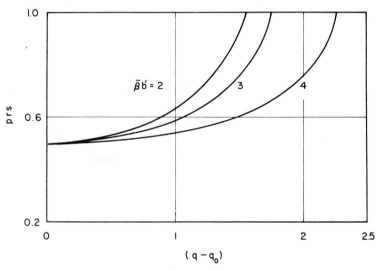

FIG. 26. Hard-kernel-bunch trajectories for a potential energy conversion region ($r = 0.5$).

c. Closed-Form Solution of M-FWA Nonlinear Equations in a Kinetic Energy Conversion Region

A kinetic energy conversion region, as illustrated in Fig. 25, might logically follow the potential energy conversion region after the electrons have moved an appreciable fraction of the distance from the sole to the anode, i.e., $prs \approx 0.8$ or 0.9. In this region the electrons are adiabatically decelerated in order to convert their kinetic energy to rf.

Specifically the conditions to be assumed in this region are that the

5. EFFICIENCY IMPROVEMENT IN CROSSED-FIELD AMPLIFIERS

potential energy is invariant with distance and also that the hard-kernel bunch remains synchronous with the rf wave as it slows down. These conditions are expressed as

$$v_{y\omega} = 0 \quad \text{or} \quad p = \text{constant}, \tag{62}$$

and

$$\frac{\partial \Phi}{\partial q} = 0. \tag{63}$$

Imposing the synchronism and constant potential energy conditions requires a particular variation of the rf structure phase velocity as in O-type devices and a prescribed variation (reduction) in the z-directed dc velocity. Thus $\partial u/\partial q$ must vary such that $p = $ constant and the additional advantage is accrued that collection on the rf structure is virtually eliminated.

Slowing of the dc velocity is accomplished either by increasing the circuit-sole spacing gradually so that the dc electric field is reduced or by slowly increasing the static magnetic field. Assume that the z-directed (dc) electric field is small compared to the transverse component and that since the electrons are being decelerated adiabatically $\partial v_{y\omega}/\partial q$ is small. If the entrance displacement plane is denoted as q, the transverse dc field is written as

$$E_{yT} = E_{y0}[1 - 2De_y]. \tag{64}$$

Under these assumptions the nonlinear equations are written as

$$\frac{dA(q)}{dq}(1 + 2Du) = \psi(p_1), \tag{65}$$

$$\frac{d\theta(q)}{dq} - b = 0, \tag{66}$$

$$\frac{du}{dq}(1 + 2Du)^2 = -\frac{\omega}{2\omega_c D}\psi(p_1)A(q), \tag{67}$$

$$e_y + u = 0, \tag{68}$$

$$\frac{d\theta(q)}{dq} = -\frac{2u}{1 + 2Du} \tag{69}$$

and

$$p = \text{constant} = p_1. \tag{70}$$

Since the kinetic energy conversion region presupposes that the bunch has moved most of the way to the circuit, it is reasonable to assume that $\psi(p_1) \approx 1$ throughout the region. If the conversion of bunch kinetic energy to rf is appreciable, then one cannot neglect $2Du$ compared to unity. For slight improvement through weak tapers (20%) this is a reasonable assumption.

The above system is first solved to find $u(q)$ and then all other quantities such as $e_y(q)$, etc., are easily calculated. Equations (65) and (67) are combined to obtain

$$X^3 \frac{d^2X}{dq^2} + 2X^2 \left(\frac{dX}{dq}\right)^2 + \frac{\omega}{\omega_c} \psi^2(p_1) = 0, \tag{71}$$

where $X(q) \triangleq 1 + 2Du(q)$. Equation (71) may be solved directly subject to the boundary condition $X = 1$ at $q = q_1$:

$$2\frac{\omega}{\omega_c}(q - q_1) = X\sqrt{C_1 - \frac{\omega}{\omega_c}\psi^2 X^2} - \sqrt{C_1 - \frac{\omega}{\omega_c}\psi^2}$$
$$- \frac{C_1}{\psi(\omega/\omega_c)^{\frac{1}{2}}} \left\{ \sin^{-1}\sqrt{\frac{1}{C_1}\frac{\omega}{\omega_c}}\psi X \right\} \cdot \left\{ -\sin^{-1}\left[\sqrt{\frac{1}{C_1}\frac{\omega}{\omega_c}}\psi\right]\right\}. \tag{72}$$

The constant C_1 is chosen such that $A(q_1) = A_1$ and from Eq. (67)

$$A_1 = -\frac{\omega_c}{\omega\psi}\frac{dX}{dq}\bigg|_{q=q_1}. \tag{73}$$

The distance-velocity profile is then found to be

$$2\psi(\omega/\omega_c)^{\frac{1}{2}}(q - q_1) = X[1 + (\omega/\omega_c)A_1^2 - X^2]^{\frac{1}{2}} - A_1(\omega/\omega_c)^{\frac{1}{2}}$$
$$-[A_1^2(\omega/\omega_c) + 1]\left\{\sin^{-1}\left[\frac{X}{\sqrt{1 + A_1^2(\omega/\omega_c)}}\right] - \sin^{-1}\left[\frac{1}{\sqrt{1 + A_1^2(\omega/\omega_c)}}\right]\right\}. \tag{74}$$

This dependence of velocity on distance is shown in Fig. 27.

Equation (74) is solved for $u(q - q_1)$ to obtain the circuit-sole spacing and circuit phase velocity functions:

$$b'(q) = \frac{b_0'}{1 - 2De_y} = \frac{b_0'}{1 + 2Du} = \frac{b_0'}{X} \tag{75}$$

and

$$v_0(q) = \frac{\bar{u}_0}{1 + Db(q)} + \bar{u}_0(1 + 2Du) = \bar{u}_0 X, \tag{76}$$

or

$$b(q) = \frac{1 - X}{DX}.$$

It is interesting to note that, as expected, the phase-focusing result for adiabatic slowing of a bunch given by Eq. (74) is exactly the same as for the O-type device, Eq. (15), if one uses the correspondence

$$2CA_0{}^2 \Leftrightarrow (\omega/\omega_c)A_1{}^2,$$
$$\sqrt{2C_0}\,y \Leftrightarrow \psi(\omega/\omega_c)^{\frac{1}{2}}(q - q_1). \tag{77}$$

Since $\psi \approx 1$ and $\omega_c/\omega \sim 0.75$ for high efficiency, the rate of taper is somewhat faster in the M-type device.

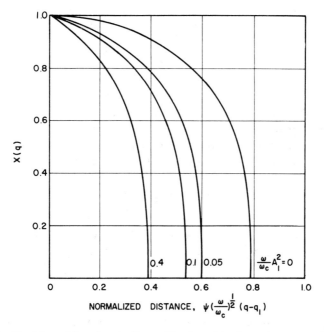

Fig. 27. Hard-kernel-bunch velocity profiles for a kinetic energy conversion region in an M-FWA.

d. Closed-Form Solution of M-FWA Nonlinear Equations in a PE-KE Conversion Region

A hard-kernel bunch is again assumed and in this PE-KE conversion region the bunch moves towards the circuit in a definite path while it is being slowed in order to convert some of its kinetic energy. The mathematical problem begins with the equations of Section 5.b and assumes bunch synchronism with the wave, i.e., $\partial \Phi/\partial q = 0$. Thus we see that tapering of the circuit-sole spacing and the rf circuit phase velocity occurs over the entire length of the region.

The nonlinear equations for this region are written as follows:

$$\frac{dA(q)}{dq} = \frac{\psi(p)}{1 + 2Du(q)}, \tag{78}$$

$$[1 + 2Du(q)]\frac{\partial[1 + 2Du(q)]}{\partial q} = -\frac{\omega}{\omega_c}\frac{\psi(p)A(q)}{[1 + 2Du(q)]}$$

$$+ \frac{s}{l}[1 + 2Du(q)]\frac{dp(q)}{dq}, \tag{79}$$

$$p = -\frac{2u(q)}{1 + 2Du(q)}, \tag{80}$$

and

$$e_y(q) = -\frac{s}{2l}\left(\frac{\omega}{\omega_c}\right)^2 [1 + 2Du(q)]\frac{dv_{y\omega}}{dq} - u(q). \tag{81}$$

One additional condition is required to close the system, namely a relationship between $u(q)$ and $p(q)$, which amounts to specifying the relative amounts of potential and kinetic energy conversion. It is convenient to write this function as follows, with $[1 + 2Du(q)] \triangleq X(q)$:

$$\frac{dX}{dp} = \frac{s}{l}\left(1 - \frac{\beta^2}{X^2}\right), \tag{82}$$

where β is the circuit phase constant. The higher the power of X, the more rapid is the taper. Solution of Eq. (79) for p yields

$$prs = 1 + rl\left[X - \beta \tanh^{-1}\frac{X}{\beta}\right], \tag{83}$$

where $prs = 1$ indicates collection on the circuit. It should be noted that the form of the above equations is quite arbitrary. For these electrons it is assumed that $X \to 0$, indicating a complete conversion of kinetic energy. Equation (83) is illustrated in Fig. 28.

The beam injection position is denoted by r and thus

$$\beta \tanh^{-1}\frac{1}{\beta} = 1 + \left(\frac{1-r}{rl}\right) \approx \frac{r+1}{2r}. \tag{84}$$

Combining Eqs. (78) and (79) gives

$$\frac{dp}{dq}\left(\frac{dX}{dp} - \frac{s}{l}\right) = -\frac{\omega}{\omega_c}\frac{\psi(p)}{X^2}\int_{q_0}^{q}\frac{\psi(p)}{X}dq. \tag{85}$$

5. EFFICIENCY IMPROVEMENT IN CROSSED-FIELD AMPLIFIERS

Thus from the solution of Eqs. (83) and (85) we can find both $X(q)$ and $p(q)$. After some manipulation the solution is

$$\left(\frac{dp}{dq}\right)^2 = \frac{2}{\beta^2}\left(\frac{\omega}{\omega_c}\right)\frac{l}{s}\psi^2[p, X(p)]\int_{p'=1/s}^{p'=p}\frac{dp'}{X(p')}$$

$$= \frac{1}{2\beta^2}\left(\frac{\omega}{\omega_c}\right)\left(\frac{l}{s}\right)^2 \psi^2[p, X(p)]\ln\left(\frac{\beta^2 - X^2}{\beta^2 - 1}\right), \qquad (86)$$

where it has been assumed that the velocity $v_{y\omega} = 0$ and $dp/dq = 0$ at the bunch entrance position, r.

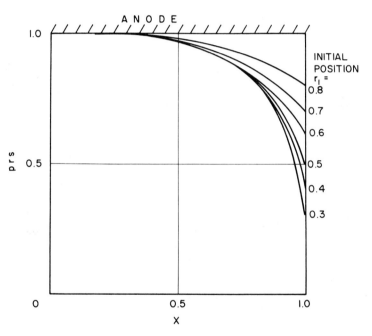

FIG. 28. Relationship between velocity taper and bunch transverse position.

Equation (86) is then integrated as follows:

$$\frac{l}{\beta s}\left(\frac{\omega}{\omega_c}\right)^{\frac{1}{2}}(q - q_0) = \int_{p'=1/s}^{p'=p} \frac{\sinh\left[\frac{\omega}{\omega_c rl}\frac{1}{X(p')}\right]dp'}{\sinh\left[\frac{\omega}{\omega_c}\frac{s}{l}\frac{p'}{X(p')}\right]\sqrt{\ln\left(\frac{\beta^2 - X^2(p')}{\beta^2 - 1}\right)}}. \qquad (87)$$

Fortunately the integrand of Eq. (87) is convergent so that the singularity at $p = p_0$ is not troublesome. Equation (87) gives p versus q and $X(q)$ is given by

$$\frac{1}{\beta}\left(\frac{\omega}{\omega_c}\right)^{\frac{1}{2}}(q-q_0) = \int_{X'=1}^{X'=X} \frac{X'^2 \sinh\left[\frac{\omega}{\omega_c rl}\frac{1}{X}\right]dX'}{(\beta^2 - X'^2)\sinh\left[\frac{\omega}{\omega_c}\frac{s}{l}\frac{p(X')}{X}\right]\sqrt{\ln\left(\frac{\beta^2-X^2}{\beta^2-1}\right)}}. \tag{88}$$

These are both shown in Fig. 29. The rf voltage amplitude function is

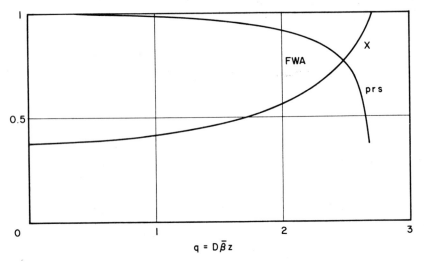

Fig. 29. Variation of bunch position and velocity versus distance [$r = 0.5$, $\beta b' = (\omega/\omega_c)(1/rl) = 4$].

found to be

$$A(q) = \beta \left[\left(\frac{\omega_c}{\omega}\frac{l}{2s}\right)\ln\left(\frac{\beta^2 - X^2}{\beta^2 - 1}\right)\right]^{\frac{1}{2}}. \tag{89}$$

This is shown in Fig. 30 for the same case as above and is compared with the amplitude for a purely potential energy conversion.

The variation of interaction gap spacing required is given by

$$b'(q) = \frac{b_0'}{X\left[1 + \frac{s}{l}\left(\frac{\omega D}{\omega_c}\right)^2\left(X\frac{d^2p}{dq^2} + \frac{dX}{dq}\frac{dp}{dq}\right)\right]}, \tag{90}$$

where b_0' is the spacing in the uniform region. A typical set of results for Eq. (90) is shown in Fig. 31.

5. EFFICIENCY IMPROVEMENT IN CROSSED-FIELD AMPLIFIERS 473

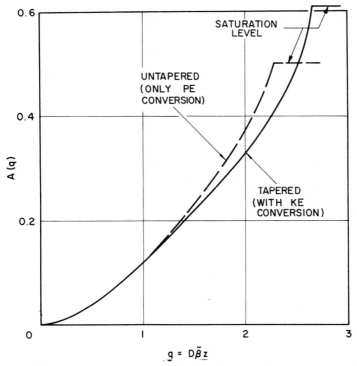

FIG. 30. Rf voltage amplitude versus distance for potential energy conversion and combined potential-kinetic energy conversion $(r = 0.5, \ A_0 = 0, \ \beta b' = 4)$.

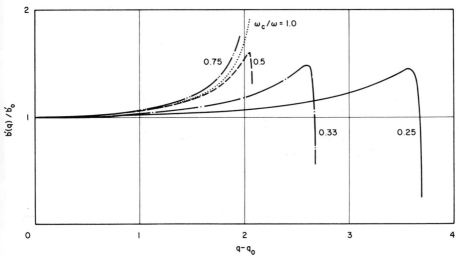

FIG. 31. Circuit-sole spacing required to maintain beam-wave synchronism in a potential-kinetic energy conversion region $(r = 0.5, \ r_1 = 0.5, \ D = 0.05, \ l = 1.95)$.

e. Nonlinear M-FWA Equations with Velocity and Interaction-Space Tapering

In order to evaluate the hard-kernel-bunch results in the presence of finite space-charge forces and the effects of fast electrons, it is necessary to solve the nonlinear M-type interaction equations for prescribed velocity and interaction-space taperings. It is also desirable to ask the computer to determine the optimum taper functions for realistic bunch configurations.

In general it might happen that the rf structure impedance and hence D will vary with distance as a result of tapering. Under these conditions the general variable-parameter (nonuniform) transmission-line equation of Chapter III, Section 7 will be required. Fortunately under most conditions these parameters remain relatively invariant with distance and thus the uniform line equations may be used. The appropriate FWA equations for this study are the following.

Circuit Equations

$$\frac{d^2A}{dq^2} - \frac{D}{1+Db}\frac{db}{dq}\frac{dA}{dq} - A\left\{\left(\frac{1}{D}+\frac{d\theta}{dq}\right)^2 - \left(\frac{1+Db}{D}\right)^2\right\}$$

$$= \mp \frac{1+Db}{\pi D}\left[\int_0^{2\pi}\int_{\frac{1}{s}-\frac{1}{2}}^{\frac{1}{s}+\frac{1}{2}} \psi(p)\cos\Phi\,\frac{1+2Du_i}{1+2Du}\,d\Phi_0'\,dp_0'\right.$$

$$\left. + 2dD\int_0^{2\pi}\int_{\frac{1}{s}-\frac{1}{2}}^{\frac{1}{s}+\frac{1}{2}} \psi(p)\sin\Phi\,\frac{1+2Du_i}{1+2Du}\,d\Phi_0'\,dp_0'\right], \qquad (91)$$

and

$$2\frac{dA}{dq}\left(\frac{1}{D}+\frac{d\theta}{dq}\right) + A\left\{\frac{d^2\theta}{dq^2} - \frac{D}{1+Db}\left(\frac{1}{D}+\frac{d\theta}{dq}\right)\frac{db}{dq} - \frac{2d}{D}(1+Db)^2\right\}$$

$$= \pm\frac{1+Db}{\pi D}\left[\int_0^{2\pi}\int_{\frac{1}{s}-\frac{1}{2}}^{\frac{1}{s}+\frac{1}{2}} \psi(p)\sin\Phi\,\frac{1+2Du_i}{1+2Du}\,d\Phi_0'\,dp_0'\right.$$

$$\left. - 2dD\int_0^{2\pi}\int_{\frac{1}{s}-\frac{1}{2}}^{\frac{1}{s}+\frac{1}{2}} \psi(p)\cos\Phi\,\frac{1+2Du_i}{1+2Du}\,d\Phi_0'\,dp_0'\right], \qquad (92)$$

where the upper sign applies to the forward-wave case, and the lower sign to the backward-wave case. The loss terms contain the parameter

5. EFFICIENCY IMPROVEMENT IN CROSSED-FIELD AMPLIFIERS

d defined by the relation $\alpha = \beta D d$. The transverse coupling function $\psi(p)$ is written as

$$\psi(p) = \frac{\sinh \beta y}{\sinh \beta b'} = \frac{\sinh\left[\frac{\omega}{\omega_c} c p \frac{s}{l}\left(1 + D \frac{d\theta}{dq}\right)\right]}{\sinh\left[\frac{\omega}{\omega_c} \frac{c}{rl}\left(1 + D \frac{d\theta}{dq}\right)\right]}$$

where $l = \bar{u}_0/\omega_c \bar{y}_0$, and $c(q) = b'(q)/b_0'$ carries the information on variable circuit-sole spacing.

Force Equations

$$\frac{\partial u}{\partial q}(1 + 2Du) = \frac{\omega}{2\omega_c}\psi(p)\left[\frac{dA}{dq}\cos\Phi - A\sin\Phi\left(\frac{1}{D} + \frac{d\theta}{dq}\right)\right]$$
$$+ \frac{1}{2D^2}\frac{s}{l}v_{y\omega} + \left(\frac{\omega_p}{\omega}\right)^2\frac{rs}{\pi D^2} F_{2-z} \qquad (93)$$

and

$$(1 + 2Du)\left[\frac{\partial v_{y\omega}}{\partial q} + \left(\frac{\omega_c}{\omega}\right)^2 \frac{l}{sD}\right] = c\frac{l}{s}\varphi(p)A\cos\Phi\left(1 + D\frac{d\theta}{dq}\right)$$
$$+ \left(\frac{\omega_c}{\omega}\right)^2 \frac{l}{sD}\left(\frac{V_a}{2V_0} rl - 1 + \frac{1}{c}\right)$$
$$- 2\left(\frac{\omega_p}{\omega}\right)^2 \frac{rl}{\pi D}\frac{\omega_c}{\omega} F_{2-y}, \qquad (94)$$

where $\varphi(p)$ is defined by

$$\varphi(p) = \frac{\cosh\left[\frac{\omega}{\omega_c} c p \frac{s}{l}\left(1 + D\frac{d\theta}{dq}\right)\right]}{\sinh\left[\frac{\omega}{\omega_c}\frac{c}{rl}\left(1 + D\frac{d\theta}{dq}\right)\right]}.$$

Dependent Variables

$$\{\text{Eq. (11) of Chapter IX}\} \qquad (95)$$

and

$$\{\text{Eq. (12) of Chapter IX}\}. \qquad (96)$$

In writing the above equations a Brillouin beam has been assumed and the varying dc electric field has been incorporated.

The equations are solved in basically the same manner as in the uniform line case except that either $b(q)$ and $e_y(q)$ must be specified or the computer must be given a criterion by which to determine optimum variations for these functions. If it is recalled that the hard-kernel velocity tapers derived for O-type devices gave nearly the same results as

Efficiency Improvement in Crossed-Field Backward-Wave Oscillators

a. Introduction

The problem of efficiency improvement in M-BWO's and M-BWA's relates to that in M-FWA's in exactly the same manner as in O-FWA's and O-BWO's. A combination of circuit velocity tapering and interaction-space tapering will lead to gain and efficiency improvement through added conversion of electron kinetic energy in an M-type device. The problem is treated by the same methods used in Section 5 and thus many of the details will be left out and only the final results given.

Recall that the only differences between the FWA and BWO nonlinear equations are changes in sign preceding the right-hand side of both circuit equations; all other equations remain unchanged. The procedures utilized follow the work of Volkholz[19] and Rowe.[20]

b. Closed-Form Solution of M-BWO Nonlinear Equations in a Potential Energy Conversion Region

The potential energy conversion in a crossed-field backward-wave oscillator may be studied using the hard-kernel-bunch model as in the case of the M-FWA. The fundamental assumptions are the same in the two cases; i.e., inter-bunch space-charge forces are neglected and the gain parameter is considered small, $D \ll 1$. The simplified nonlinear equations (47)–(52) of Section 5.b are applicable after appropriate sign changes are made in the circuit equations to account for backward energy flow.

The M-BWO nonlinear circuit equations are

$$\frac{dA(q)}{dq} = -\frac{1}{2\pi} \int_0^{2\pi} \int_{\frac{1}{s}-\frac{1}{2}}^{\frac{1}{s}+\frac{1}{2}} \psi(p) \sin \Phi(p_0', \Phi_0', q) \, d\Phi_0' \, dp_0' \tag{97}$$

and

$$\frac{d\theta(q)}{dq} - b = -\frac{1}{2\pi A(q)} \int_0^{2\pi} \int_{\frac{1}{s}-\frac{1}{2}}^{\frac{1}{s}+\frac{1}{2}} \psi(p) \cos \Phi(p_0', \Phi_0', q) \, d\Phi_0' \, dp_0'. \tag{98}$$

In addition to the above, Eqs. (49)–(52) are used to complete the system.

6. EFFICIENCY IMPROVEMENT IN CROSSED-FIELD OSCILLATORS

We again assume a purely potential energy conversion region in which beam-wave synchronism is maintained and the hard-kernel bunch is considered focused at the maximum decelerating field position of $\pi/2$ radians. These constraints are specified by Eqs. (53). Again the acceleration terms are neglected since $D \ll 1$, and the nonlinear equations simplify to

$$\frac{dA(q)}{dq} = -\psi(p), \tag{99}$$

$$v_{y\omega} = \frac{\omega D}{\omega_c}\left(\frac{l}{s}\right)\psi(p)A(q), \tag{100}$$

and

$$p = p_0 + \int_0^q \frac{v_{y\omega}}{D}\,dq. \tag{101}$$

The above nonlinear system may be solved in the same manner as previously used (M-FWA), resulting in

$$q - q_0 = \frac{1}{\sqrt{-2C_1}}\int \frac{d\pi(q)}{\sin \pi(q)\sqrt{[\alpha\pi(q)/C_1] + 1}}, \tag{102}$$

where $\pi(q)$ and α are defined in Section 5.b. C_1 is an integration constant yet to be determined.

Assume that the initial bunch position is denoted by p_0, q_0 and that

$$A(q_0) \triangleq A_0. \tag{103}$$

The integration constant is then

$$C_1 = \alpha\left[\frac{\omega}{\omega_c}\frac{s}{l}p_0 - \frac{A_1^2}{2}\left(\frac{\omega}{\omega_c}\right)^2\right] \tag{104}$$

and the y-directed velocity is

$$v_{y\omega} = D(l/s)\,\psi(p)\sqrt{A_0^2\left(\frac{\omega}{\omega_c}\right)^2 - 2\frac{\omega}{\omega_c}\frac{s}{l}(p - p_0)}. \tag{105}$$

The rf level as a function of the p-position of the bunch is then found from Eqs. (55) and (105).

$$A(p) = \frac{\omega_c}{\omega}\sqrt{A_0^2\left(\frac{\omega}{\omega_c}\right)^2 - 2\frac{\omega}{\omega_c}\frac{s}{l}(p - p_0)}. \tag{106}$$

Substitution of Eq. (104) into Eq. (102) and simplifying yields

$$q - q_0 = \frac{s}{l} \int_{p'=p_0}^{p'=p} \frac{dp(q)}{\psi(p') \sqrt{A_0^2 \left(\frac{\omega}{\omega_c}\right)^2 - 2\frac{\omega}{\omega_c}\frac{s}{l}(p' - p_0)}}. \quad (107)$$

The amplitude versus distance is calculated from a substitution for A_0 in Eq. (106) into Eq. (107). This variation is illustrated in Fig. 32.

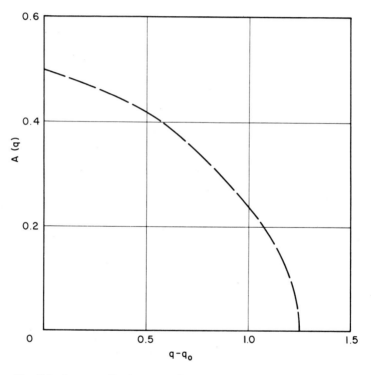

FIG. 32. Rf voltage amplitude versus distance for an M-BWO with potential energy conversion ($r = 0.5$, $\beta b' = 4$).

c. *Closed-Form Solution of M-BWO Nonlinear Equations in a Kinetic Energy Conversion Region*

Following a potential energy conversion region in an M-BWO in which the hard-kernel bunch moves through an appreciable fraction of the sole-circuit distance, we will consider a kinetic energy conversion region in which the p-position of the bunch remains invariant. Again the bunch will be maintained synchronous with the rf wave while being adiabat-

6. EFFICIENCY IMPROVEMENT IN CROSSED-FIELD OSCILLATORS

ically decelerated in order to convert some of the bunch kinetic energy to rf. The constraints are given by Eqs. (62) and (63). As in the M-FWA case this is accomplished by both slowing the rf wave phase velocity and tapering the circuit-sole spacing. Alternately, of course, the magnetic field could be appropriately tapered rather than changing the circuit-sole spacing.

Again the circuit equations must be modified to account for backward energy flow on the circuit. For the M-BWO, Eq. (65) becomes

$$\frac{dA(q)}{dq}(1 + 2Du) = -\psi(p_1). \tag{108}$$

The remainder of the system is made up of Eqs. (66)–(70). The boundary conditions invoked are that

$$A(q_1) \triangleq A_1 \tag{109}$$

and

$$1 + 2Du(q_1) = X(q_1) \equiv 1. \tag{110}$$

The initial bunch position is denoted by p_1, q_1 and since $v_{y\omega} = 0$ then $p \equiv p_1$ independent of q.

The above system is solved in a similar manner to the corresponding FWA equations. Equations (108) and (67) are combined to give

$$X^3 \frac{d^2X}{dq^2} + 2X^2 \left(\frac{dX}{dq}\right)^2 - \frac{\omega}{\omega_c} \psi^2(p_1) = 0. \tag{111}$$

Note the similarity to Eq. (71) of this chapter. Proceeding in a manner similar to that used previously yields the following velocity-distance relation:

$$2\psi \sqrt{(\omega/\omega_c)}(q - q_1) = \sqrt{(\omega/\omega_c) A_1^2 - X \sqrt{[(\omega/\omega_c) A_1^2 - 1] + X^2}}$$

$$+ [(\omega/\omega_c) A_1^2 - 1] \ln \frac{X + \sqrt{[(\omega/\omega_c) A_1^2 - 1] + X^2}}{1 + (\omega/\omega_c) A_1^2}. \tag{112}$$

It is again interesting to note that the result is identical with that for kinetic energy conversion in an O-BWO if the following transformations are introduced:

$$(\omega/\omega_c) A_1^2 \Leftrightarrow 2C_0 A_0^2 \tag{113}$$

and

$$\psi \sqrt{(\omega/\omega_c)}(q - q_1) \Leftrightarrow \sqrt{2C_0} y. \tag{114}$$

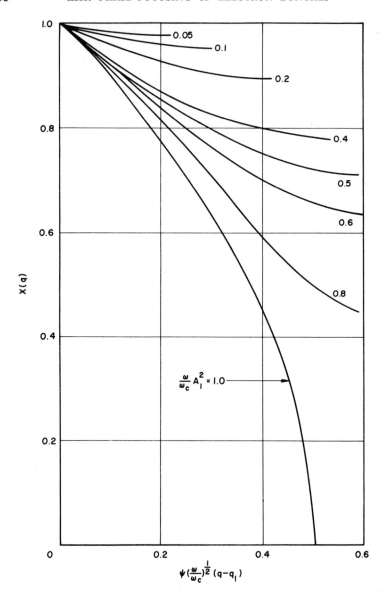

FIG. 33. Hard-kernel-bunch velocity profiles for a kinetic energy conversion region in an M-BWO.

6. EFFICIENCY IMPROVEMENT IN CROSSED-FIELD OSCILLATORS

The velocity-distance relation given by Eq. (112) is shown in Fig. 33. The total y-directed electric field is given by

$$E_{yT} = E_{y0}[1 - 2De_y] \tag{115}$$

and the electric field decrement e_y is related to the velocity by

$$e_y = -u = \frac{1-X}{2D}. \tag{116}$$

Thus both the required circuit phase velocity variation and the circuit-sole spacing are obtained from Fig. 33.

d. Closed-Form Solution of M-BWO Nonlinear Equations in a PE-KE Conversion Region

The constraints of the previous two sections may be invoked simultaneously on the nonlinear system and solution forms obtained. We consider a region in which both potential energy and kinetic energy are simultaneously converted. Again a hard-kernel bunch is assumed to be in synchronism with the rf wave. In such a PE-KE conversion region the circuit phase velocity and the circuit-sole spacing are varied over the entire length of the region.

The M-BWO circuit equation is

$$\frac{dA(q)}{dq} = -\frac{\psi(p)}{1 + 2Du(q)}. \tag{117}$$

In addition, Eqs. (79)–(81) complete the system. Again the relative amounts of potential and kinetic energy converted are specified by the arbitrary function (again the exact form is somewhat arbitrary):

$$\frac{dX}{dp} = \frac{s}{l}\left(1 - \frac{\beta^2}{X^2}\right), \tag{118}$$

where again β is the circuit phase constant.

Equations (79) and (117) are combined to obtain

$$\frac{dp}{dq}\left(\frac{dX}{dp} - \frac{s}{l}\right) = \frac{\omega}{\omega_c} \frac{\psi(p)}{X^2} \int_{q_0}^{q} \frac{\psi(p)}{X} dq. \tag{119}$$

Equations (118) and (119) are combined to give

$$\frac{dp}{dq} = -\left(\frac{\omega}{\omega_c}\right)\frac{l}{s}\frac{\psi(p)}{\beta^2}\int_{q_0}^{q}\frac{\psi(p)}{X} dq. \tag{120}$$

Solving Eqs. (83) and (119) after some manipulation yields

$$\left(\frac{dp}{dq}\right)^2 = \frac{\omega_c}{\omega}\frac{l}{s}\psi^2(p)\left[C_1 - 2\gamma^2 \int \frac{dp}{X(p)}\right], \tag{121}$$

where $\gamma \triangleq \omega/\beta\omega_c$. The integration constant C_1 is determined from the initial conditions.

In Section 6.b on potential energy conversion it was assumed that $A(q_0) \triangleq A_0$ at $p = p_0$, which occurs at $q = q_0$. Recall from the potential energy conversion section that

$$\frac{dp}{dq} = \frac{v_{y\omega}}{D} = \frac{l}{s}\psi(p)\sqrt{A_0^2\left(\frac{\omega}{\omega_c}\right)^2 - 2\frac{\omega}{\omega_c}\frac{s}{l}(p - p_0)}. \tag{122}$$

Assume that $p_0 = 1/s$; thus from Eq. (122)

$$\left(\frac{dp}{dq}\right)^2 = \left(\frac{l}{s}\right)^2 \psi^2(p) A_0^2 \left(\frac{\omega}{\omega_c}\right)^2. \tag{123}$$

Then C_1 from Eq. (121) is given by

$$C_1 = \frac{l}{s}\left(\frac{\omega}{\omega_c}\right)^3 A_0^2. \tag{124}$$

The integral of Eq. (121) is easily evaluated as

$$\int_{p'=1/s}^{p'=p} \frac{dp'}{X(p')} = \frac{l}{2s}\ln\left(\frac{\beta^2 - X^2}{\beta^2 - 1}\right). \tag{125}$$

Substituting Eqs. (124) and (125) into Eq. (121) and simplifying yields

$$(q - q_0)\frac{l}{s}\left(\frac{\omega}{\omega_c}\right)^{\frac{1}{2}} = \int_{p'=1/s}^{p'=p} \frac{dp'}{\psi(p')\left[\frac{\omega}{\omega_c}A_0^2 - \frac{1}{\beta^2}\ln\left(\frac{\beta^2 - X^2(p')}{\beta^2 - 1}\right)\right]^{\frac{1}{2}}}, \tag{126}$$

where

$$\psi(p') = \frac{\sinh\left[\frac{\omega}{\omega_c}\frac{1}{rl}\frac{1}{X(p')}\right]}{\sinh\left[\frac{\omega}{\omega_c}\frac{s}{l}\frac{p'}{X(p')}\right]}.$$

From Eq. (118), Eq. (126) may be rewritten as

$$(q - q_0)\left(\frac{\omega}{\omega_c}\right)^{\frac{1}{2}} = \int_{X'=1}^{X'=X} \frac{X'^2 \, dX'}{\psi(p')(X'^2 - \beta^2)\left[\frac{\omega}{\omega_c}A_0^2 - \frac{1}{\beta^2}\ln\left(\frac{\beta^2 - X'^2}{\beta^2 - 1}\right)\right]^{\frac{1}{2}}}. \tag{127}$$

6. EFFICIENCY IMPROVEMENT IN CROSSED-FIELD OSCILLATORS

Equation (126) gives the p-q trajectory information, while Eq. (127) yields q versus X (velocity) information. From a knowledge of X versus q, e_y may be calculated from Eq. (81). The amplitude function is evaluated from Eq. (79).

The velocity-distance and amplitude-distance functions may also be determined by using an alternative boundary condition in which the rf amplitude is assumed to be zero at the displacement plane, corresponding to collection of the hard-kernel bunch on the circuit. As a starting point rewrite Eq. (121) as

$$\left(\frac{dp}{dq}\right)^2 = 2\gamma^2 \frac{\omega_c}{\omega} \frac{l}{s} \psi^2(p) \left[C_2 - \int \frac{dp}{X(p)}\right]. \tag{128}$$

The amplitude is given by

$$A = \frac{\beta^2(s/l)}{(\omega/\omega_c)\,\psi(p)} \frac{dp}{dq}. \tag{129}$$

Combining Eqs. (128) and (129) yields

$$A^2 = 2\beta^2 \frac{\omega_c}{\omega} \frac{s}{l} \left[C_2 - \int \frac{dp}{X(p)}\right]. \tag{130}$$

Now choose

$$C_2 = \int \frac{dp}{X(p)} \quad \text{at} \quad p = \frac{1}{rs},$$

which corresponds to collection of the hard-kernel bunch on the circuit and makes $A = 0$ at q corresponding to $p = 1/rs$. Then

$$\int_{p'=p}^{p'=1/rs} \frac{dp}{X(p)} = \frac{l}{2s} \int_{X'=X}^{X'=0} \frac{2X\,dX}{X^2 - \beta^2}$$

$$= \frac{l}{2s} \ln \frac{\beta^2}{\beta^2 - X^2} \tag{131}$$

and from Eq. (128)

$$\frac{dp}{dq} = \left[\frac{\omega}{\omega_c}\left(\frac{l}{s}\right)^2 \frac{\psi^2(p)}{\beta^2} \ln \frac{\beta^2}{\beta^2 - X^2}\right]^{\frac{1}{2}}. \tag{132}$$

The bunch trajectory equation is then written as

$$\frac{1}{\beta} \frac{l}{s} \left(\frac{\omega}{\omega_c}\right)^{\frac{1}{2}} (q - q_0) = \int_{p'=p_0}^{p'=p} \frac{dp}{\psi(p)\left[\ln \frac{\beta^2}{\beta^2 - X^2}\right]^{\frac{1}{2}}}. \tag{133}$$

The amplitude function is then

$$A = \left[\beta^2 \frac{\omega_c}{\omega} \ln \frac{\beta^2}{\beta^2 - X^2}\right]^{\frac{1}{2}}. \tag{134}$$

In this alternate description the initial signal level A_0 is eliminated and replaced by the condition that the hard-kernel-bunch trajectory terminates on the circuit with zero kinetic energy. Plots of Eqs. (133) and (134) are shown in Figs. 34 and 35. The amplitude is also compared with that for only potential energy conversion.

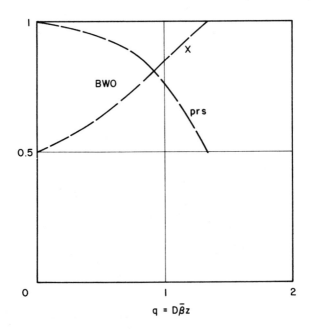

FIG. 34. Variation of bunch position and velocity versus distance for an M-BWO $[r = 0.5, \quad \beta b' = (\omega/\omega_c)(1/rl) = 4]$.

e. Nonlinear M-BWO Equations with Velocity and Interaction-Space Tapering

Tapers in nonlinear M-BWO's may be evaluated including the effects of fast electrons and space-charge forces in much the same manner as in the case of the M-FWA. The M-FWA equations are given in Section 5.e and it is recalled that these are applicable to the M-BWO if one changes the signs on the right-hand sides of Eqs. (91) and (92).

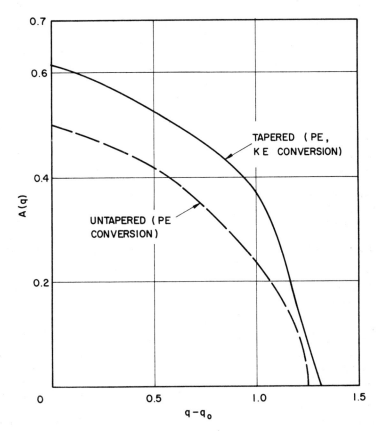

Fig. 35. Rf voltage amplitude versus distance for an M-BWO ($r = 0.5$, $\beta b' = 4$).

References

1. Pierce, J. R., "Traveling Wave Tubes." Van Nostrand, Princeton, N.J., 1950.
2. Slater, J. C., "Microwave Electronics." Van Nostrand, Princeton, N.J., 1950.
3. Cutler, C. C., "Increasing Traveling Wave Tube Efficiency by Variation of Phase Velocity." Bell Telephone Labs. Internal Memo No. MM-53-1500-28 (July 1953).
4. Geppert, D. V., Analysis of traveling-wave tubes with tapered velocity parameter. *Proc. IRE* **46**, No. 9, 1658 (1958).
5. Haddad, G. I., and Bevensee, R. M., "Start-oscillation conditions of tapered backward-wave oscillators," *IEEE Trans. Electron Devices*, Vol. ED-10, 389-393 (November 1963).
6. Haddad, G. I., and Rowe, J. E., "Start-oscillation conditions in nonuniform backward-wave oscillators," *IEEE Trans. Electron Devices*, Vol. ED-11, No. 1, 31-37 (January 1964).

7. Filimenov, G. F., Isochronous traveling-wave tube. *Radio Eng. Electron.* (*USSR*) (*Engl. Transl.*) **3**, No. 1, 124-135 (1958).
8. Vainshtein, L. A., Theorie non linéaire du tube à propagation d'onde. *Congr. Intern. Tubes Hyperfrequences, Paris, 1956* pp. 375-381.
   Vainshtein, L. A., Electron waves in delaying systems. Non-linear equations of traveling wave valves. *Radio Eng. Electron.* (*USSR*) (*Engl. Transl.*) **2**, No. 6, 22-32 (1957).
9. Hess, R. L., "Traveling-Wave Tube Large-Signal Theory, with Application to Amplifiers Having D-C Voltage Tapered with Distance." Ph.D. Dissertation, Electrical Engineering Dept. Univ. of California (July 1960).
10. Meeker, J. G., "Phase Focusing in Linear-Beam Devices." Univ. of Michigan Electron Phys. Lab. Tech. Rept. No. 49 (August 1961).
11. Meeker, J. G., and Rowe, J. E., "Phase focusing in linear-beam devices," *IRE Trans. Electron Devices* **9**, No. 3, 257-266 (1962).
12. Clarke, G. M., "Tapered Travelling Wave Tubes." Ferranti Ltd. Rept. No. 117 (November 1960); See also "Constant Impedance Tapers for Travelling Wave Tubes." Rept. No. 121 (April 1961); "Tapered Structures for Travelling Wave Tube Amplifiers." Rept. No. 127 (November 1961); "Tapered Structures for Travelling Wave Amplifiers." Rept. No. 131 (November 1962), with A. Knox, G. McFadden, and J. McKenzie.
13. Haddad, G. I., "Efficiency and Start-Oscillation Conditions in Nonuniform Backward-Wave Oscillators." Univ. of Michigan Electron Phys. Lab. Tech. Rept. No. 61 (March 1963).
14. Haddad, G. I., and Rowe, J. E., Efficiency of tapered backward-wave oscillators. *IEEE Trans. Electron Devices* **12**, No. 1, 20-30 (1964).
15. Spangenberg, K. R., "External-Circuit Traveling-Wave Tubes." Stanford Univ. Electron Devices Lab. Consolidated Quarterly Status Rept. No. 7 (1958).
16. Ruetz, J., Robinson, D., and Pavkovich, J., The effect of tapered circuits on efficiency for high power traveling wave tubes. Paper presented at *Electron Devices Meeting* (*IRE-PGED*), *Washington, D.C. 1960*.
17. Rowe, J. E., and Brackett, C. A., Efficiency, phase shift and power limitations in variable-pitch traveling-wave amplifiers. *Proc. Nat. Electron. Conf. 1962* **18**, 97-103.
18. Haddad, G. I., and Rowe, J. E., General velocity tapers for phase-focused forward-wave amplifiers. *IEEE Trans. Electron Devices* **10**, No. 3, 212 (1963).
19. Volkholz, K. L., "Energy Exchange in Crossed-Field Systems." Univ. of Michigan Electron Phys. Lab. Tech. Rept. No. 73 (June 1964).
20. Rowe, J. E., Univ. of Michigan Electron Phys. Lab. Unpublished material (1963).

CHAPTER

XIV | Prebunched Electron Beams

1 Introduction

In the previous chapters the electron beam entering the interaction region in both O- and M-type devices was generally assumed to be unmodulated and unbunched. The effects of thermal velocities and random current fluctuations in the entering beam were also generally neglected although the means for their inclusion was outlined. Since it is certainly possible either to velocity modulate and/or to density modulate a beam before entrance into a traveling-wave interaction region, one wonders about the effects of the modulation and/or bunching on the gain and energy conversion in the following interaction region. In particular, what advantages can accrue from such pre-operations on a beam? We have, of course, already considered this problem in the case of the klystron when multiple-cavity interactions are utilized. In that case an advantage accrued due to the increased i_n/I_0 values appearing at the output gap. Many configurations are possible such as cavities, short-length traveling-wave or standing-wave interaction regions, etc., and these may be utilized in either linear-beam or crossed-field devices.

In evaluating the effectiveness of various premodulation and prebunching schemes one must determine whether the increase in ultimate gain and/or efficiency more than offsets the power cost of creating the bunched beam. In this chapter various special premodulation and prebunching possibilities are evaluated, and two methods of calculating the required bunching power are developed. Modulation of the beam potential and current in the primary interaction region is treated separately in Chapter XVI. All composite interaction schemes such as illustrated in Fig. 1 may be thought of simply as serial systems of general intearction regions with arbitrary signal and beam inputs.

An interesting lesson to recall here is that learned as a result of the severed-circuit calculations of Chapter VI and the phase-focusing studies of Chapter XIII. In those studies it was found that the prebunching improvement was lost if the bunched beam was allowed to drift over any extended length with a circuit field less than the space-charge debunching field. The reason is that if the space-charge field far exceeds the containing field then the bunch will spread apart (first at the edges) and thus the rf current amplitude driving the circuit will decrease. Velocity spread within the bunch will also lead to defocusing and bunch breakup.

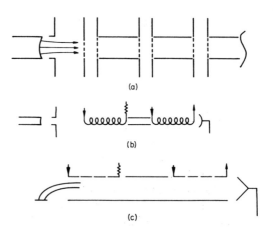

FIG. 1. Serial interaction configurations. (a) Multicavity klystron. (b) O-type traveling-wave serial interaction. (c) Crossed-field serial interaction.

2 Mathematical Formulation of the Lagrangian Equations

The Lagrangian nonlinear equations for any device are developed in much the same manner for prebunched beam situations as others, and hence only the differences are considered in detail.

The continuity equation as shown in Chapter VI requires some modification. The appropriate form for an arbitrary multidimensional entering beam is

$$\rho(0, 0, 0, t) = \rho(0, 0, 0)f(t), \qquad (1)$$

where $f(t)$ denotes the time dependence of the charge density, which is separable. Thus the already derived nonlinear equations for uniform entering beams remain unchanged, and we need only specify the entering phase positions for the electrons or charge groups relative to a cycle of the modulating waveform. For a uniform entering beam recall that the initial phases are specified as

$$\Phi_{0,j} = \frac{2\pi j}{m} \qquad = 0, 1, 2, ..., m, \qquad (2)$$

2. LAGRANGIAN EQUATIONS

where m simply indicates the number of charge groups being considered. Each represents an equivalent fraction of the beam charge. In the prebunched beam case the same amount of beam charge is injected and the phase positions depend both upon the bunch width and upon its position relative to the rf wave at the input. This description is illustrated in Fig. 2.

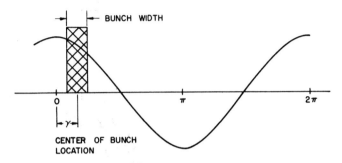

FIG. 2. Bunched-beam description.

The remaining mathematical description of the entering premodulated, prebunched beam pertains to the initial velocities. Velocity modulations are specified through the dependent variables

$$u_y(0, x_0, \Phi_0, \varphi_0) \tag{3a}$$

$$u_x(0, x_0, \Phi_0, \varphi_0) \tag{3b}$$

$$u_\varphi(0, x_0, \Phi_0, \varphi_0) \tag{3c}$$

for the general class of O-type devices and through

$$u(0, p_0, \Phi_0) \tag{4a}$$

$$v_{y\omega}(0, p_0, \Phi_0) \tag{4b}$$

for M-type devices.

As a means of simplifying the calculations pertaining to prebunched beams, only one-dimensional systems are considered for O-type devices and two-dimensional systems for M-type devices. It should be pointed out that initial distributions such as Maxwellian distributions are particular cases of a general beam input.

3 Results for Klystrons

To illustrate the spreading due to space-charge forces in a drift region, it is interesting to inject a prebunched beam without velocity modulation into a drift region. The bunch axial velocity versus phase position is shown in Fig. 3, where a $\lambda/20$ bunch has been injected at a phase position

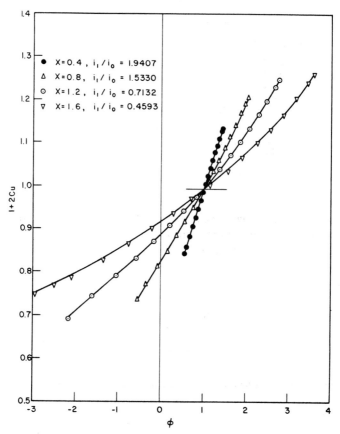

FIG. 3. Velocity-phase plot for a $\lambda/20$ bunch in a drift region ($B = 1$, $\omega_p/\omega = 0.125$, $\alpha_i = 0$).

of 1 radian. No initial modulation of the electron velocity has been considered. As expected the bunch spreads while drifting and the harmonic current amplitudes gradually decrease. The harmonic current amplitudes go through successive maxima and minima, decreasing with distance. The amplitudes at successive peaks are shown in Fig. 4. In

3. RESULTS FOR KLYSTRONS

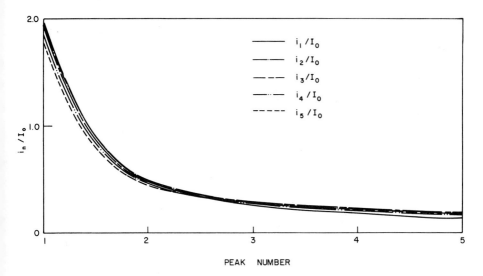

FIG. 4. Harmonic current amplitude versus peak number for a $\lambda/20$ bunch in a drift region ($B = 1$, $\omega_p/\omega = 0.125$, $\alpha_i = 0$).

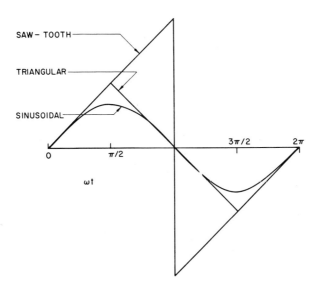

FIG. 5. Velocity modulation waveforms.

the absence of an rf circuit field there is nothing to prevent bunch defocusing.

It is also interesting to speculate on the effects of various premodulation waveforms on bunched and unbunched beams in klystron drift regions. An interesting problem not considered here is that of synthesizing an optimum premodulation waveform with the objective of maximizing i_1/I_0 at the output gap. The waveforms considered here are illustrated in Fig. 5.

The resultant normalized amplitude of the fundamental component of current versus distance is shown in Fig. 6 both for the case of an

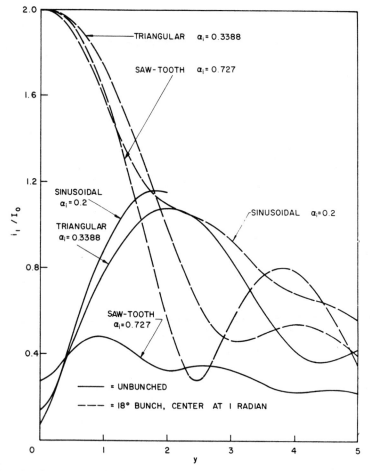

Fig. 6. Current amplitude versus distance for premodulated and prebunched drifting beams $(\omega_p/\omega = 0, \; B = 1)$.

unbunched beam input and for a bunched beam input ($\lambda/20$ bunch width). These were calculated without space-charge forces and it is seen in the prebunched beam case that the results are somewhat independent of the velocity modulation form. Note from Fig. 5 that the initial slopes of all three waveforms are identical.

In the unbunched beam cases the sinusoidal velocity modulation yields the largest value of i_1/I_0 although as expected the triangular waveform results are only slightly reduced. The presence of moderate space-charge forces should not greatly influence the results nor change the relative positioning of the various curves.

In synthesizing an optimum premodulation waveform, it is desired to obtain a condition in which the bunch-edge electrons are initially driven towards the bunch center so as to counteract partially the defocusing space-charge forces. Unfortunately the synthesis problem is complicated by the fact that performance depends on so many independent parameters.

4 Results for Traveling-Wave Amplifiers

Even in the absence of space-charge forces in amplifiers and oscillators there will be some spreading of an injected bunched beam due to the variation of rf circuit field across the bunch. Other parameters which affect the performance of bunched-beam devices are the injection phase and the relative injection velocity, b. Fortunately, the dependence on bunch injection phase, γ, is not critical, as shown in Fig. 7.

Figures 8 and 9 illustrate the effect of prebunched beam injection on traveling-wave amplifier operation in both the growing-wave and beating-wave regimes. The entrance phase of the bunch was taken as 60° and since $u_0/v_0 > 1$ the bunch proceeds to advance in phase relative to the rf wave. As energy is given to the rf wave the bunch slows down and drops back in phase until saturation is reached. Comparison of the results with those for the unbunched beam reveals that an improvement in gain, efficiency and over-all length is achieved.

The value of injection velocity chosen was arbitrary and does not represent an optimum situation. For a given bunch configuration, injection phase and initial rf signal level one must select an injection velocity value so as to maintain the bunch in a decelerating phase for as long as possible. This is illustrated in Fig. 8, where a relatively large value of b is taken. Note the improved interaction efficiency and the somewhat improved gain. The operating conditions of Fig. 8 correspond to those usually encountered in the high-power Crestatrons described in Chapter VI.

Injection of a bunched beam into a high-gain amplifier with a velocity parameter adjusted for maximum output is illustrated in Fig. 9. The value of b selected maintains phase focusing of the bunch over a considerable distance, with the result that the efficiency is significantly improved along with increased gain and a considerable reduction in length. The bunch spreading results from the nonuniform sinusoidal

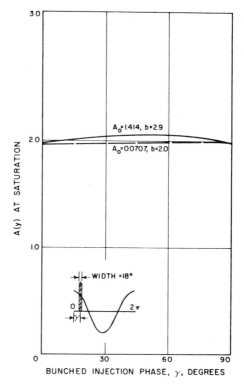

FIG. 7. Effect of bunch injection phase on saturation amplitude $(C = 0.1, B = 1)$.

field across the bunch, since space-charge forces in the bunch are neglected. The inclusion of space-charge forces, i.e., $\omega_p/\omega \neq 0$, results in additional bunch spreading as illustrated in Fig. 10, where a relatively high value of QC or ω_p/ω exists. The same space-charge defocusing results independent of the bunch injection angle. Even though the injection velocity and phase angle are adjusted to maintain phase focusing of the center of the bunch, the space-charge spreading prevents the edges of the bunch from staying in phase, with a resulting loss in fundamental current amplitude. The only means for counteracting this effect is to

4. RESULTS FOR TRAVELING-WAVE AMPLIFIERS

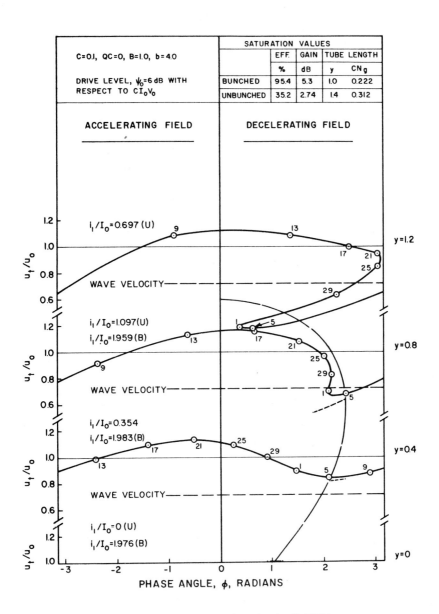

FIG. 8. Velocity versus phase for the O-TWA.

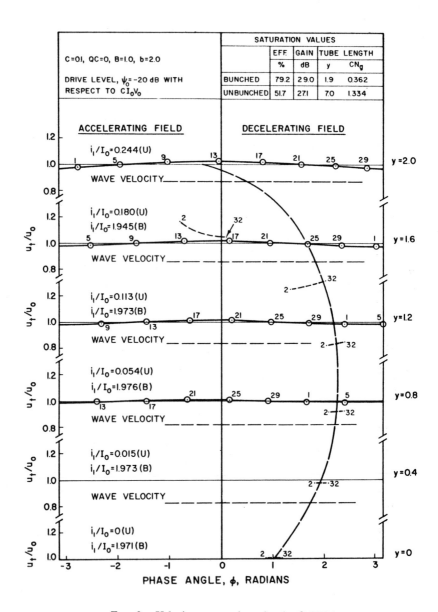

FIG. 9. Velocity versus phase for the O-TWA.

4. RESULTS FOR TRAVELING-WAVE AMPLIFIERS

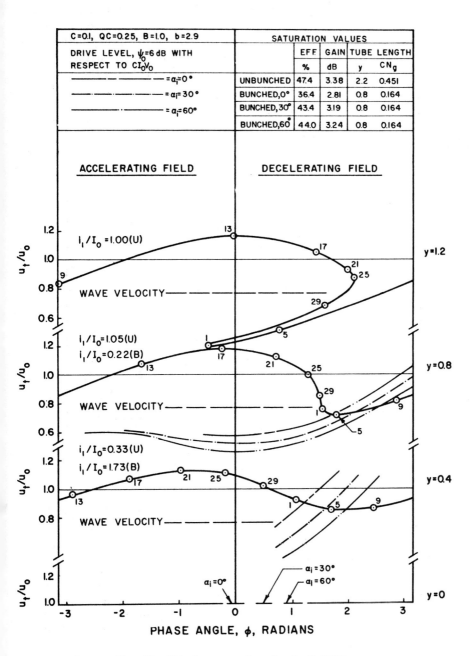

FIG. 10. Velocity versus phase for the O-TWA.

have a very strong rf signal on the circuit to balance the space-charge field.

The advantage of length reduction obtained by prebunching the electron beam is summarized for a range of parameters in Fig. 11. As the signal level at the circuit input is increased, notice that the optimum circuit length is reduced due to the counteraction of space-charge defocusing.

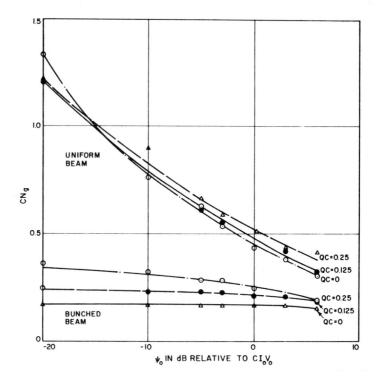

FIG. 11. Saturation tube length for an O-TWA ($C = 0.1$, $B = 1$, $b = 2$).

The balance or lack of balance between circuit field and space-charge field is shown in Figs. 12 and 13, where the efficiencies of bunched-beam (B) and unbunched-beam (U) devices are compared as a function of drive level and injection velocity. For low drive level and high space-charge forces, the efficiency falls off rapidly due to bunch spreading and a loss of phase focusing. Notice that the unbunched-beam devices exhibit rather constant efficiency versus drive level, with gain increasing as ψ_0 is reduced. At high ω_p/ω prebunching is advantageous only when ψ_0 is large and

4. RESULTS FOR TRAVELING-WAVE AMPLIFIERS

consequently the gain is low: not always desirable. As expected, intermediate values of space-charge forces are desirable to obtain high efficiency whether prebunching is used or not.

In the above calculations the entering stream or bunches were assumed to have no velocity modulation, i.e., to be uniform. The effects of

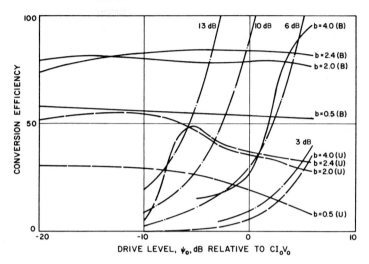

FIG. 12. Efficiency and gain versus drive level for an O-TWA $(C = 0.1, \; B = 1, \; QC = 0)$.

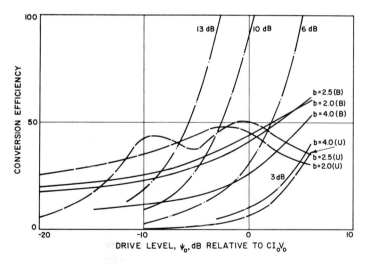

FIG. 13. Efficiency and gain versus drive level for an O-TWA $(C = 0.1, \; B = 1, \; QC = 0.125)$.

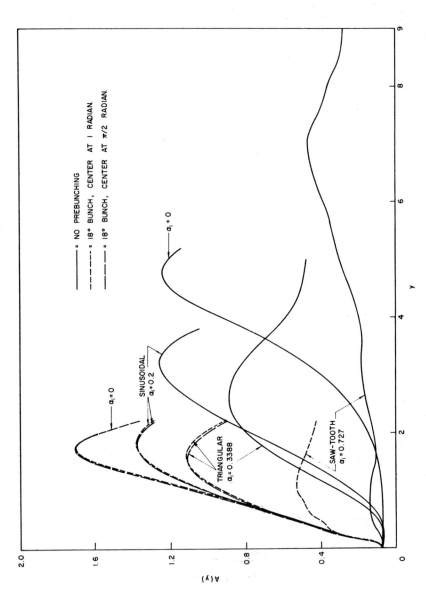

FIG. 14. Rf amplitude versus distance in a TWA with premodulation and prebunching ($C = 0.1$, $b = 0.5$, $\omega_p/\omega = 0$, $\psi_0 = -20$).

initial velocity modulation are easily incorporated as outlined previously, and these may sometimes be used to advantage. Since the circuit field acts nonuniformly across the bunch, resulting in spreading, and the bunch space-charge forces produce spreading, it is reasoned that a tailored initial velocity modulation, possibly directing the edge electrons toward the center, could reduce the rate of bunch spreading, resulting in phase focusing over a longer length and a higher efficiency. The synthesis of "the optimum" waveform is a considerable problem and is not pursued further here. Lichtenberg[2] has carried out experimental work on this idea with some success.

To illustrate the effects of velocity modulation in addition to prebunching in a traveling-wave amplifier, the waveforms of Fig. 5 are investigated. In view of the rf circuit fields in the O-TWA available for bunch maintenance it is expected that more favorable results will be realized than in the klystron case. For a particular TWA case ($C = 0.1$, $b = 0.5$, $\omega_p/\omega = 0$, $\psi_0 = -20$) the results of both premodulation and prebunching are shown in Fig. 14. For comparison purposes the rf amplitude curve for no premodulation or prebunching is also shown. The prebunched-beam results are also not critically dependent upon the bunch injection phase as illustrated for both 1 and $\pi/2$ radians. Both the premodulation and prebunching result in the rf voltage amplitude maximum occurring sooner and the prebunching, of course, yields a greater output. Here again, the sinusoidal waveform seems to yield better results than the others studied. The effects of moderate space-charge forces again are not expected to be significant.

5 Results for Crossed-Field Amplifiers

In the previous sections of this chapter we have been concerned with the general treatment of prebunched electron beams and with the characteristics of prebunched linear-beam devices. One wonders about the desirability of prebunching in crossed-field amplifiers. To what extent can the gain, efficiency and other characteristics of the family of crossed-field devices be improved through various prebunching schemes? Some indication of the expected performance was seen in Chapter XIII, where the device performance was evaluated using a hard-kernel-bunch theory.

Fortunately the handling of a prebunched beam in crossed fields proceeds in much the same manner as in linear-beam devices and we will assume a model as depicted in Fig. 2; the fundamental to dc current ratio for a $\lambda/20$ bunch is approximately 1.98. The interaction equations are those of Chapter IX for the amplifier with appropriate initial conditions.

Since the maximum decelerating phase of the circuit wave occurs at $\gamma = \pi/2$ it is logical to expect that the optimum bunch injection phase would also be $\pi/2$. The dependence of output on bunch injection phase for an M-FWA is shown in Fig. 15 for three values of injection phase. As

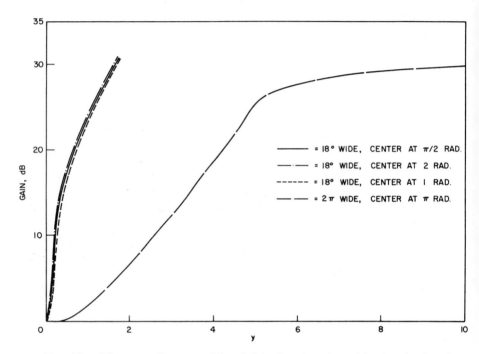

FIG. 15. Gain versus distance and bunch injection phase for a delta-function bunch in an M-FWA ($D = 0.1$, $r = 0.5$, $s = 0.1$, $b = 0$, $\omega_c/\omega = 0.5$, $\omega_p/\omega = 0$, $\psi_0 = -30$).

expected the dependence on phase is not critical although interestingly enough the output for 2 radians is slightly above the others. For comparison purposes the corresponding gain curve for the amplifier without prebunching is also shown. The striking effect is the large reduction in length required to reach saturation. The calculations were terminated at a displacement plane corresponding to collection of 90% of the entering stream current. Notice that the ultimate level reached is approximately the same in all cases.

The focusing of the bunches at or near the $\pi/2$ point is depicted in Fig. 16, where the trajectory (distance-phase) coordinates for the bunches are displayed. Since three charge layers were used in the computation, three bunches were inserted. Space-charge forces were neglected here. Along with the indicated change in phase there is a conversion of approximately 4–6% of the bunch kinetic energy to rf.

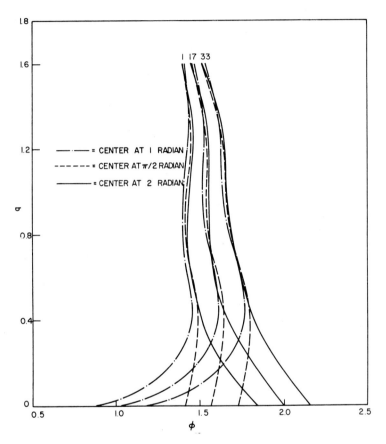

FIG. 16. Distance-phase plots for delta-function bunches in an M-FWA. (See Fig. 15, Chapter XIV.)

The prebunched M-FWA characteristics are most dependent upon ω_c/ω and r, as shown in Figs. 17 and 18. The unbunched gain curves are also shown for comparison. The saturation levels are again virtually the same with and without bunching in the $r = 0.5$ cases independent of ω_c/ω. In the $r = 0.25$ case, where the beam is launched near to the sole electrode, the rf levels achieved in the prebunched amplifier are considerably higher than those for the corresponding unbunched amplifier.

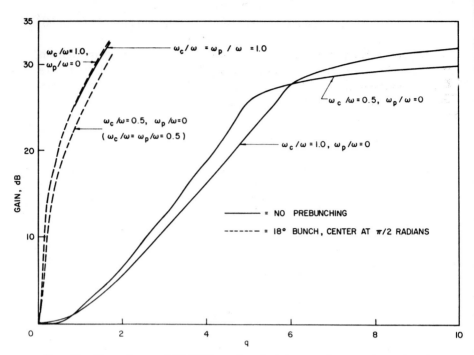

Fig. 17. Dependence of M-FWA prebunched beam gain on ω_c/ω and ω_p/ω at $r = 0.5$ ($D = 0.1$, $r = 0.5$, $s = 0.1$, $b = 0$, $\psi_0 = -30$).

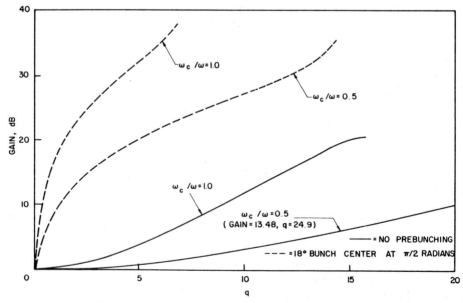

Fig. 18. Dependence of M-FWA prebunched beam gain of ω_c/ω at $r = 0.25$ ($D = 0.1$, $r = 0.25$, $s = 0.1$, $b = 0$, $\omega_p/\omega = 0$, $\psi_0 = -30$).

5. RESULTS FOR CROSSED-FIELD AMPLIFIERS

The reason for this is that in the unbunched-beam case the electrons move away from the sole quite slowly, with the result that a large number are focused in unfavorable phase positions and are eventually collected on the sole electrode.

The influence of space-charge forces, i.e., $\omega_p/\omega \neq 0$, on performance is illustrated in Fig. 17. As shown in Chapter IX for large D, i.e., > 0.05, the inclusion of space-charge forces does not significantly alter the results. Since we have assumed a Brillouin beam the space-charge forces in the delta-function bunches are the same as those in the unbunched beam. To some extent this is a function of the strength of the circuit field, since these forces tend to counteract the defocusing coulomb forces in the bunch and tend to hold the bunch together. The trajectories for the cases shown in the previous figures are given in Figs. 19 and 20.

As mentioned previously the delta-function bunch considered here is much like the hard-kernel bunch investigated in Chapter XIII. One

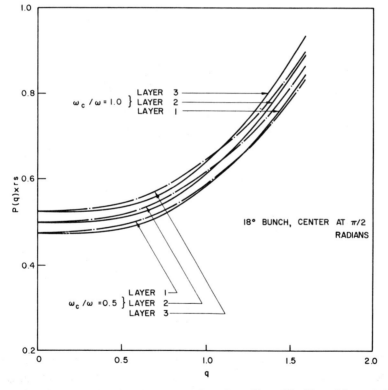

FIG. 19. Delta-function-bunch trajectories for Fig. 17 ($D = 0.1$, $r = 0.5$, $s = 0.1$, $b = 0$, $\omega_p/\omega = \omega_c/\omega$, $\psi_0 = -30$).

expects the results obtained in the two cases to be quite similar. A comparison is shown in Fig. 21, where both the hard-kernel-bunch predicted amplitude and that obtained for two injection phases of the delta-function bunch are plotted versus distance.

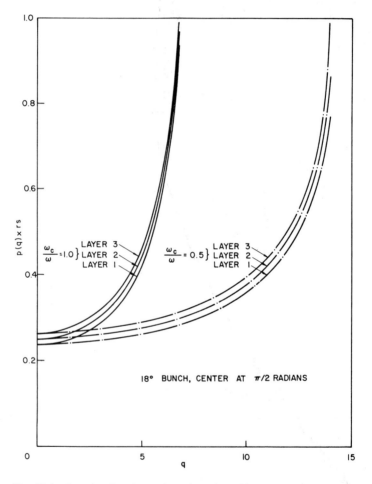

Fig. 20. Delta-function-bunch trajectories for Fig. 18 ($D = 0.1$, $r = 0.25$, $s = 0.1$, $b = 0$, $\omega_p/\omega = \omega_c/\omega$, $\psi_0 = -30$).

On the basis of the above general studies and results, it is expected that significant improvements in crossed-field device performance are achievable through beam prebunching. The most significant advantage seems to be the shortened interaction length required.

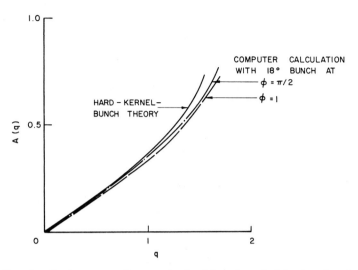

Fig. 21. Amplitude versus distance for hard-kernel and delta-function bunches ($r = 0.5$, $\omega_c/\omega = 0.5$, $\psi_0 = -30$).

6 Rf Power Required to Bunch an Electron Beam

It was pointed out earlier that the evaluation of the effectiveness of a prebunched beam on device operation must include a consideration of the rf power necessary to bunch the beam effectively, since this affects the over-all efficiency. The calculation of the required power is carried out in this section using both approximate and nearly exact methods. The approximations all relate to the manner in which the electric field of the bunch is determined.

A simple method of calculating the bunching power is to use the model outlined in Fig. 22, in which one wavelength λ_s of charge is taken from the stream and analyzed. The process involves converting this cylinder of charge to an equivalent sphere of the same charge density, compressing the sphere to the desired density (bunching), and then reconverting the compressed sphere to a cylindrical shape which is approximately a thin disk. In the limit, of course, one obtains a delta-function bunch in which $i_1/I_0 = 2$. If now we assume that the energy required to accomplish steps A → B and C → D is small compared to that required to compress the sphere (B → C) we may proceed directly to calculate the energy needed to accomplish charge compression.

Consider subdivision of the sphere into concentric shells, calculate the work done in compressing each shell, and then add (integrate) the work done over all shells. The work necessary to compress an elemental shell by dr_i is given by

$$dW_i = -q_i E_i \, dr_i, \qquad (5)$$

where q_i is the charge within the elementary shell and E_i is the electric

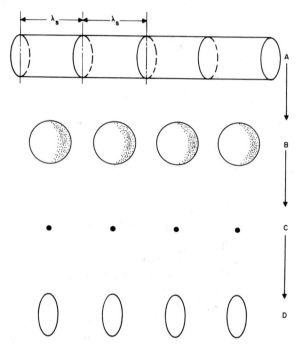

FIG. 22. Sequence in beam-bunching model (approximate method).

field over the shell. Thus the work necessary to compress the outer radius r_0 to a final radius r_f is

$$W = \int_{r_0=r_0}^{r_0=r_f} dW = \int_{r_0=r_0}^{r_0=r_f} \int_{r_i=0}^{r_i=r_0} (-q_i E_i) \, dr_i. \qquad (6)$$

Consider that the charge in an elementary shell is

$$q_i = 4\pi r_i^2 \rho \, dr_i \qquad (7)$$

and the total charge is

$$Q = \tfrac{4}{3}\pi r_0^3 \rho; \qquad (8)$$

6. RF POWER REQUIRED TO BUNCH AN ELECTRON BEAM

then the electric field over the shell is obtained from Poisson's equation as

$$E_i = -\frac{\rho}{3\epsilon_0} r_i. \tag{9}$$

Assuming that the compression throughout the sphere is uniform and hence that

$$\frac{dr_i}{dr_0} = \frac{r_i}{r_0} \tag{10}$$

and substituting Eqs. (7)–(9) into Eq. (6) yields

$$W = \frac{-3Q^2}{4\pi\epsilon_0} \int_{r_0=r_0}^{r_0=r_f} \int_{r_i=0}^{r_i=r_0} \frac{r_i^4}{r_0^7} dr_i \, dr_0$$

$$= \frac{3Q^2}{20\pi\epsilon_0} \left(\frac{1}{r_1} - \frac{1}{r_0}\right). \tag{11}$$

The number of bunches compressed per second is directly the operating frequency, and hence the bunching power in watts is simply

$$P_b = \omega W$$

$$= \frac{3Q^2\omega}{20\pi\epsilon_0} \left(\frac{1}{r_f} - \frac{1}{r_0}\right). \tag{12}$$

Equation (12) is easily expressed in terms of the dc stream parameters as

$$\frac{P_b}{I_0 V_0} = 1.577 \times 10^{-4} \frac{P_\mu V_0^{\frac{1}{4}}}{(b'/\lambda_0)^{\frac{2}{3}}} (k^{\frac{1}{3}} - 1), \tag{13}$$

where

$P_\mu \triangleq$ stream microperveance,

$V_0 \triangleq$ dc stream voltage,

$b' \triangleq$ stream radius,

$\lambda_0 \triangleq$ free-space wavelength, and

$k \triangleq$ the linear compression ratio in fractions of a stream wavelength.

Equation (13) is shown graphically in Fig. 23 in terms of familiar linear-beam device parameters. It is seen from these figures that 5–10% of the dc beam power is required for compression by a factor of 20, which gives $i_1/I_0 \approx 1.98$, a nearly delta-function bunch.

Inspection of the method of calculation indicates that the greatest approximation is involved in determining the appropriate electric field associated with the actual bunch. The limitation on obtaining particular compression ratios with a minimum of bunching power is primarily the velocity spread in the bunch.

This same procedure may be applied to the calculation of bunching power for a strip beam between parallel plates, characteristic of the injected-beam crossed-field device.

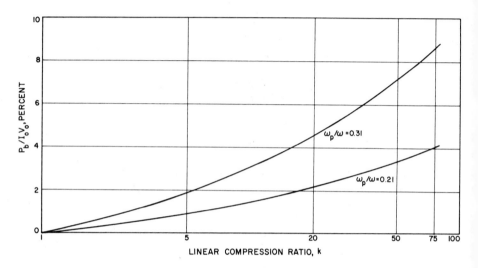

FIG. 23. Bunching power as a percentage of beam power $(B = 1.5, b = 0)$.

It is recalled that in Chapter IV on space-charge-field expressions the electric field associated with a rectangular bunch in a drift tube was evaluated from a solution of Poisson's equation using Green's function techniques. Since this is a nearly exact calculation of E, it may be used to advantage in computing bunching power. The electric potential within a rectangular bunch, neglecting the effects of neighboring bunches, is calculated as

$$V(r, z) = \frac{\rho_0}{\pi \epsilon_0 b'} \sum_n^\infty \frac{2a}{(\mu_l)^3} \left[\frac{J_0\left(\mu_l \frac{r}{a}\right) J_1\left(\mu_l \frac{b'}{a}\right)}{J_1^2(\mu_l)} \right]$$
$$\cdot \left[1 - \exp\left(-\mu_l \frac{\delta z}{2a}\right) \cdot \cosh\left(\mu_l \frac{z}{a}\right) \right]. \quad (14)$$

The charge density ρ_0 may be written in terms of the beam current and the rf period as

$$\rho_0 = \frac{I_0 T}{\delta z}. \quad (15)$$

6. RF POWER REQUIRED TO BUNCH AN ELECTRON BEAM

Equation (14) for the potential is then written as

$$V(r, z) = \frac{I_0 \sqrt{\mu_0/\epsilon_0} \lambda_0}{\pi b'} F(r, z), \qquad (16)$$

where

$$F(r, z) = \sum_{n}^{\infty} \frac{2a}{\mu_l^3 \delta z} \left[\frac{J_0\left(\mu_l \frac{r}{a}\right) J_1\left(\mu_l \frac{b'}{a}\right)}{J_1^2(\mu_l)} \right]$$

$$\cdot \left[1 - \exp\left(-\frac{\mu_l \delta z}{2a}\right) \cosh\left(\frac{\mu_l \delta z}{2a} \frac{2z}{\delta z}\right) \right]. \qquad (17)$$

The variation of the potential within the bunch in both the radial and axial directions is shown in Figs. 24 and 25, where the normalized potential is plotted. As expected the principal variation occurs near each edge, dropping off by approximately a factor of two from the center to the edge in both the radial and axial directions. This will show up later in the power required to form the bunch.

The bunch length δz may also be conveniently written in terms of the time width of the bunch relative to the rf period as

$$\delta z = \frac{\tau}{T} \lambda_0 \qquad (18)$$

providing that $(u_0/c)^2 \ll 1$.

Since the charge has been considered uniform within the bunch the work done in forming the bunch is given by

$$W = QV, \qquad (19)$$

where Q = the bunch charge. Since the rf angular frequency ω denotes the number of bunches per second, the power necessary to create the bunch is given by

$$P = \omega W = \omega QV. \qquad (20)$$

After some manipulation the bunching power over the dc beam power is written as

$$\frac{P}{I_0 V_0} = \frac{754 \hat{P} V_0^{\frac{1}{2}}}{(b'/\lambda_0)} F(r, z), \qquad (21)$$

where $\hat{P}$ is the beam perveance and V_0 the dc beam voltage. Note that $(\delta z/\lambda_0)$ is a linear compression factor indicating the bunch width relative to the rf wavelength.

The required bunching power is shown in Fig. 26 for a representative range of characteristic parameters. The calculations are made at the

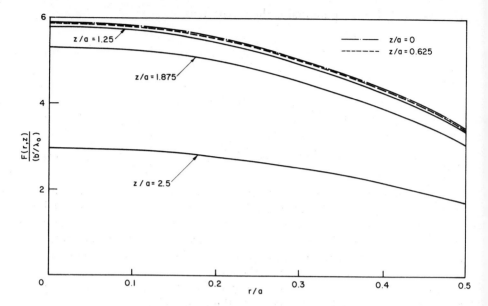

Fig. 24. Bunch potential versus radius ($b'/a = 0.5$, $\delta z/\lambda_0 = 0.1$, $a/\lambda_0 = 0.02$).

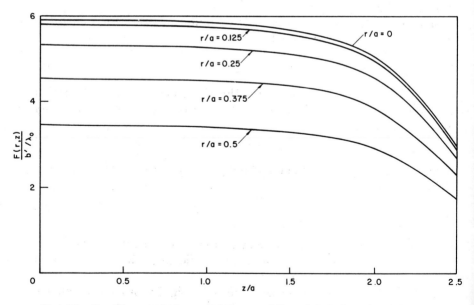

Fig. 25. Bunch potential versus axial distance ($b'/a = 0.5$, $\delta z/\lambda_0 = 0.1$, $a/\lambda_0 = 0.02$).

6. RF POWER REQUIRED TO BUNCH AN ELECTRON BEAM

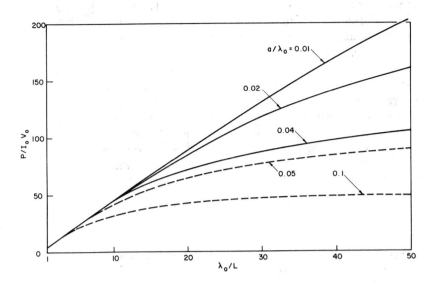

FIG. 26. Bunching power versus bunch length ($b'/a = 0.5$, $z/a = r/a = 0$, $I_0/V_0 = 10^{-2}$).

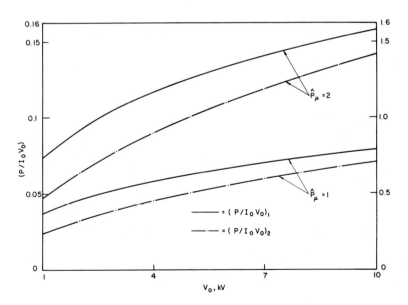

FIG. 27. Comparison of theoretical bunching power versus voltage ($b'/\lambda_0 = 0.002$, $\lambda_0/\delta z = 20$, $b'/a = 0.5$, $z/a = r/a = 0$).

center of the bunch, which, of course, yields the highest values. The values of power required to bunch the beam are reduced by a factor of four if one uses the potential at the corners (a factor of two reduction in moving to each edge).

A comparison of the power required for bunching as calculated by Eqs. (13) and (21) is shown in Fig. 27. The dependence on voltage and perveance is approximately the same and if the results of Eq. (21) are reduced by a factor of four as suggested above then the two sets of theoretical results are within approximately a factor of two of one another. The results obtained from Eq. (21) are considered quite accurate and should compare favorably with experiments providing thermal velocity effects are negligible.

REFERENCES

1. Rowe, J. E., and Meeker, J. G., Interaction of premodulated electron streams with propagating circuits. *J. Elec. and Control* **9**, No. 6, 439-467 (1960).
2. Lichtenberg, A. J., Prebunched beam traveling-wave tube studies. *IRE Trans. Electron Devices* **9**, No. 4, 345-351 (1962).

CHAPTER

XV | Collector Depression Techniques

1 Introduction

Not an incidental by-product of the nonlinear interaction calculations[1] is the information obtained on velocity and phase at various displacement planes of all devices. In Chapter XIII the velocity-phase curves at intermediate planes along the structure were consulted in developing phase-focusing criteria and techniques. It was seen that considerable kinetic energy remains in both O-type and M-type beams at the output plane of the structure and that it is possible to reduce the remaining kinetic energy by the application of phase-focusing techniques such as rf circuit tapers and/or the utilization of dc gradients along the beam path.

Due to the inherent velocity spread in the spent beam, which is occasioned by bunch space-charge forces and nonuniform circuit fields acting across the beam, it is never possible to remove all the beam kinetic energy (100% interaction efficiency). A simple theoretical plan for achieving almost 100% efficiency is radial collection of electrons after they have given up all their kinetic energy, thus removing them from the interaction region.

In this chapter we deal with the electron beam at the output plane of the rf structure, where it has considerable kinetic energy with or without the use of phase-focusing techniques. Kinetic energy information typical of both O-type and M-type beams at the output displacement plane of the rf structure is shown in Fig. 1.

We see from this figure that in both types of devices there is a considerable amount of kinetic energy remaining in the beam at the output plane of the rf structure. An additional complication appears in the M-type device in that a large percentage of the beam has been collected on the rf structure, as illustrated in Figs. 18 and 19 of Chapter IX. This is generally not a problem in O-type devices, where strong focusing fields are used and thus interception is small. The y-directed velocity distribution is also shown in Fig. 1c for a typical M-type interaction.

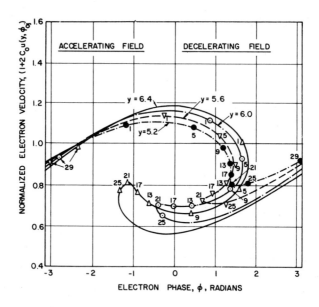

FIG. 1. Kinetic energy for O- and M-type beams. (a) Normalized z-velocity ($C = 0.1$, $\omega_p/\omega = 0$, $b = 1$).

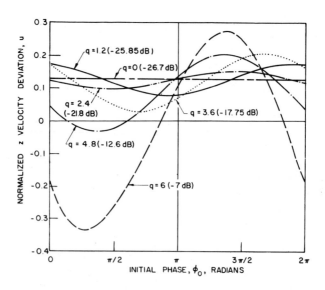

FIG. 1b. Normalized z-velocity ($D = 0.1$, $\omega_p/\omega = 0.25$, $r = 0.5$).

The remaining kinetic energy carried by the electrons at the output plane is usually dissipated in the collector electrode when this electrode is operated at the rf circuit potential. A considerable reduction in collector dissipation is afforded by segmenting the collector electrode and operating each segment at an optimum reduced potential so as to maximize the plate circuit or over-all efficiency. Collector segmentation will be viewed in this chapter solely in terms of efficiency improvement, without regard to technological problems encountered in high-beam-power devices.

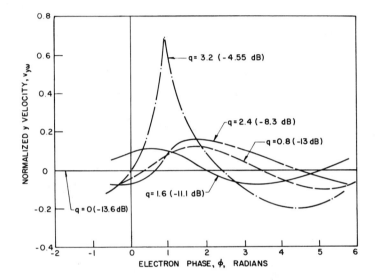

FIG. 1c. Normalized y-velocity ($D = 0.05$, $\omega_p/\omega = \omega_c/\omega = 0.25$, $r = 0.5$).

2 Graphical Evaluation of Depressed Collectors

Prior to the development of an elaborate mathematical analysis it is useful to view the segmented, depressed collector graphically. The output electron kinetic energy characteristic of O-type microwave devices is plotted in Fig. 2 versus charge group number in terms of fractions of the total beam current. The rf power and the power to be dissipated in the collector are clearly visible in the figure.

The problem is now to determine the extent of collector subdivision necessary and the optimum operating potentials. In Fig. 2, constructed for a two-segment collector, the problem is seen to be one of maximizing the areas of the two rectangles. The same process may be carried through for any number of collector segments. The area of the rectangle bounded by 1.0 on the ordinate and m on the abscissa represents the total dc power initially in the beam. The wavy line divides the energy converted to rf and delivered to the circuit and that energy still remaining in the beam to be dissipated in the collector. The energy to be dissipated in the collector is minimized by forcing the spent electrons to "work against" the field produced by depressing the collector. This minimization process is complicated by the fact that all electrons do not have the same velocity upon exit and thus several collector segments operating at different potentials are required.

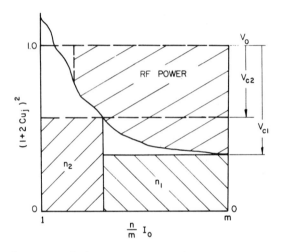

FIG. 2. Kinetic energy of electron groups at the output versus group number. Collector potentials optimized for a two-segment collector. n_1 electrons collected at V_{c1} and n_2 collected at V_{c2}.

The Fig. 2 type of presentation indicates clearly the advantages to be gained through multiple segmentation. Considerations other than efficiency, namely technological problems encountered in sorting and collecting of electrons and segment cooling, limit the number of segments to be utilized. These remarks apply both to M-type and O-type devices. It will be shown later that the processes of velocity sorting and collecting of electrons are more difficult in the case of spent M-type beams than in the case of O-type beams due to the presence of the magnetic field and the highly nonuniform velocity distribution.

The two-segment collector case illustrated in Fig. 2 results in an efficiency enhancement factor of 2.2 for the O-type forward-wave amplifier. Had a three-segment collector been used with all voltages optimized for maximum efficiency, the enhancement factor would have been 2.6.

3 Analysis of Output Energy Distribution for Collector Depression in O-Type Devices

In Chapters V, VI, and VII on O-type devices and Chapters VIII, IX, and X on M-type devices the energy distribution curves, $(1 + 2Cu_j)^2$ or $(1 + 2Du_j)^2$ versus electron group phase, were shown for various operating conditions at the output plane of the rf structure. The fact that all electron groups have a considerable amount of remaining kinetic energy indicates the desirability of collector depression. Some reduction of the residual kinetic energy has been achieved by various "phase-focusing" techniques. However, not even these have reduced the residual kinetic energy of all electrons to zero.

In view of the inherent velocity spread in spent beams as a result of nonuniform field action on charge groups and space-charge spreading, it is not anticipated that the total kinetic energy could be removed from a significant fraction of the total beam charge, the actual fraction being limited by focusing difficulties and circuit interception at the output end. One- and two-stage depressed collectors appear entirely practical. As practical problems do not limit the present discussion, three- and four-stage collectors are also considered here. It is for the experimentalist to tread the crooked path to the promised land.

The vorticities of the velocity curves indicate tight electron bunching with only a small velocity spread around the core of the bunch. Fast electrons and bunch-edge charge cause the deleterious wide velocity spread. Since experimental evidence is alleged to lend credance or disrepute to abstract theory, it is noted that Cutler has observed similar energy-phase curves with a velocity analyzer at the output of a TWA. These are shown in Fig. 3. Comparison of these data with theoretical curves of Chapters VI, VII, IX, and X reveals the expected similarity.

As the nonlinear interaction theory considers individual charge groups entering the interaction region, we conveniently make use of their properties in treating the collector depression problem[2]. Electrode interception in the electron gun region is of no consequence and hence only the charge entering the rf interaction region and its subsequent transmission to the collector or collection on the rf circuit need to be considered. In O-type devices confined flow results in near-perfect beam transmission, whereas circuit interception of a portion of the dc beam charge is inherent in unphase-focused M-type devices. Various minimal-stiffness focusing systems for O-type beams also give rise to circuit interception, although this condition is to be avoided.

Entering charge is defined on a per cycle of rf basis and thus it is convenient to define charge group energies on a per cycle basis. The spent beam characteristics are obtained in terms of charge group velocities and these are converted to energies as follows. The change in energy experienced by a particular group is

$$\frac{mu_{tj}^2}{2} - qV_0 \triangleq -qV_j \triangleq \mathscr{E}_j, \tag{1}$$

where V_j defines an equivalent rf potential and $\mathscr{E}_j$ the corresponding energy in joules. For energy given up by the charge group (deceleration) V_j is positive, while for energy absorbed (acceleration) V_j is negative. Thus V_j and $\mathscr{E}_j$ represent changes with respect to the initial average energy.

That charge group which remains unaffected by the rf fields satisfies the relation

$$\frac{mu_0^2}{2} - qV_0 = 0. \tag{2}$$

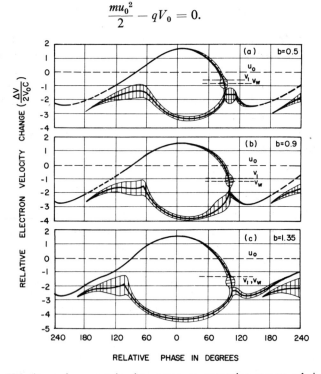

Fig. 3. Velocity and current in the stream at saturation versus relative electron phase with injection velocity as the parameter ($C = 0.075$, $QC = 0.22$, $\beta b' = 0.388$).

3. ANALYSIS OF OUTPUT ENERGY DISTRIBUTION

Combining Eqs. (1) and (2) gives the rf energy in terms of the velocity as

$$\left(\frac{u_{tj}}{u_0}\right)^2 = 1 - \frac{V_j}{V_0}. \qquad (3)$$

Since the total charge group velocity is defined as

$$u_t(y, \Phi_{0j}) \triangleq u_0[1 + 2Cu(y, \Phi_{0j})], \qquad (4)$$

Eq. (3) may be rewritten as

$$\left(\frac{u_{tj}}{u_0}\right)^2 = [1 + 2Cu(y, \Phi_{0j})]^2 = 1 - \frac{V_j}{V_0}, \qquad (5)$$

where the index j indicates, as before, the particular electron group in question. Equation (5) indicates that a knowledge of the normalized output plane velocity $[1 + 2Cu(y, \Phi_{0j})]$ of any O-type device allows the calculation of the plate circuit efficiency for a multisegment collector system.

Recall that in the nonlinear calculations of Chapter VI, electron charge groups are injected at the input to the interaction region over one cycle of the rf wave; hence it is convenient to define various characteristic energies per cycle of the beam. The beam transmission efficiency is assumed to be 100%. The defined energies are

$\mathscr{E}_{dc} \triangleq$ dc beam input energy to the interaction region,

$\mathscr{E}_i \triangleq$ rf input energy to the circuit,

$\mathscr{E}_d \triangleq$ rf energy dissipated on the circuit,

$\mathscr{E}_c \triangleq$ dc energy given to the collector, and

$\mathscr{E}_0 \triangleq$ rf energy at the circuit output.

Characteristic efficiencies are defined as follows:

$\eta_e \triangleq$ electronic conversion efficiency, and

$\eta_0 \triangleq$ over-all efficiency.

These efficiencies are now written in terms of the defined energies, assuming that m charge groups are injected into the interaction region. The electronic efficiency is thus

$$\eta_e = \frac{\mathscr{E}_0 - \mathscr{E}_i + \mathscr{E}_d}{\mathscr{E}_{dc}} = 1 - \frac{\mathscr{E}_c}{\mathscr{E}_{dc}}, \qquad (6)$$

and the over-all efficiency

$$\eta_0 = \frac{\mathscr{E}_0 - \mathscr{E}_i}{\mathscr{E}_{dc}}. \tag{7}$$

The rf energy given up by the beam may be related to the electron rf potential as follows:

$$\mathscr{E}_0 - \mathscr{E}_i + \mathscr{E}_d = \sum_{j=1}^{m} qV_j, \tag{8}$$

where generally (for high gain) $\mathscr{E}_i \ll \mathscr{E}_0$.

If the device is operated with the collector potential at the rf structure potential, the energies and efficiencies are

$$\mathscr{E}_{dc} = mqV_0, \tag{9a}$$

$$\mathscr{E}_c = \sum_{j=1}^{m} q(V_0 - V_j), \tag{9b}$$

$$\eta_e = 1 - \frac{\sum_{j=1}^{m} q(V_0 - V_j)}{mqV_0} = 1 - \frac{1}{m}\sum_{j=1}^{m}\left(1 - \frac{V_j}{V_0}\right) = \frac{1}{m}\sum_{j=1}^{m}\frac{V_j}{V_0}, \tag{9c}$$

and

$$\eta_0 = \frac{\mathscr{E}_0}{mqV_0}, \tag{9d}$$

providing that $\mathscr{E}_i \ll \mathscr{E}_0$. In view of the relationship between the equivalent rf potential V_j and the charge group velocity, Eq. (9c) may be rewritten as

$$\eta_e = \frac{1}{m}\sum_{j=1}^{m}\{1 - [1 + 2Cu(y, \Phi_{0j})]^2\}. \tag{10}$$

The efficiency expressions for multisegment-collector tubes may be developed systematically from the above relations if it is assumed that specific fractions of the output beam are collected at given potentials with one segment of the collector at the structure potential. Suppose now that the collector is segmented and that the segment operating dc potentials are such that $V_0 > V_{01} > V_{02} > V_{03} \ldots > V_{0r}$; that is, there are r + 1 segments. A schematic collector configuration illustrating the potential arrangement is shown in Fig. 4.

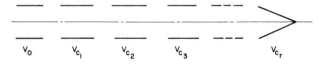

FIG. 4. Schematic diagram of segmented collector.

3. ANALYSIS OF OUTPUT ENERGY DISTRIBUTION

The energy given up to the individual collector segments is given by

$$\mathscr{E}_{c,r+1} = \sum_{i=1}^{p_1} q(V_{c,r} - V_i) + \sum_{j=1}^{p_2} q(V_{c,r-1} - V_j) + \sum_{j=1}^{p_3} q(V_{c,r-2} - V_k)$$

$$+ \cdots + \sum_{l=1}^{p_{r+1}} q(V_0 - V_l), \quad (11)$$

where

$$p_1 + p_2 + p_3 + \cdots + p_{r+1} = m,$$

p_i indicating the number of charge groups collected on the ith segment. The above can be simplified as follows:

$$\mathscr{E}_{c,r+1} = q\left[p_1 V_{c,r} + p_2 V_{c,r-1} + p_3 V_{c,r-2} + \cdots + p_{r+1} V_0 - \sum_{i=1}^{m} V_i\right]. \quad (12)$$

Thus

$$\mathscr{E}_{\text{dc},r+1} = \mathscr{E}_0 + \mathscr{E}_d + \mathscr{E}_{c,r+1} = q[p_1 V_{c,r} + p_2 V_{c,r-1} + \cdots + p_{r+1} V_0]. \quad (13)$$

The electronic efficiency for this $(r+1)$-segment collector is

$$\eta_{e,r+1} = \frac{\mathscr{E}_0 - \mathscr{E}_i + \mathscr{E}_d}{\mathscr{E}_{\text{dc},r+1}} \approx \frac{\mathscr{E}_0 + \mathscr{E}_d}{\mathscr{E}_{\text{dc},r+1}} \quad (14)$$

and

$$\eta_{0,r+1} = \frac{\mathscr{E}_0 - \mathscr{E}_i}{\mathscr{E}_{\text{dc},r+1}} \approx \frac{\mathscr{E}_0}{\mathscr{E}_{\text{dc},r+1}}. \quad (15)$$

It is convenient to refer the efficiencies for the device with $r+1$ collector segments to the efficiencies for the device with a single collector at the rf circuit potential. Thus

$$\frac{\eta_{e,r+1}}{\eta_e} = \frac{\eta_{0,r+1}}{\eta_0} = \frac{\mathscr{E}_{\text{dc}}}{\mathscr{E}_{\text{dc},r+1}}$$

$$= \frac{1}{\frac{p_1}{m}\left(\frac{V_{c,r}}{V_0}\right) + \frac{p_2}{m}\left(\frac{V_{c,r-1}}{V_0}\right) + \cdots + \frac{p_{r+1}}{m}}. \quad (16)$$

The over-all interaction efficiency for the $(r+1)$-segment-collector device is

$$\eta_{0,r+1} = \frac{\mathscr{E}_0 - \mathscr{E}_i}{\mathscr{E}_{\text{dc},r+1}}, \quad (17)$$

and thus

$$\frac{\eta_{0,r+1}}{\eta_0} = \frac{\mathscr{E}_{\text{dc}}}{\mathscr{E}_{\text{dc},r+1}} \equiv \frac{\eta_{e,r+1}}{\eta_e}. \quad (18)$$

The electronic efficiency for the $(r+1)$-segment-collector device may be written in terms of the equivalent rf potentials as

$$\eta_{e,r+1} = \frac{\sum_{i=1}^{m} \left(\frac{V_i}{V_0}\right)}{p_1 \left(\frac{V_{c,r}}{V_0}\right) + p_2 \left(\frac{V_{c,r-1}}{V_0}\right) + \cdots + p_{r+1}}. \quad (19)$$

Equation (19) may also be written in terms of particle velocities as

$$\eta_{e,r+1} = \frac{\sum_{i=1}^{m} \{1 - [1 + 2Cu(y, \Phi_{0i})]^2\}}{p_1 \left(\frac{V_{c,r}}{V_0}\right) + p_2 \left(\frac{V_{c,r-1}}{V_0}\right) + \cdots + p_{r+1}}. \quad (20)$$

In terms of the rf energy delivered to the circuit the over-all efficiency is written as

$$\eta_{0,r+1} = \frac{2C(A_{\max}^2 - A_0^2)\left(1 - C\frac{d\theta(y)}{dy}\right) \Big/ (1 + Cb)}{\left[\frac{p_1}{m}\left(\frac{V_{c,r}}{V_0}\right) + \frac{p_2}{m}\left(\frac{V_{c,r-1}}{V_0}\right) + \cdots + \frac{p_{r+1}}{m}\right]}. \quad (21)$$

Equations (16)–(21) may now be used to calculate the "partial" and "over-all" efficiencies for depressed-collector devices for an arbitrary number of collector segments but with one segment at the rf structure potential.

4 Results of Calculations for O-Type Devices

4.1 Traveling-Wave Amplifier

It is easily seen from Eqs. (18)–(21) that the efficiency enhancement achievable through depressed-collector operation is directly related to the velocity spread in the spent beam. As pointed out previously, the greater the velocity spread in the output beam the greater the number of collector segments needed to achieve a given over-all efficiency. The amount of velocity spread existing in the beam is directly dependent upon the electronic interaction efficiency.

The previous theory may be used to calculate improvement factors for any operating conditions. The most significant information is to be gleaned from the calculations on devices operating at maximum gain conditions and at maximum saturation output (maximum efficiency). A

4. RESULTS OF CALCULATIONS FOR O-TYPE DEVICES

comparison of these two sets of results is most interesting. The results have been calculated for various cases from the efficiency curves computed with the aid of the one-dimensional nonlinear theory described in Chapter VI. These are shown in Figs. 5 and 6 for one-, two-, and three-stage collectors along with the efficiency for a device without

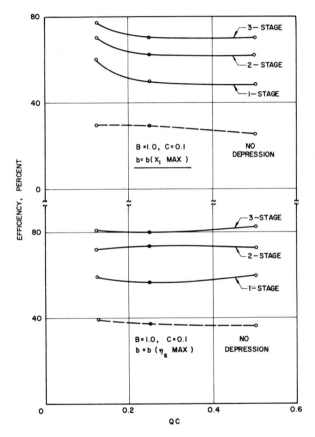

FIG. 5. O-type amplifier efficiency with collector depression versus QC.

collector depression. In Fig. 5 it is apparent that the efficiency improvement factor decreases as QC or ω_p/ω increases, as is to be expected since increased coulomb repulsion forces lead to greater velocity spread in the electron beam. The improvement factor is also seen in Fig. 6 to decrease with an increase in the gain parameter C. This also is reasonable since an enhanced interaction achieved through tighter coupling between the beam and the circuit should naturally lead to greater velocity spreads.

Close scrutiny of these results indicates that in general if the tube is operated at a condition for maximum small-signal gain to achieve linearity of operation at the expense of efficiency, the same "over-all" efficiency may be achieved as in the maximum-power-output (saturation-efficiency) case by utilizing one additional collector segment. In a

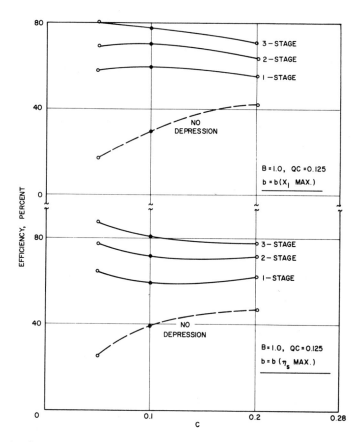

FIG. 6. O-type amplifier efficiency with collector depression versus gain parameter.

particular instance technological problems may define an upper limit on the number of collector segments permitted.

In order to achieve these maximum efficiency enhancement factors the various collector segments must be operated at the optimum potentials. Typical calculated efficiency improvement factors as a function of the various segment potentials are shown in Figs. 7 and 8 for specific operating points.

4. RESULTS OF CALCULATIONS FOR O-TYPE DEVICES

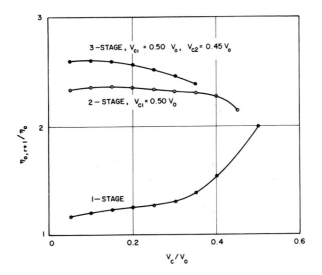

FIG. 7. O-type amplifier efficiency improvement factor versus collector segment potential ($C = 0.1$, $B = 1$, $QC = 0.125$, $b = 0.65$, $d = 0$, $y = 6.8$).

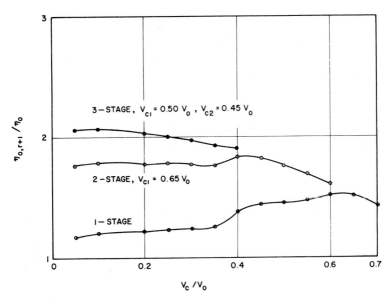

FIG. 8. O-type amplifier efficiency improvement factor versus collector segment potential ($C = 0.1$, $B = 1$, $QC = 0.125$, $b = 1.5$, $d = 0$, $y = 7.2$).

Examination of the various cases indicates that the optimum collector segment voltages are approximately as given in the following table.

TABLE I

Optimum O-Type Amplifier Collector Segment Voltages

No. of stages	V_{c1}/V_0	V_{c2}/V_0	V_{c3}/V_0
1	0.5	—	—
2	0.6	0.25	—
3	0.6	0.4	0.1

The theoretical development and calculations are independent of V_0 and thus apply equally well to relativistic beams. To illustrate the

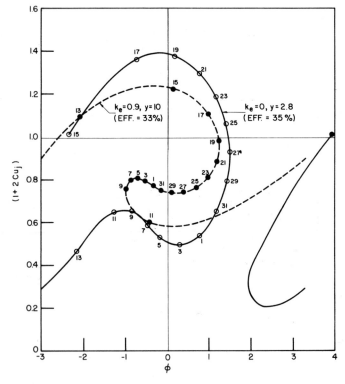

Fig. 9. Velocity versus phase for relativistic and nonrelativistic TWA's ($C = 0.3$, $b = 0$, $QC = 0$, $\psi_0 = -30$).

4. RESULTS OF CALCULATIONS FOR O-TYPE DEVICES

universal nature of the theoretical development and to illustrate relativistic effects on the efficiency and velocity spread, a selected case has been investigated. The effect of relativistic particle and wave velocities on klystron and TWA operation was considered in Chapters V and VI and some typical efficiencies were calculated. For a case in which $C = 0.3$ and $\omega_p/\omega = 0$, the velocity distributions and efficiency improvement factors for $k_e = 0$ and 0.9 are shown in Figs. 9 and 10.

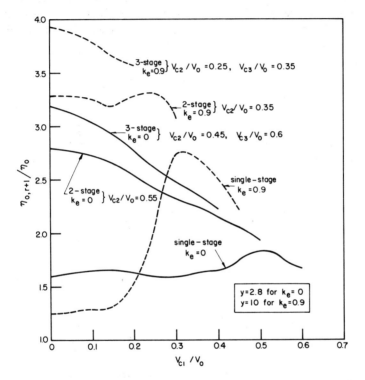

FIG. 10. Effect of relativistic factor on efficiency improvement by collector depression ($C = 0.3$, $b = 0$, $QC = 0$, $\psi_0 = -30$).

4.2 Backward-Wave Oscillator

It was explained in Chapter VII that the relatively low efficiency characteristic of the backward-wave oscillator is due to the fact that the high rf circuit field and the high rf beam current occur at opposite ends of the device. This is a natural consequence of the backward-wave mode of interaction, and as in the case of the forward-wave amplifier this process can be helped through various phase-focusing techniques. To obtain significant operating efficiencies from any device the electron beam must be tightly bunched and then prevented from spreading due to bunch space-charge forces with a strong circuit field. The lack of a large confining circuit field at the collector end of a BWO materially reduces the maximum efficiency.

In view of this low interaction efficiency, the backward-wave oscillator (both O- and M-type) is a likely candidate for depressed-collector operation. The over-all efficiency versus degree of collector segmentation is shown in Fig. 11 for several typical oscillator cases. The basically low

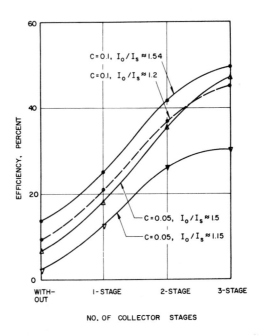

FIG. 11. Efficiency versus collector depression in an O-type backward-wave oscillator.

interaction efficiency without collector depression, and thus the low beam velocity spread, account for the large enhancement factors. Since backward-wave oscillators are usually low or moderate average power devices, there are possibly fewer problems in using multistage collectors as compared to similar forward-wave amplifiers. The optimum collector segment voltages for two typical cases are shown in Fig. 12.

4. RESULTS OF CALCULATIONS FOR O-TYPE DEVICES

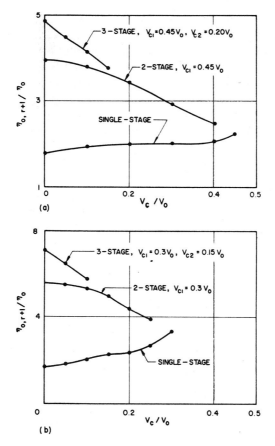

FIG. 12. Efficiency improvement factor and optimum collector segment voltages for typical O-type backward-wave oscillators. (a) $C = 0.1$, $I_0/I_s = 1.2$. (b) $C = 0.05$, $I_0/I_s = 1.5$.

4.3 Phase-Focused Oscillators and Amplifiers

In Chapter XIII it was shown theoretically, and experimental verification cited, that phase-focusing techniques could appreciably improve the saturation efficiency of linear-beam amplifiers and oscillators. Improvement of the fundamental interaction efficiency necessarily leads to a greater velocity spread in the beam and thus question arises as to the merits of collector depression on phase-focused tubes. It would be expected that the efficiency improvement factor for phase-focused devices would be less than for the unphase-focused counterparts.

The results of such computations for a phase-focused amplifier ar.d its unphase-focused counterpart are shown in Fig. 13. The efficiency is seen to be improved by a factor of 1.36 as a result of phase focusing in the forward-wave amplifier, and furthermore the efficiency of the phase-focused tube is always greater than that of the unphase-focused counterpart, independent of the degree of collector segmentation. The use of a three-stage depressed collector on the device without phase focusing gives an over-all efficiency improvement factor of 2.2, whereas the improvement factor for a three-stage collector on the phase-focused tube is only 1.7. However, the over-all efficiency is higher by a factor of 1.05.

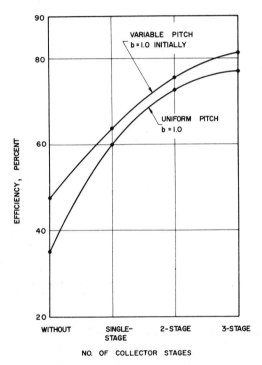

Fig. 13. Efficiency improvement for an O-type phase-focused amplifier ($C = 0.1$, $B = 1$, $\omega_p/\omega = 0$, $d = 0$).

In the case of the backward-wave oscillator the effect of phase focusing on the operation is summarized in the following table.

In comparing tapered backward-wave oscillators with untapered oscillators one must select as a common basis either the same value of velocity parameter b (voltage) or the same rf output (efficiency). For each tapered oscillator case given in Table II the two corresponding untapered

4. RESULTS OF CALCULATIONS FOR O-TYPE DEVICES

cases are given. Thus one sees that for the same voltage tapering increases the output (efficiency) and for the same output the velocity parameter increases. In both cases the starting current or length is increased.

TABLE II

PHASE-FOCUSING EFFECTS IN O-TYPE BACKWARD-WAVE OSCILLATORS

($\omega_p/\omega = 0$, $B = 1$, $d = 0$)

	Untapered	Tapered
(1)	$C = 0.1$ $A_0 = 0.7$ (eff. = 9.8 %) $b = 1.925$ $CN_s = 0.31$	
		$C = 0.1$ $A_0 = 0.825$ (eff. = 13.6 %) $b = 1.925$ $CN_s = 0.362$
(2)	$C = 0.1$ $A_0 = 0.825$ (eff. = 13.6 %) $b = 2.06$ $CN_s = 0.334$	
(3)	$C = 0.05$ $A_0 = 0.5$ (eff. = 2.5 %) $b = 1.625$ $CN_s = 0.314$	
		$C = 0.05$ $A_0 = 0.85$ (eff. = 14.4 %) $b = 1.625$ $CN_s = 0.401$
(4)	$C = 0.05$ $A_0 = 0.85$ (eff. = 14.4 %) $b = 1.825$ $CN_s = 0.355$	

In determining the efficiency enhancement possible with collector depression on tapered backward-wave oscillators one need only refer to the corresponding untapered case of Fig. 11 for the same output. Thus it is seen that the improvement results are similar to those for forward-wave amplifiers shown in Fig. 13.

4.4 Amplifier Operation below Saturation

In many situations it is desirable to operate a power amplifier several decibels below the saturated output in order to obtain greater phase linearity. If efficiency is also an important consideration then one wonders whether or not the over-all efficiency can be recovered by an additional collector depression. To be certain, this is inefficient use of a power device from both an electronic efficiency and a weight standpoint.

The velocity spread at various output rf levels below the full saturation capability of the beam is of course reduced relative to that at saturation and thus collector depression is quite effective. A rather typical case has been analyzed using the previously outlined theory and the results are presented in Fig. 14 as a function of the degree of segmentation. As

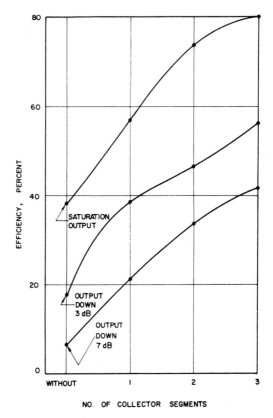

FIG. 14. Efficiency versus collector segmentation for O-type amplifier power outputs at and below saturation $(C = 0.1, \ b = 1.5)$.

generally expected, it is seen that the efficiency lost by operation below the saturated output can be regained by one additional stage of depressed collector. In view of the phase linearity near saturation obtained in phase-focused devices, the most efficient package is one incorporating both phase focusing and some degree of collector depression.

4.5 Collector Depression on Voltage-Jump Devices

It was found in Chapter XIII that the incorporation of a dc gradient variety of phase-focusing scheme in an amplifier could significantly increase the saturation gain of an amplifier although not necessarily enhance the efficiency, depending upon the details of the focusing scheme. The efficiency could be significantly increased, however, if the dc gradient were applied in the form of a voltage jump in order to reduce the percentage velocity spread. This reduction in percentage velocity spread arises from the fact that a given amount of incremental energy raises the velocity of a slow electron more than it does for a fast one. Such gradients tend to lead to a two-velocity-class condition, which of course suggests using depressed collectors.

4.6 Depressed Collectors on Voltage-Jump Devices

In the discussion of klystrons with dc voltage jumps in the output cavity it was shown that the efficiency could be considerably improved by such means. It was also shown that the percentage velocity spread was reduced as a result of voltage jumping. In a TWA, however, voltage stepping does not lead to any efficiency improvement.

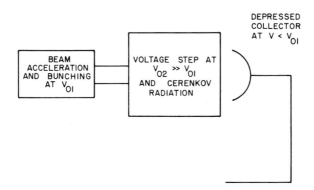

FIG. 15. Cerenkov radiator with depressed collector.

Another device in which voltage stepping may be used is a Cerenkov radiator in which the beam bunching is done at V_{01} and then the Cerenkov coupler is operated at $V_{02} \gg V_{01}$, where radiation occurs. Such a scheme is illustrated in Fig. 15 and is amenable to collector depression. A considerable improvement in efficiency can be obtained by stepping the voltage back down and collecting at a low potential. The present theory may be used directly to calculate enhancement factors after the velocity information at the output of the Cerenkov coupler is known.

5 Beam Current Flow Limitation in Collector Depression

The calculations presented in the previous sections for depressed collectors assumed rather ideal flow in the retarding field region and have not accounted for space-charge limitations. Assume that the flow in the retarding field region may be described in terms of a cylindrical beam in a concentric tube as illustrated in Fig. 16. The beam may or may not fill the drift tube and the axial magnetic field may be that for Brillouin flow or greater; in the limit, it must be sufficient for confined

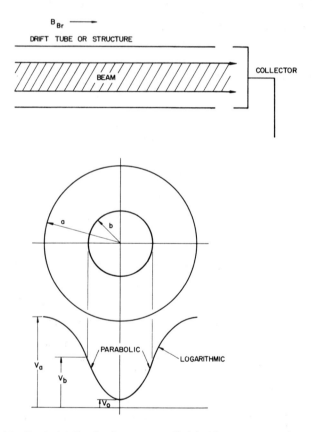

FIG. 16. Potential distribution over a cylindrical beam cross section.

flow. The potential distribution over the cross section assuming a constant charge density is also illustrated in Fig. 16.

Physically the space-charge limitation shows up in that for a given collector potential in a particular device there is a maximum value of current or perveance which can be transmitted; any increase in beam or cathode current results in a reduction in collector current and a marked increase in rf structure interception current. This is a well-known effect in power tetrodes due to the low suppressor grid potential. The maximum current flow as a function of collector potential due to space-charge depression has been studied in general by Pierce[3] and by Smith-Hartman[4] and the results were applied to the collector depression problem by Wolkstein.[5] The characteristics of both Brillouin flow beams and confined flow beams are examined to determine the limiting beam flow perveance in the drift region for each.

a. Brillouin Flow

Pierce[3] has studied Brillouin flow and shown that the maximum current which can be sustained in the geometry of Fig. 17 for a uniform axial magnetic field and a beam-edge potential of V_b is given by

$$I_{\max} = 25.4 \times 10^{-6} \, V_b^{\frac{3}{2}}, \tag{22}$$

where the beam may or may not fill the drift tube.

We are assuming idealized Brillouin flow, in which there are no radial velocities and the angular and axial electron velocities are invariant with radius. The axial velocity in such ideal flow is determined solely by the axial beam potential, and the angular velocity is equal to $\sqrt{2}$ times the axial velocity. For such ideal flow conditions Laplace's equation can easily be solved in the region between the beam boundary and the drift tube. The potential difference is given by

$$V_a - V_b = \frac{I_{\max}}{2\pi\epsilon_0 \left(\frac{dz}{dt}\right)} \int_b^a \frac{dr}{r}$$

$$= \frac{I_{\max} \ln(a/b)}{2\pi\epsilon_0 (2\eta V_b/3)^{\frac{1}{2}}}, \tag{23}$$

when $a \neq b$. The wall potential is then related to the maximum current by

$$V_a = \left(\frac{I_{\max}}{25.4 \times 10^{-6}}\right)^{\frac{2}{3}} + \frac{I_{\max} \ln(a/b)}{2\pi\epsilon_0 \left(\frac{2\eta}{3}\right)^{\frac{1}{2}} \left[\frac{I_{\max}}{25.4 \times 10^{-6}}\right]^{\frac{1}{3}}}$$

$$= [1157 + 1544 \ln(a/b)] I_{\max}^{\frac{2}{3}}. \tag{24}$$

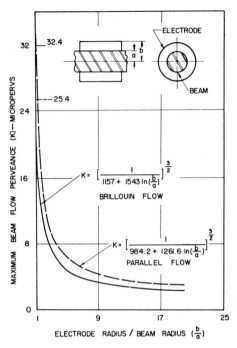

FIG. 17a. Maximum current for a solid cylindrical beam in Brillouin flow (Wolkstein[5]).

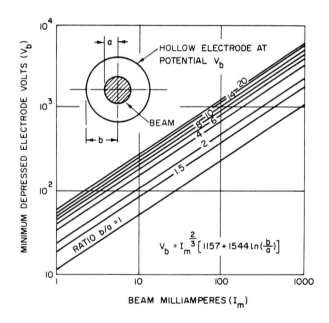

FIG. 17b. Maximum perveance for a solid cylindrical beam in Brillouin flow (Wolkstein[5]).

5. BEAM CURRENT FLOW LIMITATION

From Eq. 24 a beam-flow perveance is defined as

$$\hat{P} \triangleq \left[\frac{1}{1157 + 1544 \ln(a/b)}\right]^{\frac{3}{2}}. \tag{25}$$

Equation 25 may be used to calculate the minimum allowable collector voltage for a given maximum current and a/b value. A graph of Eq. 25 is shown in Fig. 17a and the perveance versus a/b is shown in Fig. 17b. The limiting perveance for confined flow is also shown in Fig. 17b. Notice that the limiting perveance is a maximum, and the minimum collector potential is lowest for $a = b$, which is desirably consistent with maximum efficiency conditions.

When the beam completely fills the drift tube, Pierce has shown that the current is given by

$$I = \tfrac{1}{2}\sqrt{2}\,\pi\epsilon_0 \eta^{\frac{3}{2}} B^2 V_0^{\frac{1}{2}} b^2, \tag{26}$$

and the axis voltage may be written as

$$V_0 = V_b - \frac{\eta B^2 b^2}{8}. \tag{27}$$

The beam boundary potential for maximum current yields

$$V_b = \frac{3\eta B^2 b^2}{16}. \tag{28}$$

Thus under the condition of maximum current flow for a beam filling the drift tube, $V_0 = V_b/3$. Any attempt to increase the current flow beyond the maximum given above will result in the formation of a virtual cathode, reversal of the flow and large interception of current on the rf structure. Thus in the interests of obtaining highest efficiency through collector depression one should make $a = b$.

b. Confined Flow

Many linear-beam tubes operate with axial magnetic fields which are considerably in excess of the Brillouin field and thus may be considered as having confined or parallel flow conditions. Pierce[3] has also determined the maximum current for such a flow originating from an unshielded gun. The maximum current is

$$I_{\max} = 32.4 \times 10^{-6}\, V_b^{\frac{3}{2}} \tag{29}$$

and the axial voltage is related to the beam-edge voltage by

$$V_0 = 0.174 V_b. \tag{30}$$

Notice that the maximum beam perveance is greater than that for Brillouin flow, as expected.

Unfortunately the electron velocity varies with radius, and hence some approximation must be used to continue the analysis. Following Wolkstein's suggestion, an average velocity may be used in terms of the average beam potential and the wall potential is then related to the maximum current by

$$V_a = (I_{max}/32.4 \times 10^{-6})^{\frac{2}{3}} + \frac{I_{max} \ln(a/b)}{0.2527 \times 10^{-4}[I_{max}/32.4 \times 10^{-6}]^{\frac{1}{3}}}$$

$$= [984.2 + 1261.6 \ln(a/b)] I_{max}^{\frac{2}{3}}. \tag{31}$$

The minimum allowable voltage for parallel flow as given by Eq. (31) is shown in Fig. 18.

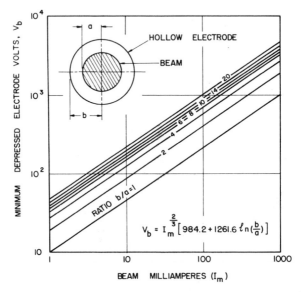

FIG. 18. Maximum beam current for parallel flow (Wolkstein[5]).

The perveance relation is found from Eq. (31) as

$$\hat{P} = \left[\frac{1}{984.2 + 1261.6 \ln(a/b)}\right]^{\frac{3}{2}}, \tag{32}$$

which has been plotted in Fig. 17. There is little difference in the results for either limiting perveance or minimum collector potential, as expected. Similar results are obtained for other types of flows and for these cases also the optimum operating conditions prevail when the beam fills the drift tube, i.e., $a = b$.

Whereas the effect of space-charge potential suppression as outlined above can limit efficiency enhancement, the effect can also be utilized. High-velocity electrons impinging on electrodes can produce secondaries which may travel back through the beam or go to the circuit. The creation of a space-charge flow with a potential minimum in front of an electrode can prevent such action by forcing the secondaries back into the electrode in question.

6 Depressed Collectors on Crossed-Field Devices

6.1 General Considerations

The nonlinear interaction problem in crossed-field injected-beam devices has been developed and studied in Chapters VIII, IX, and X following methods of analysis similar to those used in O-type beam wave devices. The efficiencies calculated for both amplifiers and oscillators agree well with experimental values and in general are higher than those for comparable O-type amplifiers and oscillators due to the fact that these (M-type) devices are potential energy conversion devices and thus are not plagued, to the first order, in the interaction process by loss of synchronism of the electron bunches. Phase-focusing techniques have been applied to both O- and M-type devices to further enhance their efficiencies.

Even with the application of various phase-focusing techniques to both classes of devices there remains a considerable amount of energy in the spent beam (at the circuit output plane) which is generally to be dissipated in the collector electrode. In previous sections of this chapter a calculation was made of the efficiency enhancement factor to be obtained by segmenting and depressing the collector on various O-type devices. In a similar fashion the over-all efficiency (plate circuit) may be improved in M-type devices by collector depression.

The problem of collector depression in M-type devices is much the same as that in O-type devices, since the objective is to decelerate the electrons of the spent beam to near-zero velocities so that as much as possible of their energy is converted to potential energy at the point of collection. By this process one recovers the potential energy of the electrons which would be lost were they collected at anode potential. The improvement in over-all efficiency is apparent. This process is somewhat more complicated in M-type devices, however, in view of the presence of the orthogonal electric and magnetic fields, which give rise to cycloidal motion of the electrons and hence highly nonlaminar motion. It would of course be possible to eliminate much of the nonlaminarity of the flow in the collector region by placing a magnetic shield around the collector so that only electric fields would be present.

The problem is thus, what is the most efficient and effective means of sorting the electrons so that their remaining energy is recovered? In addition to shielding the collector from the magnetic field as suggested above, one might consider decelerating the electrons quasi-adiabatically by gradually decreasing the electric field.

Another means would be to impose a rapid change of electric field at the entrance to the collector region, which would have the effect of forcing the electrons into cycloidal paths. Thus electrons could reach near-zero velocities over one cycloidal path and hence a sequence of collector segments at successively higher potentials would accomplish the result. Both of these systems are illustrated in Fig. 19.

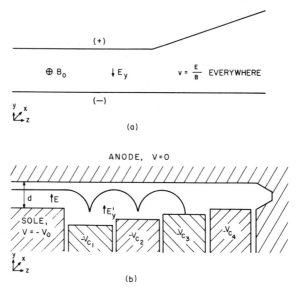

Fig. 19. Depressed collector schemes for crossed-field devices. (a) Quasi-adiabatic slowing through a gradual change in electric field. (b) Rapid slowing through an abrupt change in electric field.

6.2 Mathematical Analysis

The exact analysis of this problem is seen to be complicated by the presence of the transition fields at the exit boundary of the rf structure and by the necessity of including space-charge forces in the calculations of detailed electron motion through the collector configuration. An exact approach would be to design a configuration on a space-charge-free single-trajectory basis and then to solve the Poisson problem for the particular configuration and set of electrode potentials, accounting for the coulomb repulsion forces.

The optimum collector segment potentials and the enhancement achievable as a function of the degree of segmentation may be selected by either graphical or analytical methods similar to those employed in studying the collection of O-type beams. Nonlinear performance calculations are made for these devices assuming the injection of m charge groups into the interaction region and then integrating along trajectories until the output plane is reached. In M-type devices, due to collection of charge groups on the anode and sole, unfortunately one does not have m charge groups appearing at the output plane. This is simply accounted for in the following equation:

$$n = m - (m_a + m_s) \tag{33}$$

where

$n =$ the number of charge groups appearing at the output plane,

$m =$ the number of charge groups injected into the interaction region,

$m_a =$ the number of charge groups collected on the anode, and

$m_s =$ the number of charge groups collected on the sole electrode.

The concept of the equivalent rf potential V_j will again be used as well as its joule equivalent, $-qV_j \triangleq \mathscr{E}_j$. The fundamental electronic interaction efficiency may then be expressed in terms of the equivalent energies as

$$\eta_e = \frac{\Delta\mathscr{E}_{\text{pot.}}}{\mathscr{E}_{\text{dc}}} = 1 - \frac{\Sigma\mathscr{E}_{\text{losses}}}{\mathscr{E}_{\text{dc}}}, \tag{34}$$

where

$\mathscr{E}_{\text{dc}} = mqV_0 =$ the input energy of the electron beam,

$\Delta\mathscr{E}_{\text{pot.}} =$ electron potential energy converted to rf, and

$\Sigma\mathscr{E}_{\text{losses}} =$ energy lost due to all causes.

The energy lost in the system due to electron charge groups being collected on either the sole or the anode is given by

$$\mathscr{E}_{\text{losses}} = -q\left[\sum_{j=1}^{m_a} V_j + \sum_{j=1}^{m_s}(V_j - V_0)\right], \tag{35}$$

where again V_j is the equivalent rf potential associated with the jth charge group collected.

As illustrated in Fig. 19 the collector consists of r segments, each collecting n_i charge groups. Since there are n charge groups at the collector entrance plane,

$$n = \sum_{i=1}^{r} n_i, \tag{36}$$

where i indicates the particular collector segment in question.

If any charge groups are collected on the ith segment with, say, $(-qV_{jk})$ remaining, there will be losses (dissipation) in the collector segment, which of course limits the efficiency improvement. These losses may be expressed as follows:

$$\left(\sum \mathscr{E}_{\text{losses}}\right)_{\text{collector}} = -q \sum_{i=1}^{r} \left[\sum_{k=1}^{n_j} V_{ik} - n_j V_{ci}\right]$$

$$= -q \sum_{k=1}^{n} V_k - \sum_{i=1}^{r} q n_j V_{ci}. \tag{37}$$

Substituting Eqs. (35) and (37) into Eq. (34) gives the following for the efficiency:

$$\eta_e = 1 - \frac{1}{m} \left\{ \sum_{j=1}^{m_a} \frac{V_i}{V_0} - \sum_{j=1}^{m_s} \left(\frac{V_j}{V_0} - 1\right) - \sum_{k=1}^{n} \frac{V_k}{V_0} + \sum_{i=1}^{r} n_i \frac{V_{ci}}{V_0} \right\}. \tag{38}$$

All of the equivalent rf potentials indicated in Eq. (38) are known from the previous nonlinear calculations. The equivalent rf potential V_k for the n charge groups (electrons) at the collector entrance plane is given by

$$V_k = V_0 \left(1 - \frac{y_k}{d}\right) + \frac{1}{2\eta} (\dot{z}_k^2 + \dot{y}_k^2), \tag{39}$$

where

y_k = the y-displacement,

d = sole-anode separation, and

$\dot{z}_k, \dot{y}_k$ = z- and y-directed velocities respectively.

At this point it is prudent to resort to a graphical picture of the collection and sorting process in order to gain a little physical insight into the process. The optimization process for a four-segment collector to be discussed below is illustrated on the electron distribution diagram of

Fig. 20. The energy which can be recovered by a multisegment depressed collector is shown shaded in Fig. 20 and is expressed as

$$\text{Energy recovered} = q \sum_{i=1}^{r} n_i V_{ci}. \tag{40}$$

Whether graphical or mathematical procedures are utilized, the object is to recover the maximum energy in the collector and thus $\sum_{i=1}^{r} n_i V_{ci}$ is to be maximized.

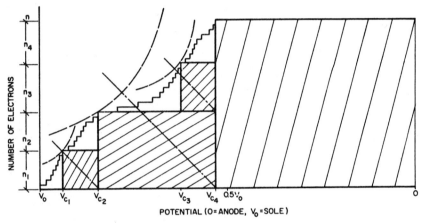

FIG. 20. Energy distribution at the output plane of an M-type device.

A graphical procedure is illustrated in which successive subdivisions of the abscissa are found by drawing rectangular hyperbolas tangent to the distribution function within each section. In order to maximize the energy recovered by the collector, the maximum potential energy achieved by the kth electron group along its trajectory through the collector region must be determined. It is readily shown from a solution of the Lorentz force equation in a parallel-plate collector region that this maximum energy V_{mk} is given by

$$V_{mk} = V_0 \left(\frac{\dot{z}_k}{2v_0} + \frac{\varDelta v_k}{2v_0} + \frac{3}{4} - \frac{y_k}{d} \right), \tag{41}$$

where

$$\varDelta v_k = \sqrt{\left(\frac{v_0}{2} - \dot{z}_k \right)^2 + \dot{y}_k^{\,2}}.$$

The efficiency enhancement thus achieved through collector segmentation may be expressed in terms of the energy losses to the sole, anode, and collector and the energy recovered in the collector region:

$$\frac{\eta_{e,r}}{\eta_e} = 1 + \frac{\sum_{i=1}^{r} n_i V_{ci}}{mV_0 - \sum_{j=1}^{m_a} V_j - \sum_{j=1}^{m_s} (V_j - V_0) - \sum_{k=1}^{n} V_k}. \quad (42)$$

6.3 Results

The above analytical method for calculating efficiency enhancement factors applied to any collector scheme, and the only problem remaining is to determine accurately the number of electron groups, n_i, which are actually collected on a particular collector segment. As mentioned earlier, this may require the calculation of actual nonlaminar motion in Poisson fields in the collector region. For the system illustrated in Fig. 19b the individual collector segment length should be greater than one cyclotron wavelength. If space charge and nonlaminar motion in the collector region are neglected the velocity curves of Fig. 1 may be used directly to calculate efficiency enhancement factors.

To indicate the advantage of depressed collectors on M-type devices, theoretical efficiency enhancement factors have been calculated using the nonlinear results shown in Chapter IX. It is assumed that in a practical device a depressed collector would be located at a plane corresponding to 50–70% electron collection on the anode. It is expected that the lower the percentage electron collection the higher will be the efficiency enhancement factor for an arbitrary number of collector stages.

Calculations were made for a variety of operating parameters which yielded significant interaction efficiencies as outlined in Chapter IX. These were made at various collection planes for one, two, three and infinitely many collector segments. The results on maximum efficiency enhancement are shown in Figs. 21 and 22, where in each case the collector potential has been optimized. The enhancement factors are seen to be significant and of course for a given velocity spread the more collector segments utilized the greater the improvement.

In order to achieve these improvements one must optimize the various collector segment potentials. These optimum operating potentials are illustrated in Figs. 23 and 24 for various numbers of collector segments.

Oscillator efficiency particularly may be improved by these techniques. The above general procedures are, of course, applicable here and the results of such calculations are quite similar to the above.

6. DEPRESSED COLLECTORS ON CROSSED-FIELD DEVICES

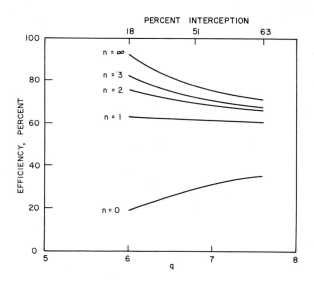

FIG. 21. Efficiency versus interaction length, for n collector segments ($D = 0.05$, $r = 0.5$, $s = 0.1$, $b = 0$, $\psi_0 = -30$, $\omega_c/\omega = 0.25$).

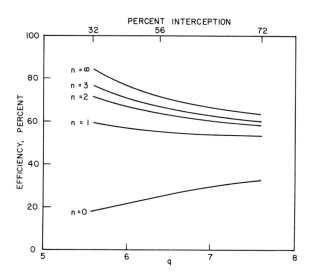

FIG. 22. Efficiency versus collector segmentation ($D = 0.1$, $r = 0.5$, $s = 0.1$, $b = 0$, $\psi_0 = -30$, $\omega_c/\omega = 0.75$).

In applying this technique of efficiency improvement to phase-focused M-FWA's and M-BWO's and to operation below the saturated output, results similar to those obtained in O-type interaction are obtained.

The preceding analyses indicate the rewards obtainable through collector depression on any device as a function of the output beam

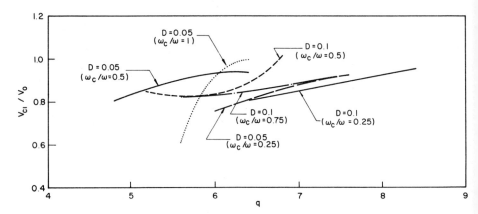

FIG. 23. Single-stage collector ($r = 0.5$, $s = 0.1$, $b = 0$, $\psi_0 = -30$).

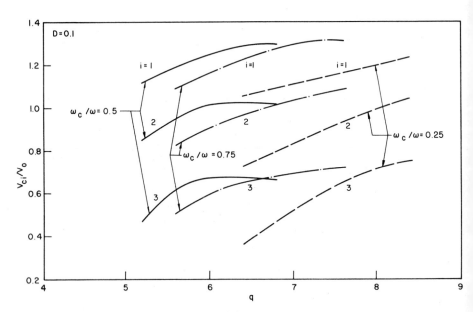

FIG. 24. Three-stage collector for $D = 0.1$ ($r = 0.5$, $s = 0.1$, $b = 0$, $\psi_0 = -30$).

velocity distribution. To be certain, technological difficulties will be met in attempting to achieve the desired results and thus, as always, there are experimental limitations to be considered. Philosophically speaking, one might avoid these problems by maximizing the fundamental interaction efficiency and dispensing with collector segmentation and its burdensome additional power supplies and power dissipation problems.

REFERENCES

1. Rowe, J. E., One-dimensional traveling-wave tube analysis and the effect of radial electric field variations. *IRE Trans. Electron Devices* **7**, No. 1, 16-22 (1960).
2. Rowe, J. E., Efficiency enhancement by phase focusing and collector depression, in *Proc. 4th Intern. Congr. Microwave Tubes, The Netherlands, 1962*, pp. 640-646.
3. Pierce, J. R., "Theory and Design of Electron Beams," pp. 153-159. Van Nostrand, Princeton, N. J., 1949.
4. Smith, L. P., and Hartman, P. L., The formation and maintenance of electron and ion beams. *J. Appl. Phys.* **11**, No. 3, 220-230 (1940).
5. Wolkstein, H. J., Effect of collector potential on the efficiency of traveling-wave tubes. *RCA Rev.* **19**, 259-282 (1958).
6. Sterzer, F., Improvement of traveling-wave tube efficiency through collector potential depression. *IRE Trans. Electron Devices* **5**, No. 4, 300-306 (1958).
7. Dunn, D. A., Luebke, W. R., and Wada, G., A low potential collector employing an asymmetrical electrode in an axially-symmetric magnetic field. *IRE Trans. Electron Devices* **6**, No. 3, 294-297 (1959).
8. Dunn, D. A., Borghi, R. P., and Wada, G., A crossed-field multisegment depressed collector for beam-type tubes. *IRE Trans. Electron Devices* **7**, No. 4, 262-268 (1960).
9. Hansen, J. W., and Susskind, C., Improvement of beam-tube performance by collector-potential depression, and a novel design. *IRE Trans. Electron Devices* **7**, No. 4, 282-289 (1960).

CHAPTER

XVI | Modulation Characteristics

1 Introduction

There are many systems applications for microwave amplifiers. Hence their modulation characteristics are of interest; i.e., how is the operation influenced by periodic variations of either the stream potential or the current? The term modulation has been used in previous chapters in discussing the action of an impressed rf signal on the velocity and bunching characteristics of the electron stream. In this chapter "modulation" will refer only to the process of placing on the electron stream a coherent periodic disturbance (signal) at some frequency different from the carrier signal frequency.

The modulating signal will generally produce sidebands around the carrier and its harmonics, and these will alter the amplitude and phase of the carrier at the output. In view of the wideband nature of such rf structures as the helix and the vane line, the harmonics and their associated sidebands may be within the passband of the device. A particular case of interest depending on two-frequency operation is the beam-type parametric amplifier, in which many frequencies are involved when the modulation is due to impressed noise.

The modulations of interest may be divided into two distinct classes:

(1) those modulations whose rates are sufficiently slow so that the period of the modulating cycle is much greater than the transit time for an electron through the device; and

(2) those modulations whose periods are comparable to the electron transit time.

Classes (1) and (2) are called respectively low-frequency modulation and high-frequency modulation. A detailed study of the operation of O-type devices in both the linear and nonlinear regimes for both classes of modulations has been given by Sobol and Rowe.[1,2] Chapter XIV on prebunched electron beams considers one type of high-frequency modulation in that pre-velocity and density modulations are at the carrier frequency.

Under low-frequency modulation a variation of the stream velocity (anode voltage) produces a phase modulation (PM) and/or a transit-time modulation (TTM). Variation of the entering electron stream current density produces an amplitude modulation (AM). Of course, the production of PM or TTM is accompanied by a certain amount of AM due to nonlinearities, just as AM is always accompanied by PM or TTM. In studying low-frequency modulations, assume that the sidebands produced around the carrier and each of its harmonics are sufficiently close that the circuit properties are identical for all. This is not the case in the BWO, since modulation changes the carrier frequency and the circuit impedance changes.

Also assume that for low-frequency modulations all space and time derivatives of the modulation function at that frequency are negligibly small compared to derivatives of the same function at the carrier frequency. Thus one may consider that all functions of the modulating frequency are constants in the interaction equations; also, quasi-stationary equations describe the behavior at the modulation frequencies. The output signal under modulation is described by the gain and phase shift relative to the input signal. These "modulation device functions" must be calculated on a point-by-point basis in the nonlinear regime.

For high-frequency modulations the various sideband components may see different circuit properties than does the carrier. In this case, for a given modulation, there will be differences in the output spectra resulting from PM and TTM. The differences occur in the amplitudes of spectral components but not in the frequency location of the components. A list of references on various modulation studies is given at the end of this chapter.

2 Mathematical Analysis for O-Type Devices

In considering the low-frequency modulation characteristics of nonlinear O-type devices, the one-dimensional Lagrangian analysis of Chapter VI is utilized and the stream potential and current are varied as follows over an rf cycle:

$$V_0 = V_{01} + \Delta V(t, t_A) \tag{1}$$

and

$$I_0 = I_{01} + \Delta I(t, t_B), \tag{2}$$

where in general ΔV and ΔI are periodic functions of different modulation frequencies. The times t_A and t_B give the coherence and phase difference between the modulations ΔV and ΔI and the rf applied signal. V_{01} and I_{01} describe the unmodulated state of the stream. All of the assumptions of the one-dimensional Lagrangian analysis are made here, along with the restriction that the modulating frequencies are very much lower than the rf signal frequency. The average values of ΔV and ΔI are small, and the following "modulation parameters" are defined:

$$M^3 = M^3(t, t_A, t_B) \triangleq \frac{(1 + \Delta V/V_{01})}{(1 + \Delta I/I_{01})}, \tag{3}$$

$$\xi_1 \triangleq (1 + \Delta V/V_{01})^{\frac{1}{2}}, \tag{4}$$

and

$$\xi_2 \triangleq \left(\frac{1 + \Delta I/I_{01}}{M^3}\right)^{\frac{1}{2}} = \frac{(1 + \Delta I/I_{01})}{\xi_1}. \tag{5}$$

These parameters will appear in the final equations and give a direct measure of the modulation amplitudes.

Since the rf circuit equation (equivalent circuit) is independent of the average beam conditions, the homogeneous portion of the nonlinear circuit equations will not change. However, since stream velocity and density are being periodically varied, we expect the force and continuity equations to be affected. Recall that charge conservation for the unmodulated (quiescent) condition is stated as

$$\rho(z_0, t_0)\, dz_0 = \rho(z, t)\, dz,$$

where z_0 and t_0 indicate the initial charge position and time. In terms of the modulation parameters the conservation equation is written as

$$\rho(z, t) = -\frac{I_{01}}{u_{01}} \left(\frac{\partial z_0}{\partial z}\right)_t \left(\frac{1 + \Delta I/I_{01}}{M^3}\right)^{\frac{1}{2}}$$

$$= -\frac{I_{01}}{u_{01}} \xi_2 \left(\frac{\partial z_0}{\partial z}\right)_t. \tag{6}$$

Recall that the Lagrangian method is to integrate over the trajectories for all electrons entering the interaction space during one rf cycle. Thus it would seem necessary in the case of the modulated device to integrate over the entire amount of charge entering during one modulation cycle. In order to avoid this undesirable complication a quasi-stationary approximation is made in which each point within the modulation cycle is treated as a stationary point and it is thus necessary to integrate over only one rf cycle. Again we refer entering charges to the rf signal at the input plane. The phase variable Φ_{0j} is used,

$$\Phi_{0j} = \omega t_{0j},$$

where the entrance time is now

$$t_{0j} = \frac{z_{01j}}{u_{01}} = \frac{z_{01}}{u_{01}(1 + \Delta V/V_{01})^{\frac{1}{2}}}.$$

z_{01j} measures the distance traveled in t_{0j} when $u = u_{01}$ and z_{0j} is traversed in t_{0j} when $\Delta V \neq 0$. For convenience all normalized variables are defined in terms of the unmodulated parameters.

Rather than integrate with time as a running variable, consider that all charge groups entering during each rf cycle enter simultaneously but with different phases relative to the signal wave at the input. Hence integration is made with respect to y as previously, with Φ_{0j} and y as independent variables. The charge group trajectory diagram for the modulated device is shown in Fig. 1. Referring to this figure reveals that

$$\left(\frac{\partial \Phi}{\partial z}\right)_{\Phi_{0j}} + \frac{d\theta}{dz} = \left(1 - \frac{u_{01}}{u_{tj}}\right)\frac{\omega}{u_{01}}.$$

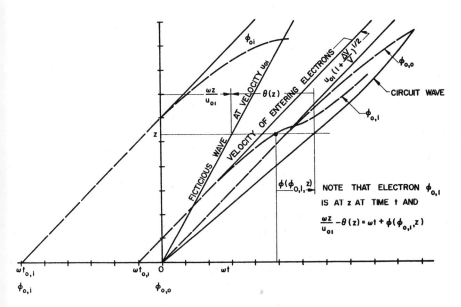

FIG. 1. Distance-phase diagram for a modulated traveling-wave device.

The charge group velocity is defined as

$$u_{tj} = u_{01}\left(1 + \frac{\Delta V}{V_{01}}\right)^{\frac{1}{2}}[1 + 2C_0 u(y, \Phi_{0j})]$$

and proceeding as in the unmodulated case gives the velocity-phase relation as

$$\left[\frac{\partial \Phi(y, \Phi_{0j})}{\partial y}\right]_{\Phi_{0j}} + \frac{d\theta(y)}{dy} = \frac{1}{C_0}\left[1 - \frac{1}{\xi_1[1 + 2C_0 u(y, \Phi_{0j})]}\right] \quad (7)$$

where the subscript 0 on C indicates the unmodulated parameter value.

In view of stream potential and current modulations, the normalized rf circuit voltage is defined by

$$V(y, \Phi) = \text{Re}\left[\frac{Z_0 I_0}{C_0} M\left(1 + \frac{\Delta I}{I_{01}}\right) A(y) e^{-j\Phi}\right]. \quad (8)$$

Since the derivative of $V(y, \Phi)$ is involved in the force equation, the modulation parameters will appear in both the force equation and the inhomogeneous circuit driving terms.

In converting the conservation equation to new variables the z_0 variable must be written in terms of the phase variable. Note that

$$\left(\frac{\partial z_0}{\partial z}\right)_t = \frac{u_{01}}{\omega}\left[1 + \frac{\Delta V}{V_{01}}\right]^{\frac{1}{2}}\left(\frac{\partial \Phi_0}{\partial z}\right)_t \quad (9)$$

and since

$$d\Phi = \frac{dy}{C_0} - \frac{d\theta}{dy} - \omega\, dt,$$

Eq. (9) may be written as

$$\left(\frac{\partial z_0}{\partial z}\right)_t = \frac{1}{[1 + 2C_0 u(y, \Phi_{0j})]} \frac{1}{(\partial \Phi/\partial \Phi_{0j})_y}. \quad (10)$$

Substituting into the conservation of charge equation yields

$$\rho(u, \Phi_{0j}) = -\frac{I_{01}\xi_2}{u_{01}} \frac{1}{[1 + 2C_0 u(y, \Phi_{0j})]} \frac{1}{(\partial \Phi/\partial \Phi_{0j})_y}. \quad (11)$$

The space-charge density of Eq. (11) is expanded into a Fourier series, as in Chapter VI, with the result

$$\rho(u, \Phi_{0j}) = \frac{I_{01}\xi_2}{u_{01}} \sum_{u=0}^{\infty}\left(\sin n\Phi \int_0^{2\pi} \frac{\sin n\Phi(\partial\Phi_{0j}/\partial\Phi)_y}{[1 + 2C_0 u(y, \Phi_{0j})]} d\Phi \right.$$
$$\left. + \cos n\Phi \int_0^{2\pi} \frac{\cos u\Phi(\partial\Phi_{0j}/\partial\Phi)_y}{[1 + 2C_0 u(y, \Phi_{0j})]} d\Phi\right) \quad (12)$$

2. MATHEMATICAL ANALYSIS FOR O-TYPE DEVICES

where again all harmonics are used in the force equation and it is assumed that only the fundamental component of ρ drives the circuit. The resulting circuit and force equations are as follows.

Circuit Equations

$$\frac{d^2 A(y)}{dy^2} - A(y)\left[\left(\frac{1}{C_0} - \frac{d\theta(y)}{dy}\right)^2 - \frac{(1 + C_0 b_0)^2}{C_0^2}\right]$$

$$= -\frac{(1 + C_0 b_0)}{\pi C_0} \frac{1}{M^{5/2}(1 + \Delta I/I_{01})^{\frac{1}{2}}}$$

$$\times \left\{\int_0^{2\pi} \frac{\cos \Phi(y, \Phi'_{0j})\, d\Phi'_0}{[1 + 2C_0 u(y, \Phi'_{0j})]} + 2C_0 d_0 \int_0^{2\pi} \frac{\sin \Phi(y, \Phi'_{0j})\, d\Phi'_0}{[1 + 2C_0 u(y, \Phi'_{0j})]}\right\}, \quad (13)$$

and

$$A(y)\left[\frac{d^2\theta(y)}{dy^2} - \frac{2d_0}{C_0}(1 + C_0 b_0)^2\right] + \frac{2\, dA(y)}{dy}\left(\frac{d\theta(y)}{dy} - \frac{1}{C_0}\right)$$

$$= -\frac{(1 + C_0 b_0)}{\pi C_0} \frac{1}{M^{5/2}(1 + \Delta I/I_{01})^{\frac{1}{2}}}$$

$$\times \left\{\int_0^{2\pi} \frac{\sin \Phi(y, \Phi'_{0j})\, d\Phi'_0}{[1 + 2C_0 u(y, \Phi'_{0j})]} - 2C_0 d_0 \int_0^{2\pi} \frac{\cos \Phi(y, \Phi'_{0j})\, d\Phi'_0}{[1 + 2C_0 u(y, \Phi'_{0j})]}\right\}. \quad (14)$$

Force Equation

$$[1 + 2C_0 u(y, \Phi_{0j})] \frac{\partial u(y, \Phi_{0j})}{\partial y}$$

$$= -\frac{1}{M^2} A(y)\left(1 - C_0 \frac{d\theta(y)}{dy}\right) \sin \Phi(y, \Phi_{0j}) + \frac{C_0}{M^2} \frac{dA(y)}{dy} \cos \Phi(y, \Phi_{0j})$$

$$- \frac{1}{(1 + C_0 b_0) M^3 \xi_1}\left(\frac{\omega_{p0}}{\omega C_0}\right)^2 \int_0^{2\pi} \frac{F_{1-z}(\Phi - \Phi')\, d\Phi'_0}{[1 + 2C_0 u(y, \Phi_{0j})]}, \quad (15)$$

where

$$b_0 \triangleq \frac{1}{C_0}\left(\frac{u_{01}}{v_0} - 1\right).$$

The space-charge-field expression used is that obtained by the harmonic method of Chapter IV with the expansion of $\rho(u, \Phi_{0j})$ given by Eq. (12). The result is

$$E_{sc-z} = \frac{2\omega u_{01}}{|\eta|(1 + C_0 b_0)}\left(\frac{\omega_{p0}}{\omega}\right)^2 \xi_2 \int_0^{2\pi} \frac{F_{1-z}(\Phi - \Phi')\, d\Phi'_0}{1 + 2C_0 u(y, \Phi'_0)}. \quad (16)$$

Thus in the absence of modulations the above equations reduce directly to those of Chapter VI. The solution of the above equations is obtained in the same manner as before and subject to the same boundary conditions.

If the multidimensional analyses of Chapter VI are utilized, the modulation analysis can proceed in the same way, provided it is assumed that the modulation affects only the longitudinal velocities and density. In fact, the appropriate equations can be written down directly following the form of the above results.

3 O-Type Nonlinear Modulation Results

Since the output signal of an amplifier under modulation is describable by the gain and phase shift relative to the input signal, we proceed to

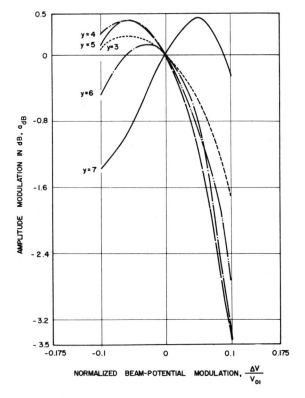

FIG. 2. Amplitude-modulation device function during beam-potential modulation, large-signal calculation ($C_0 = 0.1$, $QC_0 = 0.125$, $d_0 = 0$, $b_0 = 0.65$, $B = 1$, $a/b' = 2$).

3. O-TYPE NONLINEAR MODULATION RESULTS

define modulation device functions as follows. The phase modulation function is

$$\Delta S = \theta(y, \Delta V, \Delta I) - \theta(y, 0, 0)$$

and the amplitude modulation function is

$$\Delta a_{db} = M\left(1 + \frac{\Delta I}{I_{01}}\right) A_{db}(y, \Delta I, \Delta V) - A_{db}(y, 0, 0).$$

Unfortunately one cannot give explicit analytic expressions for ΔS and Δa_{db} above due to the nonlinear nature of the equations. Therefore various specific computer calculations must be utilized.

Phase and amplitude device functions for a range of quiescent O-TWA operating parameters are summarized in Figs. 2 and 3. Large-signal operation refers to the carrier in all cases.

Several characteristics are revealed in the above results. For short

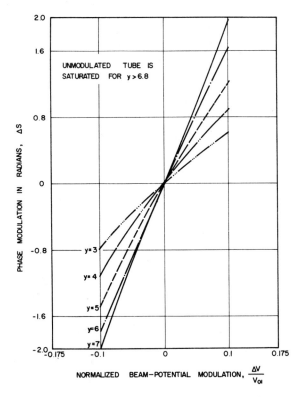

FIG. 3. Phase-modulation device function during beam-potential modulation, large-signal calculation ($C_0 = 0.1$, $QC_0 = 0.125$, $d_0 = 0$, $b_0 = 0.65$, $B = 1$, $a/b' = 2$).

lengths (small y values) the device is still operating linearly, since a low-level input signal was assumed and the results agree well with those from the linear theory.[1,2] Furthermore at sufficiently long lengths that saturation is impending, the phase modulation experienced during a periodic voltage modulation is more nearly linear than at low signal levels. Amplitude modulation, on the other hand, tends to be limited at saturation and also the voltage for maximum gain is shifted to higher voltages, as expected. Linear amplitude modulation is obtained for small-amplitude current modulations.

In all cases the velocity parameter b was chosen for maximum gain and the input drive level selected to be 30 dB below CI_0V_0. Satura ion occurs at a level from 3–5 dB above CI_0V_0. Experimental data on modulated large-signal traveling-wave amplifiers are in quite good agreement with the theoretical predictions, thereby justifying the assumptions made in the low-frequency modulation theory.

The large linear phase modulation characteristic during a beam potential modulation is desirable for certain applications such as the serrodyne. The inclusion of circuit loss effects tends to extend the linear range of operation and the inherent AM can be minimized by operating near saturation. One difficulty, however, which arises when working near saturation is that if there is any AM of the carrier this is converted directly to produce an additional phase modulation. Simultaneous application of ΔV and ΔI can result in linear PM with little AM.

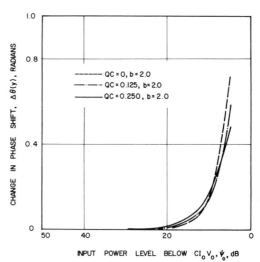

FIG. 4. Change in phase shift versus input power ($C = 0.1$, $d = 0$, $N_g = 5.75$, $B = 1$).

4. MATHEMATICAL ANALYSIS FOR M-TYPE DEVICES

Noise modulation of a large-signal device can be treated by assuming that the entering stream is divided into a number of velocity and current classes and treating these as premodulations and prebunchings. The methods of analysis outlined in Chapters VI and IX then apply directly. Multiple-frequency analyses, though straightforward, are quite formidable due to the fact that the time dependence of the modulating signal must also be considered. Such analyses can be carried out directly using the basic method of Chapter VI.

The nonlinear calculations can also be used to determine the amount of phase shift (PM) produced during an amplitude modulation (AM) of the carrier. These results are shown in Figs. 4 and 5 for structures of two different lengths. Note that no PM is produced when the carrier amplitude is relatively small. However, for a change in ψ_0 from -9 dB to 5 dB a linear change in phase shift of 0.5 radian is obtained. Thus the phase deviation is 0.5 radian with an inherent AM of 4 dB.

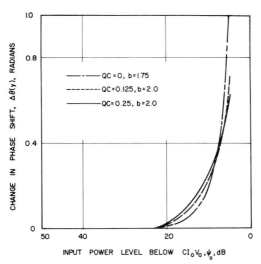

FIG. 5. Change in phase shift versus input power $(C = 0.05, \ d = 0, \ N_g = 13, \ B = 1)$.

4 Mathematical Analysis for M-Type Devices

In carrying out an analysis of the low-frequency modulated crossed-field amplifier, M-FWA, a two-dimensional analysis must be utilized since energy conversion in these devices results from the y-directed movement of the electrons. The nonlinear crossed-field Lagrangian equations of Chapter IX are the basis of the analysis. The mathematical procedure developed here follows that used in Section 2 for the one-dimensional O-FWA. The results are quite similar and in fact correspond directly to those for the two-dimensional O-FWA.

The stream potential and current are varied over one rf cycle as follows:

$$V_0 = V_{01} + \Delta V(t, t_A)$$

and

$$I_0 = I_{01} + \Delta I(t, t_B).$$

It will be assumed that the entrance time for charge groups is given by

$$t_{0j} = \frac{z_{01}}{\bar{u}_{z1}(1 + \Delta V/V_{01})^{\frac{1}{2}}}$$

and the modulation (*beam potential*) is assumed to affect only the z-directed electron (charge group velocity). Thus

$$u_{zj} = \bar{u}_{01}(1 + \Delta V/V_{01})^{\frac{1}{2}}[1 + 2D_0 u(p_0, \Phi_0, q)],$$

where u_{01} = the average dc stream velocity at the input and D_0 indicates the unmodulated value of the gain parameter.

Following a procedure similar to Section 2 gives the velocity-phase equation as

$$\left(\frac{\partial \Phi(p_0, \Phi_0, q)}{\partial q}\right)_{\Phi_{0j}} - \frac{d\theta(q)}{dq} = \frac{1}{D_0}\left[1 - \frac{1}{\xi_1[1 + 2D_0 u(p_0, \Phi_0, q')]}\right] \quad (17)$$

and the equation for the y-position variable as

$$p(p_0, \Phi_0, q) = p_0 + \int_0^q \frac{v_{y\omega} dq'}{D_0 \xi_1[1 + 2D_0 u(p_0, \Phi_0, q')]}. \quad (18)$$

The potential function is defined by

$$V(p, \Phi, q) = \mathrm{Re}\left[\frac{Z_0 I_0}{D_0} M\left(1 + \frac{\Delta I}{I_{01}}\right) A(q)\psi(p) e^{-j\Phi}\right]$$

and the space-charge density components are written as

$$\rho_{ns} = -|\rho_0| h w \xi_2 \int_0^{2\pi} \int_{\frac{1}{s}-\frac{1}{2}}^{\frac{1}{s}+\frac{1}{2}} \psi(p) \frac{1 + 2D_0 u_i(p_0', \Phi_0', 0)}{1 + 2D_0 u(p_0', \Phi_0', q)} \sin n\Phi \, d\Phi_0' \, dp_0'$$

and

$$\rho_{nc} = -|\rho_0| h w \xi_2 \int_0^{2\pi} \int_{\frac{1}{s}-\frac{1}{2}}^{\frac{1}{s}+\frac{1}{2}} \psi(p) \frac{1 + 2D_0 u_i(p_0', \Phi_0', 0)}{1 + 2D_0 u(p_0', \Phi_0', q)} \cos n\Phi \, d\Phi_0' \, dp_0'.$$

4. MATHEMATICAL ANALYSIS FOR M-TYPE DEVICES

Again we assume that only the $n = 1$ component of ρ is effective in driving the circuit, although all harmonics are utilized in the force equation. The rest of the working equations are written as follows.

Circuit Equations

$$\frac{d^2 A(q)}{dq^2} - A(q)\left[\left(\frac{1}{D_0} + \frac{d\theta(q)}{dq}\right)^2 - \left(\frac{1 + D_0 b_0}{D_0}\right)^2\right]$$

$$= -\frac{(1 + D_0 b_0)}{\pi D_0} \frac{1}{M^{5/2}(1 + \Delta I/I_{01})^{\frac{1}{2}}} \left[\int_0^{2\pi} \int_{\frac{1}{s}-\frac{1}{2}}^{\frac{1}{s}+\frac{1}{2}} \psi(p) \cos \Phi \right.$$

$$\cdot \frac{1 + 2D_0 u_i(p_0', \Phi_0', 0)}{1 + 2D_0 u(p_0', \Phi_0', q)} d\Phi_0' \, dp_0' + 2d_0 D_0$$

$$\left. \cdot \int_0^{2\pi} \int_{\frac{1}{s}-\frac{1}{2}}^{\frac{1}{s}+\frac{1}{2}} \psi(p) \sin \Phi \frac{1 + 2D_0 u_i(p_0', \Phi_0', 0)}{1 + 2D_0 u(p_0', \Phi_0', q)} d\Phi_0' \, dp_0'\right] \quad (19)$$

and

$$2\frac{dA(q)}{dq}\left(\frac{1}{D_0} + \frac{d\theta(q)}{dq}\right) + A(q)\left[\frac{d^2\theta(q)}{dq^2} + \frac{2d_0}{D_0}(1 + D_0 b_0)^2\right]$$

$$= \frac{(1 + D_0 b_0)}{\pi D_0} \frac{1}{M^{5/2}(1 + \Delta I/I_{01})^{\frac{1}{2}}} \left[\int_0^{2\pi} \int_{\frac{1}{s}-\frac{1}{2}}^{\frac{1}{s}+\frac{1}{2}} \psi(p) \sin \Phi \right.$$

$$\cdot \frac{1 + 2D_0 u_i(p_0', \Phi_0', 0)}{1 + 2D_0 u(p_0', \Phi_0', q)} d\Phi_0' \, dp_0' - 2d_0 D_0$$

$$\left. \cdot \int_0^{2\pi} \int_{\frac{1}{s}-\frac{1}{2}}^{\frac{1}{s}+\frac{1}{2}} \psi(p) \cos \Phi \frac{1 + 2D_0 u_i(p_0', \Phi_0', 0)}{1 + 2D_0 u(p_0', \Phi_0', q)} d\Phi_0' \, dp_0'\right]. \quad (20)$$

Force Equations

$$[1 + 2D_0 u(p_0, \Phi_0, q)] \frac{\partial u(p_0, \Phi_0, q)}{\partial q} = \frac{\omega}{2\omega_c} \frac{\psi(p)}{M^2}$$

$$\times \left[\frac{dA(q)}{dq} \cos \Phi(p_0, \Phi_0, q) - \frac{A(q)}{D_0} \sin \Phi(p_0, \Phi_0, q)\left(1 + D_0 \frac{d\theta(q)}{dq}\right)\right]$$

$$+ \frac{1}{2\xi_1^2 D_0^2} \frac{s}{t} v_{y\omega} + \frac{1}{M^3 \xi_1}\left(\frac{\omega_p}{\omega}\right)^2 \frac{rs}{\pi D_0^2} F_{2-z} \quad (21)$$

and

$$[1 + 2D_0 u(p_0, \Phi_0, q)] \left[\frac{\partial v_{y\omega}}{\partial q} + \left(\frac{\omega_c}{\omega}\right)^2 \frac{l}{sD_0}\right]$$

$$= \frac{l}{sM^2}\left(1 + D_0 \frac{d\theta(q)}{dq}\right) \frac{\cosh\left[\frac{\omega}{\omega_c}\frac{ps}{l}\left(1 + D_0 \frac{d\theta(q)}{dq}\right)\right]}{\sinh\left[\frac{\omega}{\omega_c rl}\left(1 + D_0 \frac{d\theta(q)}{dq}\right)\right]}$$

$$\times A(q) \cos \Phi + \left(\frac{\omega_c}{\omega}\right)^2 \frac{rl^2}{sD_0 \xi_1^2}\left(\frac{V_a}{2V_0}\right) - \frac{2rl}{M^3 \xi_1}\left(\frac{\omega_p}{\omega}\right)^2 \left(\frac{\omega_c}{\omega}\right) F_{2-y}, \tag{22}$$

where F_{2-y} and F_{2-z} are the space-charge integrals.

In the event that the sole-circuit voltage is also modulated by $k(1 + \Delta V/V_{01})$, the second term on the right of Eq. (22) is multiplied by k.

With the modulation parameters placed equal to one, the above equations reduce to those of Chapter IX. Numerical solutions are obtained as previously with the addition of specifying $M_1 \xi_1$ and ξ_2 as system parameters.

5 Output Spectra for Low-Frequency Modulations

In addition to the change in gain and phase shift (measured by Δa_{dB} and ΔS) produced by low-frequency modulations, it is desired to know the makeup of the output spectrum. It is always assumed that the various modulations are applied to quiescent cw tubes. Calculation of the output spectra as shown by Sobol[1] necessitates explicit expressions for the modulation device functions varying with ΔV and ΔI. Unfortunately, as we saw previously, this is not possible in the nonlinear regime and therefore some approximation must be introduced.

A convenient means is to assume that the modulation device functions can be fitted with and described in terms of polynomials. Although it is not a limitation on the basic method, we will consider modulation of either v_{01} or I_{01} but not both simultaneously. Since all the modulations are assumed periodic, the output can conveniently be Fourier analyzed.

Characterize the output by

$$\frac{E}{E_0} = G(\xi) \cos [\omega_c t + \Phi(\xi)], \tag{23}$$

5. OUTPUT SPECTRA FOR LOW-FREQUENCY MODULATIONS

where

$E \triangleq$ the output electric field with modulation,
$E_0 \triangleq$ the output electric field without modulation
$\Phi \triangleq \Phi_0(1 + p_1\xi + p_2\xi^2)$, the phase function,
$\Phi_0 \triangleq$ the phase function without modulation,
$G(\xi) \triangleq (1 + h_1\xi + h_2\xi^2)$, the amplitude function, and
$\omega_c \triangleq$ carrier radian frequency.

The modulating signal given by ξ perturbs either $G(\xi)$ or $\Phi(\xi)$ at ω_a, and the output in general will contain harmonics of ω_c and ω_a and combinations represented by $n\omega_c + m\omega_a$. Assume that the rf circuit impedance at harmonics of the carrier is low and therefore they may be neglected. Thus the output is also conveniently written as a doubly periodic series as follows:

$$\frac{E}{E_0} = \sum_{n=-\infty}^{\infty} \sum_{m=-\infty}^{\infty} A_{nm} e^{j(n\omega_c + m\omega_a)t}.$$

Under the assumption of low-frequency modulations the above is written approximately as

$$\frac{E}{E_0} \approx \sum_{m=-\infty}^{\infty} A_{1m} e^{j(\omega_c + m\omega_a)t} + A_{-1m} e^{j(-\omega_c + m\omega_a)t}$$

and comparison with the Eq. (26) reveals that

$$A_{1m} = \frac{1}{4\pi} \int_0^{2\pi} G(\xi) e^{j[\Phi(\xi) - m\omega_a t]} \, d(\omega_a t). \tag{24}$$

The amplitudes of the spectral lines are found from $2\mathrm{Re}(A_{1m})$; thus Eq. (27) must be integrated for various modulation functions. After a great deal of mathematical labor, the following output-spectra phase and amplitude results are obtained for the modulation waveforms of Fig. 6:

(*i*) *Sinusoidal Modulations*: $\xi = q \sin \omega_a t$.

$$\{\text{Sideband phase shift}\} = \Phi_0 + \Phi_0 \frac{q^2 p_2}{2}$$

$$\{\text{Sideband amplitude}\} = 2 \, \mathrm{Re} \, A_{1m}$$

$$= \sum_{k=-\infty}^{\infty} (-1)^k J_{2k}\left(\frac{-q^2 \Phi_0 p_2}{2}\right) \left\{ \left(1 + \frac{h_2 q^2}{2}\right) J_{4k-m}(-\Phi_0 p_1 q) \right.$$

$$- q^2 \frac{h_2}{4} [J_{4k-n-2}(-q\Phi_0 p_1) + J_{4k-n+2}(-q\Phi_0 p_1)] \bigg\}$$

$$- J_{2k+1} \frac{h_1 q}{2} [J_{4k-n+1}(-\Phi_0 p_1 q) - J_{4k-n+3}(-\Phi_0 p_1 q)].$$

Two degenerate cases of the above are of interest.

(1) Small-amplitude modulation allows the phase shift to be approximated by a linear function. Then $p_2 = 0$ and only the $k = 0$ terms persist.

$$2 \operatorname{Re} A_{1m} = (-1)^m \left(1 + \frac{h_2 q^2}{2}\right) J_m(-\Phi_0 p_1 q)$$

$$- \frac{q^2 h_2}{4} [(-1)^{m+2} J_{m+2}(-\Phi_0 p_1 q) + (-1)^{m-2} J_{m-2}(-\Phi_0 p_1 q)].$$

(2) Condition of zero AM in the output.

$$2 \operatorname{Re} A_{1m} = (-1)^m J_m(-\Phi_0 p_1 q).$$

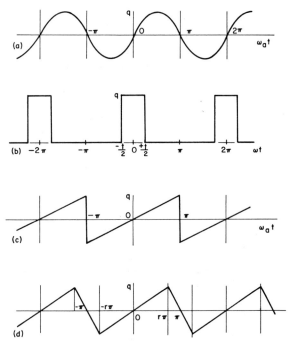

FIG. 6. Modulation wave forms. (a) Sinusoidal. (b) Pulse. (c) Ideal saw-tooth. (d) Saw-tooth with finite flyback time.

(*ii*) *Pulse Modulation.* Again the device is considered as being *cw* in the unmodulated state and then the pulses of Fig. 6b are applied to either the potential or current. The duty cycle is sufficiently low that the

5. OUTPUT SPECTRA FOR LOW-FREQUENCY MODULATIONS

average potential and current are unaffected. The modulation waveform is given by

$$\xi = 0 \quad \text{for} \quad -\pi < \omega_a t < -\frac{\pi\tau}{T}$$

$$= q \quad \text{for} \quad -\frac{\pi\tau}{T} < \omega_a t < \frac{\pi\tau}{T}$$

$$= 0 \quad \text{for} \quad \frac{\pi\tau}{T} < \omega_a t < \pi.$$

The sideband amplitudes are given by

$$2A_{1m} \exp[-j\Phi_0]$$
$$= \frac{\tau}{T}\left(\frac{\sin \pi m\tau/T}{\pi m\tau/T}\right)\{1 + (1 + h_1 q + h_2 q^2)\exp j(\Phi_0 p_1 q + \Phi_6 p_2 q^2)\}.$$

This expression indicates that the spectrum is composed of two sets of lines. One is the usual $\sin x/x$ distribution arriving at the output at the same phase as the unmodulated signal. The second set of lines also has a $\sin x/x$ distribution but arrives with a delay dependent upon the phase-modulation characteristics, and an amplitude dependent upon the amplitude-modulation characteristics.

(iii) Saw-Toothed Modulation. A scheme to transfer energy from the carrier frequency to one of the sidebands with little energy in other sidebands or the carrier is called the "serrodyne" and utilizes a saw-toothed modulation function. Several particular cases are considered.

(1) Consider the ideal saw-toothed modulation of Fig. 6c with zero flyback time applied to an ideal amplifier which has a linear phase characteristic and no amplitude variation, $p_2 = h_1 = h_2 = 0$. The saw-toothed function is described by

$$\xi = \frac{q}{\pi}\omega_a t \quad -\pi < \omega_a t < \pi.$$

The sideband amplitudes are calculated as

$$2A_{1m}e^{-j\Phi_0} = \frac{\sin\left(\frac{p_1 q \Phi_0}{\pi} - m\right)\pi}{\left(\frac{p_1 q \Phi_0}{\pi} - m\right)\pi}.$$

The desired frequency shift is obtained by adjusting the tube length and modulation signal amplitude so that

$$\frac{p_1 q \Phi_0}{\pi} = n' = \text{an integer.}$$

Then the sideband for $m = n'$ has a unit amplitude and all others vanish.

(2) Consider the saw-tooth of Fig. 6a with a finite flyback time, again applied to the ideal amplifier with a linear phase shift and no amplitude variation. The modulation function is described by

$$\xi = \frac{-q}{1-r}\left(1 + \frac{\omega_a t}{\pi}\right) \quad \text{for} \quad -\pi < \omega_a t < -r\pi$$

$$= \frac{q}{r\pi}\omega_a t \quad \text{for} \quad -r\pi < \omega_a t < r\pi$$

$$= \frac{q}{1-r}\left(1 - \frac{\omega_a t}{\pi}\right) \quad \text{for} \quad r\pi < \omega_a t < \pi.$$

In this case the sidebands have amplitudes given by

$$2A_{1m}e^{-j\Phi_0} = r\frac{\sin\left(\frac{p_1\Phi_0 q}{r\pi} - m\right)r\pi}{\left(\frac{p_1\Phi_0 q}{r\pi} - m\right)}$$

$$+ \frac{\sin\left[\left(\frac{p_1\Phi_0 q}{(1-r)\pi} + m\right)\pi - \frac{qp_1\Phi_0}{1-r}\right] - \sin\left[\left(\frac{p_1\Phi_0 q}{(1-r)} + m\right)r\pi - \frac{qp_1\Phi_0}{1-r}\right]}{\left(\frac{p_1\Phi_0 q}{(1-r)\pi} + m\right)\pi}.$$

Comparison of the above relations reveals that the effect of a finite flyback time is to create a large number of sidebands in the output spectrum.

(3) Consider the ideal saw-toothed modulation applied to a nonideal tube. The sideband amplitudes are given by

$$A_{1m}\exp[-j\Phi_0] = \frac{e^{-j\alpha^2}}{\beta}\left(\frac{\pi}{2}\right)^{\frac{1}{2}}\exp[-j\Phi_0][C(P_1) - C(P_2) + jS(P_1) - jS(P_2)]$$

$$+ \frac{qh_1\exp[-j\alpha^2]}{2\pi^2\beta}\left\{\frac{j}{2\beta}(e^{jP_2} - e^{jP_1}) - \alpha\left(\frac{\pi}{2}\right)^{\frac{1}{2}}[C(P_1) - C(P_2) + jS(P_1)\right.$$

$$\left. - jS(P_2)] + j\exp[jP_1]\left[\frac{\alpha}{\beta} - \frac{P_1^{\frac{1}{2}}}{\beta^2}\right] - j\exp[jP_2]\left[\frac{\alpha}{\beta} - \frac{P_2^{\frac{1}{2}}}{\beta^2}\right]\right\},$$

where $C(P)$ and $S(P)$ are Fresnel integrals given by

$$C(P) = \frac{2}{\pi} \int_0^{P^{\frac{1}{2}}} \cos t^2 \, dt,$$

$$S(P) = \frac{2}{\pi} \int_0^{P^{\frac{1}{2}}} \sin t^2 \, dt,$$

$$\alpha = \frac{(q\Phi_0 p_1/\pi) - m}{2q^2(\Phi_0 p_2/\pi^2)}, \qquad \beta = \frac{q^2 \Phi_0 p_2}{\pi},$$

$$P_1 = \beta(\pi + \alpha)^2, \quad \text{and} \quad P_2 = \beta(\alpha - \pi)^2.$$

The output spectra for transit-time modulations and phase modulations were studied in detail by Cumming[3] and he found that the output spectra for the two are different for the same input. The difference arises from the fact that a transit-time modulation changes the time delay which follows the generation of a periodic time function, while phase modulation changes the phase during the time of generation of the periodic function.

The normalized spectral amplitude of the nth sideband for a TTM is

$$I = M_n[g, \; (1 + \rho_c) r_t \omega_c]$$

where

$q = $ the depth of AM,

$r_t = $ peak deviation in transit time during the modulation cycle, and

$\rho_c = \omega_a/\omega_c$.

The spectral amplitude for the nth sideband with PM is

$$I = M_n(q, r_p)$$

where $r_p = $ the peak deviation.

For low-frequency modulations $\rho_c \ll 1$ and thus the output spectra for TTM and PM are the same since $r_p = r_t \omega_c$.

6 Modulation by Multiple High-Frequency Signals

Another interesting modulation case is one in which several high-frequency signals modulate the beam so that the time dependencies of all signals must be considered. Also, it may be that each modulation signal propagates at a different velocity and thus each signal may see a different rf circuit impedance.

The general nonlinear problem may be treated in the Lagrangian frame using the basic method outlined in Chapter VI. A specific form of the coupling can be assumed to simplify the calculations. The stream equations are easily written for multiple signals and the resulting system of equations can be solved as before.

REFERENCES

A. General Modulation

1. Sobol, H., "Modulation Characteristics of O-Type Electron Stream Devices." Univ. of Michigan Electron Phys. Lab. Tech. Rept. No. 33 (October 1959).
2. Sobol, H., and Rowe, J. E., Analysis of modulated traveling-wave devices and beam-type parametric amplifiers. *J. Elec. and Control* **8**, No. 5, 321-350 (1960).
3. Cumming, R. C., "Frequency Translation by Modulation of Transit-Time Devices." Stanford Univ. Appl. Electron. Lab. Tech. Rept. No. 39 (August 1955).
4. Learned, V., The klystron mixer applied to T-V relaying. *Proc. IRE* **38**, No. 9, 1033-35 (1950).
5. Bray, W. J., The traveling-wave valve as a microwave phase-modulator and frequency shifter. *Proc. IRE (London)* **99**, Pt. III, 15-20 (1952).
6. Steele, G. F., The modulation of traveling-wave tubes. *Electron. Eng.* **29**, 429-433 (1957).
7. Mendel, J. T., "Grid-Modulated Traveling Wave Tube for Low-Pass Amplification." Stanford Univ. Electron. Res. Lab. Tech. Rept. No. 47 (July 1952).
8. Beam, W. R., and Blattner, D. J., Phase angle distortion in traveling-wave tubes. *RCA Rev.* **17**, 86-99 (1956).
9. Putz, J. L., "Nonlinear Phenomena in Traveling-Wave Amplifiers." Stanford Univ. Electron. Res. Lab. Tech. Rept. No. 37 (October 1951).
10. Nation, A. W. C., and Harrison, A. E., "Cross Modulation in Traveling-Wave Tube Amplifiers." Univ. of Washington Dept. Elec. Eng. Rept. No. 15 (November 1954).
11. DeGrasse, R. W., and Wade, G., "Microwave Mixing and Frequency Dividing." Stanford Univ. Electron Tube Lab. Stanford Electron. Labs. Tech. Rept. No. 386-1 (November 1957).
12. DeGrasse, R. W., "Frequency Mixing in Microwave Beam-Type Devices." Stanford Univ. Electron Devices Lab. Stanford Electron. Lab. Tech. Rept. No. 386-2 (July 1958).
13. Louisell, W. H., and Quate, C. F., Parametric amplification of space-charge waves. *Proc. IRE* **46**, No. 4, 707-716 (1958).

B. Noise

14. Haus, H. A., "Analysis of Signals and Noise in Longitudinal Electron Beams." M.I.T. Res. Lab. Electron. Tech. Rept. No. 306 (August 1955).
15. Smullin, L. D., and Haus, H. A., "Noise in Electron Devices." Wiley, New York, 1959.
16. Haus, H. A., and Robinson, F. N. H., A minimum noise figure of microwave-beam amplifiers. *Proc. IRE* **43**, No. 8, 981-991 (1955).

C. Parametric Amplification

17. Louisell, W. H., and Quate, C. F., Parametric amplification of space-charge waves. *Proc. IRE* **46**, No. 4, 707-716 (1958).
18. Haus, H. A., The kinetic power theorem for parametric, longitudinal electron-beam amplifiers. *IRE Trans. Electron Devices* **5**, No. 4, 225-232 (1958).
19. Manley, J. M., and Rowe, H. E., Some general properties of nonlinear elements, Part I. *Proc. IRE* **44**, No. 7, 904-914 (1956).

APPENDIX

A | Rf Structure Impedance Variations

1 Helical Line for O-FWA

In solving the nonlinear equations of Chapter XIII for the O-FWA it is necessary to prescribe an impedance variation $Z_0(y)/Z_0$. Of course to do this a particular rf structure form must be considered. In view of the wide use of the helical line, its impedance variation when used with a solid cylindrical electron beam has been chosen. Pierce has calculated the helix impedance versus γa and his results are the basis of these calculations.

The parameter γa increases with decreasing structure velocity at constant frequency. For $\gamma a = 1.5$ in the uniform region, the impedance variation in the variable-pitch section is shown in Fig. 1 as a function of a velocity parameter (Meeker[1]). The phase velocity curve for the forward fundamental space harmonic is shown in Fig. 7, Chapter III.

The impedance variation may be approximated, for many conditions, as

$$\frac{Z_0(y)}{Z_0} \approx \left(\frac{v_0(y)}{v_0(0)}\right)^{\frac{3}{2}} = \left(\frac{1 + C_0 b_0}{1 + C_0 b(y)}\right)^{\frac{3}{2}} \tag{1}$$

and

$$\frac{Z_0(y)}{Z_0} \approx \left(\frac{v_0(y)}{v_0(0)}\right)^2 = \left(\frac{1 + C_0 b_0}{1 + C_0 b(y)}\right)^2. \tag{2}$$

The assumption of a constant helix impedance in the variable-velocity region is seen to be somewhat impractical. $Z_0(y)$ may be held constant, however, by decreasing the diameter as the velocity is tapered in such a way as to maintain γa constant.

2 Helical Line for O-BWO

The tape helix operated at the first backward space harmonic has found wide application in backward-wave oscillators due to its inherent broad frequency range of operation. In view of its popularity the characteristics of such a variable-pitch helix will be examined here.

2. HELICAL LINE FOR O-BWO

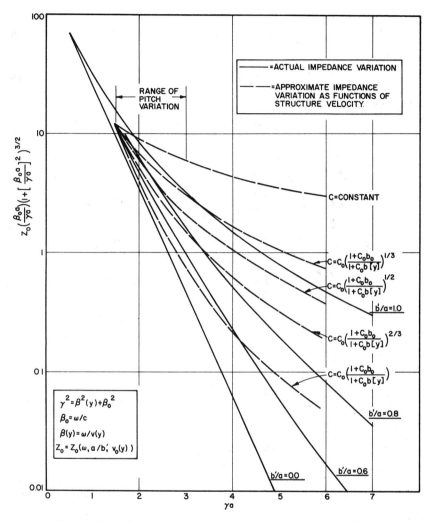

FIG. 1. Impedance variation versus γa for solid-beam helix TWA.

The impedance and phase velocity characteristics of these structures have been calculated by Watkins and Ash.[2] The phase velocities versus ka for various forward and backward modes as indicated by the index m are shown in Fig. 2. The impedance of the mth space harmonic component is defined as

$$K_m \triangleq \frac{E_m^2}{2\beta_m^2 P}, \tag{3}$$

where
$$\gamma_m = (\beta_m^2 - k^2)^{\frac{1}{2}} \qquad (4)$$
and
$$\beta_m = \beta_0 + \frac{2\pi m}{p}. \qquad (5)$$
For slow waves
$$\gamma_m \approx \beta_m. \qquad (6)$$

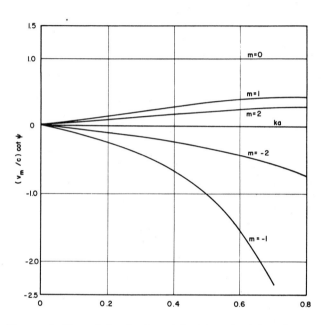

FIG. 2. Phase velocities of the fundamental, $+1$, -1, $+2$, -2 space harmonic components of a wave on a developed helix with a positive group velocity as a function of ka (Watkins and Ash[2]).

The impedance of the -1 space harmonic is shown in Fig. 3 versus ka for single and bifilar tape helices. Under the slow-wave approximation we may write (at the tape radius)

$$\frac{E_{mz}^2}{2\beta_m^2 P_m} = \frac{30}{|m + ka|} \qquad (7)$$

and thus we see that the impedance versus frequency relationship is independent of the pitch.

The impedance at the mean beam position is, of course, reduced from that at the helix plane and may be expressed as

$$R_m = \left[\frac{I_m(\gamma r_0)}{I_m(\gamma a)}\right]^2. \tag{8}$$

R_m is shown versus γa in Fig. 4 for the first backward space harmonic. Recall that $I_{-1}(\nu) \equiv I_1(\nu)$.

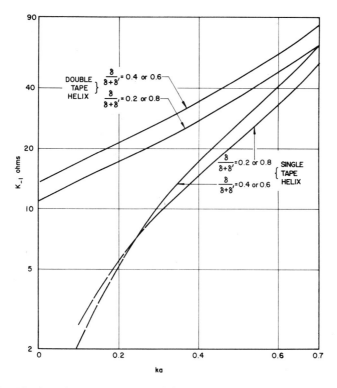

FIG. 3. The impedance parameter $E_{-1}^2/2\beta_{-1}^2 P$ at $r = a$ as a function of ka for single and bifilar tape helices (Watkins and Ash[2]).

Using the above results we proceed to write the impedance of the untapered helix, operated at $m = -1$ at the beam position, as

$$Z_{r_0}^u = Z_a^u \left[\frac{I_1(\gamma_{-1}^u r_0)}{I_1(\gamma_{-1}^u a)}\right]^2. \tag{9}$$

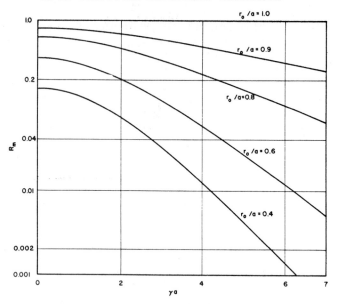

FIG. 4. The impedance reduction factor R_{-1} for a thin hollow beam of radius r_0 as a function of $\lambda_{-1}a$ for several values of r_0/a. When applied to a thick hollow beam, r_0 is taken as the mean radius. (Watkins and Ash[2])

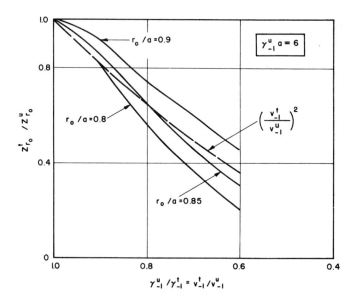

FIG. 5. Impedance velocity variation for a backward-wave mode.

The analogous relation for the tapered helix is

$$Z_{r_0}^t = Z_a^t \left[\frac{I_1(\gamma_{-1}^t r_0)}{I_1(\gamma_{-1}^t a)} \right]^2. \tag{10}$$

Taking the ratio of Eqs. (9) and (10), noting that $Z_a^t \equiv Z_a^u$ in view of Eq. (7), yields

$$\frac{Z_{r_0}^t}{Z_{r_0}^u} = \left[\frac{I_1(\gamma_{-1}^t r_0)}{I_1(\gamma_{-1}^t a)} \right]^2 \left[\frac{I_1(\gamma_{-1}^u a)}{I_1(\gamma_{-1}^u r_0)} \right]^2. \tag{11}$$

The propagation or phase constants of the tapered and untapered helices are related as

$$\frac{\gamma_{-1}^t}{\gamma_{-1}^u} = \frac{v_{-1}^u}{v_{-1}^t}. \tag{12}$$

Equation (11) is thus easily evaluated if specific values of $\gamma_{-1}^u a$ and v_{-1}^u/v_{-1}^t are assumed. A particular set of results is shown in Fig. 5 (Haddad[3]). Also shown is the approximate relation

$$\frac{Z_{r_0}^t}{Z_{r_0}^u} \approx \left(\frac{v_{-1}^t}{v_{-1}^u} \right)^2. \tag{13}$$

3 Tapered Interdigital Line Characteristics

In considering variable-pitch rf structures for crossed-field devices it is again necessary to consider specific structure types in order to calculate the effect of tapering on phase velocity, group velocity and rf impedance. The structures most commonly used are the vane opposite a flat conducting plate (sole) and the interdigital line opposite a flat conducting plate. The interdigital line is examined here for application in a kinetic-energy conversion region of an M-FWA. This structure is illustrated schematically in Fig. 6.

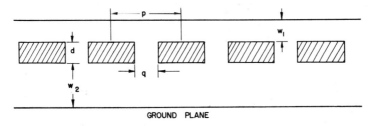

FIG. 6. Schematic of interdigital line.

The phase velocity is conveniently varied by variation of p and q. In order to realize similar variations of circuit phase and group velocities for wideband operation, it is necessary also to vary w_1 and w_2. Of course, w_2 must be varied in order to slow the beam bunches adiabatically. The interdigital-line impedance has been calculated by Walling[4] as a function of the section phase shift under the approximation that $w/p \ll 1$. This approximation is considered valid here.

$$K(\theta) = \frac{\mu}{\epsilon} \left[\frac{4d}{q} \sin^2 \frac{\theta}{2} + p(1 - \alpha) \left(\frac{1}{w_1} + \frac{1}{w_2} \right) \right]^{-1}, \tag{14}$$

where

$p \triangleq$ pitch distance,

$q \triangleq$ finger gap width,

$d \triangleq$ finger thickness,

$\alpha \triangleq q/p$, and

$\theta \triangleq$ phase shift per finger section.

For operation at the first forward space harmonic, the variation of $K(\theta)$ with pitch for a typical set of dimensions is shown in Fig. 7 (Volkholz[5]).

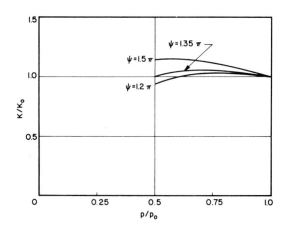

Fig. 7. Variation of interdigital line impedance with pitch ($w_1/p_0 = 0.4$, $w_2/p_0 = 0.5$, $\alpha = 0.6$, $4d/q_0 = 1.2$).

It is interesting to note that, unlike that of the helical line rf structure, the interdigital-line impedance varies very little for a considerable variation in pitch. This certainly is advantageous since a reduction of impedance would result in a lowering of efficiency and gain per unit length.

3. TAPERED INTERDIGITAL LINE CHARACTERISTICS

The dispersion relation

$$\frac{K(\theta + \pi)}{K(\theta)} = \tan^2 \frac{kh}{2} \tag{15}$$

may be used to evaluate the variation of group velocity with pitch and hence the effect on bandwidth. The approximate impedance variation given by Eq. (14) is combined with the dispersion equation, Eq. (15), to obtain v_p and v_g as

$$v_p = \frac{\omega}{\beta} = \frac{2cp}{h\theta} \tan^{-1} \left[\frac{\frac{4d}{q} \sin^2 \frac{\theta}{2} + p(1-\alpha)\left(\frac{1}{w_1} + \frac{1}{w_2}\right)}{\frac{4d}{q} \cos^2 \frac{\theta}{2} + p(1-\alpha)\left(\frac{1}{w_1} + \frac{1}{w_2}\right)} \right]^{\frac{1}{2}} \tag{16}$$

and

$$v_g = \frac{d\omega}{dp} = \frac{2cd}{\alpha h}$$

$$\cdot \frac{\sin \theta}{\left[\frac{4d}{q} \cos^2 \frac{\theta}{2} + p(1-\alpha)\left(\frac{1}{w_1} + \frac{1}{w_2}\right)\right]^{\frac{1}{2}} \left[\frac{4d}{q} \sin^2 \frac{\theta}{2} + p(1-\alpha)\left(\frac{1}{w_1} + \frac{1}{w_2}\right)\right]^{\frac{1}{2}}}$$
(17)

The velocity variations are shown in Fig. 8 as functions of pitch and phase shift. The increasing variance of the v_p and v_g curves with pitch change indicates a decreasing bandwidth. These have been calculated assuming w_1 and w_2 constant. In order to realize low dispersion, w_1

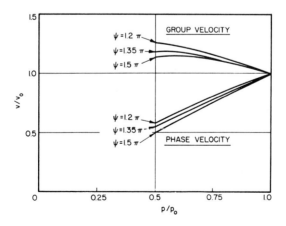

FIG. 8. Variation of interdigital-line group and phase velocities with pitch ($w_1/p_0 = 0.4$, $w_2/p_0 = 0.5$, $\alpha = 0.6$, $4d/q_0 = 1.2$).

must be varied to maintain $v_g \approx v_p$. The required distance variations may be calculated for any specific case.

The lower cutoff frequency is increased with a decreased pitch according to

$$\tan^2\left(\frac{\pi h f_c}{c}\right) = 1 + \frac{4d}{p^2 \alpha (1-\alpha)\left(\frac{1}{w_1} + \frac{1}{w_2}\right)}. \qquad (18)$$

The upper cutoff frequency is also increased and the net result is an approximate maintenance of the bandwidth.

Similar calculations are easily carried out on the vane line or any other type of structure for either forward or backward space harmonic operation.

References

1. Meeker, J. G., "Phase-Focusing in Linear-Beam Devices." Univ. of Michigan Electron Phys. Lab. Tech. Rept. No. 49 (August 1961).
2. Watkins, D. A., and Ash, E. A., The helix as a backward-wave circuit structure. *J. Appl. Phys.* **25**, No. 6, 782-790 (1954).
3. Haddad, G. I., "Efficiency and Start-Oscillation Conditions in Nonuniform Backward-Wave Oscillators." Univ. of Michigan Electron Phys. Lab. Tech. Rept. No. 61 (March 1963).
4. Walling, J. C., Interdigital slow-wave structures. *Proc. 1ᵉ Congr. Intern. Tubes Hyperfréquences, Paris, 1956* pp. 454-464.
5. Volkholz, K. L., "Energy Exchange in Crossed-Field Systems." Univ. of Michigan Electron Phys. Lab., Tech. Rept. No. 73 (June 1964).

APPENDIX

B | O-TWA Kompfner-Dip Conditions

The phenomenon now known as the Kompfner-dip condition was first discovered experimentally by Kompfner[1] in 1950 and was subsequently studied by Johnson[2] in 1955. For a particular electron beam

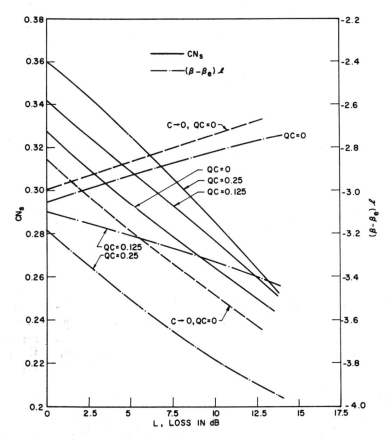

FIG. 1. Kompfner-dip conditions ($C = 0.05$).

velocity which is less than the characteristic circuit phase velocity a nonzero rf input to a traveling-wave amplifier is completely transferred to the electron beam, giving a zero rf output. It can be shown using a coupled-mode analysis that the energy transferred to the beam is put into the fast space-charge wave, which carries positive ac energy.

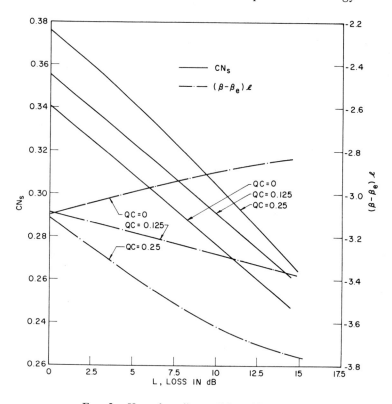

FIG. 2. Kompfner-dip conditions ($C = 0.1$).

One can show from a small-signal analysis that this complete transfer of rf energy from circuit to beam occurs for a critical combination of beam velocity and normalized circuit length. These quantities are conveniently measured in terms of $(\beta - \beta_e)l = \theta b$ and CN_s respectively. Johnson[2] has investigated the conditions for the Kompfner dip as a function of space charge, QC, and circuit loss, d, when the interaction parameter, C, approaches zero.

It is interesting to investigate the existence of Kompfner-dip conditions in TWA's whose interaction parameter, C, may not be small. This phenomenon could be investigated using the nonlinear theory

B. O-TWA KOMPFNER-DIP CONDITIONS

(see Chapter XI), which would indicate that the rf circuit energy goes into accelerating the electrons.

The appropriate small-signal equations for the finite-C traveling-wave amplifier are the determinantal equation

$$\delta = f(C, QC, d, b) \tag{1}$$

and the rf circuit voltage equation

$$\frac{V(\theta)}{V} = \exp\left[-j\frac{\theta}{C}\right] \sum_{i=1}^{3} \left(\frac{V_{ci}}{V}\right) \exp[\delta_i \theta], \tag{2}$$

where $\theta \triangleq 2\pi CN_s$.

We wish to find the combinations of b and CN_s which make the left-hand side of Eq. (2) zero for $V \neq 0$ and $\theta > 0$. These equations are easily solved using the "downhill" method on a digital computer and the results are shown in Figs. 1–3. In Fig. 1 the results for the $C \to 0$

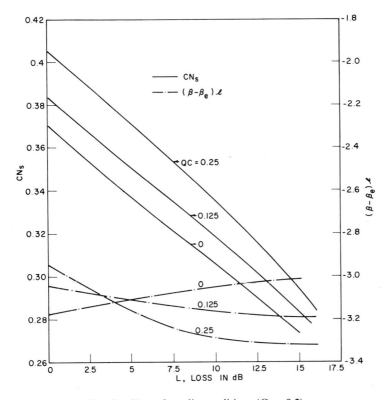

FIG. 3. Kompfner-dip conditions ($C = 0.2$).

case are plotted along with those for $C = 0.05$ for comparison purposes. It is seen that a substantial difference in the results occurs when $C \neq 0$. The variations in the Kompfner-dip conditions with loss and space charge, however, are similar for all values of C. These results now serve as a basis for investigating energy transfer from the circuit to the beam in the nonlinear case.

REFERENCES

1. Kompfner, R., On the operation of the traveling-wave tube at low level. *J. Brit. IRE* **10**, 283-289 (1950).
2. Johnson, H. R., Kompfner-dip conditions. *Proc. IRE* **43**, 874 (1955).

APPENDIX

C | M-FWA Kompfner-Dip Conditions

The dispersion equation for near-synchronous waves in injected-beam crossed-field amplifiers was obtained by Gould[1] using a small-signal field-theory analysis. A thin but finite laminar Brillouin beam was assumed to be replaced by rippled surface charge densities in the manner of Hahn. Cyclotron waves were shown to be negligibly excited and thus only a third-order dispersion equation was obtained. The input boundary-value problem was then solved approximately using these three waves. Later Dombrowski[2] presented a circuit-type analysis, including the effects of cyclotron waves, which reduces to the same near-synchronous third-order dispersion equation.

The dispersion equation for a Brillouin beam is

$$(\delta + jb + d)(\delta^2 + 2jgS\delta - \delta^2) = \delta, \tag{1}$$

where S is the space-charge parameter and g the beam location parameter. The incremental propagation constants of the system are obtained from solutions of Eq. (1) for particular values of b, d, S, and g. The voltage along the circuit may be written as

$$\frac{V_t(\theta)}{V} = \exp\left[-j\frac{\theta}{D}\right] \sum_{n=1}^{3} \frac{V_n}{V} \exp\left[-\theta \delta_n\right], \tag{2}$$

where $\theta \triangleq 2\pi DN_s$. Equations (1) and (2) must now be solved simultaneously to determine the lowest-order combination of b, DN_s which will cause $V_t(\theta)/V = 0$ for $\theta > 0$. These computations have been carried out for $d = 0$ and are presented in Fig. 1. It is interesting to note that the required DN_s decreases with increasing space charge and is essentially independent of g. The critical value of b is always negative (beam below synchronism) and increases negatively with increasing space charge.

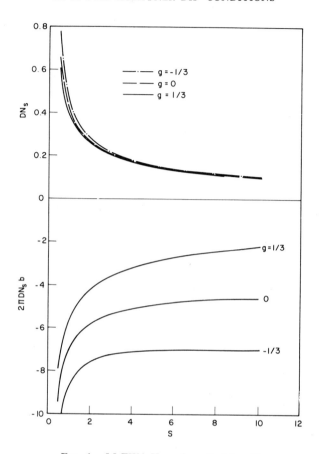

Fig. 1. M-FWA Kompfner dip ($d = 0$).

REFERENCES

1. Gould, R. W., Space charge effects in beam-type magnetrons. *J. Appl. Phys.* **28**, 599-604 (1957).
2. Dombrowski, G. E., "A Small-Signal Theory of Electron-Wave Interaction in Crossed Electric and Magnetic Fields." Univ. of Michigan Electron Tube Lab. Tech. Rept. No. 22 (October 1957).

Author Index

Numbers in parentheses are reference numbers and indicate that an author's work is referred to although his name is not cited in the text. Numbers in italic show the page on which the complete reference is listed.

Akhiezer, A. I., 397, *422*
Allen, M. A., 397, *421*, *422*, *423*
Anderson, J. M., 397, *423*
Anderson, J. R., 381, 382, 383, *384*
Arnaud, J., 54(6), *68*
Ash, E. A., *15*, *281*, 571, 572, 573, 574, *578*

Beam, W. R., *568*
Beck, A. H. W., *12*
Benham, W. E., *14*
Bernashevsky, G. A., *15*, 396, *421*
Bernier, J., *13*, *280*
Bernstein, I. B., 397, *422*, *423*
Bevensee, R. M., 426, *485*
Birdsall, C. K., *13*
Blattner, D. J., *568*
Bloch, F., *68*
Bloom, S., 47(2), *68*, 396, *421*
Bobot, D., *14*
Bogdanov, E. V., *15*, 396, *421*
Bohm, D., 397, *422*
Borghi, R. P., *549*
Bowen, A. E., *12*
Boyd, G. D., *15*, 396, 397, *421*
Brackett, C. A., 426, *486*
Brangaccio, D. J., 209, *262*
Bray, W. J., *568*
Brewer, G. R., *13*
Brillouin, L., *13*, *14*, 54, *68*, 178, *261*
Brossart, J., *14*
Buneman, O., 2(42), *14*, *15*, 397, *422*, *423*

Case, K. M., 397, *422*
Chodorow, M., *13*
Chorney, P., *15*, 397, *421*, *423*

Chu, C. M., 35(4), *68*
Chu, L. J., *13*, 30, *68*
Clarke, G. M., 426, *486*
Coleman, P. D., *171*
Condon, E. U., *12*
Crawford, F. W., *15*, 397, *423*
Cumming, R. C., 567, *568*
Cutler, C. C., 204, 208, 209, *262*, 425, *485*

DeGrasse, R. W., *568*
Demirkhanov, R. A., 397, *423*
Doehler, O., 1(51, 53), *13*, *14*, 54(6), *68*, 121, *176*, 178, *261*, *281*
Dombrowski, G. E., 375, *384*, 583, *584*
Dunn, D. A., *549*

Epstein, D. W., 69(1), *119*
Epsztein, B., *14*, 54(6), *68*
Ettenberg, M., 397, *423*

Fainberg, Y. B., 397, *421*, *422*
Feenberg, E., 120, 121, *176*
Feinstein, J., 306, 355, 357, 358, 359, *363*, 397, *422*
Field, L. M., *15*, 396, 397, *421*
Filiminov, G. F., *15*, 397, *423*, 426, *486*
Fried, B. D., 397, *423*
Frieman, E. A., 397, *423*

Gabor, D., *15*
Gandhi, O. P., 282(1), 283, *303*, 306, 342, 355, *363*, 364, 365, 374, *384*
Geppert, D. V., 426, *485*
Gevorkov, A. K., 397, *423*
Ginzton, E. L., *13*

585

Goldberger, A. K., *280*
Gorbatenko, M. F., 397, *421*
Gordon, E. I., 397, *423*
Gould, R. W., *14*, *15*, 375, *384*, 396, 397, *421*, *423*, 583, *584*
Greene, J. M., 397, *422*
Gross, E. P., 397, *422*
Grow, R., 264, 277, *280*
Guenard, P. R., *13*, *14*, 120, 140, *176*, *281*

Haddad, G. I., 426, 429, 431(13, 18), 455, 461, *485*, *486*, 575, *578*
Haeff, A. V., *13*, 395, *421*
Hahn, W. C., 1, *12*, 120, *176*
Hamilton, D. R., *12*
Hansen, J. W., *549*
Hansen, W. W., 1, *12*
Harman, W. A., *281*
Harrison, A. E., *13*, *568*
Hartman, P. L., 537, *549*
Haus, H. A., *568*, *569*
Hay, H. G., *14*
Heagy, M. S., *14*
Hebenstreit, W. B., *13*, 395, *421*
Heffner, H., *13*, *281*
Hefni, I., 173(19), 176, *176*
Heil, A., *13*
Heil, O., *13*
Hernqvist, K. G., *15*, 397, *423*
Hess, R. L., 426, 446, 448, *486*
Hok, G., *262*
Huber, H., 1(53), *14*
Hull, A. W., *14*
Hull, J. F., 306, 355, 360, 363, *363*
Hutter, R. G. E., *12*

Iziumova, T. I., 396, *421*

Jackson, J. D., *13*, 30, *68*
Johnson, H. R., *13*, *281*, *394*, 579, 580, *582*

Kino, G. S., *15*, 35, *68*, 306, 355, 357, 358, 359, *363*, 381, *384*, 397, *421*, *422*, *423*
Kirstein, P. T., 381, *384*
Kislov, V. J., *15*, 396, *421*
Kleen, W., 1(51, 53), *13*, *14*, 121, *176*, 178, *261*
Knipp, J. K., *12*
Kompfner, R., *13*, *281*, 579, *582*
Kooyers, G. P., 306, 355, 360, 363, *363*

Kruskal, M. D., 397, *422*
Kulsrud, R. M., 397, *423*
Kuper, J. B. H., *12*

Lampert, M. A., 397, *421*
Langmuir, I., *15*
Lawson, J. D., 397, *423*
Learned, V., *568*
Lerbs, A., 1(53), *14*
Lichtenberg, A. J., 501, *514*
Lim, Y. C., 397, *423*
Llewellyn, F. B., *12*
Louisell, W. H., *568*, *569*
Luebke, W. R., 277, *281*, *549*

MacFarlane, G. G., *14*
Maloff, I. G., 69(1), *119*
Manley, J. M., *569*
Maxwell, J. C., 28, *68*
Meeker, J. G., 65, *68*, 121, 125, *176*, 426, 431(10, 11), 444(11), 446, *486*, *514*, 570, *578*
Mendel, J. T., *568*
Metcalf, G. F., 1(3), *12*
Midford, T. A., 381(9), *384*
Mihran, T. G., 120, 131, 132, *176*
Morrison, J. A., 397, *422*
Mourier, G., *14*, 54(10), *68*
Muller, M., *13*, *281*
Myers, L. M., 69(2), *119*

Nation, A. W. C., *568*
Nergaard, L. S., *13*, 395, *421*
Nordsieck, A. T., 178, 191, 218, *261*, 264, *281*

Paik, S. F., 35, 51, *68*
Palluel, P., *68*, *280*
Panofsky, W. K. H., 72, *119*
Paschke, F., 121, *176*
Pavkovich, J., 426(16), 427(16), 428(16), *486*
Peter, R. W., 47(2), *68*, 396, *421*
Phillips, M., 72, *119*
Piddington, J. H., 397, *422*
Pierce, J. R., 2(8), *12*, *13*, 44, 58(13), *68*, *281*, 395, 397, *421*, *422*, *423*, 425, *485*, 537, 539, *549*
Pines, D., 397, *422*
Popov, A. P., 397, *423*

AUTHOR INDEX

Poulter, H. C., 106, *119*, 178, 254, 258, *261*
Putz, J. L., 277, *281*, 568

Quate, C. F., *568*, *569*

Ramo, S. I., 1, *12*, 62, *68*, 120, *176*
Rapoport, G. N., *281*
Revans, R. W., 397, *423*
Richtmyer, R. D., 1(13), *12*
Robinson, D., 426(16), 427(16), 428(16), *486*
Robinson, F. N. H., *568*
Rowe, H. E., *569*
Rowe, J. E., 62, *68*, 106, *119*, 173(20), *176*, 178, 213, 216, 217, 218, 250(18, 19), *261*, *262*, 264(19), *281*, 283, *303*, 306, 355, *364*, 426, 429(14), 431(11, 18), 444(11), 446(11), 463, 476, *485*, *486*, *514*, 515(1), 519(2), *549*, 550, 558(2), *568*
Ruetz, J., 426, 427, 428, *486*
Rydbeck, O. E. H., *13*
Rynn, N., *281*

Schelkunoff, S. A., 30, *68*
Schumann, W. O., 397, *422*
Scott, A. W., 213, 214, 215, *262*
Sedin, J. W., 89, *119*, 264, *281*, 306, 350, 355, *363*, 365, 375, 377, 378, 379, 380, *384*
Sen, H. K., 397, *422*
Sensiper, S., 35(17), 37, *68*
Shulman, C., *14*
Sirkis, M. D., *171*
Slater, J. C., *12*, 178, *261*, 425, *485*
Smith, L. P., 537, *549*
Smullin, L. D., *15*, 397, *421*, *568*
Smythe, W. R., 72, 89, 92, 114, *119*
Sobol, H., 550, 558(1, 2), 562, *568*
Solymar, L., 121, *176*
Spangenberg, K. R., 426, *486*
Steele, G. F., *568*
Sterzer, F., *549*
Sturrock, P. A., *119*, 397, *421*
Suhl, H., *68*
Sumi, M., 397, *423*
Susskind, C., *549*

Targ, R., 397, *423*
Tchernov, Z. S., *15*, 396, *421*
Thomas, J., 385, *394*
Thomson, Sir William (Lord Kelvin), 30, *68*
Tidman, D. A., *15*
Tien, P. K., 35, *68*, 103, 105(8), 106, *119*, 178, 208, 209, 254, 258, *261*, *262*, *281*
Trivelpiece, A. W., *15*, 397, *421*, *422*
Turner, C. W., 121, *176*
Twiss, R. Q., 397, *422*, *423*

Vainshtein, L. A., 179, *262*, 426, *486*
Van Kampen, N. G., 397, *422*
Varian, R. H., 1, *13*
Varian, S. F., 1, *13*
Vlasov, A. A., 397, *422*
Volkholz, K. L., 463, *486*, 576, *578*
Voronov, Z. S., 396, *421*

Wada, G., *549*
Wade, G., *568*
Walker, L. R., 103(8), 105(8), 106(8), *119*, 178(9), 208(9), 209(9), *261*, *281*, 397, *422*
Walling, J. C., 576, *578*
Wang, C. C., 178, *261*
Warnecke, R., 2(51, 53), *14*, 120, 140, *176*, *281*
Watkins, D. A., 264, 277, *280*, *281*, 571, 572, 573, 574, *578*
Webber, S. E., 103, *119*, 121, 125, 132, 137, 138, 139, 140, 141, 142, 143, 144, 145, 172, *176*
Webster, D. L., *13*, 120, 122, 130, 132, *176*
Weglein, R. D., *281*
Wehner, G., 397, *423*
Weiss, G., *15*
Whinnery, J. R., *13*
Williams, N. T., *13*, *281*
Wilson, R. N., 92, 114, 121, *176*
Wolkstein, H. J., 537, 538, 540, *549*
Wolontis, V. M., 103(8), 105(8), 106(8), *119*, 178(9), 208(9), 209(9), *261*

Zitelli, L., 120, *176*

Subject Index

A

Adiabatic systems, drift region, 302
  M-BWO results, 370–375
  M-FWA results, 346
  slowing in, 469
Amplifiers, M-type, 304–363
  O-type, 177–261
Axially symmetric beams, physical models, 69
  space-charge potential in, 92–113

B

Backward-wave systems, amplifier. 5
  circuits, 64, 265–266
  equivalent circuits, 62–65
  group velocities, 62
  oscillators, 2, 5–7, 529–531
    carcinotron, 263
    closed-form solution, 455
    crossed-field, 476–485
    efficiency improvement, 476–485
    O-type, 263–303
    starting current, 426
    velocity tapering in, 454–461
    voltage-tunable, 7
  phase velocities, 62
  telegraphist's equations for, 64
Beam current flow, limitation in collector depression, 536–541
  perveance limitation, 539, 540
Beam input conditions, 128, 148, 153, 154, 187–189, 230, 267, 268, 291, 312, 368
Beam-plasma interactions, 395–423
  amplifiers, 7, 9, 396
  collision effects, 397
  double-beam system, 410
  drift-space interaction, 414
  one-dimensional effects, 398
  oscillators, 7
  two-dimensional effects, 419
  velocity distributions, 417
Beating-wave phenomenon, 216, 338
Bessel function, 93, 107
Biperiodic rf structures, 54
Brillouin flow, 71, 159, 233, 282, 286, 291, 292, 312, 339, 475, 536, 537
Busch's theorem, 151, 233

C

Cartesian coordinate system, space-charge field components, 77
Cerenkov radiator, 535
Charge continuity, 19, 127, 149, 154, 180, 181, 227, 233, 285, 286, 399, 488
Circuit loss, 189, 210, 313, 325, 368
Closed-form solutions, 3, 424–476
Collector depression, 515–549
  graphical evaluation, 517
  M-type devices, 541–549
    analysis, 542
    results, 546
  O-type devices, 214, 519–540
    analysis, 519
    current limitation, 536
    operation below saturation, 533–534
    phase-focused amplifiers, 531–533
    phase-focused oscillators, 531–533
    results, 524
    voltage-jump devices, 535
Confined flow, 159, 229, 539
Crestatrons, 216, 217, 493
Crossed-field interaction, 11, 282–384
  backward-wave oscillator, 9, 365–384
    adiabatic equations, 375
    closed-form solution, 476–485
    cyclotron waves, 380
    rf conversion efficiencies, 372

588

two-dimensional negative sole, 366
two-dimensional positive sole, 374
drift-space, 282–303
  adiabatic motion, 302
  gap modulation, 292
  three-dimensional equations, 299
  two-dimensional equations, 283
  two-dimensional results, 295
forward-wave amplifier, 9, 304–363
  beating-wave amplification, 338
  circuit sever, 340
  closed-form solution, 461–476
  comparison of calculations, 350
  cyclotron waves, 348
  depressed collectors, 541–549
  efficiency improvement, 461–476
  gain and efficiency, 316
  Kompfner-dip conditions, 583–584
  nonlaminar streams, 339
  phase shift, 337
  prebunching results, 501–507
  three-dimensional negative sole, 348
  two-dimensional negative sole, 306
  two-dimensional positive sole, 342

D

D'Alembertian, 169, 250
Density modulation, 17, 120, 551
Depth of modulation index, 122
Dirac delta function, 96
Disk electron space-charge-field, 125
Double-beam interactions, 397, 410, 414
  amplifiers, 7, 395, 396
  circuit-wave solutions, 410–414
Downhill method, 270, 271, 369, 388, 459

E

Efficiency, improvement in M-BWO, 476–485
  improvement in M-FWA, 461–475
  in O-BWO, 454–460
  in O-TWA, 429–454
  klystron, 140
  M-BWO, 372
  M-FWA, 314–316
  O-BWO, 272
  O-TWA, 190, 203, 208, 235
  prebunched-beam devices, 499

traveling-wave energy converter, 390
Electron-beam bunching, 487–514
  klystrons, 490
  M-FWA, 501
  O-TWA, 493
  power required, 507
  rf power required, 507–514
Electron plasma frequency, 170, 397
Elliptic integral, 452
Emitting-sole crossed-field devices, 2, 11
  Amplitron, 11
  emitting-sole amplifiers, 11
Equivalent circuit, 19, 33, 56, 65, 311
  concept, 33
  E-type devices, 62
  forward-wave amplifier, 20
  impedance, 51
  w-β diagram, 56
  phase constant, 51
  spatially varying line parameters, 65
  transmission line, 20, 51
Eulerian analysis, 16–27
  backward-wave oscillator, 19
  klystron, 19
  traveling-wave amplifier, 19

F

Flight-line diagram, 23, 24, 123, 124, 197
Floquet's theorem, 25, 62, 263
  generalization due to Bloch, 25

G

Gain parameter, 189, 203, 313, 368, 406
Gap modulation, 292–295
  coupling coefficient, 128
  crossed-field, 292
Gauss' theorem, 75, 94
Gibbs phenomena, 110
Green's function, 73, 75, 93, 96, 104, 106, 117, 510
  delta function ring of charge, 96
  dyadic, 74
  one-dimensional disk, 104

H

Hamilton's variational principle, 168
Hard-kernel bunch, 12, 431, 444, 474, 501

SUBJECT INDEX

Harmonic current amplitudes, 128, 132, 149
Helical circuit, bifilar, 572
  Cutler impedance variation, 436
  for O-BWO, 570
  for O-FWA, 570
  with solid beam, 436
Helical waveguide, 44, 177
Hertzian vector potentials, 36
  TE waves, 36
  TM waves, 36
Hollow beams, concentric rings of charge, 102

I

Impedance sheet, 89
Inhomogeneous wave equation, 29
Injection velocity parameter, 189, 313, 368
Integral equation analysis, 253–261
  circuit equations, 257
  circuit voltage, 254
  equivalence, 259
  traveling-wave-amplifier equations, 256

K

Kelvin theory, equivalence, 34
Kirchhoff equations, 65
Klystrons, 2, 4, 7, 11, 120–176, 490–493
  bunching parameter, 138
  multicavity analysis, 140
  one-dimensional analysis, 121–129
  one-dimensional results, 130–144
  radial and angular effects, 155–168
  relativistic analysis, 168–172
  three-dimensional analysis, 150–154
  two-cavity, 4
  two-dimensional analysis, 144–150
  voltage stepping, 172–176
Kompfner-dip conditions, 393, 579–584
  M-FWA, 583–584
  O-TWA, 579–582
  O-TWEC, 387

L

Lagrangian analysis, 16–27, 121–154, 179–234, 266–268, 283–294, 299–302, 306–314, 342–348, 366–371, 374–380

Laminar Brillouin flow, 284
Laplace's equation, 75, 93, 107, 403, 537
Linear compression ratio, 509
Lorentz force equation, 19, 29, 122, 147, 151, 180, 229, 233, 284, 299, 307, 348, 402
Lorentz transformation, 250
Low-frequency modulations, 551
  output spectra, 562–567

M

Maxwellian distributions, 167, 489
Maxwell's equations, 28, 30, 179
Method of images, 89
Modulation studies, 550–568
  analysis for M-type devices, 559–562
  analysis for O-type devices, 551, 555
  device functions, 551
  high-frequency modulation, 567
  low-frequency modulation, 562–567
  modulation index, 125
  modulation parameters, 552
  O-type device results, 556–559
  pulse modulation, 564
  saw-toothed modulation, 565
  sinusoidal modulation, 563
Multidimensional propagating structures, 54–62
  equivalent transmission lines, 54
Multiple bunch focusing, 448

N

N-beam TWA analysis, 217–224
  circuit equations, 221
  computations, 223
  force equation, 221
  weighting functions, 222

O

Oscillators
  M-type, 365–384
  O-type, 263–280

P

Periodic structures, 28–68, 265
Phase focusing, 3, 19, 387, 424, 485

SUBJECT INDEX 591

Planar symmetric beams, physical models, 69
Plasma, 395–421
  amplifiers, 35
  column, 51–54
  frequency, 175
  reduction factor, 108, 138
  transmission-line equivalent, 51
Poisson's equation, 69, 72, 107, 169, 250, 403, 509, 510
Poynting vector, 44, 190
Prebunching, 487, 514

R

Ramo's induced current theorem, 57
Rebatron-harmodotron, 172
Relativistic devices, BWO, 279
  klystron, 251
  TWA, 250
Rf to dc converters, 3, 385–394

S

Schelkunoff's generalized telegraphist's equations, 30
Space-charge fields, 69–119
  beam-plasma systems, 403
  crossed-field drift space, 290
  klystron, 125, 147, 151, 170
  M-BWO, 367
  M-FWA, 309, 352
  O-BWO, 267
  O-TWA, 186, 229, 234, 251
Start-oscillation conditions, 270, 274, 371

T

Tapered devices, 461
  helical lines, 570
  interdigital lines, 575
  M-BWO, 476–485
  M-FWA, 461–476
  O-BWO, 454–461
  O-TWA, 429–454
Telegraphist's equations, 31

Transmission-line equivalent circuits, 28–68, 431
  backward-wave circuits, 62
  helical structure, 34
  multidimensional structures, 54
  plasma column, 51
  variable parameter circuits, 68
Traveling-wave amplifiers, 1–3, 5–7, 16, 35, 177–261, 429–454, 493–501, 551–558, 579–582
  integral equation analysis, 253–261
  Kompfner-dip conditions, 579–582
  modulation of, 551–558
  multidimensional results, 235–250
    circuit interception, 244
    circuit loss, 244
    effect of stream diameter, 237
    harmonic current, 242
    stream subdivision, 236
  N-beam analysis, 217
  one-dimensional analysis, 179–193
  one-dimensional results, 193–217
    beating-wave operation, 216
    effect of circuit loss, 208
    efficiency, 203
    flight-line diagram, 197
    gain, 193
    harmonic current, 200
    phase shift, 193
    severed circuit, 210
  prebunching in, 493–501
  relativistic analysis, 250–253
  tapered operation, 429–454
    closed-form solution, 433
    computer solution, 437, 445
    experimental, 444
    general equations, 431
    multiple bunch, 448
  three-dimensional analysis, 231–234
  two-dimensional analysis, 224–231

V

Velocity tapering
  M-type amplifier, 461–476
  M-type oscillator, 476–485
  O-type amplifier, 429–454
  O-type oscillator, 454–461